So Many Brave Men

So Many Brave Men

A History of the Battle at Minisink Ford

Mark Hendrickson
Jon Inners
Peter Osborne

Pienpack Publishing
Pennsylvania

So Many Brave Men:
A History of the Battle At Minisink Ford

By Mark Hendrickson, Jon Inners, Peter Osborne

Published by
Pienpack Publishing
An imprint of
The Pienpack Company, Inc.
7 Bunker Hill Court
Easton, Pennsylvania 18040-1037
Easton, Pennsylvania

ISBN 978-0-615-34658-8
Library of Congress Control Number: 2010900306

HISTORY /United States /Revolutionary Period (1775-1800)

Cover design by Rick Hibberd
Book design by Rick Hibberd and Peter Osborne
Cover photograph by Peter Osborne
Manuscript read by Mary Tiebout, Frank Salvati,
Fred and Barbara Johnson Weissman, and Nancy Vocci
Index by Peter Osborne

First Edition
10 9 8 7 6 5 4 3 2 1

Printed in the United States of America

The Pienpack Company
Providing Journeys of Discovery Since 1979

Visit us on the Internet at *www.pienpack.com*

Dedicated To

Lt. Col. Benjamin Tusten

The bravest of the brave on that fateful day in July, 1779

We were about one third Tories, and (one) third timid and one third true blue.
John Adams

To
Martha and Kat

Contents

Darley
A. Bollett Sc.

Introduction

To the Heroes of 1779
May their memory be gratefully cherished so
long as the sun rises in the east and sets in the west.
A Toast From
A Nineteenth Century July 4th Commemoration

On July 20, 1779, British Indian Department Captain Joseph Brant and almost ninety Loyalists and Indians descended upon the Neversink Valley from the north and then through Minisink (present-day Port Jervis, New York), conducting a destructive raid on the settlements in their path. They retreated back up the Delaware River and towards their homes in central upstate New York. The raid, the second by Brant into the area in less than a year, set off a series of events that culminated in a military catastrophe at the battle at Minisink Ford on July 22nd. Minisink Ford is about twenty-seven miles north of Port Jervis following the route of the Delaware along New York State Route 97.

Militiamen who chased Brant up the river were drawn from neighboring Sussex County, New Jersey, central and western Orange County, New York and southern Ulster County, New York. Approximately one third of those who were involved in the battle were killed. It was, according to Pike County historian George Fluhr, the most lethal battle of the Revolutionary War in terms of the number of men killed versus the number of participants. It is a chapter in the Upper Delaware River's history that is told every time a book is written about the valley. It was the only Revolutionary War battle that took place in the Upper Delaware. Unfortunately, it is often misunderstood or the details are incorrectly relayed.

It is worth remembering that the years leading up to and including the American Revolution were passionate, dangerous, and anxious times for those living on the frontier along the Delaware River. The French and Indian War (1754-1763) was still a recent memory for many, and some of those early settlers may have even served in local militia units twenty years before. Certainly the settlers knew neighbors whose families had suffered the loss of a loved one. Another contentious issue had been the boundary conflict between New York and New Jersey, a dispute that almost erupted into warfare. It had only been resolved in 1774 after more than one hundred years of court battles, fights, arrests and lawsuits.

As relations with Britain began to dissolve, neighbors along the Delaware frontier were pitted against each other. Oaths of allegiance and loyalty were required as a Committee of Safety monitored activities to ascertain that there were no Loyalists among them. Indian raids into the area, or just the threat of them, forced those hardy settlers to remove to safer places.

Today, it is hard to imagine what the raid and subsequent battle at Minisink Ford meant to that generation. One local militiaman's widow, who applied for a pension in the years after the battle, remembered Brant's raid into the Neversink Valley in the following account:

That on the nineteenth day of July 1779 long to be remembered by the people living in the valley of the Neversink the Indians again made their appearance on the frontiers and burned a number of houses and barns and destroyed much property.

For the men who served in the militias of Orange, Ulster and Sussex Counties during the Revolution there were often difficult and dangerous assignments along the frontier. Sergeant Jacob Aldrich, a member of the Orange County Militia, described those times in his pension application with the following statement:

But to return to the western frontiers against the Indians I have performd many hard marches and dangerous fatiegues . . . I was out when so many brave men were lost att Lackaway.

The men who fought at Minisink Ford were part of a civil war that pitted a relatively small group of colonists against the greatest empire in the world at the time. In 1778 and 1779 those living in the Minisink region were faced with a deadly threat from the skilled and feared strategist Joseph Brant. Most of these early settlers had been loyal subjects to the King of England only a decade before.

This book is the first major work written on the Battle at Minisink, or the "Delaware Battle," as it was described by participants, in more than thirty years. It includes contemporary documents, pension applications and memoirs that have never been published before. In many cases these documents have not been seen by anyone since they were submitted to government agencies more than one hundred and sixty years ago, and in some cases more than two hundred years ago.

The first chapter is an adaptation of a paper that Jon Inners and I delivered in 2001 to a geology conference about how geology often impacts military history. Jon was doing research on the battle at the Minisink Valley Historical Society where I worked at the time, and we joined forces to write the paper.

There have been some minor updates to that original paper, but it was and remains the only effort to take the letters of the two commanding officers, Colonel John Hathorn of the Orange County Militia and Captain Joseph Brant, and place them side by side, and include maps to interpret how each saw the battle unfolding. It is also the first time in recent memory that an effort has been made to explain how the battle developed, and how such a terrible military disaster was visited upon the American militia units that day.

The after-action reports, which are reproduced in their entirety in Chapter 2, are remarkable for many reasons. First and foremost is their immediacy and the freshness with which they are written. They are the earliest surviving records of the battle. Second, they mirror each other in many ways.

There are only a few points of major disagreement between Hathorn and Brant - the number of men who participated in the battle, and Hathorn's belief that Brant had been reinforced, when Brant makes no mention of it. Brant's casualties were also exaggerated. From the letters and maps that we have created one can follow the battle's progress and its terrible conclusion.

A third reason that the reports are so intriguing is the perspective that each commander had. Brant's letter has a matter-of-fact sense to it, just another day on the job if you will. Hathorn's letter is much more defensive; and with good reason, he had just commanded a military expedition that had concluded in disaster. He had placed his command in great jeopardy and had left his home area unprotected.

The remaining chapters of this book rely on the impressive and scholarly research that fellow author Mark Hendrickson has undertaken over the last six years. I believe that there is no one living who knows more about the battle than Mark, or where the records are, and indeed what records still even exist. This book is largely a result of his work, and I know I speak for many area historians when I say that we will be forever grateful for his sound research.

Aside from the sheer volume of material that Mark has uncovered, he has created a new, and we hope, more accurate list of the participants. It is based upon the pension application records and other manuscripts that are buried deep within our country's National Archives in Washington, D.C. We have used a different methodology than previous authors have in organizing our listing of names, and have attributed all of our sources for future researchers to use.

Because of Mark's extensive research, we also have important new perspectives about the battle. Chapter 3 offers his reflections on the research that was compiled, and provides an important overview of the battle and its participants. The chapter is the result of many conversations and e-mails between us about his research.

There are three chapters, organized by the state from where the battle veteran had served, which include the applications and paperwork the men filed to receive a federal pension in the decades after the battle. In some cases the individual has extensive recollections of the battle, another may offer just one sentence.

We have chosen to include the entire pension application of each man and not just excerpts, because it reveals how difficult the process was in getting a pension and how much military service a patriot might have had before and after the battle in 1779. For some, the process of obtaining and filing an application was an ordeal itself, and often resulted in a veteran or his widow and family being given a small pension. Or, in a number of cases, veterans were denied a small stipend for brave and dedicated service earned decades before in a time when there was not a lot of paperwork to confirm and prove one's service.

While one can understand the government clerk's efforts of being frugal with the people's money, and wanting to be sure that the pension was merited, it seems like a great injustice from this late date that pensions were denied to men who were in the autumn of their lives, and in genuine need, and who did not have the ability to track down commanders, get affidavits or have a paper trail demonstrating their service some fifty years before.

The only material that has been deleted are the standard forms that were attached and not filled out, such as the regulations to obtain a pension. A citation notes the order that it was found in the application. The form is printed in its entirety in Chapter 4. The spelling and capitalization in the original documents is maintained, as is the grammar and organization of the application. The original layout of documents, for example, is not duplicated in an effort to save pages. We have inserted a divider after each document to indicate that it is a separate item.

Additionally, we have included the occasional letter that was placed in the file long after the pension had been filed and ruled upon. Family historians and members of the Daughters of the American Revolution were often trying to find information about their ancestors, and were requesting material from the government.

The method that we have used provides a better context in understanding and describing the experience of the men who fought on the hill overlooking the Delaware River. The pensions indicate that most of the men who fought at Minisink Ford were not inexperienced or cowards, as has sometimes been suggested, but dedicated patriots. This dispels one of the enduring misconceptions about the battle.

If not for the accidental firing of a rifle by a member of the militia before Hathorn had his troops in place to attack Brant and his forces, the battle might have ended quite differently. The American militia had a major strategic advantage in both where they were situated and in numbers. However, once the battle began it was hard to keep a unified command given that the battle was spread across such a wide area, and the units were physically separated in a dense forest setting. The units ultimately split apart and the remaining men at the top of the hill were surrounded and killed or barely able to escape.

Many of the militiamen served in militia units before and after the battle. Many of the participants at the tragedy at Minisink Ford came back to fight time and time again, proving their courage and their loyalty to the cause. A disastrous day with its horrible memories did not stop their commitment to the cause.

All three of us are mindful of previous efforts to tell the history of the battle at Minisink Ford beginning in the early nineteenth century and continuing with Vernon Leslie's pioneering work *The Battle of Minisink: A Revolutionary War Engagement in the Upper Delaware Valley* and George Fluhr's research and publications as well. Messrs. Leslie and Fluhr looked at the battle using a more scholarly approach, and both historians inspired our interest in the battle. We are grateful for their research, and have built on their work and that of others who have gone before us.

We are sure our book is going to encourage further research and create new controversy. Such is the case when one steps into the history of such a pivotal event in our region's history. For all three of us, that is just fine. We have compiled this material with the notion that we will be challenged. This work may become the basis on which the next round of research begins and that is the way it should be.

Finally, we want to honor the service of the militiamen who fought that day on a remote hilltop far from their homes and families. They fought for ideals that pertained to their very lives. They fought for liberty and freedom.

Peter Osborne
March 31, 2010
Easton, Pennsylvania

Chapter 1
Blood and Mayhem At The Minisink

...[He] has endeavoured to bring on the Inhabitants of our Frontiers, the merciless Indian Savages, whose known Rule of Warfare, is an undistinguished Destruction, of all Ages, Sexes, and Conditions.

The Declaration of Independence

July 4, 1776

These ringing words of "the Unanimous Declaration of the Thirteen United States of America" had a special meaning for the people of the upper Delaware River valley in the late 1770's and early 1780's - for this was the frontier during the American Revolutionary War, and it was subject, on numerous occasions, to the depredations of Indians allied to the British - though the raiding parties often contained as many white Loyalists, or Tories, as Indians. (Whether these Indian and Tory raiders were any more indiscriminate than their American counterparts in their slaughter of innocents is highly questionable.)

The most significant of these "partisan" raids was the second Minisink raid led by Joseph Brant, an educated Mohawk warrior serving as a captain in the British Indian Department in July of 1779. The actual attack on the settlement of Minisink (present-day Port Jervis, New York) was destructive enough, but it was the ensuing Battle of Minisink (in which nearly fifty New York and New Jersey militiamen lost their lives) that sent shock waves of loss and grief though the frontier population along the Delaware.

The Minisink battle has become part of the folklore of the region encompassed by Sullivan County, New York, Sussex County, New Jersey, and Pike County, Pennsylvania. Each year in July, dozens of people gather by the grave of an unknown militiaman at Lackawaxen, Pennsylvania, and on a nearby hilltop at Minisink Ford, NY, to commemorate the patriots who died in the battle.

General Physiography Of The Minisink Country

The term "Minisink" has a variety of geographical meanings *(e.g., see Witte, 2001)*. For our purposes "Minisink Country" refers to the area centered on Port Jervis and extending up and down the Delaware River, perhaps twenty miles, and up the Neversink River valley an equal distance. (The historical Indian Minisink village was apparently located on the New Jersey side of the Delaware River opposite Minisink Island, two miles below Milford and nine miles below Port Jervis (*Donehoo, 1928*)).

Defined in this fashion, the Minisink country's most striking topographic features are the broad and deep valleys of the Delaware and Neversink Rivers, which unite at Tri-States Point (where the southeast-flowing "upper" Delaware River makes a right-angle bend to flow southwest); the Minisink escarpment, a steep southeast-facing bluff that extends for many miles along the northwest side of the Neversink and "lower" Delaware River valleys in the vicinity of Cuddebackville, Port Jervis, and Milford; and Shawangunk Mountain, the broad northwest slope of which forms the southeastern boundary of the Minisink country.

Northwest of the Minisink escarpment on both sides of the "upper Delaware" River is a broad region of rugged, hilly terrain that constitutes the area where the Indian-Tory and Patriot forces maneuvered immediately prior to the Battle of Minisink. In the midst of these hills about twenty-seven miles north of Port Jervis following the route of the Delaware along New York State Route 97, the tributary Lackawaxen River, flowing from the west, has deposited a broad gravel bar in the Delaware just downstream of their confluence. This shallow bar marks the site of Minisink Ford ("Lackawack" in *Hathorn's Dispatch to Gov. George Clinton, Draper Manuscripts; 20F35)*, the river crossing where the battle took place.

Brant's Minisink Raids

The First Raid

Brant's first incursion into the Minisink country was in October of 1778. On the 13th of that month he led a band of about one hundred Indians on a raid on "Peenpack," an old name for the area around present-day Cuddebackville, about eight miles up the Neversink River from Port Jervis. In their initial surprise attack the raiders killed several settlers, including Thomas Swartwout and three of his sons *(Stickney, 1867)*.

The other settlers managed to gain refuge in nearby forts at Gumaer's and DeWitt's, both of which the enemy force attacked in desultory fashion. Brant and his men then retired to the west that same day,

> *...after having burned all the houses, barns, &c, they found in their course; a circumstance that afterwards distressed the inhabitants much for the want of hay, grain and articles of furniture (Stickney, 1867).*

Peaceful Interlude

The October raid roused the settlers to strengthen their defenses. The forts were repaired and application made to the general government for assistance. The result of this petition was the assigning of Count Casimir Pulaski's Legion of mixed cavalry and infantry to patrol the area around Minisink. Thus encouraged, the inhabitants set about rebuilding the burned houses and barns and prepared for the spring planting.

The cavalry was withdrawn in January, 1779 and in February, Pulaski himself and the infantry were ordered to South Carolina. The war effort had shifted south and his presence there was deemed more important than on the then quiet Minisink frontier. A contingent of troops under Brigadier General Edward Hand remained behind for a few months, but this final regular force was withdrawn in mid-April.

Though the settlers were initially disturbed at being left in a relatively defenseless position, they soon put their fears behind them and life at the confluence of the Neversink and Delaware returned to its normal routine. Alas, the first raid had been but a prelude to a much more destructive incursion the following summer *(Stickney, 1867, Leslie, 1976)*.

Second Raid and Militia Pursuit

Brant's second Minisink raid was designed to capture foodstuffs, particularly cattle, for his hungry followers. It began on July 8, 1779 when he left Chemung (near Elmira, NY) with a war party of sixty Indians and twenty-seven Tories *(Stone, 1865, Kelsey, 1984)*. He apparently moved eastward to the upper Delaware at Cushetunk (Cochecton, New York), then southeastward across the "Catskill foothills" of present-day Sullivan and Orange Counties to the lower Neversink River valley several miles northeast of the Minisink settlement.

Brant planned to strike at dawn on July 20th, but his party did not reach its destination until noon *(Leslie, 1976, Williams, 2005)*. Even so, the raiders caught the settlers napping. Brant's men attacked down the Neversink valley, burning a succession of houses, barns, and mills located mainly on the flats between the river on the northwest and the elongate "Trilobite Mountain" on the southeast. His only setback came at "Fort Van Auken" near the present-day country club on Neversink Drive, which he surrounded for an hour but could not capture.

Before nightfall, the raiders had also burned Maghaghkamik Church - not far from Tri-States Point at the confluence of the Neversink and Delaware - and Martinus Decker's fort on the Delaware River flats to the west of the steep, rocky bluffs at Point Peter on the Minisink escarpment.

In his report to Lieutenant Colonel Mason Bolton after the raid, Brant described the action of July 20th as follows:

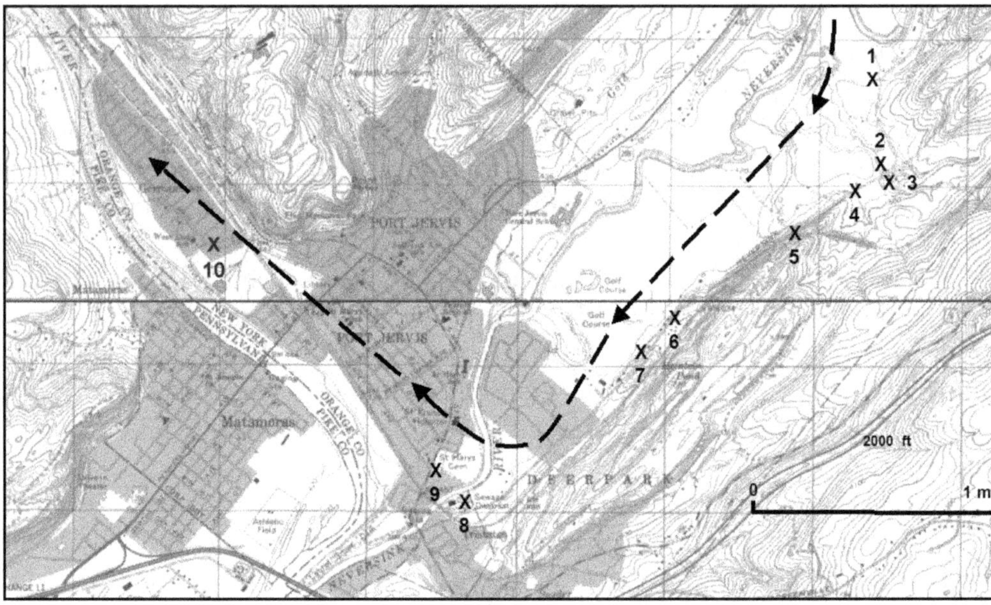

Map Courtesy Gary Fleeger
Pennsylvania Geological Survey

Figure 1

Map of the Port Jervis area showing some of the important sites related to Brant's second Minisink raid.
1 = Johannes Decker's stockaded fort (burned);
2 = Anthony Van Etten's blacksmith shop (spared);
3 = Solomon Davis' grist mill (burned);
4 = schoolhouse (burned);
5 = Wihlahoosa Cave (where Johannes Decker hid from the raiding party);
6 = Jacobus Van Fleet's house and barn (burned);
7 = Van Auken's fort (unsuccessfully attacked by the raiders);
8 = Solomon Kuykendall's house (burned);
9 = Maghaghkamik Church (burned);
10 = Fort Decker (Lt. Martinus Decker's fortified house, burned).

The arrows are an approximation of Brant's path of destruction through the settlement.

We have burnt all the Settlement called Minisink, one Fort excepted, round which we lay before about an hour, & had one man Killed and one wounded. We destroyed several small stockaded Forts, and took four scalps & three Prisoners; but did not in the least injure Women and Children. The reason that we cou'd not take more of them, was owing to the many Forts about the Place, into which they were always ready to run like ground hogs. (Sir Frederick Haldimand Papers)

Later that day, Brant's war-party, now encumbered with considerable plunder, but apparently not nearly as many cattle as their leader had hoped for, headed upstream along the Delaware on the old Delaware River trail, past the spectacular Hawk's Nest cliffs, which ascend abruptly from the east bank of the river three -and-a-half miles upstream of Port Jervis. That evening they camped at the mouth of the Mongaup River.

On the next morning, July 21st, they marched towards the mouth of the Halfway Brook (Barryville, NY) where they camped overnight. The next day, the 22nd, the raiders continued to proceed along the river to the Minisink ford, where that morning, the militia under Colonel John Hathorn caught up with them *(Leslie, 1976)*.

The pursuing Patriot militia force, initially eighty men from Goshen, New York, and Sussex, New Jersey, under the command of Lieutenant Colonel Benjamin Tusten, had left the ravaged Minisink settlement toward the evening of July 21st, and marched several miles up the river. Somewhere in the vicinity of Sparrowbush, New York, Colonel John Hathorn and forty men from Warwick, New York caught up with the main party.

Hathorn, being senior in rank, took command, and led the combined force back into the hills along the "Keshethton path" with the design of marching around and surprising Brant's war party farther up the river. About midnight, they camped at or near present-day Lochada Lake about three-and-a-half miles from Brant's camp at the mouth of the Halfway Brook. The next morning, Hathorn's men, leaving their horses and (in Hathorn's words) *"disengaging of everything heavy...Marched on with the Intention to make the Attack the Moment an opportunity offered" (Draper Manuscripts; 20F35)*.

The Battle At Minisink Ford
First Phase: Ambush at the Ford

Brant's war party and Hathorn's militia approached the Minisink Ford by different "roads"- the former along the old Indian trail that followed the east shore of the Delaware River, the latter by the ill-defined Keshethton (Cushetunk) path through the hills. Coming up about mid-morning, Hathorn observed the Indians and Tories in the process of transporting their plunder across the Delaware at the ford and dispersed his men accordingly.

After sending Captain Bezaleel Tyler ahead with a small intercepting force, he divided his force into three "divisions" - Lieutenant Colonel Benjamin Tusten on the right (on top of "Ambush Hill"), Hathorn himself in the center, and Lieutenant Colonel Wisner (misspelled "Wesner" in Hathorn's report) on the left and somewhat to the rear. But Captain Tyler, or one of his men, fired too soon on the front of the enemy force, prematurely giving away Hathorn's planned ambush.

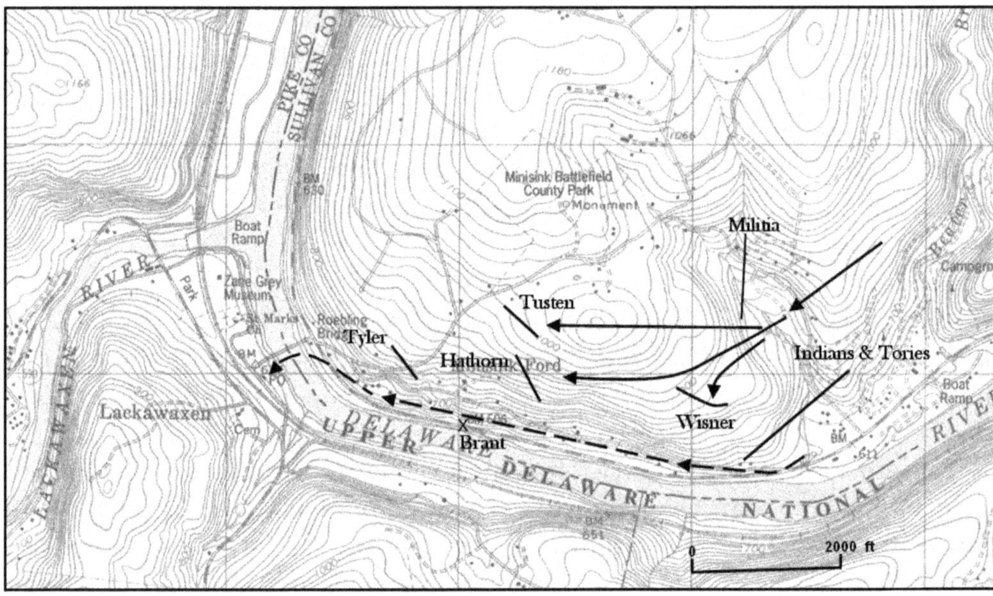

Map Courtesy Gary Fleeger
Pennsylvania Geological Survey

Figure 2
Map showing the initial action of the "ambush at the ford."

Even now it is not clear what happened, although Hathorn lays the blame clearly on Tyler. Hathorn's division then pushed forward, firing vigorously on the Indians and Tories who were by this time partly across the river.

Brant himself was about a quarter-mile downstream of the ford when the firing began. He quickly gathered about forty men and struck up the hill from the river (probably not directly up the spring-fed ravine called "Dry Brook" by Leslie as reported in Wilson *(Wilson, 1822)* and Stone *(Stone, 1865)*, but along its western slope). This movement brought Brant's party around behind Hathorn's and Tusten's divisions, and completely split off Wisner's division (which may have been slow in coming up on Hathorn's left). Wisner and his men - accounting for at least a third of the militia force - retired from the field, never to be heard from again.

When Brant's flanking party fired on the rear of the remaining militia, which was also being engaged by the Indians and Tories rallying in front, more of the militia apparently broke and fled. This left Hathorn and about forty men (as he reported) hard-pressed on two or three sides, their only escape route to the north across the one thousand foot-wide topographic swale between "Ambush Hill" and "Rocky Hill."

The following is how each commander described the first phase of the battle

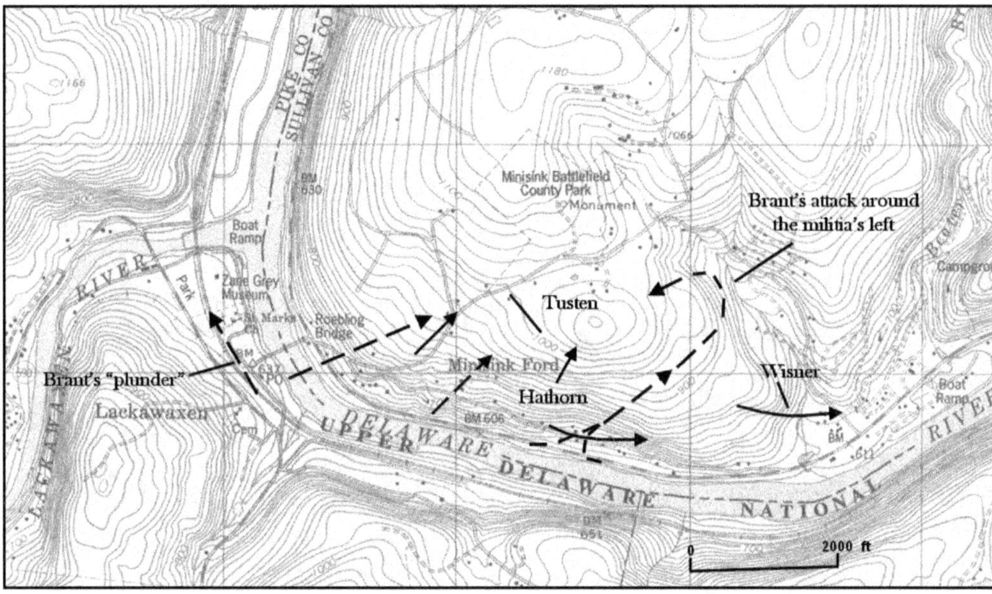

Map Courtesy Gary Fleeger
Pennsylvania Geological Survey

Figure 3
Map showing Brant's flank movement ending the first phase of the battle.

in their after-action reports:

Brant:
. . . there are two Roads, one thro' the Woods, the other along side of the River; We were coming up this Road next morning, and I sent two Men to examine the other Road, the only way the Rebels cou'd come to attack us; these Men found the Enemys path not far from our Camp, and discovered they had got before to lay in Ambush. The two Rascals were afraid when they saw the Path, and did not return to inform us, so that the Rebels had fair play at us. They fired on the front of our People when crossing the River. (Sir Frederick Haldimand Papers)

Hathorn:
. . . the Indians Probable from some discovery they had made of us, marched with more Alacrity than Usual, with an Intention to get their Prisoners, Cattle & plunder taken at Minisink over the River, they had almost Effected getting their Cattle & baggage across when we discovered them at Lacawack 27 Miles from Minnisink some Indians in the River and some had got over, it was determined in Council to make an attack at this place I therefore disposed of the Men into three Divisions, ordered Col Thurston [Tusten] to Command the one on the Right and to take post about three

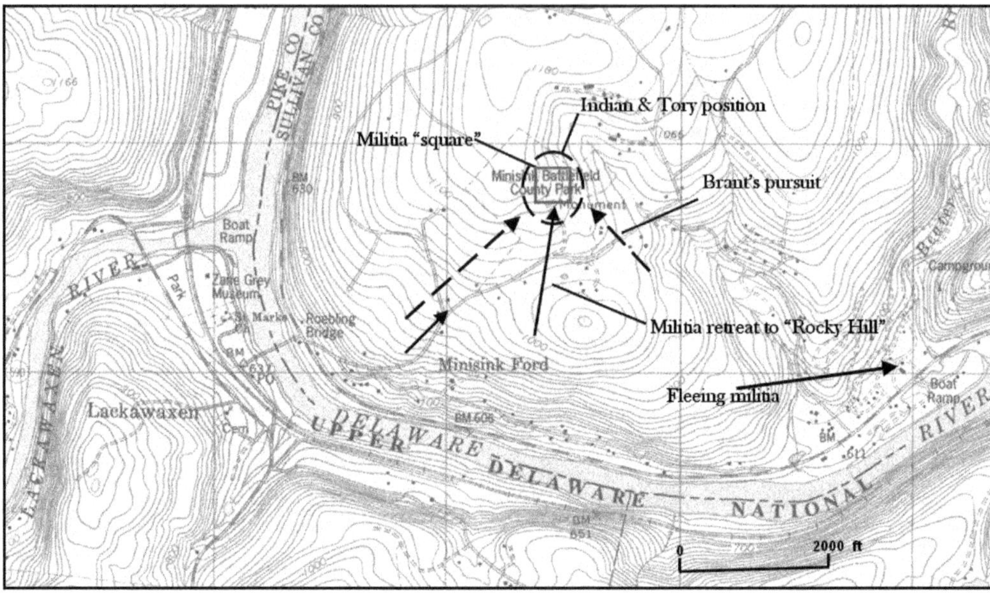

Map Courtesy Gary Fleeger
Pennsylvania Geological Survey

Figure 4
Map showing the militia retreat and encirclement on the "Rocky Hill" to start the second phase of the battle.

hundred yards distance on an Eminence to secure our Right, Lt.Col Wesner [Wisner], with another Division to file of to the Left, and to dispose of himself in the like manner, In order to prevent the Enemy from gaining any Advantage on our flanks the other Division under my Command to attack them with that Vigour necessary to Strike Terror in such a foe. (Draper Manuscripts; 20F35)

Second Phase: "Last Stand" on the "Rocky Hill"

Both commanders describe the second phase of the battle in their after-action reports and both demonstrate how the strategic advantage once held by the militia was reversed by Brant at this point:

Brant:

. . . the Rebels soon retreated and I pursued them, untill they stopt upon a Rocky Hill, round which we were employed and very busy, near four hours . . . (Sir Frederick Haldimand Papers)

Hathorn:

. . . we returned the fire and kept up a Constant Bush firing up the hill from the River, in which the brave Capt. Tyler fell, Several were wounded, the People being

Exceedingly fatigued Obliged me to take post on a heighth which proved to be a piece of Strong and Advantagious ground the Enemy Repeatedly approached upon us within from 40 to 100 yds distance and were as repeatedly repulsed, I had now but about 45 Men (Officers Included) who had lost their Command naturally drew towards me, the Spirits of these few, notwithstanding their fatigue, Cituation & unallayed thirst adding to that, the Cruel Yellings of those bloody monsters, the seed of Anak in Size Exceed thought or description, we defended this Ground near three hours and half, during the whole time one blaze without Intermission was kept up on both Sides . . . (Draper Manuscripts; 20F35)

Hathorn's small remaining militia force, which included parts of his and Tusten's divisions, conducted a fighting retreat for about one-half mile until they reached the high ground bounded by the sandstone ledges of "Rocky Hill." Here they took up a position about an acre in extent with "Hospital Rock" on the west, the high Monument ledge (and its stony extension) to the east and south, and relatively open ground to the north. Brant's Indians and Tories completely encircled them at a distance of about one hundred yards. Both sides probably had relatively clear "fields of fire" from behind trees, rock ledges and boulders, as the primeval hemlock woodlands would have been rather dark, but clear of underbrush beneath the great forest canopy.

The Americans were probably outnumbered nearly two-to-one during this final phase of the battle, though some of Brant's men were undoubtedly on the other side of the Delaware guarding the plunder from the Minisink raid. For three-and-a-half or four hours, the battle raged. Though few appear to have been killed outright, many on both sides were wounded.

Toward sundown, the beleaguered militia began running low on ammunition. As their fire slackened, Brant urged his men on to a final assault. Although tradition and battleground signs say that the Indians and Tories broke through the American "square" from the southwest at "Sentinel Rock" after a particularly stalwart militiaman was killed, Hathorn clearly states in his report that the enemy cracked his line on the northeast.

The militia now had little choice but to run for their lives. Those who were able retreated precipitously downhill toward the river, scattering through the woods to save themselves as best they could. Some were killed as they ran away, but some escaped - swimming and wading across the Delaware - and made their weary way back through the wilderness to the Minisink settlement.

According to tradition, the American wounded were less lucky. During the militia's "last stand" Lieutenant Colonel Tusten, a physician in private life, had been treating the wounds of as many as seventeen men under the protection of the overhanging "Hospital Rock" ledge.

With the rout of the Americans from their hilltop stronghold, the Indians and

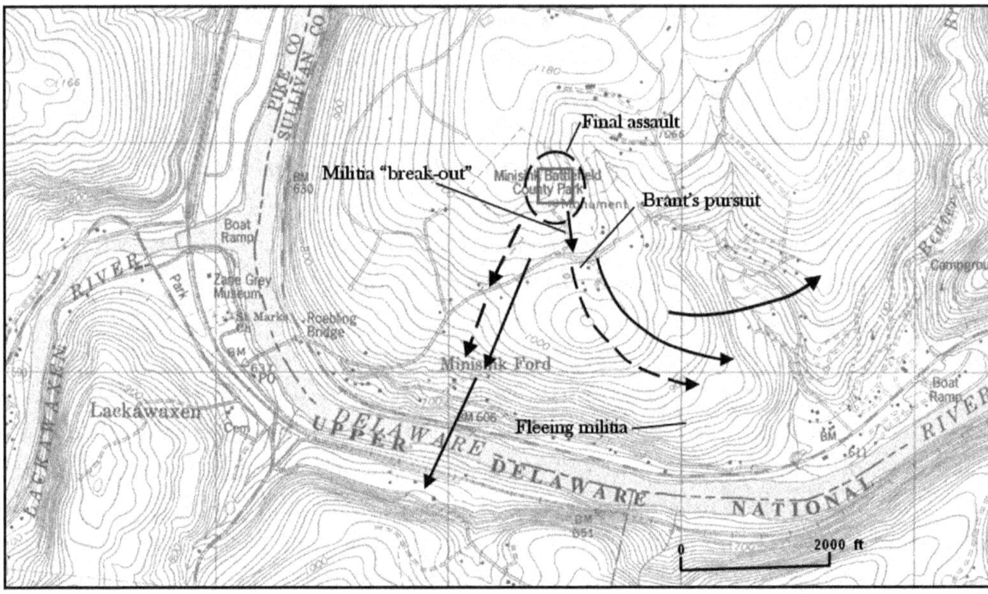

Map Courtesy Gary Fleeger
Pennsylvania Geological Survey

Figure 5
Map showing the "last stand" and the final rout of the militia, ending the battle.

Tories fell upon Tusten and the wounded men, quickly dispatching them with tomahawks *(Wilson, 1822; Stone, 1865)*. At least one pension application, that of John Daniels, states that some of the men were tortured. Brant does not mention this incident in his report, although Hathorn expressed his concern about the wounded.

Another reputed eyewitness, John Knapp of the militia, stated in his pension application fifty-two years later:

"I heard the wounded beg of us not to leave them and ere I was out of hearing I heard some begging of the Indians to spare their lives which I believe was those wounded men." (Draper Collection, see Leslie, 1976)

After the battle ended, both commanders concluded their reports as follows:

Brant:
. . . before we cou'd drive them out. We have taken 40 odd Scalps and one Prisoner, a Captain. I suppose the Enemy have lost near half of their Men & most of their Officers; they all belonged to the Militia and were about 150 in number. (Sir Frederick Haldimand Papers)

Hathorn:

I found myself under the necessity of Ceasing the fire our Ammunition from the Continued fire of more than five hours, naturally Suggested that it must be Exhausted, ordered no person to Shoot, without having his Object Sure that no Shot might be lost, this gave Spirit to the Enemy who. . . formed their whole strength and forced the North East part of our Lines, here we gave them a Severe Gaul, our people, not being able to Support, the lines retreated down the hill precipitately towards the River the Enemy kept up a Constant fire on our Right which was returned, the people this time was so Scattered I found myself unequal to rally them again consequently every man made Choice of his own way, thus Ended the Action. . . In the whole Twenty one Men [are missing.] (Draper Manuscripts; 20F35)

After the battle, the rest of Brant's war party crossed over to Pennsylvania and camped that night on the Lackawaxen River, about a mile upstream from its junction with the Delaware River.

Aftermath

The next morning Brant and his party continued up the river to Cushetunk (Cochecton, New York), and then carried their plunder over the divide to the North Branch Susquehanna River and Onaquaga (Ouaquaga, New York) where he wrote the dispatch to Lieutenant Colonel Bolton *(Kelsey, 1984)*. By July 29th, he knew that a large American force was at Wyoming and that it intended to invade Indian country.

On July 31st, the Sullivan-Clinton Expedition, led by Major General John Sullivan and Brigadier General James Clinton and consisting of about 2,300 regulars and militia, got a belated start from Wyoming (present-day Wilkes-Barre, Pennsylvania) and moved up the North Branch of the Susquehanna via Wyalusing and Tioga (Athens, Pennsylvania). Meanwhile, Brant and some of Brant's Volunteers conducted a small scouting expedition north to the Mohawk, returning in time to join Butler's Rangers, and approximately four hundred to four hundred fifty Indians, mainly Six Nations warriors, with approximately twenty Delawares. They were to be defeated by Sullivan at the only pitched battle of the campaign - at Newtown, New York, on August 29 *(Fischer, 1997; Williams, 2005)*.

Sullivan marched on into western New York, destroying numerous villages and crop fields all the way to the Genesee River before returning to Wyoming early in October. Sullivan's march did not put an end to the Indian menace on the frontier as he and Gen. George Washington intended *(Fischer, 1997)*, and in fact there were more raids into the Minisink region in the years to come. But it did effectively destroy the economic and social fabric of Iroquoia.

Recovery of the Bodies and Later Commemoration

Although at least two rather half-hearted attempts to recover the bodies of the militiamen slain at the Minisink Ford are recorded as having taken place in the weeks following the battle, it was not until 1822, forty-three years after the event, that the remaining bones were gathered up and returned to Goshen, New York, twenty-two miles east of Port Jervis.

The bones were interred beneath a monument on the anniversary of the battle, July 22, 1822, with the Reverend James R. Wilson delivering the funeral eulogy and the aged Colonel Hathorn speaking a few words *(Wilson, 1822)*. The monument erected at that time was subsequently defaced and a new monument, still standing in Goshen's village square, was erected in 1862.

A monument at the battleground site was dedicated on the centennial of the battle in 1879. For many years, annual ceremonies have been held at either the battlefield or at Goshen, commemorating the patriot dead of Minisink. Many of the orations delivered at these ceremonies are preserved in the archives of the Minisink Valley Historical Society in Port Jervis, the Orange County Historical Society in Arden, New York, the Pike County Historical Society in Milford, and in numerous local newspapers.

The preceding chapter is based on a paper entitled ***Blood and Mayhem on the Upper Delaware: Military Geology of Brant's Minisink Raid and the Battle of Minisink, July 1779*** *that was written by Jon Inners and Peter Osborne. It was delivered at the 66th Annual Field Conference of Pennsylvania Geologists at Shawnee-on-the-Delaware, Pennsylvania, in 2001.*

Chapter 2
The Loss of So Many Brave Men

We have burnt all the Settlement called Minisink, one Fort excepted, round which we lay before, about an hour, and had one man killed and one wounded. We destroyed several small stockade Forts, and took four Scalps and three Prisoners; but did not in the least injure Women or Children.

Captain Joseph Brant to Lieutenant Colonel Bolton
July 29, 1779

. . . altho we Suffered by the Loss of so many brave men, the best, for the Number without Sensible Error in the Precinct, its beyond doubt, the Enemy Suffered much more.

Colonel John Hathorn to Governor George Clinton
July 27, 1779

The after-action reports of both of the commanders at the Battle at Minisink Ford, reprinted in their entirety here, provide researchers and historians with the best eyewitness accounts of the battle available because they were both filed within a week of the encounter. As noted, they provide interesting insights into the commander's views of what happened and what their dispositions were.

These two letters have a freshness and a clarity to them, especially when compared to the pension applications, filed decades after the battle, by the soldiers who fought at Minisink Ford. On July 27, 1779, less than a week after the battle, a shaken Colonel Hathorn wrote Governor George Clinton from Warwick to report on the latest action that his regiment was involved in.

The two men were friends, but Clinton did not let that stand in the way of his criticism of Hathorn and Tusten after the battle. In a letter to the Reverend Nathan Kerr, the minister of the Presbyterian Church in Goshen, dated July 30, 1779, Clinton said that it was very unfortunate that he had not been advised of the calling out of the militia to oppose Brant. Tusten and Hathorn had not informed the Governor, as required by the Militia Law, that they were taking the field with their regiments after the raid at Minisink. Had Clinton been notified he might have been able to send Pawling's Regiment of Levies on a forced march to the Delaware, and thereby cut off Brant's avenue of retreat.

The original Hathorn report can be found in the Draper Manuscripts *(Draper Manuscripts; 20F35)* at the University of Wisconsin. The letter was rediscovered by Vernon Leslie when he was writing *The Battle of Minisink: A Revolutionary War Engagement in the Upper Delaware Valley.* A copy of the letter can be seen at *www.albertwisnerlibrary.org/Factsandhistory/History/Minisink.htm.* It also appeared in *The Weekly Register,* a Newburgh, New York newspaper on July 23, 1879.

---

by Express
To His Excellency George Clinton Esqr.
Governor of the State of New York
on public Service

Pougkeepsie

Sir/

Warwick 27th July 1779

In Conformity to the Militia Law, I Embrace the first oppertunity to Communicate to your Excellency my proceedings on a late Tour of duty with my Regiment, on the Evening of the 21st of this Instant I recieved an Order from his Excellency General Washington, together with a requisition of the Commissary of Prisoners to furnish one hundred men of my Regiment for to Guard the British Prisoners on their way towards Easton, at the same time recieved an Express from Minisink that the Indians were ravageing and burning that place. I Ordered Three Companies of my Regiment - Includeing the Exempt Company, to parade for the purpose of the Guard the Other three Companies to March Immediately to Minisink on the 22nd. I arrived with a part of my people at Minisink, where I found Col Thurston [Tusten] & Major Meeker of New Jersey with parts of their Regiments who had marcht with about Eighty Men up the River a few Miles I Joined this party with about Forty men, the whole amounting to one hundred and Twenty men Officers Included, a Spy came in and Informed me the Enemy lay about four hours before at Mungaup six miles distant from us, our people appeared in high Spirits, we marched in pursuit, with an Intention Either to fall on them by surprize, or to gain in front, and Ambush them, we was soon informed that, they were on their March up the River, I found it Impracticable to surprize them on the Grounds they now were, and, took my Rout along the Old Keshethton path, the Indians Encamped at the Mouth of the halfway brook, we encamped, at 12 o'Clock at night at, Skinners Saw Mill three Miles and a half from the Enemy, where we lay the Remainder of the night, the Mountains were so Exceeding rugged & high we could not possibly get at them, as they had passed the Grounds the most favourable for us to attack them on before we could possibly, overtake them. Skinners, is about Eighteen Miles from Minisink, at day light on the morning of the 23rd after Leaving our Horses and disengageing of every thing heavy we Marched on with Intention to make the Attack the Moment, an oppertunity offered, the Indians Probable from some discovery they had made of us, marched with more Alacrity than Usual, with an Intention to get their Prisoners, Cattle & plunder taken at Minisink over the River, they had almost Effected getting their Cattle & baggage across when we discovered them at Lacawack 27 Miles from Minnisink some Indians in the River and some had got over, it was determined in Council to make an attack at this place I therefore disposed of the Men into three Divisions, ordered Col Thurston [Tusten] to Command the one on the Right and to take post about three hundred yards distance on an Eminence to secure our Right, Lt.Col Wesner [Wisner], with another Division to file of to the Left, and to dispose of himself in the like manner, In order to prevent the Enemy from gaining any Advantage on our flanks the other Division under my Command to attack them with that Vigour necessary to Strike Terror in such a foe. Capt Tyler with the advance Guard, unhappily discharged his piece before the Divisions could be properly posted, which put me under the necessity of bringing on the Action. I ordered My Division to fix their Bayonets and push forcibly on them, which order being resolutely Executed put the Indians in the utmost Confusion, great numbers took into the River who fell from the well directed fire of our Rifle Men, and Incessant Blaze from our Musketry without returning any fire, the Divisions in the Rear, not Subject to Order broke some advanced down the hill towards me others fled, into the Woods, I soon percieved the Enemy Rallying on our Right and Recrossing the River to gain the heighths, I found myself under the necessity to Rally all my force which by this time was much less than I Expected, the Enemy by this time had Collected in force & from the best Accounts can be collected recieved a Reinforcement from Koshethton began to fire on our Left, we returned the fire and kept up a Constant Bush firing up the hill from the River, in which the brave Capt. Tyler fell, Several were wounded, the People being Exceedingly fatigued Obliged me to take post on a heighth which proved to be a piece of Strong and Advantagious ground the Enemy Repeatedly approached upon us within from 40 to 100 yds distance and were as repeatedly repulsed, I had now but about 45 Men (Officers Included) who had lost their Command naturally drew towards me, the Spirits of these few, notwithstanding their fatigue, Cituation & unallayed thirst adding to that, the Cruel Yellings of those bloody monsters, the seed of Anak in Size Exceed thought or description, we defended this Ground near three hours and half, during the whole time one blaze without Intermission was kept up on both Sides, here we had three men killed and nine Wounded, among the Wounded were Lieut. Col Thurston [Tusten], in the hand Major Meeker in the Sholder, Adjutant Finch in the Leg Capt. Jones in the foot and Ensign Wood in the Wrist, the Chief of our people was wounded by Angle Shotts from the Indians from behind Rocks & Trees, our Rifles here were very Usefull, I found myself under the necessity of Ceasing the fire our Ammunition from the Continued fire of more than five hours, naturally Suggested that it must be Exhausted, ordered no person to Shoot, without having his Object Sure that no Shot might be lost, this gave Spirit to the Enemy who formed their whole strength and forced the North East part of our Lines, here we gave them a Severe Gaul, our people, not being able to Support, the lines retreated down the hill precipitately towards the River the Enemy kept up a Constant fire on our Right which was returned, the people this time was so Scattered I found myself unequal to rally them again consequently every man made Choice of his own way, thus Ended the Action.

The following are missing in the whole from the Last Accounts
Col Ellisons Regiment
Adjutant Finch
Lieut. Col. Thurston [Tusten] one private of New Jersey
Ensign Wood &
Capt. Jones
one private of
Capt. Wood
my Regiment
Capt. Little
Capt. Duncan
and
Twelve privates
In the whole Twenty one Men

Several Wounded Men are come in I hope numbers of Others will yet be found. I recieved a Wound in my head, one in my Leg and one in my thigh Slightly the one in my thigh from Inattention is a little Troublesome Several Spies that lay near the Enemy the night following the Action inform us that they moved of their Wounded in Canoes, on the day following that on the Ground where they Lay there was great Quantities of Blood their whole Encampment was marked with wounded men, great Numbers of Plaisters & bloody Rags was found, altho we Suffered by the Loss of so many brave men, the best, for the Number without Sensible Error in the Precinct, its beyond doubt, the Enemy Suffered much more, from the Various parts of the action can be collected a greater number, of Indians Dead than we lost besides their Wounded, the Number of Indians & Tories is not ascertained, some accounts says 90 others 120 others 160. Col Seward of New Jersey with 93 men was within five or six Miles of the Action on Pensylvania Side did not hear the firing, approached and Lay near the Indians all night following, and from their Conduct and, Groanings of the Wounded, gave rise to a belief that they had been in some action where they had suffered and would have attacked them round their fires but a mutiny arose among some of his People which prevented: a very unfavourable and to be Lamented Circumstance if in their Cituation he had attacked them with the common Smiles of Providence he must have succeeded and put them to a Total rout.

Dear Governor its not in my Power to point out to you the disagreeable Situation I was in surrounded by a foe with a handfull of such Valuable Men not only as Soldiers but, as fellow Cittizens and members of Society and nothing to be Expected but the Hatchet Spear, and Scalping Knife. The Tremendous Yells and Whoops all the fiends in the Confines of the Infernal Regions with one United Cry could not Exceed it, add to this the Cries the Entreaties and feeling petitions of the Wounded around me not to leave them was - is beyond parrallel or Idea, my heart Bleeds for the unfortunate who fell wounded into their hands, however one Circumstance gives me a little Consolation, Mr. Roger Townsend of Goshen received a Wound in his Thigh being Exceeding Thirsty makeing an attempt to go to find some Water, was met by an Indian who very friendly took him by the hand, said he was his Prisoner, and would not hurt him, a well directed Ball from one of our Men, put the Indian in a dose and Mr. Townsend ran back into the Lines. I hope some Little humanity may be Yet found in the breast of those Savages.

I should be at the Greatest Loss was I to attempt to point out one Officer or soldier that Exceeded another in bravery during the time of the general Action, too much praise cant be given to them for their attention in Recciving orders and alacrity in Executing them.

I have acquisced with Col Woodhull in ordering 1/8th of our Regiments to Minisink as a Guard Temporary Guard, untill your Excellency's pleasure is known on the subject.

The Indians were under the Command of Brant who was Either Killed or Wounded in the Action they Burnt Major Deckers House and Barn Samuel Davis's House Barn & Mill Jacobus Van Vlecks House & Barn, Daniel Vanokers Barn, here was Two Indians Killed from a Little Fort round the house - which was saved, Esquire Cuykindalls house & barn Simon Westfalls house and barn, the Church Peter Cuykindalls house and Barn Mertinus Deckers Fort, house, Barn and Saw Mill and Nehemiah Pattersons Saw Mill. Killed & Scalped Jeremiah Vanoker Daniel Cole Ephraim Ferguson & one Travern, took with them Several Prisoners, most Children with a great Number of Horses, Cattle, & Valuable Plunder, some of the Cattle we resqued and returned to the owners. I hope your

Excelly. will make allowance for the Imperfect, Stile, Razures & Blotts of the Line whilst I have the honour to Subscribe myself with the most perfect Esteem in hast

Your Excellencies
Most Obt Servt.
(signed) John Hathorn Col.

On July 29th, one week after the battle, Joseph Brant sent a letter to Lieutenant Colonel Mason Bolton reporting his activities at Minisink. Bolton was the commander at Fort Niagara at the time and, though he was mistrustful of Indians, he held Brant in high regard because of his humane behavior during the Cherry Valley Massacre on November 11, 1778 at Cherry Valley, New York.

The original letter can be found in the papers of Sir Frederick Haldimand, then the acting Governor of the Province of Quebec. *(Sir Frederick Haldimand: Unpublished Papers and Correspondence 1758-1784; B.100; 21,760; 173-174.)*

Oghwage July 29th 1779

I beg leave to acquaint you, that I arrived here last Night from Minisink, and was a good deal disappointed that I cou'd not get into that Place at the time I wished to do, a little before day; instead of which I did not arrive 'till noon, when all the Cattle was in the Woods, so that we cou'd get but a few of them. We have burnt all the Settlement called Minisink, one Fort excepted, round which we lay before, about an hour, and had one man killed and one wounded. We destroyed several small stockade Forts, and took four Scalps and three Prisoners; but did not in the least injure Women or Children. The Reason that we cou'd not take more of them, was owing to the many Forts about the Place, into which they were always ready to run like Ground-Hogs. I left this Place about 8 O'Clock next day, and marched 15 miles, there are two Roads, one thro' the Woods, the other along side of the River; We were coming up this Road next morning, and I sent two Men to examine the other Road, the only way the Rebels cou'd come to attack us; these Men found the Enemys path not far from our Camp, and discovered they had got before to lay in Ambush. The two Rascals were afraid when they saw the Path, and did not return to inform us, so that the Rebels had fair play at us. They fired on the front of our People when crossing the River. I was then about 400 yards in the Rear, as soon as the firing began I immediately marched up a Hill on their Rear with 40 men, and came round on their backs, the rest of my Men were all scattered on the other side; however, the Rebels soon retreated and I pursued them, untill they stopt upon a Rocky Hill, round which we were employed and very busy, near four hours before we cou'd drive them out. We have taken 40 odd Scalps and one Prisoner, a Captain. I suppose the Enemy have lost near half of their Men & most of their Officers; they all belonged to the Militia and were about 150 in number.

I am informed by the Prisoners, that the King's Troops had taken a Post below the Highlands on the North River, called King's Ferry, in which were 50 men, and had built a Fort on each side of the River: That after this Genl. Clinton sent a part of his Army into new England, took several Towns, and destroyed a great deal of Stores &c. - that Genl. Washington in the mean time sent part of his Army in the Night and surprized one of his Forts, in which 500 men were taken Prisoners - this affair happened some time ago.

The Night after we left Minisink, I received another piece of Intelligence - that Genl. Clinton at the head of a great Army was coming up the North River, and had drove Genl. Washington and his Army before him, and obliged him to retreat up the River in a hurry; this news I received from the Rebels, who also said the Country were extremely alarmed.

I find the Enemy certainly intends an Expedition into the Indian Country, and have built strong Forts - by the last accounts they were at Wioming perhaps by this time they may be at Shimong, where I have sent my Party to remain 'till I join them; I am now seting off with 8 men to the Mohawk River in order to discover the Enemys motions. In the last Skirmish, we had 3 Men killed & 10 wounded. John the Mohawk dangerously wounded and 3 more almost in as bad a Situation. I am afraid they will not recover.

I am, Sir, Your Obedient Servant
(signed) Joseph Brant

Lt. Col. Bolton

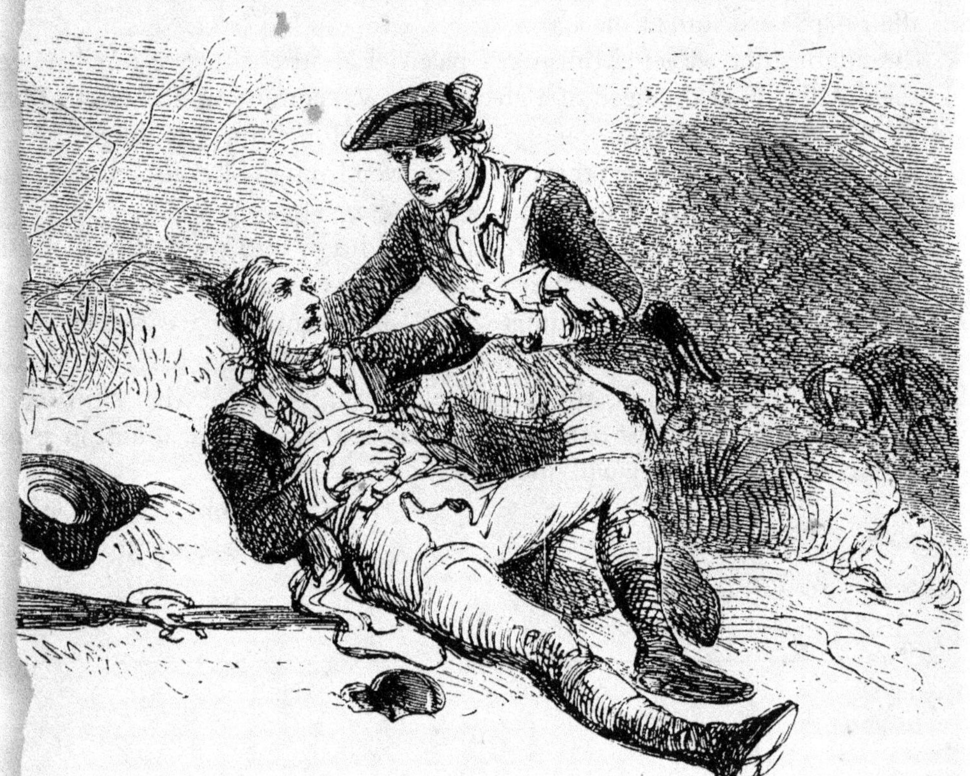

Chapter 3
The Patriot's Deed is Honored

Our duty is now done, the Patriot's deed is honored
One hundred years ago our father did it well.
Excerpts From a Poem read by Ezra Conklin
At the 1879 Minisink Battle Centennial Commemoration

Revisiting Old Traditions

In the years that followed the Delaware battle, a history of that horrible day was built upon memories (some faded, some clear), family tradition, stories and accounts written by authors who attempted to make some sense of what had happened July 22, 1779. A long line of writers contributed to the inconsistencies, inaccuracies and poor history that have since plagued the efforts to remember the events of that dramatic day.

Our effort has attempted to correct many of the misconceptions as best we can from this late date. In this chapter we deal with some of the most glaring issues raised by writers past and speakers at the commemorative programs in decades gone by. Many times those who were trying to remember the bravery of the men who fell did not understand the larger issues at work during that time.

This project was originally undertaken to locate, transcribe and publish the records of the service of the men who were in the Battle of Minisink. Once those documents, which included individual service records and pension applications, were rediscovered, reviewed and transcribed, a significant amount of information about the battle was recovered, much of it presented here for the first time.

In this chapter we will look at many of the traditions about the battle through the prism of knowledge gained from the pensions and various archival materials. In some cases, we have confirmed what has long been accepted about the battle's history and in other cases we are presenting materials that should alter our views.

Militia and Pension Records

All of the records presented here are of the militiamen serving under Colonel John Hathorn at the time of the battle. All of the records come from *Series M-804 Revolutionary War Pension and Bounty Land Warrant Application Files; M804*; 2,670 rolls; owned by the National Archives and Records Administration.

Select items from the *Lyman Copeland Draper Manuscript Collection*, owned by the State Historical Society of Wisconsin have been included as well. The *Draper Manuscripts* is a collection of correspondence, newspaper clippings, notes and other material collected by Lyman C. Draper (1815-1891). Draper, the corresponding secretary of the State Historical Society of Wisconsin, was a student of history and he carried on an extensive amount of correspondence, collecting material with the intent to write several books on American history. Unfortunately, Draper died after completing just one book. Upon his death his entire collection was bequeathed to the Wisconsin Historical Society, whose staff organized it into 491 volumes.

Two of the volumes in the Brant Manuscripts section of the collection deal specifically with the Revolutionary War in the Minisink area, and much attention is given to the Battle of Minisink. Additional material on the battle and its participants can be found scattered throughout other volumes of the Brant Manuscripts.

Much of the correspondence is with descendants of men who fought in the war, or with people who had heard accounts about the battle told long after the war had ended. As one reads some of the letters in the collection, one will see that many of the letters are filled with errors and hearsay. However, that is not to say that there is not valuable material in the *Draper Manuscripts;* there is. For example, Jonathan Bailey (included on the Leslie list but not ours), recounts stories of his father's service directly to Draper. They have a striking similarity to what we do know about the battle.

Loyalist Records

An attempt to locate records of men serving under Joseph Brant at the battle was undertaken by searching the *British Headquarters (Sir Guy Carleton) Papers, the Haldimand Papers, Audit Office Series 12 Loyalist Claims (AO12), Audit Office Series 13 Loyalist Claims (AO13)*, and the *Draper Manuscripts*. Nothing new was found that would prove a person's involvement in the battle.

In *Volume 5*, page 163 of *The Public Papers of George Clinton, First Governor of New York*, there is a listing of thirteen men by name who were with Brant on July 8th at Chemung, as stated by Moabary Owen. Owen is known to have been with Brant during the raid on Minisink and, after being captured by the militia, pretended to be forced into the service of the British. He gave the names of those men under interrogation, and they are suspected to have been with Brant at the battle.

However, they cannot be placed there with certainty. Of these thirteen men, only a very few of them were found to have filed claims to the Crown for losses during the war and none of them mentioned anything that would place them at the battle. They cannot be placed in the engagement any other way than by Moabary Owen's statement, which listed those men with Brant about two weeks previous to the battle. Any one of these men could have taken sick or been dispatched on temporary duty between July 8th and the time of the battle.

Moabary Owen, an Indian named John Tayojaronsere (also called Captain John, John the Mohawk and Tuscarora John), and a man named Cole are the only three men that can be placed in Brant's party during the battle. It is to be noted that in the *Revolutionary War Pension and Bounty Land Warrant Application Files (M-804)*, there can be found an application for a pension from a Mowberry Owens (W.19,950). This is more than likely the same Moabary Owen mentioned above.

According to *Volume 5*, pages 162-164 of *The Public Papers of George Clinton, First Governor of New York*, Owen was a deserter from a State Regiment and this Mowberry Owens claims he served from 1777 to 1783 in Colonel Philip Van Cortland's Regiment (Second New York Regiment). It has also been said that Moabary Owen was from Warwick, which is where the applicant states he enlisted into the service. There is no mention of desertion or service under Joseph Brant in the file, and it is doubtful that he would mention it either when trying to secure a pension. In the absence of irrefutable proof that this is the same man, his pension application has not been included in this work.

Errors in the Pension Records

There are a number of errors to be found in the records and a number of items that can be misinterpreted. The first item is the microfilm *Series M-881 (Compiled Service Records of Soldiers Who Served in the American Army During the Revolutionary War); Roll #742; Hathorn's Regiment, Militia*. Within this reel of microfilm are found the names of Peter Bertholf, William Blane (Blain), Jeremiah Curtis, George Luckey, David McCamley (McCambly), Andrew Miller, John Minthorn and Colvill (Colville) Shepherd who were in service on an alarm at Minisink in July, 1779. This is not proof positive that the men were in the battle. As will be shown later, not everyone got to Minisink in time to march against the enemy and of those who did, some were detailed to other duties rather than to chase Brant.

The men did respond to the alarm, though, and therefore were entitled to pay, which is what this shows. The same thing applies to *Roll #751; McClaughry's Regiment, Militia*, where a number of names appear to have been called out in response to an alarm in July, 1779. They too are entitled to pay, but cannot be placed in the battle solely upon these records.

Second, there is a mistake in the pension file of Johannes Miller. In his file you see mention of John Miller being wounded in the battle, and his name being submitted to the House of Representatives as an applicant for an invalid pension on April 25, 1794, and printed in the *American State Papers, class 9, page 94*. It is believed that records for John Miller were mistakenly placed in the file of Johannes Miller. As one reads the file for Johannes Miller, you will see that there is no mention of him being wounded at all during the war or being either an applicant for, or in receipt of, an invalid pension.

Every single source consulted says that the John Miller, who was wounded at the battle, was from the Warwick area. The people testifying in support of Johannes Miller's claim give the impression he is from Montgomery or the surrounding area, and not from Warwick. It is believed that the document should have been filed under another John Miller instead, most likely the file of John Miller (S.28,128). He is from the Warwick/Florida area, and was wounded according to information found in his file. Not only that, but he was placed on the pension list by Act of March 23, 1792. This date is near to the cited information printed in the *American State Papers*. The file is, however, short, and nothing positively identifying this John Miller as the correct one is found within it.

Third, in the application of Peter Westbrook, Sarah Van Auken, his daughter, claims that he was killed in the Battle of Minisink, but lists the date of the Conesaugh battle where he was really killed. This is an example of people giving depositions for someone's claim, and making a mistake. Attention is brought to it because someone will eventually find it and think that maybe he was really killed at the Battle of Minisink when every other pension and record shows he was killed the following year.

Fourth, not everything found in the book *New York in the Revolution* by Berthold Fernow can be trusted as fact. On page 550, for example, where it lists those killed at Minisink, it lists Tusten as having been in Hathorn's Regiment, an obvious mistake. Tusten commanded Colonel William Allison's Regiment while Allison was a prisoner in New York at the time of the battle.

In Fernow's book the names of men were taken from different records kept throughout the war at different dates, and the names were placed in the units accordingly. For example, there may be a record of a John Graham having served in the Third Regiment of the Orange County Militia in 1777. John Graham may have moved after the date of that record and been serving in the Fourth Regiment of the Orange County Militia the following year, but no record was preserved to show that. *New York in the Revolution* would correctly list him in the Third Regiment, but has no way of showing that the man served in other units throughout the war, leaving one with the false impression that the man served the entire war in the same unit.

As seen by pension records, a great many of the men served in several different units throughout the war. Men could, and did, move from one location to another. Recruiting Officers roamed the country looking for willing men to enlist into the unit they were recruiting for, which may not have been the local regiment of militia. Drafts of men were made for service in different units, and it was common to see someone hire a substitute to take their place while they returned home. In addition, there were many different variations of how names were spelled.

For example, Jonathan Knap and John Knapp should not automatically be considered two separate people. It has been seen where someone with the last name Pawling would also be mentioned in records as Spalding or Paulding. Other examples are the last name Seward, which has been seen as Soward, and Shimer has been seen as Chambers. While the book does provide some good evidence and great leads, it is not 100% accurate.

Fifth, in the application of Lawrence Kortright (S.13,676), those giving depositions for his application say that he was out with the rest of the company, said company being engaged in the Battle of Minisink. Again, this is not good enough to place him in the battle because not everyone arrived in time to take part in the engagement. However, the War Department thought it good enough, and they listed him as being in the battle when replying to requests for his military records. He may or may not have been; more details of his service during that time are needed, something specifically saying he was there. His is not the only case of this, either. The same is true for many others.

Sixth, the file of John English (S.22,756) says that he took part in the Battle of Peenpack. Because so many pension files refer to Minisink and Peenpack as the same either by mistake or intent, it is impossible to determine if he was in the Battle of Minisink, or if they are talking about the skirmishing that occurred during the raid on Peenpack in October, 1778.

Seventh, some files mention the person engaged in a skirmish with Indians at the mouth of the Lackawaxen, and nothing more. One has to remember that there was a skirmish with Indians in August, 1778 in the same vicinity, when the militia was marching to Cochecton to deal with threats from that quarter. Therefore, something a little more detailed would be required to identify the engagement as the Battle of Minisink.

Eighth, there is the file of William Winans (R.11,919). Here we have one person stating that he and Winans were in the Battle of Minisink together, while another person states that he and Winans were marching to Minisink on foot together and did not arrive in time to be in the battle. In view of contradictory testimony, it is best not to include Winans or the person who said they were in the battle together for the obvious reason of not knowing which person is right.

Lastly, in the file of John R. Whitman (R.11,475), he states that in 1779 he was under Captain Samuel Jones at Fort Dewitt, and that they had an engagement with the Indians, whereupon the enemy retreated to the Delaware River and they followed them.

An engagement took place, and the enemy proved too numerous for them. They then retreated back to the frontiers. While it is very tempting to include him as being in the battle, one has to remember that there was a constant presence of Indians in the area, and that he may be getting the dates mixed up.

There was skirmishing at Fort Dewitt the previous year, but not in the raid on Minisink. While it sounds like the Battle of Minisink, it could have been that battle, or any number of small skirmishes that occurred along this section of the frontier, and were not recorded. The account is too vague to be conclusive regarding Whitman's service.

Known Units in the Battle

Vernon Leslie, in his book *The Battle of Minisink: A Revolutionary War Engagement in the Upper Delaware Valley*, did an excellent job identifying the units that were at the battle based upon available information at that time. However, with access to additional records, we can now add to the number of known units in the battle. As is already known, Brant's Volunteers, along with Indians, were present on behalf of King George.

No known Pennsylvania military units participated in the battle. The pension application of Reuben Place states that he belonged to a company of Rangers under Captain Henry Shoemaker and Colonel Jacob Stroud, and that the Rangers were engaged alongside the New York Militia in the fighting at Beaver Brook. However, reason to question his claim will be addressed in the section covering men whose participation in the battle can be proved.

Most of the men from New Jersey who participated in the battle were from the Second Regiment of Sussex County Militia, commanded by Colonel John Seward (formerly Colonel Aaron Hankinson's Regiment). However, three men (William De Shay, Ezekiel Lane and Samuel Vangorder) were from Captain Peter Westbrook's Company, according to their testimony, and it appears that they were able to get onto the ground before the rest of the company and therefore were able to join those who marched off in pursuit of Brant with Meeker and Tusten. Westbrook's company was from the Third Regiment of Sussex County Militia, commanded by Colonel John Rosencrantz. There is no evidence showing that the company as a whole was in the battle. Only a few of the men were engaged in the fighting.

Men from New York who participated in the battle were from Colonel John Hathorn's Fourth Regiment and Colonel William Allison's Third Regiment, both militia, and both from Orange County. The Third Regiment was commanded by Lieutenant Colonel Benjamin Tusten, Colonel Allison then being a prisoner of the British in New York City. Also found at the engagement are men from Colonel James McClaughrey's Second Regiment of Ulster County Militia as represented by Captain Abraham Cuddeback's and Captain Benjamin Vail's Companies.

There was present at the battle one company of Associated Exempts. Leslie lists the units involved in the battle and treats these men as a separate entity from the regiments listed. It was determined from a number of sources during the course of our research for this book that the single company of Associated Exempts at the battle was a company that belonged to the Third Regiment of Orange County Militia, commanded by Captain John Wood.

From information found in several sources, we can now add one more unit, Pawling's Levies, to those known to have taken part in the battle from the State of New York. On March 13, 1779, an order was issued to raise 1,000 men for the defense of the frontiers of the State of New York. On May 3, 1779, it was decided to divide these 1,000 men into two separate corps, one commanded by Lieutenant Colonel Henry K. Van Rensselaer, the other by Lieutenant Colonel Albert Pawling.

According to *Volume 5*, page 75 of *The Public Papers of George Clinton, First Governor of New York*, Pawling's Regiment was stationed from Great Shandaken to Peenpack. John Wood, though he was elected Captain of the Exempt Company in Goshen (see *Volume 4*, pages 627-628 of *The Public Papers of George Clinton, First Governor of New York*), raised a company for Pawling's Levies. A certified copy of Wood's appointment as Captain in this company can be found in the pension application of William Cuddeback (R.2,557).

Abraham Westfall (W.4,386), according to his own testimony in his application for a pension, was appointed a Lieutenant in Captain John Wood's Company of Colonel Albert Pawling's Regiment in April, 1779 and both men's names can be found on the roster of officers for this regiment on page 256 of *New York in the Revolution* by Berthold Fernow.

Though Wood raised the company, Westfall probably commanded it, or a portion of it, in the field. Wood's company of "Nine Months Men," according to *Volume 5*, page 107 of *The Public Papers of George Clinton, First Governor of New York*, was stationed at Peenpack on July 1, 1779. For further reading on the act to raise these men for the defense of the frontier, and the officers appointed to the regiments, refer to pages 255-261 of *New York in the Revolution* by Berthold Fernow.

Further evidence of this company raised by John Wood can be found in the pension files of Joel Bower, Benjamin Carpenter, William Cuddeback (R.2,557), and others. Bower states that the company he belonged to (Wood's) was the only company of the regiment in the battle. This is correct, as Pawling's Regiment was divided by companies and posted along the frontier stretching sixty miles from Great Shandaken to Peenpack. It also needs to be remembered that John Wood, according to the certificate from New York State found in William Cuddeback's application, was only in charge of the company in Pawling's Regiment from May 3rd until the time of the battle, less than three months, and then his Lieutenants, including Abraham Westfall, commanded the company for the remainder of the year's service.

In 1780, Westfall was made captain of the company, and his name can be found in Fernow's *New York in the Revolution* in that rank in the years after the battle. Men claiming to have been in Captain Westfall's Company at the battle were in fact in Wood's Company. They probably saw more of Westfall than Wood during the brief time he commanded the company. After Wood's capture at the Battle of Minisink, Westfall commanded the same for the rest of the war. It is no surprise that many years later they would remember it as Westfall's Company and not as Wood's.

Number of Men in the Battle

Traditional accounts and other records indicate that anywhere between 120 and 150 men actually left the smoldering ruins of Minisink in pursuit of Brant's force. That number range does appear to be accurate. Several of the men who applied for pensions give numbers in that range, among whom are Jeremiah Kinner, who places Hathorn's force at 133 men, and Levi Totten, who puts the number at 130. Adonijah Middaugh, whose father fell at the battle, estimates that there were 140 to 150 men. All of these accounts fall within the 120 - 150 range handed down by tradition.

Several other pension applications, however, such as those of William De Shay, Ezekiel Lane and Reuben Place, give much higher numbers. When reading their statements concerning the battle, it becomes rather obvious that their recollections about the engagement are somewhat confused, leaving reason to doubt their numbers.

In the end, it is best to trust Hathorn and Brant. Both men wrote reports shortly after the battle, and are therefore less likely to confuse the numbers at that time. Also, both commanders would be in the best position to know how many men they and their opponent took to the field with. Hathorn, in his report to Governor George Clinton *(Draper Manuscripts; 20F35)*, places his command at "one hundred and Twenty men Officers Included" while Brant, in his report to Lieutenant Colonel Bolton *(Sir Frederick Haldimand: Unpublished Papers and Correspondence 1758-1784; B.100; 21,760; 173-174)*, places his enemy at "about 150 in number."

While a few of the pension applications show that they were detailed to guard the horses or were dispatched to find powder, the rest claim involvement in the actual battle itself. Remembering that only a handful made it off the hill with Hathorn, it should be noted that the majority of these pension applications are from men who were not in that stage of the battle.

Of the one hundred and twenty men who left Minisink, one of them, Samuel Vanfleet, took sick and returned home before the battle. Three men were dispatched for powder or supplies, two of whom were Jacob Aldrich and Benjamin Carpenter. The number of men left to guard the horses the morning of the battle is unknown, but among them were William Curry, Isaac Jennings and John Poppino (son of Major John Poppino).

It has been asked why so few men from the regiments responded to the alarm. This has been attributed to casualties from previous campaigns, especially the invasion of Canada in 1775. While casualties of course contributed to reduced numbers, that by no means explains the small number of men initially on the ground when the pursuit began. As seen in the pension files, there were constant requests for men to be furnished to guard mountain passes, ferry crossings and jails, among other places. Calls for men to build and repair fortifications along the Hudson River were common as well.

Joseph McConnel, when giving testimony in the application of William Curry, explained it best when he stated that "calls for troops in various places in the said County of Orange & the adjoining country were of almost continual occurrence during the greater part of the war." In addition, the pension files are filled with details about the militia being classed, with one class always on duty to be relieved by another at a set time and date. Therefore, the men available to respond to alarms of any sort would never be the total strength of the regiment.

Most importantly, though, are the pension applications of Richard Clark (S.12,509), Daniel Gilbert (S.13,133), Benjamin Davis (R.2,702), and others, that show not everyone was able to get to Minisink in time to march up the Delaware River in pursuit of the enemy. These men met the survivors on their retreat.

In the case of William Vail (R.10,816) and William Winans (R.11,919), they marched to Minisink on foot. A deposition by John Hall in the William Winans pension application clearly points out that those who went to Minisink by horse did not wait for those on foot before marching against the enemy. In his application for a pension, Alexander Miller (S.23,320) states that he and others were sent back from Minisink to Florida in order to guide the rest of the company who had not been ready to march when the rest had, and that he therefore, was not in the battle.

Joel Coleman (S.22,680), in his own application, claims that he was stationed at Neversink under Lieutenant Tryon to guard some horses while the rest of the militia pursued the Indians. Adonijah Middaugh (R.7,157) claims in his pension application that some men had been sent to Fort Van Tuyle before they marched up the Delaware. In the pension application of William Wells (S.6,360), Thomas Knight states that he and Wells were left at Major Decker's Fort when the rest of the militia marched up the Delaware after Brant.

Lastly, as seen in Hathorn's report to Governor George Clinton *(Draper Man-*

uscripts; 20F35), Hathorn received orders to furnish one hundred men from his regiment to escort the prisoners taken at Stony Point, and at the same time received the news of the attack upon Minisink, forcing him to divide his command. All of this combined shows that some militiamen were not able to get to Minisink in time to march with the others up the Delaware and participate in the battle, while others were assigned to other duty when they arrived at the settlement.

Known Participants in the Battle

One of the most difficult parts of this project was to compile a complete listing of the participants in the battle. There was no official effort to compile a detailed list of participants, or of those killed in the immediate aftermath. In the two centuries that have passed since that fateful day various attempts have been made to organize a listing, albeit in a piecemeal fashion. Prior to 1976, the only listing that was recognized by all were the names of those inscribed on the monument commemorating the battle in Goshen, New York. Forty-six names are listed there.

In his pioneering 1976 work *The Battle of Minisink: A Revolutionary War Engagement in the Upper Delaware Valley (2nd edition)*, Vernon Leslie compiled a list of men whom he thought had taken part in the battle. It was the first time a serious effort had been made to document each of the participants using all of the known sources at hand including the monument listing. For thirty-five years it has been generally accepted as the best list that could be compiled.

Leslie compiled his names using additional sources and came up with a list of ninety-two men whom he believed took part in the engagement. An important source for inclusion on his list was the material he found in the *Draper Manuscripts*. In other instances he does not offer any documentation for listing a name although given his thoroughness he must have seen some evidence to include them.

Our own listing came by using a different methodology and because we had access to records that Vernon Leslie did not, most notably the federal pension records which, while available on microfilm since 1941, did not have an index of the 88,000 names of Revolutionary War pension applications until after 1976. Using our methodology we have created a list that includes the names of 103 men who participated in some aspect of the battle. Surprisingly, and we expect controversy to accompany our list because of this, of the 130 names on the two lists that were created by Vernon Leslie and us, sixty-three of the names appear on just one or the other list and not both.

As to why there are not at least 120 names on our list there are several answers. It may be that some of the families may not have been in need of assistance in the form of a pension; others might not have thought they qualified. Some, like James Reeve, might have thought the government owed them nothing. Reeve's file reveals that his sons pressed for a pension after he died though Reeve never applied for one on his own. Perhaps he felt that he didn't need one.

Some, like the widows of Nathan Wade and Stephen Mead, might only have applied to their home state (New Jersey, in their cases) before federal pensions were available. Another answer lies in the possibility that there may be a few who never mentioned the battle in their pensions or did so in such a way that it is not clear what they were referring to.

In some cases men may have just joined up with the militia on their way to the battle and were not known to the others. When they died on that battleground, so would the memory of their service. Some may have died and simply left no record of service or of family behind. Another factor is that there were roaming recruiting officers in Goshen, for example, recruiting people to fill the ranks of other regiments. In addition, people were able to hire substitutes. Men easily moved from unit to unit and rank to rank within that unit. It can cause great confusion to present day researchers.

The testimony of veterans struggling to recall their service fifty years earlier in a number of different units and ranks throughout the war leaves present-day researchers little conclusive evidence regarding who was at the battle. This is our best guess as to why there are not 120 names on our list. In the end, only forty-three Minisink veterans applied for pensions.

For the purpose of this work and the creation of our list, it was decided that anyone who left the ruins of Minisink under Hathorn, Meeker or Tusten in pursuit of Brant should be counted as a participant because these men would have the most to say about the battle, and could influence events based upon their actions. Deciding who fell into this category proved a very difficult task.

There were problems of vague statements, names of places and dates being confused, failing memory, and the possibility of fraud. There were several cases where a person said that he was at Minisink under an officer known to be at the battle, and that the company was engaged in it, but the person never came right out and placed himself in the action. The War Department considered that proof enough for his participation in the battle, and placed the man in the engagement when responding to requests for his war records. However, that is not good enough evidence.

As will be seen in the section dealing with the assembly of the militia, not everyone who responded to the alarm got to Minisink in time to march after Brant. Others were left behind, or sent to other forts, before the men marched up the Delaware. In addition, there were a number of mistakes made by either the War Department or by persons providing evidence for a claim. A number of applications, where it was concluded that someone was at the Battle of Minisink Ford, were not so straightforward as to come right out and say the man was engaged in the action. In these cases it was a matter of placing them there based upon a deposition they gave for someone else's file or taking one or two things they said in a deposition that could prove they were there even though they made no mention of it.

There has been some confusion about who exactly was in the battle, who was killed or wounded, and the ranks of the men involved. Even the applications for pensions contain mistakes, which is to be expected since these men were applying so many years after the event. The *Draper Manuscripts* in particular are filled with errors and hearsay. In many instances it is the son of a participant describing what his father used to tell him about the battle, or in other cases, people repeating stories that had been told in their neighborhood about the event.

In order to eliminate as much confusion as possible, and with access to so many sources, it was decided to list only the men whose participation could be proved and allow future researchers to expand upon the list as more evidence is found. Military documents such as service records, legal documents such as pension applications, newspaper articles from the time, and even obituaries that identified a person's participation in the battle were considered as proof of one's involvement.

The following list will show how a person can be placed in the battle or with the men that left Minisink in pursuit of Brant. With regard to rank, if it is not shown, if there is any reason to doubt, or if it cannot be proved that a person served in a certain rank at the time of the battle, then their rank is not listed, and no explanation is given for why it is left out.

As can be seen in the applications for pensions, men served in different units, and at different ranks throughout the war. A person could have been listed in records as a Sergeant in 1778, but he may not have been serving in that rank on July 22, 1779. One must also consider faulty memory in pension applications, and possible errors by the War Department. It should also be mentioned that on page 550 of Fernow's *New York in the Revolution*, there is a listing of some of the dead from the battle, along with their ranks. None of the information found on that page has been used, as there is doubt about its accuracy.

Some of the ranks are correct, but in the case of Bezaleel Tyler, for example, he is listed as a private on that page when he was actually acting as a captain at the battle. Another error found on that page is the listing of Ephraim Masten as a private. He is listed as an ensign on the monument in Goshen.

It is thought that the errors are in Fernow's book, and not on the monument because among the casualties listed on this page in Fernow is Lieutenant Colonel Tusten, who is incorrectly listed as belonging to Colonel Hathorn's Regiment. Tusten was in fact in command of the regiment from Goshen. In some of the files included in this work you will see veterans of the battle identify other soldiers who were in the engagement with them. In many instances there is reason to disagree with the veteran's statements and in those cases we will discuss why in the narrative that follows.

We have provided three lists of the participants in different formats. The first is a list of participants in a table format for easy reference. The table includes the known regiment and rank. The units that were known to have been at the battle are abbreviated in the table as follows: Third and Fourth Regiments of the Orange County Militia (OCM3 and OCM4), the Second Regiment of Ulster County Militia (UCM2), the Second Regiment of the Sussex County Militia (SCM2), and Pawling's Levies (PL). We have also included the participant's age where it is known or can be approximated.

The second list, and it is the first time that this has been done, is a unit roster that includes the known units of participants. It does not, however, include the names of participants for whom there is no data on the unit in which they served. The third list offers a comprehensive explanation of why each participant was included on the list and in some cases why men were not included on the list.

Alphabetical Listing of Participants At The Battle At Minisink Ford

Number	Last Name	First Name	Regiment (If known)	Rank (If known)	Age
1	Aldrich	Jacob	OCM3	Sgt.	23 or 24
2	Bailey	Gamaliel	?	?	?
3	Barber	Timothy	UCM2	Private	?
4	Barker	William	?	?	?
5	Barney	David	?	?	?
6	Baylis	Nehemiah	OCM3	?	19 or 20
7	Bedell	Isaac	SCM2	Private	15 or 16
8	Bennett	Benjamin	?	?	?
9	Bertholf	Henry	OCM4	Ensign	28
10	Bower	Joel	PL	Private	16
11	Brink	John	OCM3	?	18 or 19
12	Brooks	John	UCM2	Private	23
13	Carpenter	Benjamin	PL	Private	?
14	Carpenter	John	?	?	?
15	Cartwright	Solomon	SCM2	Private	27
16	Chapman	Uriah	PL	Private	?
17	Cuddeback	Abraham	UCM2	Captain	?
18	Curry	William	OCM4	Private	31

Number	Last Name	First Name	Regiment (If known)	Rank (If known)	Age
19	Daniels	John	OCM3	?	18 or 19
20	Dean	David	OCM3	Private	16
21	Decker	Joel	?	?	?
22	De Shay	William	?	Private	16
23	De Witt	Francis	UCM2	?	33
24	Dewitt	Moses	SCM2	Private	17
25	Duncan	John	OCM3	Captain	?
26	Dunning	Benjamin	?	Lieutenant	?
27	Dunning	Jacob	?	?	?
28	Embler	Adam	?	?	?
29	Finch	Nathaniel	OCM4	Adjutant	?
30	Forgerson (Ferguson)	Ephraim	?	?	?
31	Gunsaluss	Henry	?	Private	19
32	Gunsolley	Benjamin	OCM3	Private	19 or 20
33	Hains (Haynes)	Benjamin	?	?	?
34	Harding	Abraham	OCM3	Lieutenant	33
35	Harker	Joseph	SCM2	Captain	34 or 35
36	Haskell	Jonathan	?	?	?
37	Hathorn	John	OCM4	Colonel	30
38	Hill	Thomas	SCM2	Captain	32
39	Horton	Jason	OCM3	?	20
40	Jennings	Isaac	?	?	?
41	Jones	Samuel	OCM3	Captain	?
42	Kemble (Kimble)	Peter	OCM3	Sergeant	?
43	Kinner	Jeremiah	OCM3	Private	20 or 21
44	Knapp	James	?	?	?
45	Knapp	?	?	?	?
46	Knapp	John	OCM3	Private	21 or 22
47	Knapp	Samuel	OCM3	Lieutenant	?
48	Lane	Ezekiel	?	Private	23
49	List	Jacob	OCM3	?	18 or 19

Number	Last Name	First Name	Regiment (If known)	Rank (If known)	Age
50	Little	James	?	?	?
51	Little	John	OCM3	Captain	?
52	Little	Samuel	?	?	?
53	Lockwood	Joshua	?	?	?
54	Mastin (Masten)	Ephraim	?	Ensign	?
55	Mead	Stephen	SCM2	Private	?
56	Meeker	Samuel	SCM2	Major	?
57	Middaugh	Ephraim	OCM3	Ensign	?
58	Middaugh	Solomon	SCM2	Lieutenant	23
59	Miers (Myers)	Daniel	OCM3	?	21 or 22
60	Miller	John	OCM4	Private	?
61	Mosher	James	?	?	?
62	Nierpos (Nearpass)	Baltus	?	?	?
63	Norris	Joseph	?	?	?
64	Owens	Eleazer	UCM2	Private	?
65	Patterson	?	?	?	?
66	Pierce	Jonathan	?	?	?
67	Place	Reuben	?	?	16
68	Poppino	John	OCM4	Major	52
69	Poppino	John	OCM4	?	28
70	Reed	Daniel	?	?	?
71	Reeve	James	OCM3	Private	17
72	Rider	Joseph	?	?	?
73	Ried (Reed)	John	OCM3	Private	24
74	Sawyer	James	OCM3	?	?
75	Schoonover	Thomas	SCM2	Private	?
76	Shepherd	Abraham	?	Lieutenant	?
77	Stewart	Nathan	OCM3	Private	20
78	Swarthout	Anthony	?	?	20
79	Talmadge	Daniel	SCM2	?	?

Number	Last Name	First Name	Regiment (If known)	Rank (If known)	Age
80	Terwilliger	Matthias	UCM2	?	?
81	Thomas	Moses	?	?	?
82	Totten	Levi	?	?	20
83	Townsend	Roger	?	?	?
84	Tusten	Benjamin	OCM3	Lieutenant Colonel	35
85	Tyler	Bezaleel	?	Captain	34 or 35
86	Vail	Benjamin	UCM2	Captain	?
87	Vail	Gilbert T.	?	?	?
88	Van Fleet	James	OCM3	Sergeant	26
89	Vanfleet	Daniel	PL	?	29 or 30
90	Vanfleet	Samuel	SCM2	Sergeant	20
91	Vangorder (Vangorden)	Samuel	?	?	15 or 16
92	Wade	Nathan	SCM2	Private	?
93	Wait (Wade)	Simon	?	?	?
94	Wallace	John	UCM2	?	?
95	Ward	Isaac	?	?	?
96	Waters	Thomas	OCM3	Private	16 or 17
97	Whitaker	Benjamin	OCM3	Private	27 or 28
98	Williams	Abraham	?	?	?
99	Wisner	Gabriel	?	?	?
100	Wisner	Henry	OCM4	Lieutenant Colonel	58 or 59
101	Wood	John	OCM3	Captain	39 or 40
102	Wood	John	OCM4	Lieutenant	?
103	Woodward	Hezekiah	UCM2	Sergeant	23 or 24

List of Participants at the Battle at Minisink Ford by Unit Roster

One of the challenges in creating this list is that family genealogists and historians may make a case that they can prove, for example, that their ancestor lived in Goshen at the time of the battle and, though they will probably be right in placing him in the Goshen Regiment, we could not make that determination with certainty. The pension applications are not always clear as to where someone resided.

In addition, even though someone may be proved to have been from Goshen, it does not necessarily mean that the soldier was serving in the Goshen Regiment. Benjamin Carpenter, for example lived within the bounds of the Goshen and Minisink Regiment, but at the time of the battle, he was serving with Pawling's Levies.

The following is our effort to create a listing by unit.

Fourth Regiment, Orange County Militia
"the Florida and Warwick Regiment"
Colonel John Hathorn
Lieutenant Colonel Henry Wisner
Major John Poppino
Adjutant Nathaniel Finch
Lieutenant John Wood
Ensign Henry Bertholf
Private William Curry
Private John Miller
John Poppino

Third Regiment, Orange County Militia
"the Goshen and Minisink Regiment"
Lieutenant Colonel Benjamin Tusten
Quarter Master James Sawyer
Captain John Duncan
Captain Samuel Jones
Captain John Little
Captain John Wood
Lieutenant Abraham Harding
Lieutenant Samuel Knapp
Ensign Ephraim Middaugh
Sergeant Jacob Aldrich
Sergeant Peter Kemble/Kimble
Sergeant James Van Fleet
Private David Dean
Private Benjamin Gunsolley
Private Jeremiah Kinner
Private John Knapp
Private James Reeve

Private John Ried/Reed
Private Nathan Stewart
Private Thomas Waters
Private Benjamin Whitaker
Nehemiah Baylis
John Brink
John Daniels
Jason Horton
Jacob List
Daniel Miers/Myers

Second Regiment, Ulster County Militia
"the Second or South End Regiment"
Captain Abraham Cuddeback
Captain Benjamin Vail
Sergeant Hezekiah Woodward
Private Timothy Barber
Private John Brooks
Private Eleazer Owens
Francis DeWitt
Matthias Terwilliger
John Wallace

Lt. Col. Albert Pawling's Regiment, New York Levies
"Pawling's Levies"
Private Joel Bower
Private Benjamin Carpenter
Private Uriah Chapman
Daniel Vanfleet

Second Regiment, Sussex County Militia
Major Samuel Meeker
Captain Joseph Harker
Captain Thomas Hill
Lieutenant Solomon Middaugh
Sergeant Samuel Vanfleet
Private Isaac Bedell
Private Solomon Cartwright
Private Moses Dewitt
Private Stephen Mead
Private Thomas Schoonover

Private Nathan Wade
Daniel Talmadge

Listing of Participants of The Battle At Minisink Ford
With explanation for Listing

The following is a detailed description of how individuals were determined to be eligible for listing on our inventory of participants.

Aldrich, Jacob (Sergeant) - In his application for a pension, he claims that he was not present at the battle because he was sent back with a file of men to get provisions. The phrase "sent back" implies that he started with the men in pursuit of Brant before being detached in search of supplies. In addition, the pension application of Benjamin Carpenter mentions that after traveling in pursuit of the enemy for a distance, several men were sent back in search of more powder. Aldrich was probably one of those men. His rank is established by statements made in his application for a pension stating that "he was an Orderly Sergeant during the whole time of his said service."

Bailey, Gamaliel - Killed, name on the monument in Goshen.

Barber, Timothy (Private) - Killed, name on the monument in Goshen. His rank is established by *National Archives Microfilm Publication M881 (Compiled Service Records of Soldiers Who Served in the American Army During the Revolutionary War); Roll #751 (McClaughry's Regiment, Militia)* indicating that he acted in this rank at the time of the battle.

Barker, William - Killed, name on the monument in Goshen.

Barney, David - Killed, name on the monument in Goshen.

Baylis, Nehemiah - His participation in the battle can be proved by statements in his application for a pension.

Bedell, Isaac (Private) - Though his pension application does not specifically state that he was in the battle, he does claim to have been in a severe skirmish at Scheholy (Shohola) under Major Meeker in August of 1779. In this engagement, Major Meeker received a musket shot to the shoulder. This is the same wound Meeker received at the Battle of Minisink.

Bedell also states that of those engaged the New York troops suffered the greatest number of casualties, which was true of the Battle of Minisink. He even places the severe skirmish in the month of August, 1779, which places it relatively close to the time of the Battle of Minisink. Therefore, it is easy to conclude that he is talking about the Battle of Minisink. His rank is established by testimony in his application for a pension stating that he "did duty for 15 days as a private upon this occasion" when talking about the "severe skirmish" discussed above.

Bennett, Benjamin - Killed, name on the monument in Goshen.

Bertholf, Henry (Ensign) - His participation in the battle can be proved by statements in his widow's application for a pension. His rank is established by numerous statements in his pension application that he served as an Ensign during the war. More importantly, though, is a letter from the Comptroller's Office in Albany, dated April 29, 1846, in which it is shown that he was serving on June 10th, and 25th, as well as July 1st of 1779, as an ensign.

A letter in his application dated February 10, 1847 from Henry W. Bertholf, grandson of Henry Bertholf, to the Commissioner of Pensions, states that Robert Carr was with Henry Bertholf at the Battle of Minisink but Carr's own statement in Bertholf's application makes no such claim, only that one or more alarms were raised, including one at Minisink in 1777. In addition, Robert Kerr (as his name is spelled on his own application for a pension) never mentions actual involvement in the battle when trying to secure his own pension, only that he responded to alarms at Minisink. Therefore, there is not enough evidence to include Robert Carr as a participant in the battle.

Bower, Joel - In his pension application, he mentions that in the spring of 1782 he was sent to reinforce Colonel Spaulding (Pawling) on the frontiers, and that he joined the company of Captain Abraham Westfall. As has been discussed, Westfall was a lieutenant under Captain John Wood in Lieutenant Colonel Albert Pawling's Regiment of Levies in 1779.

John Wood raised the company, and after his capture, his lieutenants commanded the company for the rest of their term of service. In the following years Westfall served as a captain, as can be seen on pages 255-261 of *New York in the Revolution* by Berthold Fernow. This is probably why Bower refers to Captain Westfall's company.

Bower states that Spaulding's Regiment was divided into companies, and guarded a distance of about sixty miles, which is exactly how Lt. Col. Pawling's regiment was posted along the frontier from Great Shandaken to Peenpack. He claims that his company was the only one of his regiment that was in the battle, which is also true of Pawling's Levies. Captain John Wood's company was the only company from Pawling's Levies at the Battle of Minisink. In addition, he mentions that a Captain Wood was taken prisoner at the battle, and that the majority of the men were militia.

He also tells of how the Indians fell upon a section of the frontier, and that the militia pursued them, eventually being defeated when they were able to engage the enemy. He describes being wounded in the engagement and spending the night under a log. The next morning he and several others floated back down the Delaware on a makeshift raft. He is clearly talking about the Battle of Minisink.

Brant, Joseph (Captain) - His participation in the battle can be proved by his letter to Lieutenant Colonel Bolton reporting the battle (*Sir Frederick Haldimand: Unpublished Papers and Correspondence 1758-1784; B.100; 21,760; 173-174)*. A letter from Lord George Germain to Sir Frederick Haldimand dated April 16,

1779 states that "The astonishing Activity & Success of Joseph Brants Enterprizes and the important consequences with which they have been attended, give him a claim to every Mark of our Regard and which you think will be pleasing to him. What has occurred to me as most likely to gratify him, has been done and inclosed herewith you will receive a commission signed by His Majesty appointing him a Colonel of Indians" *(Sir Frederick Haldimand: Unpublished Papers and Correspondence 1758-1784; B.043; 21,703; 119-125).*

Though Sir Frederick Haldimand, the acting Governor of the Province of Quebec, received this letter, he suppressed it because it had the potential to harm Brant's position among the Indians. He was still young compared to other Indian war leaders, and had only recently gone to war.

Furthermore, he was not yet recognized by the senior war chiefs as being a seasoned warrior. Because of this, it was feared that the other Indians would become jealous and hateful of him. Therefore, his commission as colonel was rejected by Haldimand and he instead received an official commission as "Captain of the Northern Confederated Indians" in 1780 *(Sir Frederick Haldimand: Unpublished Papers and Correspondence 1758-1784; B.085.1; 21,745; 60).* He is the only Indian known to have received an official commission, and that not until after the Battle of Minisink. Therefore, any suggestion that he served as a colonel at the time of the battle must be dismissed.

Brink, John - His participation in the battle can be proved by statements in his widow's application for a pension. Henry Van Etten provides a statement for her application, and one would be tempted to include Van Etten as a participant in the battle as well, except that the statement is not detailed enough. The statement only says that Van Etten was among those called out on the 21st to fight the Indians and Tories but it does not place him in the battle. He must therefore be excluded because he could have been sent on an errand, taken sick, or any number of things before the march up the river began.

In the May 23, 1839 declaration of Dianna Brink, the names of Vantyle and Major John Decker must be ignored for the same reasons, especially since Decker was wounded in the raid two days previous. James Van Fleet's statement, unlike Van Etten's, clearly places himself with Brink in the battle.

Brooks, John (Private) - His participation in the battle can be proved by statements in his application for a pension. His rank is established by testimony found in his pension application from Richard Varick and Richard Platt, two persons appointed by New York State to examine people applying for invalid pensions in order to determine their entitlement. In their testimony, dated April 18, 1787, Brooks is listed as a private at the time he was wounded in the Battle of Minisink. This is thought to be the most accurate document in his file concerning his rank.

In Brooks' statement of March 6, 1833, he names Major Thomas Dewitt as being at the battle. However, no other information has been found placing Dewitt at the battle. Dewitt did serve as a major on the frontiers in Pawling's Levies, but only a single company of that regiment was in the action. Therefore, there is reason to think that Brooks may be mistaken. Brooks claims to have kept garrison at Fort DeWitt from 1780 until the close of the war, and Dewitt, acting as a Major along the frontier, would have had plenty of business at Fort Dewitt. Unless new information comes to light, Dewitt must be excluded from the list of known participants in the battle. Brooks also mentions a Brazilla Tyler at the battle who is obviously a distortion of Bezaleel Tyler.

Carpenter, Benjamin - In his widow's application for a pension, she claims that he started in pursuit of Brant, but after traveling about four miles, he and two other men were sent in pursuit of powder and were unable to rejoin the regiment.

Carpenter, John - Killed, name on the monument in Goshen.

Cartwright, Solomon - His participation in the battle can be proved by statements in his application for a pension.

Chapman, Uriah - His participation in the battle can be proved by statements in his widow's application for a pension.

Cole, ? - His participation in the battle under Joseph Brant can be proved by testimony given in John Knapp's account of the battle *(Draper Manuscripts; 8F144)*. It is reported by Knapp that an exchange took place between Daniel Myers and a Mr. Cole in the heat of the battle. Myers supposedly knew Cole and it is to be noted that traditions about the raid and battle have long placed a man named Cole as being with Brant. However, there have been many variations of what his first name was.

Tories named Cornelius Cole, Case Cole and Daniel Cole have all been credited with being under Brant during the raid on Minisink and the ensuing battle. (It is worth nothing that Case is a nickname for Cornelius and Cornelius and Case may in fact be the same person.) In *Volume 5*, page 163 of *The Public Papers of George Clinton, First Governor of New York*, Moabary Owens lists a Daniel and Johs. (Johannes) Cole as being with Brant at Chemung on July 8th.

Cuddeback, Abraham (Captain) - In the pension application of Daniel Vanfleet, William A. Cuddeback states that he saw Vanfleet start in pursuit of Brant the day before the Battle of Minisink in company with Captain Abraham Cuddeback. His rank is established by National Archives Microfilm Publication *M881 (Compiled Service Records of Soldiers Who Served in the American Army During the Revolutionary War); roll #751 (McClaughry's Regiment, Militia)* indicating that he acted in this rank at the time of the battle.

Curry, William (Private) - In his application for a pension, he states that he was left to take care of the horses the morning of the battle, five miles from the scene of action, so he clearly marched with Hathorn in pursuit of Brant. His rank is established by testimony he gives in his pension application that specifically shows he was serving in that rank at the time of the battle.

Daniels, John - His participation in the battle can be proved by statements in his application for a pension.

De Shay, William (Private) - Though his pension application does not specifically state that he was in the battle, he does claim that he was with a large party that was hastily formed in pursuit of Indians within the bounds of New York State and near the Delaware River during one summer. He says that this party marched in pursuit all day with the intention of surprising the enemy at night, but that towards night they themselves were attacked and surrounded, all being killed except eight or ten men who saved themselves by running. He states that the enemy were commanded by Brant, and that the engagement occurred some miles from the settlements. This is unmistakably the Battle of Minisink. His rank is established by testimony in his application for a pension that makes it clear that he only served in that rank during the war.

Dean, David - In his application for a pension, he said that John Ried was with him at Beaver Brook where Colonel Tusten was killed in 1779. This is sufficient evidence to include his name.

Decker, Joel - Killed, name on the monument in Goshen.

DeWitt, Francis - His participation in the battle can be proved by statements in his application for a pension.

Dewitt, Moses (Private) - His participation in the battle can be proved by statements in his application for a pension. His rank is established by testimony in his pension application submitted by Solomon Middaugh that states that Dewitt served as a private in the same company during the entire war.

Duncan, John (Captain) - Killed, name and rank on the monument in Goshen.

Dunning, Benjamin (Lieutenant) - Killed, name on the monument in Goshen. In *The Royal Gazette* of August 18, 1779, he is listed as a lieutenant. The fact that this paper correctly named seventeen men who were killed in the battle less than a month after the action, along with their ranks or positions of importance in the community, leads us to believe that Dunning was acting as a lieutenant at the time of the battle and is listed as such in this work.

Dunning, Jacob - Killed, name on the monument in Goshen.

Embler, Adam - Killed, name on the monument in Goshen.

Finch, Nathaniel (Adjutant) - Killed, name and rank on the monument in Goshen.

Forgerson (Ferguson), Ephraim - Killed, name on the monument in Goshen. It should be mentioned that Hathorn, in his report of the battle to Governor Clinton, lists a man by the same name as having been killed in the raid a few days previous to the battle. Whether this is the same man mistakenly listed on the monument or another man with the same name, cannot be determined.

Gunsaluss, Henry (Private) - He is considered a participant based upon a statement he gave in his pension application. He stated that "during this period of service he was engaged in several skirmishes with the Indians, in one of which Capt. Little of the said regiment was killed." He confuses the year, but that can be explained by faulty memory. It was concluded that he must be referring to the Battle of Minisink because there is no other known skirmish with Indians in present day Orange, Sullivan or Ulster Counties where a Captain Little was killed.

He also places the skirmish in Ulster County, which the Battle of Minisink was fought in, as Sullivan County was not yet formed. He mentions belonging to Colonel Newkirk's Regiment which was often posted on the frontiers and Fort Dewitt, which places him in the right area. However, according to other pension applications, and on page 550 of Fernow's *New York in the Revolution*, under "At Minisink on the Delaware, July 22, 1779," where it talks about men killed in action, it lists Little as belonging to Colonel Allison's Regiment when he was killed, which appears to be correct. His confusion about which regiment Little belonged to could be explained by faulty memory after so many years. His rank is established by testimony in his pension application stating that he served only as a private.

Gunsolley, Benjamin (Private) - His participation in the battle can be proved by statements in his application for a pension. His rank is established by testimony in his application for a pension that indicates that he only served as a private.

Hains (Haynes), Benjamin - His participation in the battle can be proved by statements in his application for a pension. In the same pension application, Aaron Westbrook testifies that he was with Hains "at the battle with the Indians at Lackawaxen." This statement can be misleading. One's first thoughts are that this is the Battle of Minisink and that Westbrook is claiming that he and Hains were in the battle together.

This is not the case. In August 1778, after receiving reports that Brant was in the Cochecton area, a force of militia from New Jersey, New York and Pennsylvania, as well as a detachment of Continental troops rendezvoused at the mouth of the Lackawaxen River and marched up to Cochecton in order to deal with the threat from that quarter. While on this expedition, at least one skirmish was had with the Indians. This could be the skirmish he is referring to.

In addition, the Battle of Minisink has been referred to as the Battle of Cochecton in pension applications, including those not transcribed for this work. In Westbrook's testimony, it is written that he was with Hains"at the battle with the Indians at ~~Cochecton~~ Lackawaxen." Clearly, he is not placing himself at the Battle of Minisink, but at a skirmish at the Lackawaxen because "Cochecton" is crossed out.

Throughout Hains' pension application the Battle of Minisink is referred to as "the Battle at Halfway Brook", "the Battle at Beaver Brook," or "the Battle at Cochecton." It is never referred to as "the Battle at Lackawaxen." The engagement with Indians at the Lackawaxen is clearly identified as a separate skirmish throughout Hains' application, with several people giving testimony for Hains pointing out that he was in both battles.

Westbrook makes no mention of either battle in his own application, adding to the confusion. Therefore, there is reason to doubt his involvement in the battle and unless more evidence is found to place him there, Westbrook must be left off the roll of known men at the Battle of Minisink.

Harding, Abraham (Lieutenant) - His participation in the battle can be proved by statements in his application for a pension. Despite claims that he served as a captain throughout the war, he is listed on page 291 of Fernow's *New York in the Revolution*, as a lieutenant in Captain John Little's Company of the Goshen Regiment on February 26, 1778. On the following page, he is listed as being elevated to the rank of captain to replace Captain Little who fell in this battle.

Harker, Joseph (Captain) - His participation in the battle and his rank can be proved by referring to *NJDARM: New Jersey State Archives Holdings: Sussex County - Clerk's Office, Courts and Miscellaneous; Series: Sussex County Court of Common Pleas/Court of General Quarter Sessions Rough [Pamphlet] Minutes, 1776-1852; Control #CSUCP005; Box 1b August term, 1784; Sussex County, Court of Common Pleas and Quarter Sessions. August Term, 1784 (Folder #22)*. In these records can be found a certificate from both Major Samuel Meeker and Colonel John Hathorn that detail how Harker served under them in the battle. His listing as a captain can also be supported by testimony in the pension application of Jason Horton who mentioned Harker in that rank.

Haskell, Jonathan - Killed, name on the monument in Goshen.

Hathorn, John (Colonel) - His participation in the battle and rank are confirmed by his letter to Governor George Clinton after the battle (see *Hathorn's Dispatch to Governor George Clinton, Draper Manuscripts; 20F35*).

Hill, Thomas (Captain) - He is considered a participant based upon a statement given by Isaac Bedell in Hill's widow's application for a pension. Bedell states that Hill was in an engagement at Sheholi where Daniel Talmage was killed and Bedell was wounded in the thigh. This is clearly the Battle of Minisink because Daniel Talmage was killed in the battle.

In addition, refer to Isaac Bedell above for further discussion on the engagement at Sheholi. Bedell also mentions that they were under Colonel Seward, but it is well known that Seward was on the march up the Pennsylvania side of the Delaware River attempting to aid Hathorn, and was not present at the battle (see *Hathorn's report to Governor George Clinton, Draper Manuscripts; 20F35*). This reference to Seward only means that Seward was the commander of their regiment, and not that he was present at the battle.

Horton, Jason - His participation in the battle can be proved by statements in his application for a pension. He lists a Captain Brown of the Continental Army commanding a company of Artillery and pursuing Brant with them. This is most likely an error. The only possibility that a company of artillerymen could have been involved is if they were passing through the area to join Sullivan's Expedition (still not yet underway) or if they were passing through on their way to somewhere else. Therefore, unless new information comes to light, Captain Brown is not included as a participant in the battle.

Jennings, Isaac - In a statement given by Daniel Poppino in the pension application of John Poppino (son of Major John Poppino and brother to Daniel), it is mentioned that John Poppino and Isaac Jennings were placed as guards over the horses of those who went into the battle. For further discussion about Jennings, see *Draper Manuscripts; 8F155 and 8F156*. Poppino's pension application and the Draper Manuscripts, when taken together, support inclusion of his name as a participant.

Jones, Samuel (Captain) - Killed, name and rank on the monument in Goshen.

Kemble (Kimble), Peter (Sergeant) - His participation in the battle can be proved by statements in his widow's application for a pension. His rank is established by testimony given in his widow's pension application that specifically shows he was serving in that rank at the time of the battle.

Kinner, Jeremiah (Private) - His participation in the battle can be proved by statements in his application for a pension. His rank is established by testimony given in his pension application that points out that he joined Captain McDowell's (McDowell was a lieutenant in July, 1779) company at the time the militia assembled to pursue Brant and that he finished his term of enlistment as a private at a fort near the residence of Captain Jacob R. DeWitt (at Peenpack). In his application he says that he rode to Minisink with two men named Bradley. There is no mention as to whether they joined in the pursuit of Brant or if they were left at the settlement when the others marched up the Delaware, so they are not listed as participants.

Knapp, James - Killed, name on the monument in Goshen.

Knapp, ? - In John Knapp's account of the battle *(Draper Manuscripts; 8F144)*, he claims that his brother, while trying to get away, "was wounded while crossing the river in his thigh but escaped and got in the next day." This is sufficient evidence to include his brother's name, whose first name is currently unknown.

Knapp, John - His participation in the battle can be proved by statements in his application for a pension.

Knapp, Samuel (Lieutenant) - Killed, name on the monument in Goshen. In *The Royal Gazette* of August 18, 1779 he is listed as a lieutenant. The fact that this paper correctly named, less than a month after the action, seventeen men who were killed in the battle along with their ranks or positions of importance in the community, leads us to believe that Knapp was indeed serving as a lieutenant at the time of the battle, and is listed as such in this work.

Lane, Ezekiel (Private) - His participation in the battle can be proved by statements in his application for a pension. His rank is established by testimony given in his pension application that states that he " Volunteer'd as a Private, in the service of the United States during the Revolutionary War."

List, Jacob - His participation in the battle can be proved by statements in his application for a pension.

Little, James - Killed, name on the monument in Goshen.

Little, John (Captain) - Killed, name and rank on the monument in Goshen.

Little, Samuel - Killed, name on the monument in Goshen.

Lockwood, Joshua - Killed, name on the monument in Goshen.

Mastin (Masten), Ephraim (Ensign) - Killed, name and rank on the monument in Goshen.

Mead, Stephen (Private) - Killed, name on the monument in Goshen. His rank is established by referring to *NJDARM: New Jersey State Archives Holdings: Department of Defense - Colonial Wars and Revolutionary War; Series: Department of Defense Adjutant General's Office (Revolutionary War) Numbered Manuscripts, 1770s-1890s; Control #SDEA1001; Items #10230A, 10230B, 10230C* which clearly show that he held the rank of private at the battle.

Meeker, Samuel (Major) - His participation in the battle and rank can be proved by numerous pension applications that list him in this rank, and by Hathorn's report to Governor George Clinton after the battle (*Draper Manuscripts; 20F35*).

Middaugh, Ephraim (Ensign) - Killed, name and rank on the monument in Goshen.

Middaugh, Solomon (Lieutenant) - His participation in the battle can be proved by statements in his application for a pension. His rank is established both by his commission dated May 16, 1777, which is included in his pension application, and by testimony in the said application that specifically says that he was serving from 1779 until the close of the war at that rank, and was present at the battle during this time.

Miers (Myers), Daniel - His participation in the battle can be proved by statements in his application for a pension. He also claims that John Davis was his captain, and Samuel Cole his lieutenant at the time. That is not enough evidence to include them as participants. They were his company officers at the time, but he does not say definitively that they were at the battle with him.

Not everyone got to Minisink in time to join in the pursuit, and therefore there is a possibility that they did not arrive in time to take part in the battle, or were left behind when the others marched. Also, the only Captain John Davis encountered in the research was from the Second Regiment of Sussex County, New Jersey (see pension applications of Moses Dewitt, Solomon Middaugh and Samuel Vanfleet).

The reference to Major John Woodhull could easily be a distortion of Captain John Wood who years later was a Major. Therefore, John Woodhull cannot be counted among the men involved in the battle.

Miller, John (Private) - In order to prove his participation in the battle, one must refer to two pension applications because it is believed that records for John Miller were mistakenly placed in the file of Johannes Miller (W.16,650). In Johannes Miller's file there is a list of applicants for invalid pension returned by the District Court for the District of New York, submitted to the House of Representatives by the Secretary of War on April 25, 1794, and printed in the *American State Papers, class 9, page 94*. This is the document that is suspected to have been filed in the wrong person's application.

As one reads the file for Johannes Miller, one will see that there is no mention of him being wounded during the war or being either an applicant for, or in receipt of, an invalid pension. Every single source consulted says that the John Miller who was wounded at the battle was from the Warwick area.

The people testifying in support of Johannes Miller's claim give the impression that he is from Montgomery, or the surrounding area, and not from Warwick. It is believed that the document should have been filed under John Miller (S.28,128) instead. This John Miller (S.28,128) is from the Warwick/Florida area, and was wounded according to information found in his file. Not only that, but he was placed on the pension list by Act of March 23, 1792. This date is near to the above-mentioned information printed in the *American State Papers*. The file is, however, short, and nothing positively identifying this John Miller as the correct one is found within it. The John Miller who fought in the battle fought as a private based upon information found in the supposedly inaccurately filed document. Both pensions have been included in this book.

Mosher, James - Killed, name on the monument in Goshen.

Nierpos (Nearpass), Baltus - Killed, name on the monument in Goshen.

Norris, Joseph - Killed, name on the monument in Goshen.

Owens, Eleazer (Private) - Killed, name on the monument in Goshen. His rank is established by *National Archives Microfilm Publication M881 (Compiled Service Records of Soldiers Who Served in the American Army During the Revolutionary War); roll #751 (McClaughry's Regiment, Militia)* indicating that he acted in this rank at the time of the battle.

Owens, Mowberry - His participation in the battle under Joseph Brant can be proved by correspondence found in *Volume 5*, pages 162-164 of *The Public*

Papers of George Clinton, First Governor of New York.

Patterson, ? - A statement given by Joel Bower when applying for a pension, indicates that after the battle he spent the night under a log and when he made his way to the banks of the Delaware River the following morning, he found five other wounded militiamen, one of whom was named Patterson. This is sufficient evidence to include his name.

Pierce, Jonathan - Killed, name on the monument in Goshen.

Place, Reuben - His participation in the battle can be proved by statements in his application for a pension. However, there are doubts about his involvement. First off, no other men from Shoemaker's Rangers claim involvement in the Battle of Minisink, but that in itself does not mean he wasn't there. Second, Peleg Place, whom it is assumed may be a relative, said in his own application for a pension that he was on the march from the Stroudsburg area under Colonel Stroud to Beaver Brook when they received news of the Battle of Minisink. He accurately names several of the casualties, and says they returned to their starting point when word of the battle was received, there being no reason to continue on.

It is possible that Reuben Place was in the area, and had opportunity to join in the pursuit, but it seems more likely that he was a little confused in old age. Perhaps he was with Colonel Stroud's column and, like Peleg Place, was familiar enough with some of the names to accurately name the casualties in his own declaration.

The description of Tusten being shot from a horse while crossing a brook isn't even remotely accurate and with his reference to the battle as being fought at Beaver Brook, it is interesting that a brook is mentioned when there is none in the vicinity of where Tusten really fell. However, he does mention an early effort to bury the bodies but that they were prevented from completing the task. Nathan Stewart, a participant, also tells of returning to the battleground to recover the dead, and having to abandon their plans. Similar statements like these concerning recovering the dead make it impossible to dismiss his claim, and therefore his name must be listed as a participant.

Poppino, John (Major) - His participation in the battle and rank can be proved by a copy of his obituary which appeared in *The New York Spectator* on April 4, 1828 (*Draper Manuscripts; 8F122*), and in *The Goshen Independent Republican* of March 24, 1828. In addition, the August 7, 1779 edition of *The Pennsylvania Packet* and the August 14, 1779 edition of *The Royal Gazette* both contain an extract of a letter from Orange County dated July 29th, which gives an account of the events at Minisink, and mistakenly lists a Major Papan as being killed in the engagement. This Major Papan is obviously a distortion of the name Poppino.

Poppino, John - This is Major John Poppino's son. His participation in the battle can be proved by statements in his widow's application for a pension. Daniel Poppino, his brother, states that John was placed as a guard over the horses of those who marched into the battle, as was Isaac Jennings.

Reed, Daniel - Killed, name on the monument in Goshen.

Reeve, James (Private) - His participation in the battle can be proved by statements in his children's application for a pension. His rank is established by referring to *New York State Comptroller's Office Revolutionary War Accounts and Claims (1775-1808); Series Number: A0200; Roll # 11; Volume 16: Applications for Pensions*. He mentions that Colonel Allison was his colonel at the time of the battle, but as is well known, Allison was a prisoner of the British in New York at the time of the battle and the regiment was led by Lieutenant Colonel Benjamin Tusten in his absence.

Rider, Joseph - Killed, name on the monument in Goshen.

Ried (Reed), John - His participation in the battle is based upon a single sentence in the pension application of David Dean in which Dean states that John Ried was with him at Beaver Brook where Colonel Tusten was killed in 1779. James Knapp, who fell at the battle, is listed as a captain according to Ried, but no other record has shown Knapp serving in that rank at the battle. Therefore, until more proof is found, Knapp's rank is still considered unknown.

Sawyer, James - His participation in the battle is based upon a statement in the pension application of Thomas Waters. In the statement, he mentions several officers, one of which was a Captain Sawyer, as having been in the action. He also lists Sawyer as having been killed. However, there is no mention of any man named Sawyer being killed at the battle in any record or previous work.

The fact that he mistakenly lists Sawyer as having been killed in the engagement leads one to think that he is convinced of Sawyer's having been there, mentioning him twice. All of the other officers he listed are known to have been in the engagement, and he lists them with their correct rank, making his testimony very credible. He does not give the first name of Captain Sawyer, but on page 291 of Fernow's *New York in the Revolution*, under the heading "Levies and Militia," is found a commissioned officer named James Sawyer serving as quartermaster at some point during the war.

In the book *Register of the Empire State Society of the Sons of the American Revolution*, James Sawyer is listed as a captain in Colonel William Allison's Regiment and it is stated that he was at the Battle of Minisink. His rank at the time of the battle cannot be determined with accuracy, despite Waters' claim that he was a captain.

Schoonover, Thomas - His widow's pension application does not mention the Battle of Minisink by name, but it is somewhat obvious when you take into account that several members of Captain Cole's Company have been identified as being present at the battle (see Moses Dewitt, Solomon Middaugh and Samuel Vanfleet). Solomon Middaugh even claims that seven men from said company were killed at the battle.

The month is also right for the battle, but the year mistaken. The battle mentioned is placed in the proper locale for the Battle of Minisink. The application states that this soldier marched with Captain Cole into that region, but there is reason to wonder if Cole actually commanded the company, or if Solomon Middaugh commanded it in his absence. Therefore, Cole cannot be listed as being present at the engagement without further evidence.

Shepherd, Abraham (Lieutenant) - Killed, name on the monument in Goshen. In *The Royal Gazette* of August 18, 1779, he is listed as a Lieutenant. The fact that this paper correctly named seventeen men who were killed in the battle less than a month after the action, along with their ranks or positions of importance in the community, leads us to believe that Shepherd was acting as a lieutenant at the time of the battle, and is therefore listed as such in this work.

Stewart, Nathan (Private) - His participation in the battle and his rank can be proved by statements in his application for a pension. He mentions marching to Minisink under Lieutenant Gale, but there is no mention of Gale being in the battle itself. (As not everyone marched against Brant when they arrived at the settlement, further proof is needed before Gale's name can be included.)

Swarthout, Anthony - His participation in the battle can be proved by statements in his application for a pension.

Talmadge, Daniel - Killed, name on the monument in Goshen.

Tayojaronsere, John (also called Captain John, John the Mohawk and Tuscarora John) - His participation in the battle is proved by Joseph Brant's report of the battle to Lieutenant Colonel Mason Bolton *(Sir Frederick Haldimand: Unpublished Papers and Correspondence 1758-1784; B.100; 21,760; 173-174),* where Brant mentions that John the Mohawk and three others were dangerously wounded and unlikely to recover.

In a letter to Daniel Claus dated August 19th, 1779, from Chemung *(Draper Manuscripts; 9F207)*, Brant writes of John Tayojaronsere's death eight days after the battle and how five others met the same fate. Captain John Wood's journal detailing his captivity *(Draper Manuscripts; 9F89)* mentions his death, and refers to him as "Tuscarora John."

Terwilliger, Matthias - Killed, name on the monument in Goshen. Vernon Leslie, in his book on the battle, lists a Nathaniel Terwilliger but the monument says Matthias Terwilliger. The first name Matthew has also been seen in other sources. Matthias is German and a variant of Matthew.

Thomas, Moses - Killed, name on the monument in Goshen.

Totten, Levi - His participation in the battle can be proved by statements in his application for a pension.

Townsend, Roger - Killed, name on the monument in Goshen.

Tusten, Benjamin (Lieutenant Colonel) - Killed, name on the monument in

Goshen. His rank is established by Hathorn's report to Governor George Clinton after the battle *(Draper Manuscripts; 20F35).*

Tyler, Bezaleel (Captain) - Killed, name and rank on the monument in Goshen.

Vail, Benjamin (Captain) - Killed, name and rank on the monument in Goshen.

Vail, Gilbert T. - Killed, name on the monument in Goshen.

Van Fleet, James (Sergeant) - Nothing in his pension application places him in the battle. However, in the application of John Brink's widow for a pension, there is a statement from James Van Fleet who states that "Brink was in the action or Battle of Beaver Brook or at the mouth of the Lackawaxen River, together with myself and many others, were in that hard fought battle." His rank is established by testimony in his application for a pension, which states that "For three years commencing in 1778 I served as a minute man and was an orderly sergeant."

Vanfleet, Daniel - His participation in the battle can be proved by statements in his application for a pension. He mentions joining the company of Captain Abraham Westfall, and joining in the pursuit of Brant, but Westfall cannot be positively placed at the battle. Even the pension application of Abraham Westfall is silent on whether he took part in the battle.

Vanfleet, Samuel (Sergeant) - In his application for a pension, he states "That he marched with his company under Capt. Cole towards Lackawack at the time of the Battle of that place, but became so lame as to be unable to proceed on, to the place of the Battle." Clearly, he started in pursuit of Brant with the others.

His rank is established by testimony in his pension application where he lists himself as a first sergeant, and others confirm that he served as a sergeant. (As discussed under Thomas Schoonover, above, there is reason to wonder if Cole actually commanded the company at the battle, or if Solomon Middaugh commanded in his absence. Therefore, Cole cannot be listed as being present at the engagement without further evidence.)

Vangorder (Vangorden), Samuel - His participation in the battle can be proved by statements in his application for a pension.

Wade, Nathan (Private) - Killed, name on the monument in Goshen. His rank is established by referring to *NJDARM: New Jersey State Archives Holdings: Department of Defense - Colonial Wars and Revolutionary War; Series: Department of Defense Adjutant General's Office (Revolutionary War) Numbered Manuscripts, 1770s-1890s; Control #SDEA1001; Items #10257A, 10257B, 10257C, 10257D* which clearly show that he held the rank of private at the battle.

Wait (Wade), Simon - Killed, name on the monument in Goshen.

Wallace, John - His participation in the battle can be proved in the pension application of Cornelius Wallace. Cornelius, his son, states that "The enemy retreated up the Delaware River & in a day or two afterwards, a party of the militia started in pursuit & came up with the enemy on the second day at Beaver Brook. Capt. Cuddeback & this Deponents father were of the number."

WARD, ISAAC - Killed, name on the monument in Goshen.

WATERS, THOMAS (Private) - His participation in the battle can be proved by statements in his application for a pension. His rank is established by testimony given in his pension application, in which he states that he served "in every case as a private."

WHITAKER, BENJAMIN (Private) - His participation in the battle can be proved by statements in his application for a pension. His rank is established by a certificate in his pension application from Doctor Jonathan Swezy that lists him as a private at the time of the battle.

WILLIAMS, ABRAHAM - Killed, name on the monument in Goshen.

WISNER, GABRIEL - Killed, name on the monument in Goshen.

WISNER, HENRY (Lieutenant Colonel) - His participation in the battle and rank can be proved in Hathorn's report to Governor George Clinton *(Draper Manuscripts; 20F35)*.

WOOD, JOHN (Lieutenant) - Killed, name and rank on the monument in Goshen. It is to be noted that in Hathorn's report of the battle to Governor George Clinton *(Draper Manuscripts; 20F35)*, Hathorn lists him as an ensign. However, in *The Royal Gazette* of August 18, 1779, he is listed as a lieutenant. The fact that this paper correctly named seventeen men who were killed in the battle less than a month after the action, along with their ranks or positions of importance in the community, leads us to believe that Hathorn erred in his report, and that Wood was acting as a lieutenant at the time of the battle, and is therefore listed as such in this work. The fact that the monument also lists him as a lieutenant furthers the case that Hathorn may have been mistaken on Wood's rank in his report to Governor Clinton.

WOOD, JOHN (Captain) - His participation in the battle and his rank can be proved by testimony he gives in the pension application of Nehemiah Baylis, the journal of his captivity *(Draper Manuscripts; 9F89)*, and many other sources.

WOODWARD, HEZEKIAH (Sergeant) - His participation in the battle can be proved by statements in his application for a pension. His rank is established by testimony given in his application for a pension which states that, with the exception of a brief period long before the Battle of Minisink, he served as a sergeant during the entire war.

CONTEMPORARY QUOTES FROM PARTICIPANTS

The only documents that contain direct quotes regarding the battle are the pension file of Nathan Stewart, and John Knapp's account of the Minisink Battle *(Draper Manuscripts; 8F144)*. In the Nathan Stewart file, he claims that Lieutenant Colonel Tusten was wounded three times, and saw the fire cease as the ammunition began to run short. He claims that Tusten said "My brave fellows, every man for himself and the Lord for us all. I never will leave this ground." Those words, however, have also been attributed to Colonel Hathorn in other sources.

When one reads Hathorn's report to Governor George Clinton *(Draper Manuscripts; 20F35)*, you see that Hathorn states he was wounded three times in the engagement while mentioning that Tusten was wounded once, in the hand, whereas Stewart's account claims that Tusten was wounded three times, not Hathorn. This suggests the possibility that Stewart is confusing who said what. It is impossible to know who called out those words, but it is possible that Stewart, after so many years, is mistaken.

It is noteworthy that he mentions Tusten's determination not to leave the field. It is thought that Hathorn and Tusten were both in the general vicinity of Hospital Rock when the rout began because Hathorn, in his report to Governor Clinton, stated that "the cries and petitions of the wounded around me not to leave them, was beyond parallel or idea" and the majority of the wounded had been collected at Hospital Rock. Therefore it is possible that Stewart may have witnessed Hathorn calling out "My brave fellows, every man for himself and the Lord for us all" while Tusten, who was also nearby, may have made it clear through words or actions that he had no intention of fleeing.

John Knapp's account of the Minisink Battle contains a quote also. Shortly before Brant's men succeeded in driving the militia from the field, there was an exchange in the heat of the battle between Daniel Myers and a Tory named Cole. According to Knapp, finding that they could not drive the militia off the hill, the enemy collected all their force in a thicket and tried to convince the militia to surrender. A Tory named Cole, known to Daniel Myers, identified himself as Captain Brant and promised quarter if the men surrendered, but it was refused.

This exchange obviously took place shortly before the militia was driven from the field because Knapp states that the enemy collected their force together before the exchange was made, and Hathorn's report to Governor George Clinton makes it clear that the battle ended when the enemy "formed their whole strength" and "forced the North East part of our Lines." There has long been debate about whether Brant did in fact attempt to get the militia to surrender, and Knapp's account of the battle should leave no doubt that an attempt was made, even if not by him personally, and, after the battle had been raging for some time.

There is another quote that speaks volumes about the eventful lives many of these men must have lived. This quote comes not from a participant in the battle, but from a judge who listened to Joel Bower, a veteran of the engagement, recite his service in the war. The judge remarked that "it seemed more like the narration of one who had risen from the dead, than one who had survived to this late period & now rehearsing to us the events of that important era as correctly as we now heard them."

Experience of the Militia

When looking at the pension files of the men who were in the battle, we see that they were much more experienced than some have argued they were. Some of the men had served in Fort Montgomery where they resisted the attack of British Regulars and Hessians before having to make their escape over the surrounding hills. Still others had faced British Regulars at the battles of Long Island, White Plains, Monmouth, Bound Brook, Trenton, Brandywine, Germantown and other places. Some had also served on the frontiers, scouting and fighting the Indians, for several years before the Battle of Minisink.

Among the most battle hardened of the men is Captain Thomas Hill who, according to his pension application, was in no less than six major engagements before the Battle of Minisink. He participated in Washington's retreat through New Jersey which came to be known as "the Mud-rounds" in late 1776, shortly before the Hessians were taken at Trenton. He then went on to fight in the Battles of Quibbletown, Short Hills, Brandywine, Germantown, and Monmouth. He was at Fort Mifflin when it was surrendered to the British, was in a skirmish at Aquakanonk Bridge, and had served on the frontiers against the Indians, all before his participation in the Battle of Minisink. Uriah Chapman, another participant in the Battle of Minisink, had been wounded in a skirmish with the Indians the previous year and still carried the ball in his body.

Joel Bower, also a veteran of the battle, was another experienced soldier. He was in the thick of the fighting at Monmouth in June 1778 and gave a detailed account of his experiences in that battle. He claimed that he "was on the right of the left wing in Clintons Brigade, I stood in the front rank where we sustained a tremendous cannonade as well as a heavy fire of musketry for a considerable time we then were impetuously charged by the Dragoons, and in their second charge a Dragoon aimed a deadly blow at my head which would have instantly killed me, but I raised my musket to shield my head from the meditated stroke, which caused his sword to glance, & cut off my two middle fingers & the little finger about half off, this was my right hand which was holding on the small of my gun, my hand has ever since been verry much disabled from this wound, as he was then passing me he made an underhanded blow at my head, and the point of his sword passed through my left eyebrow which made a deep incission & considerably stunded me, at this instant his horse was shot through the head & fell dead at my feet, as I recovered from the last blow I received, I drew my bayonet & would have terminated his existance had not Adjutant Johnson catched me by the shoulder & bade me desist saying 'no murder here you young rogue the man cries for quarter.'"

Joel Bower went into as much detail describing his other services during the war. When giving sworn testimony for his pension application, a Judge remarked that "it seemed more like the narration of one who had risen from the dead, than one who had survived to this late period."

How the Militia Were Equipped

The traditional claim that the pursuing militia was ill-equipped does in fact stand up when compared with the records. Benjamin Carpenter's pension application points out that he and two others were sent in quest of powder, but that they failed to find the regiment after procuring the powder and Hathorn's men were no better off as a result. In the pension application of Jason Horton, Jeremiah Horton states that the defeat of the militia was caused by the lack of ammunition.

According to information found in Item #7258 in *Department of Defense Adjutant General's Office (Revolutionary War) Numbered Manuscripts, 1770s-1890s* at the New Jersey State Archives, Clofas Letteer loaned a musket, powder horn, shot bag, one pound of powder, thirty-three balls and eighteen cartridges to Daniel Talmadge so that he could respond to the alarm at Minisink.

Hathorn's Hesitation to Follow the Indians

In the pension application of John Wisner (R.11,744) is a deposition by Jasper Armstrong. Armstrong says that after the war he used to hear his father Robert Armstrong, John Wisner and Colonel Hathorn talk about the Battle of Minisink. Hathorn used to say to the two men that "If you had been with me I never would have followed the Indians so far into the woods," and that he was pressured or "overpersuaded" by the other officers. Based upon this statement, Hathorn also appears to have had misgivings about following Brant.

Of note in the same pension application is testimony from Anna Post, daughter of John Wisner, that describes how, when word came of Nathaniel Finch's death at the hands of the Indians, his wife went into fits for several days, and everyone thought that she might die as a result. Anna's mother stayed with Mrs. Finch during this time until she recovered.

Assembly and Pursuit

In the pension application of Eliud Tryon's widow (W.16,183), William Wells gives testimony that when Lieutenant Colonel Tusten received news of the attack upon Minisink, they marched the same night from Goshen, and arrived at Dewitt's Fort in the morning, having traveled about eighteen miles. From Fort Dewitt, he claims that they marched to Van Etten's Fort on the Horn Road (present day Neversink Drive). This is incorrect.

In the pension application of Ebenezer Dunning (W.21,008), Levi Van Etten claims that after the Battle of Minisink those of Captain Wood's company that were neither killed nor captured returned and "took up their quarters at the deponents mother's house which had been spared from the flames and of which the soldiers made a kind of a fort."

Therefore, Van Etten's Fort could not have been their destination when they left Fort Dewitt because it was built after the battle. It is therefore believed that Wells really means that they either marched to Van Auken's Fort, which was the only fort held during the raid, or to the burning remains of Major Decker's Fort, both on the Horn Road.

Evidence exists to support either place as the starting point for the men who went in pursuit of Brant, and it is impossible to tell from which location the men actually began their march. Jeremiah Kinner, in his application for a pension, makes a case for Van Auken's Fort being the place where the militia assembled prior to their pursuit of Brant. He states that he and others rode to the fort at Minisink where, after dispersing some Indians who besieged the place, he "and his associates entered the Fort." He also claims that "the militia commanded by Colonels Thurston, Haythorn and Wisner entered the Fort on the same day." The fort he refers to must be Van Auken's Fort as it is obviously still standing, and as stated before, Van Auken's Fort is the only fort that Brant did not destroy during his raid on Minisink.

John Hall, who gave testimony in the pension application of William Winans and Adonijah Middaugh, in his own application, both make a case for Major Decker's Fort as the assembly point. Hall states that those who arrived on horseback at Major Decker's did not wait for those coming on foot before they pursued the enemy while Middaugh states that when they reached the remains of Major Decker's Fort, men were sent back to Fort Van Tuyle, and the main body of men pursued the enemy. Clearly, the force that followed after Brant, assembled and began their march from one of these two points on the Horn Road.

The Cochecton Trail and Skinner's Sawmill

In his report to Governor George Clinton *(Draper Manuscripts; 20F35)*, Colonel Hathorn states that he set out in pursuit of Brant "along the Old Kesheth-ton path." Unfortunately, the exact route of this trail has not been preserved in historical records and it will probably forever remain a mystery. That said, a few opinions can be formed based upon the research done for this work.

We now know that the militia assembled at the fort of either Major John Decker or Daniel Van Auken on the Horn Road (present-day Neversink Drive) and that they began their march from there. With this in mind, it is thought that the militia marched from their rendezvous on the Horn Road to the Peenpack Trail in Huguenot, which may have been one end of the "Old Cochecton Path." The present day road bearing that name in Huguenot was probably the original Peenpack Trail that Brant is said to have descended when he attacked Peenpack in 1778 and Minisink in 1779.

According to tradition, when Brant fell upon Peenpack, he first struck the home of a man named Westfall. Near the junction of the Peenpack Trail and present day Route 209 is an historic marker indicating the residence of Abraham Westfall. This being the original Peenpack Trail is supported by the marker. (Many old Indian paths were widened as they were used by settlers coming into frontier areas, and eventually made into wagon roads, and later, in the age of automobiles, paved.)

From Huguenot, the militia then marched up the Peenpack Trail to the junction with present-day Route 42. This conclusion is based upon an historic marker that stands near the intersection of these two roads. It is thought that when the marker was placed there by the State Education Department in 1936, it was placed based upon some knowledge of the route that the militia had actually followed.

From this point, the exact route is unknown, but the militia crossed the Mongaup River and more than likely continued their march along present-day County Route 31 to Glen Spey, and from there along present-day County Route 32 to the vicinity of Lochada Lake. Skinner's Sawmill, the campsite of the militia, is said to be near this place (see *History of the Minisink Country* by Horace E. Twichell).

There are several reasons why it is believed that the Lochada Lake area is the general location of Skinner's Sawmill. First, Lochada Lake is located three-and-a-half to four miles from Brant's campsite at the mouth of the Halfway Brook, and approximately eighteen miles from Minisink (see *Hathorn's Dispatch to Governor George Clinton in Draper Manuscripts; 20F35*). Second, in an unnamed newspaper clipping attributed to Horace E. Twichell dated October 7, 1922 and found within the files of the Minisink Valley Historical Society, it is said that the militia encamped "at midnight at Lochada, then called Skinner's Mills."

Third, according to William Curry's testimony in his pension application, the militia left their horses and baggage under a guard five miles from the scene of action, while Hathorn makes it clear that they left their horses and baggage under a guard before resuming their march from Skinner's. Therefore, because Hathorn is making it clear that the horses and baggage were left at Skinner's, Curry's claim that this was done about five miles from the scene of action means that Skinner's Sawmill was that distance from the scene of action. Lochada Lake is approximately that distance from where the battle unfolded. For all of these reasons, the Lochada Lake area is a very good candidate for the campsite of the militia.

From Lochada Lake, the militia probably first marched along present-day Haring Road, then Hillside Road to the mouth of the Halfway Brook where Brant had encamped the night before the battle. Here the militia saw the campfires of their enemy, as tradition states. It is interesting to note that Twichell, in the unnamed newspaper article mentioned earlier, claims that the militia left Skinner's Sawmill and followed down a stream from "Tallwood" to the mouth of the Halfway Brook. The exact location of "Tallwood" or the stream that Twichell claims the militia followed from there cannot be determined. Whether by the roads mentioned above or by a stream from "Tallwood," the militia probably marched from Skinner's Sawmill to the mouth of the Halfway Brook.

From Brant's camp, the militia probably continued their march up today's Route 97 along the Delaware River, the same path Brant was on further ahead, until somewhere near or just after the Beaver Brook, where they left the path and cut across the mountain to beat the enemy to the ford.

LENGTH OF THE BATTLE

According to Hathorn's report to Governor George Clinton *(Draper Manuscripts; 20F35)*, when describing his defense of the hill which lies within the present-day Minisink Battleground Park boundaries, he says that they defended that ground for nearly three hours. Later in the same report, he says that he was under the necessity of ordering the men to cease their fire as they had been firing for more than five hours. This implies that it was about an hour and a half from when they first exchanged fire near the ford to when they took their position on the hill in the present-day park.

Brant's letter to Mason Bolton reporting the battle *(Sir Frederick Haldimand: Unpublished Papers and Correspondence 1758-1784; B.100; 21,760; 173-174)*, dated July 29th, gives the duration of the battle as about four hours. John Knapp's account of the battle *(Draper Manuscripts; 8F144)* states that the "whole conflict it was believed lasted four hours." A number of pension files of men who were in the battle also place the engagement in the three-to-five hour range (see the files of Ezekiel Lane, Nathan Stewart and Thomas Waters). The discrepancies could be based upon failing memory or whether the person is counting from the first shots at the ford or from the time the militia became surrounded on the hill.

John Wood, however, gives a different length of time in his journal (see *Draper Manuscripts; 9F89*). He states that he was taken after a "severe conflict which lasted all most the whole day." He is the only person found to give a different length of time for the battle. It is to be remembered that Hathorn and Brant both wrote their reports within a week of the engagement, so their memories are much clearer than those giving depositions many years later. Therefore, based upon similar lengths of time given by the commanders of the opposing armies, it is safe to conclude that the engagement lasted within that three-to-five hour range, and not all day as some previous authors have suggested.

BREASTWORKS ON THE BATTLEFIELD

In several books, it has been written that the militia created breastworks from which they were able to beat back the attacks of the enemy. There is no evidence of this, and it is highly unlikely. Neither commander mentioned anything about breastworks. Hathorn only states that he held an advantageous piece of high ground, and that the enemy had to resort to ricocheting rounds off the surrounding rocks in an effort to dislodge his men.

One must also consider how, in the middle of a running fight up the hill, the men would have had time or energy to build breastworks. With the enemy all around, exposing yourself for even a few seconds could bring about your own death. What most likely happened was that the militia took what cover they could find, and probably moved a few things around, or as John Knapp claims in his account of the battle *(Draper Manuscripts; 8F144)*, "flat stones that were on the place were in many instances set up edgewise for our defence." This is the extent of any defensive works created on the hill.

One cannot account for what authors from the nineteenth century saw to lead them to believe that breastworks had been built, but it should be pointed out that many of them were writing in the period just after the Civil War when breastworks would have been something they would have been accustomed to reading or hearing about. In fact, it is possible that this influenced their interpretation of things they may have seen in the vicinity of the battleground, or how they imagined the battle unfolding. Another possible scenario is that what those nineteenth century writers saw was stone fencing that had been built in the years after the battle.

Desertion of the New Jersey Militia

Tradition has long held that the troops from New Jersey left their comrades from New York to bear the brunt of the fighting. Nearly every previous author has repeated the stories of desertion by men from New Jersey. Vernon Leslie, in his book *The Battle of Minisink: A Revolutionary War Engagement in the Upper Delaware Valley*, was the first person to seriously question that assertion.

While some men from New Jersey may have fled on their own during the engagement, nothing could be found that would support the common belief that those troops deserted en masse. There are actually fragments of evidence in different records that suggest that the tradition may not be accurate at all.

According to some authors, Major Meeker, the senior ranking officer from New Jersey, led his men away from the scene of action as soon as the fighting began. Still other authors claim that Brant, whether by accident or design, cut off the part of the militia commanded by Major Meeker, and that those men from New Jersey, as a result, were not in the battle. The men from New Jersey were never posted by themselves in the opening stage of the battle, and therefore could not have been cut off and forced from the field without a portion of the New York militia also being driven off.

Colonel Hathorn, in his report to Governor George Clinton *(Draper Manuscripts; 20F35)*, states that he "disposed of the Men into three Divisions, ordered Col Thurston to Command the one on the Right and to take post about three hundred yards distance on an Eminence to secure our Right, Lt. Col Wesner, with another Division to file of (sic) to the Left, and to dispose of himself in the like manner, In order to prevent the Enemy from gaining any Advantage on our flanks the other Division under my Command to attack them with that Vigour necessary to Strike Terror in such a foe."

Clearly, Wisner commanded the division closest to the Dry Brook. It is thought that Brant came up onto the plateau to the west of the brook and cut off a part of the militia. Though Brant's exact line of march onto the rear of the militia is disputed, it is obvious that Wisner's men were the ones who were cut off and fled, not Major Meeker's men. Tradition holds many stories on the higher-ranking officers who were in the battle, but not a single word on Wisner nor his escape from the battlefield. In the absence of any accounts concerning him on that day, it seems even more plausible that Henry Wisner's group was the one put out of the fight, and not Meeker's.

In addition, Vernon Leslie correctly argues that the New Jersey men did sustain a number of casualties that would rule out the likelihood of their fleeing the ground before the battle got under way. Also of note is the fact that of all the pension applications read during the research for this book, including those not transcribed in this work, there is not one that mentions the New Jersey men deserting their comrades, nor does there appear to be any animosity exhibited towards the New Jersey troops or Major Meeker. That alone should speak volumes.

As can be seen in records found in the New Jersey State Archives, and included in this work, Colonel Hathorn supplies Captain Harker of the New Jersey Militia with a certificate in which he says that Harker "Evinced his Bravery as a Soldier" during the battle. The wording in the certificate is interesting, as it leaves one under the impression that Hathorn was very pleased with Harker's actions in the battle. However one could be reading too much into the document's wording.

As evidenced by John Knapp's account of the battle *(Draper Manuscripts; 8F144)*, Meeker urged the men from New York to follow the Indians even though they were inclined to wait for reinforcements. It seems very unlikely that someone who was so anxious to get at the enemy would flee at the opening of the engagement. Meeker, like most of the other participants in the battle, was an experienced soldier; he was not new to combat. Pension applications that were read during the research for this book have shown him to be a very active officer. In the pension applications of Jonathan Catterlin (S.12,444) and Elias Clark (S. 12,479), Meeker is mentioned as having been present at the Battle of Germantown (October 4, 1777).

Benjamin Coykendall (S.12,607), in his application, claims that Major Meeker met up with two companies that Coykendall was leading to Monmouth on June 28, 1778, and took the command. They arrived at the site of the Monmouth battle the day after the fighting had ended.

In summary, there is no evidence to support accusations of cowardice levied against Major Samuel Meeker or the soldiers from New Jersey.

The Death of Benjamin Tusten

Tradition has long held that Lieutenant Colonel Benjamin Tusten, a doctor, fell a hero in the Battle of Minisink. According to Colonel Hathorn's report to Governor George Clinton *(Draper Manuscripts; 20F35)*, Tusten was wounded in the hand. Despite such wound, he obviously did his best to tend to the other wounded as seen in the testimony of John Knapp *(see Draper Manuscripts; 8F144)*.

Knapp, in a detailed account of the engagement prepared for his application for a pension, states that "Col. Thurston with us was a surgeon to him the wounded repaired in the centre and he dressed their wounds as well as he could." It is said that Lieutenant Colonel Tusten fell defending the wounded, and based upon some of the testimony given by applicants for pensions, that may actually be the case.

Tusten did not fall before Brant led his men en masse into the militia lines and, drove them from the field because Nathan Stewart claims in his pension application that he heard Tusten say that he would never leave the ground while Hathorn (Stewart mistakenly attributes these words to Tusten) supposedly called out words to the effect of "Every man for himself and the Lord for us all!"

Apparently, Tusten did fall as the enemy swept across the plateau because Jeremiah Kinner, who also applied for a pension, claims to have been next to Tusten when he fell. Benjamin Gunsolley, in his application, states that "he saw said Col Tustan lying dead on the ground." Because Kinner and Gunsolley both claim to have seen Tusten dead, yet made their own escape, he must have been killed before the fleeing militia exited the battlefield.

Though Tusten probably did fall determined not to leave the ground as tradition suggests, not all of the wounded under his care appear to have been killed with him. John Daniels, in his application for a pension, states "that from the appearance of several dead bodies, which were seen the next day on said battle ground, which were fastened to the ground, with stakes, & had the appearance of being burnt, this deponent, has no doubt but they were put to death in that barbarous manner, according to the savage custom sometimes adapted at that day."

If true, this clearly shows that some of the wounded were kept alive long enough to be tortured to death. They were, however, tortured and killed before nightfall because Brant's force had to have time to cross back over the Delaware River, and set up camp on the Pennsylvania side.

It is worth noting however that Brant is not known to have allowed torture during his campaigns and raids. It is also not known if any militia men returned to the battleground the next day because most appear to have wanted to get back to the Minisink settlement as quickly as they could. What Daniels may have seen, or heard later, was that the bodies, which had become badly decomposed in the heat of the summer day, may have had a horrible appearance that seemed to indicate that they were tortured. Another option is that Daniels may have been recalling an event from an entirely different battle.

Those Missing and Captured

Though at least two men claim to have seen Tusten dead upon the battleground, there was apparently some question as to his fate and that of as many as thirty others. In *Volume 5*, page 217 of *The Public Papers of George Clinton, First Governor of New York,* Governor Clinton asks Colonel Goose Van Schaick, on August 27, 1779, to instruct the commanding officer at Fort Schuyler, formerly Fort Stanwix, to "make the strictest Inquiries respecting them and how the Survivors are disposed of." It was hoped that through his scouts and their dialogue with friendly Indians, they could discover the fate of those missing.

In *Volume 5*, page 218 of *The Public Papers of George Clinton, First Governor of New York*, a similar letter with the same date was sent to General Philip Schuyler, who had the "Direction of the Indian Affairs," to make the same attempt to learn their fate. In this case, Governor Clinton also instructed Schuyler that "If there was a Practicability of send'g a Flag to some of the western Posts for that Purpose I should be happy if it co'd be done." In both letters, Clinton indicates that Lieutenant Colonel Tusten is among those missing. He is not considered killed in action though several men claimed in their pension applications that they had seen him dead.

Perhaps that information was never passed up the chain of command, or maybe some hoped that reports of his death were wrong. On September 13, 1779, Schuyler followed his instructions and wrote to Major John Butler asking to be given the names of those who were captured, their places of captivity, and the number of prisoners still in the hands of the Indians "as the duties of humanity can never interfere with those of Office" *(Sir Frederick Haldimand: Unpublished Papers and Correspondence 1758-1784; B.100; 21,760; 254).*

Colonel Guy Johnson, Superintendent of Indian Affairs, responded to Schuyler's letter on December 7, 1779. He stated that he had "readily enquired of the proper persons respecting the Colonel Thurston you mention, but have not been able to find him in any List of prisoners, and the partys concerned in the affair near the Minisink inform me that to the best of their recollection he fell in Action, neither do I find any prisoners here from that Settlement." He also informed Schuyler that "Major Butler who took pains on the occasion, has informed me that an Old Militia Capt. whose name he thinks was Hoghtelling or Wood was taken & sent to Canada" *(Papers of the Continental Congress; Roll #173; Item 153; Letters from Maj. Gen. Philip Schuyler, 1775-85; Jan. 1777-June 1785 (Vol. III);*

Pages 495 - 498).

The "Old Militia Captain" John Wood, whom the British listed as "a Man devoid of every Principle of Humanity, having carried his Persecution of Loyalists to such lengths as to have caused the Execution of many of them" *(Sir Frederick Haldimand: Unpublished Papers and Correspondence 1758-1784; B.183; 21,843; 334-335),* was being held at Chambly by this time and managed to get a letter to Governor Clinton dated February 16, 1780. He informed Clinton of his being a prisoner, and asked him to make arrangements for an exchange so that he could return as soon as possible. He also asked the favor of having his wife informed of his whereabouts *(Volume 5, page 498 of The Public Papers of George Clinton, First Governor of New York).*

On February 23, 1780, Van Schaick, who was by now aware of Captain John Wood being held prisoner at Chambly, wrote to Brigadier General H. Watson Powell, then commanding at Fort Niagara, entreating him to question Captain Wood as to who was captured and killed in the engagement and where any survivors might be held *(Sir Frederick Haldimand: Unpublished Papers and Correspondence 1758-1784; B.175; 21,835; 3-4).* General Powell responded to Van Schaick's letter on March 15, 1780, informing him that Captain Wood had been questioned and that the answer was being returned with his letter *(Sir Frederick Haldimand: Unpublished Papers and Correspondence 1758-1784; B.175; 21,835; 7).*

Unfortunately, Captain Wood's reply has not been found. Powell also responded to another letter he received from Van Schaick on the same day delivered by Colonel Yates, who was accompanying some women and children to Canada under a flag of truce. By the wording in Powell's reply, it is evident that Van Schaick requested Captain Wood's release as part of a prisoner exchange.

However, Powell was surprised both by the number of Loyalist families being delivered and the fact that there was an attempt to exchange women and children for combatants when no exchange of any combatants had been agreed upon. He informed Van Schaick that since neither of the Mrs. Butlers "have ever been in Arms, it is therefore unreasonable to expect Prisoners of War in exchange for them, particularly as I advised you that it was not His Excellency the Commander in Chiefs intention to enter into an Exchange of prisoners, further than dictated by humanity in the case of Women and Children, it is not therefore in my power to permit Captain Wood to return with the Flag" (The women referred to here are the wife and daughter of Major John Butler, commander of Butler's Rangers.) *(Sir Frederick Haldimand: Unpublished Papers and Correspondence 1758-1784; B.175; 21,835; 9).*

Governor Clinton, responding to a letter from Issac Nicoll, who wrote at the urging of Captain Wood's wife, was able to inform them on March 19, 1781, that he authorized someone in Albany to travel to Canada, and offer themselves up in exchange for Captain Wood *(Volume 6, pages 647-648 of The Public Papers of George Clinton, First Governor of New York).*

This exchange did not go as hoped. Though Doctor George Smith was released from confinement in order to effect the release of Captain Wood, no exchange took place. Smith did make it to Canada, but tried to arrange for the exchange of his son for Captain Wood, ignoring the arrangement that had resulted in his own release. John Wood had appealed to his captors on July 25, 1781 suggesting that he should be exchanged for Doctor Smith's son as well.

Much to his disappointment, Wood received a reply in early August that his letter had not made it beyond the Commander-in-Chief's secretary because it was thought "unnecessary to make your Request known to His Excellency, as Doctor Smyth's application on the same Subject had been unsuccessful" *(Sir Frederick Haldimand: Unpublished Papers and Correspondence 1758-1784; B.183; 21,843; 187-188).*

Wood's request for parole if the exchange was not allowed, was also denied. British approval of the matter was unnecessary as Governor Clinton, disappointed in Doctor Smith's conduct, had no intention of letting the younger Smith go until the father had made good on his agreement. Failing to exchange his son for Captain Wood, it is found that a Captain Brunson was suggested in place of Captain Wood. General John Stark hoped to arrange the exchange that was suggested by the enemy, and communicated with Governor Clinton on the subject, but was disappointed.

Clinton, in a letter dated August 23, 1781, listed several reasons for denying the release of Smith's son. Among them was the desire to "have this matter settled before young Smith's Exchange is consented to, as by his detention we have additional Security for the Doctor's Complyance with the Conditions of his Bonds" *(Volume 7, pages 228-230 of The Public Papers of George Clinton, First Governor of New York).* On October 6, 1781, Doctor Smith tried again to gain the release of his son, and sent a letter to Captain Wood suggesting that he write to a Captain Bevier from New York who was at Crown Point under a flag with orders to "have an exchange settled on for that State." He told Captain Wood to suggest to the officer that if Terence Smith (his son) was freed and sent to Canada, that he would be released in exchange *(Draper Manuscripts; 9F92-93).*

Captain Wood rejected Smith's suggestion and replied that "I have no objections of being exchanged for Mr. Smyth or any other gentleman; but I shall leave that entirely to Capt. Bevier, as I am not the proper person to make proposals, & am not uneasy about my being neglected by my countrymen" *(Draper Manuscripts; 9F93).* Despite attempts to secure his release in the years after his capture, John Wood was not released until July 31, 1783 *(Draper Manuscripts; 8F145).*

Anxious eyes were not only cast northward in hopes of finding news of their loved ones. Several months after the battle, Colonel William Allison, who was still a prisoner of the British on Long Island, was approached by Briant Hammell, a Tory from Orange County *(Volume 5, pages 563-564* of *The Public Papers of George Clinton, First Governor of New York).* Hammell was serving in the Indian Department, and told him that he could get a listing of the men taken prisoner "on the Dilewar in July Last." He claimed that he could get the names from a Superintendent of Indian Affairs living in New York.

Hammell at the same time questioned Allison about the possibility of his wife and children being sent to join him, no doubt trying to win favor by offering the names of the prisoners. Allison informed him that he thought it very possible for Mrs. Hammell and their children to be granted permission to relocate to New York. The men parted ways, with Hammell asking Allison to write to Governor Clinton on his family's behalf and promising to look into the prisoners.

The issue of the men killed and missing in the battle was something that Allison had "much at Heart." To his disappointment, he received word from Hammell on March 26, 1780, that the superintendent had carefully searched his papers, but in vain.

Colonel Allison wrote Governor Clinton on the subject ,and Henry Wisner, whose son Gabriel was missing in the action, heard about it *(Volume 5, pages 604-605* of *The Public Papers of George Clinton, First Governor of New York).* Wisner wrote to Clinton stating that he was "very desirous of Being informed what Reasons Coll. Allison had to Expect, to Be able to git a list of the names of the prisnors taken at the Battle on Delawar last summer." No further attempts to learn the fate of those missing is known to have taken place through New York City.

Attempts to Reinforce the Orange County Militia

There were two attempts to reinforce or cooperate with Hathorn's command. The first is the well known column led by Colonel John Seward of the Second Regiment, Sussex County Militia. As Hathorn's defeated men made their individual ways home, Seward led a force of ninety-three men up the Pennsylvania side of the Delaware River and reached the mouth of the Lackawaxen after the guns had fallen silent.

According to the pension application of Barnardus Swartwood (S.22,522) who was with Seward, on arriving at the mouth of the Lackawaxen, signs of Indians were discovered in the evening and "after night the Indian Fires were discovered about one mile from the last mentioned Place, It being the same Party of Indians which was ascertained on our return home that had defeated and killed nearly all the New York Militia that same day oposite the mouth of the Lackawaxen River in the State of New York, a counsel was held by the commanding officers of his company as to the propriety of attacking the Indians that night considering the weakness of our Forces and not knowing the strength of the Indians it was decided to be impracticable to make an attack that night, when the day dawned the next morning the Indians had marched off after reconoitering the country along the Delaware River deponent returned home."

Hathorn's report to Governor George Clinton (Draper Manuscripts; 20F35) mentions that a mutiny prevented Colonel Seward from attacking the enemy around their campfires that night. Seward was obviously in favor of a night attack based upon Hathorn's report, but other officers must have been opposed, holding out until he relented.

In the pension application of Rice Price (W.5,605), there is an example of this. On April 25, 1781, a party of Tuscarora Indians and Tories attacked in the area of Captain Abraham Shimer's Fort/Blockhouse in Montague, New Jersey. After meeting some resistance at that place, they withdrew. Receiving news of the attack, Colonel Seward led a portion of his regiment to Shimer's and a council was held there amongst the officers as to whether they should cross the Delaware River into Pennsylvania and pursue the Indians. Stephen Truesdell, one of those with Colonel Seward and who gave testimony in Price's application, stated that "when they arrived at the Delaware Col Seward proposed crossing the River in persuit of the Indians but was opposed by Capt Beckwith the oldest Captain and the expedition was given up."

This is probably what also occurred at the mouth of the Lackawaxen the night of July 22, 1779, though who may have opposed attacking the enemy around their campfires is unknown. John Knapp, in his account of the battle (*Draper Manuscripts; 8F144*), also claims that Seward failed to attack because of a "cowardly subaltern." Barnardus Swartwood states that when the morning came, they watched the Indians march off and Seward's men reconnoitered the surrounding country before returning to the settlements where they learned of the defeat of Hathorn's command.

Hathorn's report to Governor George Clinton does not say that the enemy withdrew up the Delaware in canoes either, but rather that the enemy "moved of their Wounded in Canoes." It cannot be determined with certainty whether they withdrew by foot or canoe, but evidence discussed in the section dealing with Brant's campsite after the battle makes a strong case that they withdrew by foot, their wounded, who were unable to march, were moved off in the canoes.

The second and lesser known column was led by Captain Abraham Shimer of the Third Regiment of Sussex County Militia. Mention of this column can be found on page 573 of *Volume VII* of the *Pennsylvania Archives.* On this page is a letter from John Van Campen to President Reed of Pennsylvania reporting the devastation at Minisink and that men were already pursuing Brant. He also reported that on the morning of the 22nd Captain Shimer marched across the Delaware River with one hundred and seventy men with an intent to head the enemy off at the mouth of the Lackawaxen.

There were Captain Shimers in both New Jersey and Pennsylvania but it can be concluded that this is Captain Abraham Shimer of Sussex County, New Jersey based upon the pension application of Jacob Lott (S.5,709) who claims that Captain Abraham Shimer's company was ordered to march to the assistance of the Orange County Militia. He says that they left Vangorden's Fort in Sussex County and after marching up the Delaware River as far as Butler's Falls (Butler's Rift near the Hawk's Nest), they encountered wounded men from Hathorn's command and returned to their station, the wounded men in tow.

Number of Brant's Casualties

When Brant made his report of the battle to Mason Bolton *(Sir Frederick Haldimand: Unpublished Papers and Correspondence 1758-1784; B.100; 21,760; 173-174)*, he reported that in the preceding raid he had one man killed and one wounded when attacking Van Auken's Fort on the Horn Road. He also listed three men killed and ten wounded at the end of the Battle of Minisink, giving a total of four men killed and eleven men wounded in his foray against Minisink. Several days later, in his letter to Daniel Claus *(Draper Manuscripts; 9F207)*, he stated that John Tayojaronsere, one of the wounded from the battle, died of his wounds, and that five of the other wounded men met the same fate. This means that six of those ten wounded men from the battle died, making a total of ten men killed and five wounded as a result of action connected with the attack upon Minisink.

Brant's Camp After the Battle

Tradition says that after defeating the militia, Brant's men made their camp on the Pennsylvania side of the Delaware River. In all of the material viewed during the research phase of this project, nothing was seen that would cause one to question the tradition about where they encamped, until we read the pension application of Barnardus Swartwood (S.22,522).

Swartwood was with Colonel John Seward, who was marching up the Pennsylvania side of the Delaware to reinforce or cooperate with Hathorn. In his testimony, he states that when they arrived near the mouth of the Lackawaxen River, "signs of Indians were discovered in the area and their campfires were spotted after dark about a mile from the junction of the two rivers." One might initially dismiss this as faulty memory. However, nothing in his testimony would make one question his recollection of the event. More importantly, he was actually there and not an author writing about the event one hundred years or more afterwards. He would know better where the Indian encampment was located.

While one map in the *Draper Manuscripts* shows Brant encamping along the Delaware and marching up its Pennsylvania shore the following morning, another shows him marching up the Lackawaxen. This second map was drawn by a resident of the area, approximately one hundred years after the battle, to be sent to Lyman C. Draper who was collecting information about the battle. Could this mean that it was told by some people in the area that Brant marched up the Lackawaxen and not the Delaware from the ford?

One has to remember that Brant was crossing the ford near the mouth of the Lackawaxen River for a reason. Research done for this work supports the theory that the trail Brant was following back towards Cochecton ran from Minisink to the ford, where one would have to first cross the Delaware, then the Lackawaxen. Once across both rivers, one would continue to march up the Lackawaxen a distance before setting out overland towards Mast Hope and on to Cochecton.

There is evidence that Brant may have done just that. The *Draper Manuscripts; 8F103*, include a description of the Indian trail from somewhere along the Lackawaxen River northward. It is not dated, but based upon wording found in it, it is from some time around the centennial of the battle. It states that "From here the Indian trail led through Tory Swamp into Tory hollow, and over Sim's Point, a high promontory five miles west of the village of Mast Hope." *Draper Manuscripts; 8F104* also states that "On the morning of the 23d Brandt broke camp and marched away toward the north, passing where the village of Mast Hope now is. Some three miles to the left a portion of the Torys stopped in a deep laurel hollow near the place now occupied by John Grogan, and it has been called Tory Hollow ever since. The Indians crossed over Sim's Point, entered the County of Wayne and bade an eternal farewell to the Valley of the Delaware."

Therefore, based upon the information presented here, it is quite clear that Brant did not literally camp on the Pennsylvania side of the Delaware River, but encamped along the Lackawaxen River, in an area that would have been large enough to accommodate his party and prisoners, a mile from the junction of the Lackawaxen and Delaware Rivers.

It is also thought that since he was left in control of the battlefield at the end of the engagement, Brant would have collected his dead, and moved them off the hill to his campsite where they could have performed whatever burial rituals were common at the time before interring them. One can easily imagine him having them buried a distance from the camp so as to prevent the enemy from finding them.

ATTEMPTS TO RECOVER THE FALLEN

Nathan Kerr, Pastor of the First Presbyterian Church in Goshen, wrote to Governor George Clinton that on Saturday, July 24, 1779, he led a party of 240 from Minisink to within two miles of the battleground in an effort to recover the fallen *(see Volume 5, pages 162-164 of The Public Papers of George Clinton, First Governor of New York)*. However, two newspapers, *The Pennsylvania Packet* of August 7, 1779 and *The Royal Gazette* of August 14,1779, both contain information indicating that Kerr and his party made it to within twelve miles of the battlefield, not two.

This would make more sense. It is hard to believe that after traveling through the wilderness to within two miles of the defeat, you would choose to return because of rain without ever having set foot on the battleground. The twelve miles mentioned in the newspapers is probably correct. What more than likely took place was that Kerr led his party up the river on Saturday, and, after traveling to within twelve miles of the battlefield, spent the night.

The following day was rainy, and, as he reported to Governor Clinton, "many were not prepared to be out after such a wet day, nor was it in our Power to keep our Arms dry." The decision was then made to return. It is of interest that within those twelve miles that were covered, they "found two who were scalped and mangled most cruelly" *(see The Pennsylvania Packet* of August 7, 1779 and *The Royal Gazette* of August 14, 1779*)*. This suggests the distance some of Brant's men may have pursued the routed militia was beyond twelve miles from the battlefield.

Kerr, in his letter to Governor Clinton, also mentioned fearing that the Indians were still in the area, which they apparently were. *The New York Gazette* and *The Weekly Mercury* of August 30, 1779 and *The Pennsylvania Packet* of August 19, 1779, all have an extract of a letter suspected to have been from Lieutenant Colonel Henry Wisner to Major General John Sullivan dated July 29th from Orange County. In this letter, it is claimed that "Their scouts appeared yesterday in this place" which shows that there were still Indians lurking in the area, whether of Brant's party, or acting on their own.

The pension files of Enoch Jackson (S.9,361), Reuben Place and Nathan Stewart also mention attempts to recover the fallen. Jackson makes it clear that he was called out a few days after the battle to guard the frontiers and recover the dead, but was unable to travel to the battleground because of the Indian threat. Nathan Stewart simply claims that they received recruits, and returned to the scene of action to bury the dead, but were forced to abandon their plans.

Whether they were part of Nathan Kerr's party cannot be determined. Reuben Place, on the other hand, claims that he made it onto the battleground, being sent back to bury the dead, but that the bodies smelled so badly, they could not perform that duty. As mentioned previously, there is reason to question his participation in the battle, so his statements should be carefully considered. What is known is that Colonel John Seward's men never set foot on the battlefield even though they were in the vicinity that same night. In his application for a pension, Barnardus Swartwood (S.22,522) makes it clear that they did not know of Hathorn's defeat until they returned back down the Delaware.

While it is apparent that no organized unit made it to the battlefield to bury the dead, one or more people may have actually been on the battlefield shortly after the engagement.

It is unknown for certain who may have been on the battleground the next day, but according to *Draper Manuscripts; 9F12*, when a committee went to the battleground in 1822 to recover the bones, they were met by a farmer who had been on the battleground the morning after the battle, and he was able to give detailed information to them about where men were known to have fallen in the engagement.

There has been much criticism over the years about the remains of the fallen not being recovered. While it is a shame that it took so long to recover the dead, it must also be remembered that the militia was not left in control of the field where they could have attended to such things right away. It was not an easy place to get to either, being twenty-seven miles from Minisink and over rough ground.

Many of the battles fought during the war were fought in populated areas where it was possible to properly recover and inter the dead but not so in this action. As evident from multiple sources, there was a constant threat of Indians who had the ability to strike anywhere on a frontier that stretched from Shandaken to Minisink.

This frontier was guarded by local militia and Pawling's Regiment of Levies. Pawling's Regiment was distributed between forts running the entire length of that frontier. *The Public Papers of George Clinton, First Governor of New York* are filled with urgent requests for more men to help guard the region from the Indian threat because the men available were never enough. Add to this the fact that there was still a war to fight along the Hudson River. The British still maintained their hold on New York City.

While it might have been possible to form a force of men to march back to the scene of action to tend to the fallen, one has to remember that there was already a strain on the troops along the frontier brought on by the impending Sullivan-Clinton Expedition against the Indians. Pawling's Regiment, scattered along the frontier, was under orders to march through the wilderness, and join General James Clinton at Oquaga, where they would take part in the expedition. Orders to march were finally received after the battle, and, though they were unsuccessful in making a junction with Clinton, their absence left the frontier even more exposed.

Farms and homesteads had been abandoned along the frontier due to the years of Indian threats, and personal businesses suffered due to the war. It is likely that when the Battle of Minisink Ford was long over, many people got caught up in the euphoria of peace in 1783 and were anxious to get back to their lives after so many years of privation and danger. There are stories, in other sources, of individual men making their way back to the battlefield to look for lost relations or friends, but nothing that can be proved. It is obvious that the thought of recovering the dead was on people's minds shortly after the battle ended, so is it still right to criticize them this far removed from the events?

Return of the Indians After the Battle

It has been mentioned in the past that, though the militia was soundly defeated, the enemy suffered to such a degree that they never returned to the area afterwards. This is not true. The Battle of Minisink did nothing to dissuade the Indians from returning to the Minisink area. According to the pension applications reviewed for this work, there was a constant presence of Indians throughout the war.

Many of the pension applications read during the research for this work, but not included here, say the same thing. In April, 1780, less than a year after the battle at the mouth of the Lackawaxen, New Jersey and Pennsylvania militia battled Indians and Tories near Milford, Pennsylvania, in what came to be known as the Conesaugh Battle. Montague, New Jersey was attacked by a party of Indians and Tories in April, 1781, and Wawarsing was attacked by a party of Butler's Rangers, and a large force of Indians in August, 1781. They were obviously not afraid to visit the area again.

For further consideration, there are several records of Indian war parties out in 1780 and 1781 that clearly show activity in the Minisink area. The first return *(Sir Frederick Haldimand: Unpublished Papers and Correspondence 1758-1784; B.109; 21,769; 64-65)* shows a party of twenty-two Mohawks and Senecas that marched on April 2, 1780 for Minisink and they killed four people. Also listed on the same return is a party of fourteen Delawares that marched on May 19, 1780 for the same place. The second return *(Sir Frederick Haldimand: Unpublished Papers and Correspondence 1758-1784; B.109; 21,769; 116-118)* shows a party of Senecas that marched February 21, 1781 for Minisink.

A party of Delawares is reported to have returned to Niagara with one prisoner and one scalp from Minisink, according to a letter dated April 23, 1781 *(Sir Frederick Haldimand: Unpublished Papers and Correspondence 1758-1784; B.107; 21,767; 175)*. When dealing with the reports of Indians returning to the area after the Battle of Minisink, one has to address another tradition that has been handed down.

It has been reported in several sources that in April, 1780 a party of eleven Indians was sent by Brant to Minisink, where they captured several men who later rose upon their captors in the night and killed most of them. No proof of this can be found. It is believed that the original source of this story is Freegift Patchin, a soldier who was captured with Captain Alexander Harper of Schoharie on April 7th.

According to Patchin, after being marched to Tioga Point, he and Harper were met by two Indians who had been sent to Minisink a few days earlier to secure scalps and prisoners. These two Indians belonged to a party of eleven and managed to capture five men that they brought to the vicinity of Tioga Point. During the night, the Indians bound their prisoners, and, thinking it safe, went to sleep. The prisoners managed to free themselves and attack their captors, killing nine of them.

It is thought that Patchin is somehow mistaken when he says that the party had been to Minisink. It is more likely that the party mentioned was actually the same party that captured the well-known Indian fighter Moses Van Campen and others below Wyoming. Consider the similarities in the stories.

According to tradition, in April, 1780, Van Campen and five others were captured by ten Indians who marched them to the vicinity of Wysox, Pennsylvania, just below Tioga Point. These men were bound, and the Indians, thinking it safe, went to sleep. During the night, the prisoners managed to free themselves, and kill nine of their captors. The stories are very similar. During the research part of this work, several letters were found that gave even more details. In these letters, it is reported that Brant was able to march against the frontiers in the middle of February, 1780 with about 220 Indians. The weather was severe, and their progress was greatly hindered. Some of his men also became sick and died.

As a matter of convenience, the party separated. Brant went to Harpersfield where he captured Captain Harper and others, while two separate parties went below Wyoming, another went to Bethlehem, and two other parties went to unnamed destinations. One of the parties that went below Wyoming captured six men. Some of their prisoners rose in the night, however, and killed three of the Indians *(Sir Frederick Haldimand: Unpublished Papers and Correspondence 1758-1784; B.106; 21,766; 7 and B.106; 21,767; 61-62)*. This is the party that captured Moses Van Campen, as mentioned before.

There is no mention in these letters of another party taking prisoners at Minisink, and the prisoners rising against their captors like Van Campen's party did. However, because parties did go to two locations that were not identified, it is possible that a party did go to Minisink. It is more than likely, though, that if another party had gone to Minisink, and the prisoners had risen up like Van Campen's group did, that this would have been mentioned in the letters found in the *Haldimand Papers*, which it is not. Therefore, while it cannot be dismissed that a party visited Minisink at this time, a strong case can be made to dismiss the tradition that prisoners were taken there, and that they rose against their captors while they slept.

The Conesaugh Battle

Several of the men who applied for pensions were not only in the Battle of Minisink, but also in an engagement in present day Pike County, Pennsylvania on April 19, 1780. This action is commonly called the Conesaugh Battle, Raymondskill Battle or the Battle of Raymondskill. It is also referred to as the Battle of McCarty's in the pension file of Benjamin Hains, a participant in the Battle of Minisink, because it took place near the residence of Philip McCarty, which was located near the Raymondskill Creek. The Raymondskill flows into the Delaware River and is directly across from Minisink Island, downstream from Port Jervis. The men who were in both of these battles are Isaac Bedell, William De Shay, Benjamin Hains, and Ezekiel Lane.

What follows are some interesting pieces of data that address various lesser issues related to the battle over the years.

Place Names

As can be seen in many of the *Series M-804 Revolutionary War Pension and Bounty Land Warrant Application* files, there are numerous references to the "Minisink Frontier" or the "Minisink Indian Frontier." Though there may have been references to the area being a frontier in past studies, this particular wording is repeated so often in the pension applications that it leaves one with the strong impression that the area must have been referred to as "the Minisink Indian Frontier" during the applicants lives.

The Tom Quick Legends

The only thing that can be found in the *Series M-804 Revolutionary War Pension and Bounty Land Warrant Application* files concerning the legends of Tom Quick, a controversial figure in the Milford area, is found in the file of Benjamin Hains, a participant in the Battle of Minisink. Within this file is a deposition by Thomas Spangenberg mentioning that Hains was reputed to have killed an Indian named Canope. If one reads *Tom Quick the Indian Slayer* by James E. Quinlan, one will see that Quinlan mentions the killing of Canope by Hains after the Revolutionary War. Spangenberg's deposition is further proof of Hains' act.

Rumored Petition to Congress from Major Samuel Meeker or his Descendants

In the *Draper Manuscripts* there are several letters from the well-known Civil War General Hugh Judson Kilpatrick. Kilpatrick was the great-grandson of Major Samuel Meeker, his mother being the Major's granddaughter *(Draper Manuscripts; 8F74)*. Writing about Meeker, Kilpatrick states that "A claim was made or placed before Congress to reimburse him for all he had lost, but he died & it was never pressed by his heirs" *(Draper Manuscripts; 8F76)*.

Research done at the Center for Legislative Archives (National Archives and Records Administration) shows only one man named Samuel Meeker who petitioned Congress in the early nineteenth century. However, this person is not the same Samuel Meeker who fought in the Battle of Minisink. In the absence of evidence that he or his family ever petitioned Congress, this rumor must be dismissed.

There is evidence that several of the men who were in the Battle of Minisink, or their families, may have petitioned Congress, and it is highly recommended that future researchers start there. According to the staff at the Center for Legislative Archives (National Archives and Records Administration), petitions to Congress for relief were more common at that time than one might think.

Captain Harker's Wounds

It has been argued by some that the wounds Captain Joseph Harker received in the battle were so severe that he never returned to duty. While his wounds do appear to have been severe, there is mention in several pension files of his commanding a company in the years after the Battle of Minisink (see the pension applications of Thomas Ayres (S.17,245), Solomon Brundage (W.20,779), William Hedglen (W.6,728), Nathan Spencer (S.14,550) and Abraham Westbrook (S.6,357)).

This company was apparently engaged in driving the Indians off when they attacked Captain Abraham Shimer's home in Montague on April 25, 1781. Harker was a blacksmith and records show that he was unable to continue in that trade because of a wound to his right arm, received in the Battle of Minisink.

In addition, Doctor Jacob Everett, in an affidavit dated August 19, 1784 *(NJDARM: New Jersey State Archives Holdings: Department of Defense - Colonial Wars and Revolutionary War; Series: Department of Defense Adjutant General's Office (Revolutionary War) Numbered Manuscripts, 1770s-1890s; Control #SDEA1001; Item #6001)* in support of Harker's claim for reimbursement for medical expenses, stated that Captain Harker "has never perfectly recovered but is still lame in one arm by reason of sd wound." This suggests that though he did return to duty, he suffered with a severe handicap.

Samuel Helm

Tradition states that Samuel Helm was wounded in both thighs during the battle, in the act of discharging his weapon at an Indian. In truth, no evidence has been found to place Samuel Helm at the Battle of Minisink. Helm, in his own application for a pension (S.4,356), mentions that he was a witness to the preceding raid, but says nothing about the battle which followed.

No person who served with Helm during the war, or who was in the battle, places him there, either. In addition, Helm's wound to both thighs is identical to the wound he received in the Conesaugh Battle (also referred to as the Battle of McCarty's or the Battle of Raymondskill) in April, 1780. Helm, describing the Conesaugh Battle in his application for a pension, states that "after discharging seven times my Rifle and in the act of discharging it again I recived a wound through both my Thighs." Vernon Leslie, in his book *The Battle of Minisink: A Revolutionary War Engagement in the Upper Delaware Valley*, also touches upon the subject. Unless new information is found to place him in the Battle of Minisink, his name must be stricken from the roll of participants.

Images of Participants

The only known images of people who took part in the battle are those of Joseph Brant and Ezekiel Lane. Lane's image is an early form of photography called a daguerreotype. This image of him was made shortly after his death in 1848, and can be found in *Publications* by the Buffalo Historical Society, V*olume 7*, published by the Society. Several portraits of Joseph Brant were made during his lifetime. George Romney made the first in 1776 and Gilbert Stuart painted two portraits of Brant in 1786. There is also a portrait of John Hathorn made after his death based upon descriptions of his appearance. The portrait can be found in the book *Days Gone By: A History In Pictures Town of Warwick.*

Our Duty Is Now Done

In conclusion, it is evident that not all of the traditions that have been handed down about the battle are true. The militia that fought Brant was not inexperienced, as has been said, and there is no evidence found that can support the accusations of cowardice made against any of the men from New Jersey, including Major Meeker. It has been repeated many times that though the militia was soundly defeated, the Indians suffered to such a degree that they never returned afterwards. This is also not true; the battle did nothing to dissuade their return to the area.

Some of the traditions about the battle, however, do stand up to scrutiny. It does appear that, as tradition has long told, the number of men Hathorn commanded was in the 120-150 range, while Brant's force consisted of 60 Indians and 27 Tories. It is also true that the militia was not properly equipped when they marched off in pursuit of Brant, and would later suffer for want of ammunition. A part of Hathorn's command does appear to have been put out of the fight in the early stages of the battle, and Lieutenant Colonel (Doctor) Benjamin Tusten, the celebrated hero of the battle, does appear to have fallen in the manner handed down over the years, apparently determined never to leave the battleground.

The preceding chapter draws extensively upon Mark Hendrickson's research and organized his extensive findings about the Battle of Minisink. His essay, written in January 2009, provides many new perspectives about the battle, its participants and some of the traditions that have long been a part of battle lore. It is bound to raise controversy as some of the material that is presented here will address some errors of omission or addition, such as the roster of the men who fought in the battle. This chapter includes material written by Peter Osborne as well.

TWENTY FOUR SHILLINGS
Issued in defence of American Liberty
MAGNA CHARTA
Ense petit placidam, sub Libertate Quietem
Aug.t 18. 1775.

Chapter 4
A Declaration of Service

Every claimant must state, with precision, the length of his service, and the different grades in which he served, in language so definite as to enable the Department to determine to what amount of pension he is entitled.
The Act Of June 7, 1832

Pensions for Services Rendered

In the decades following the American Revolution, there were efforts undertaken to honor the service of those who had fought for the young nation's independence. The men who had taken part in the military campaigns that created the United States were venerated, and one can often find mention of their service in their obituaries, on their tombstones, and in monographs. One of the most important ways to honor the soldiers and officers of the Revolution was to grant pensions for services rendered.

The idea of offering pensions was not new to veterans of military service. The British Colonies had provided pensions for disabled servicemen well before the first shots at Lexington and Concord were fired. The idea of offering pensions continued during the Revolution in the hopes of encouraging enlistment and curtailing desertions. Through different acts during the Revolutionary War, and after, Congress authorized additional pensions for invalids who had performed military service, and for widows and orphans of soldiers killed in battle.

The first pensions began in 1778 and half pay was given to disabled officers and enlisted men were awarded $80. On August 24, 1780, widows could collection pensions based on meeting the necessary requirements. In 1789, the first Congress took over the financial responsibility for pensions that had been previously paid by the states.

As the years went by the pensions became more generous and began to include more recipients until finally Congress passed a law on June 7, 1832 that allowed every man who had served for two years to collect his full pay. If he had served less than two years he could receive a reduced pension. Widows benefits were expanded continuously until 1855 when all restrictions were removed. An important factor in the gradual increase in the scope and cost of the pension program was that by the 1830s many of the veterans were old men, and a significant number of them, or their widows had no means of support.

Of the 103 men who we have been able to prove took part in the battle, 42%, or some forty-three individuals applied for pensions afterward. Of those forty-three applicants, which included both veterans and widows, sixteen, or 37% were rejected. Twenty-seven, or 63% were approved. Of course there were many variables as to why they might have been approved or rejected. Most applied after the passage of the 1832 legislation, although at least three veterans or widows applied for pensions under earlier legislation, and even had state pensions before giving those up to get the federal pensions. By the 1830s, many of the veterans or their widows were of course quite elderly, and many had either very limited resources or no resources at all.

The paperwork that was filed, while onerous for the applicants, has proved to be a gold mine of information for later historians and researchers and an invaluable resource in preserving and interpreting the history of the Revolutionary War. Genealogists have been writing to the federal government for almost one hundred years requesting access to records of their ancestors. Today they are microfilmed, and available for researchers to borrow.

As a matter of clarification, the identification number that follows each veteran's or widow's name - S.43,576 or W.23,765 or R.78,465 - for example, are the pension file number, while the letters indicate which of three series the application is found in. Before 1910, Revolutionary War pension applications were placed in one of three series. The first series, identified by the "S" before the file number indicates that it is an approved application from the Revolutionary War veteran himself, the "S" meaning "Survivor." The second series, identified by the "W" before the file number indicates that it is an approved application from the widow of a Revolutionary War veteran. The third series, identified by the "R" before the file number indicates that it is a rejected application of either the Revolutionary War veteran or the widow of a Revolutionary War veteran. In some cases "B. L. Wt." is seen, followed by a set of numbers. This indicates that a person applied for bounty land for their services during the war.

Proving One's Service

In order to qualify for a pension, an applicant had to appear in a Court of Record in their county of residence. They had to give a declaration in which they stated their age, the period of the war they served, the length of their service, and the officers under whom they served. They were required to be as detailed as possible because the amount of their pension was based upon rank and length of service. In addition, if an applicant was already in receipt of a pension, such as an invalid pension from their state, they were required to give up that claim in order to be eligible for one from the federal government.

They also had to answer a prepared set of questions or interrogatories. The court had to state its opinion in the matter, and the clerk had to certify that the judge was in fact an officer of the court. The clerk also had to seal the proceedings with a ribbon so as to prevent tampering. If the applicant could not be found in military records in possession of the federal government, he was allowed to prove his service by two credible witnesses who could list his service, and they were required to be as specific as the applicant himself.

He was also required to obtain a letter from a local clergyman who would give his opinion as to the truthfulness of the statements given by the former soldier. A clergyman was suggested as one of an applicant's witnesses, because of the high regard men of the cloth were held in those days. He tended to the people of a local community, and may have known whether a person had a reputation as a former soldier during the Revolutionary War before the latest pension law was enacted.

This lengthy process was required because many of the men who had fought had not been paid, and were not listed on any rolls. They had simply volunteered for the cause of liberty. Another intent of the process was to prevent fraud. Despite these measures, fraud did occur. Men are known to have lied about their ages in order to appear old enough to have served in the war, when in fact they were too young to have borne arms in the conflict. Others created fictitious units that they claimed to have served in. Some soldiers might have served just short of the minimum required to qualify for a pension, and would therefore stretch the truth about their length of service. Widows and children of former soldiers were also capable of stretching the truth from time to time.

Rules and Regulations

The 1832 act provided regulations that had to be followed by the applicants in order to gain approval for a pension. Many of the Minisink participants had these regulations appended to their applications because if they had not included all the information a note was made on the regulations, and sent back. For the purposes of saving space, the regulations are reprinted in their entirety here, and not in each application file. We have noted where the regulations were inserted in the paperwork for each veteran.

REGULATIONS UNDER THE ACT OF JUNE 7, 1832.

The following regulations have been adopted:

This law has been construed to extend, as well to the line, as to every branch of the Staff of the Army, and to include under the terms "Continental Line," "State troops," "militia," and "volunteers," all persons enlisted, drafted, or who volunteered and who were bound to military service, but not those who were occasionally employed with the army upon civil contracts, such as Clerks to Commissaries and to Store Keepers, &c. Teamsters, Boatmen, &c. Persons who served on board of Private Armed Vessels are also excluded from the benefits of the law, as well as persons who turned out as patrols, or were engaged in guarding particular places at night, and were not recognized as being in actual military service.

Four general classes of cases are embraced in this law:

1. The Regular Troops.
2. The State Troops, Militia, and Volunteers.
3. Persons employed in the Naval service.
4. Indian Spies.

As rolls of the regular troops in the Revolutionary War exist in this Department, all persons, claiming the benefit of this law as officers, non-commissioned officers, musicians or privates, will in the first instance, make application by transmitting the following declaration, which will be made before a Court of Record of the County where such applicant resides. And every Court having by law a seal and Clerk is considered a Court of Record.

DECLARATION,

In order to obtain the benefit of the Act of Congress of the 7th of June, 1832.

State, Territory, or District of

County of ss.

On this day of personally appeared (a) before A. B. a resident (b) of in the county of and State, Territory or District of aged (c) years, who being first duly sworn, according to law, doth on his oath make the following declaration, in order to obtain the benefit of the provision made by the act of Congress, passed June 7th, 1832. That he enlisted in the Army of the United States in the year (d) , with and served in the (e) regiment of the line, under the following named officers:

[Here set forth the names and rank of the Field and Company Officers; the time he left the service; (and if he served under more than one term of enlistment, he must specify the particular period, and rank and names of his officers;) the town or county and State in which he resided when he entered the service; the battles, if any, in which he was engaged, and the country through which he marched. * This form is to be varied so as to apply to the cases of officers and persons who belonged to the militia, volunteers, navy, &c.]

He hereby relinquishes every claim (f) whatever to a pension or annuity, except the present, and he declares that his name is not on the Pension Roll of any Agency in any State, or (if any) only on that of the Agency in the State of

Sworn to and subscribed the day and year aforesaid.

[And then will follow the certificate of the Court.]

And the said Court do hereby declare their opinion (g) that the above named applicant was a revolutionary soldier, and served as he states.

I of the Court of do hereby certify (h) that the foregoing contains the original proceedings of the said Court in the matter of the application of for a pension.

In testimony whereof I have hereunto set my hand and seal of office (i) this day of &c.

If, on examination of the proper record the names of applicants, making such declaration, cannot be found, they will produce such proof as the rule given in note (j) points out.

Every applicant who claims a pension by virtue of service in the State Troops, Volunteers or Militia, except those who belonged to the New Hampshire Militia and State Troops of Virginia, will make and subscribe a declaration similar to the foregoing, with the following additions, viz:

(k) We, A. B., a Clergyman, residing in the and C. D. residing in (the same) hereby certify, that we are well acquainted with , who has subscribed and sworn to the above declaration; that we believe him to be years of age; that he is reputed and believed, in the neighborhood where he resides, to have been a soldier of the Revolution, and that we concur in that opinion.

Sworn and subscribed the day and year aforesaid.

[And then will follow the certificate of the Court.]

And the said Court do hereby declare their opinion, after the investigation of the matter, and after putting the interrogatories prescribed by the War Department, that the above named applicant was a Revolutionary Soldier, and served as he states. And the Court further certifies, that it appears to them that A. B. Who has signed the preceding certificate, is a clergyman, resident in the and that C. D., who has also signed the same is a resident in the and is a credible person, and that their statement is entitled to credit.

I, Clerk of the Court of do hereby certify that the foregoing contains the original proceedings of the said Court, in the matter of the application of for a pension.

In testimony whereof, I have hereunto set my hand and seal of office, this day of &c.

Every applicant will produce the best proof in his power. This is the original discharge or commission; but if neither of these can be obtained, the party will so state under oath, and will then procure, if possible, the testimony of at least one credible witness, stating in detail his personal knowledge of the services of the applicant, and such circumstances connected therewith, as may have a tendency to throw light upon the transaction.

If such surviving witness cannot be found, the applicant will so state in his declaration (l) and he will also, whether he produce such evidence or not, proceed to relate all the material facts, which can be useful in the investigation of his claim, and in the comparison of his narrative with the events of the period of his alleged service, as they are known at the Department. A very full account of the services of each person will be indispensable to a favorable action upon his case. The facts stated will afford one of the principal means of corroborating the declaration of the applicant, if true, or of detecting the imposition, if one be attempted; and unless, therefore, these are amply and clearly set forth, no favorable decision can be expected. All applicants will appear before some Court of Record in the County, in which they reside, and there subscribe and be sworn to, one of the declarations above provided, according to the nature of his case.

The Court will propound the following (m) interrogatories to all applicants for a pension, on account of service in the Militia, State troops, or Volunteers, except the Militia of New Hampshire and the State troops of Virginia.

1. Where and in what year were you born?

2. Have you any record of your age, and if so, where is it?

3. Where were you living when called into service; where have you lived since the Revolutionary war, and where do you now live?

4. How were you called into service; were you drafted, did you volunteer, or were you a substitute? And if a substitute, for whom?

5. State the names of some of the Regular Officers, who were with the troops, where you served; such Continental and Militia Regiments as you can recollect, and the general circumstances of your services.

6. To a Soldier. Did you ever receive a discharge from the service, and if so, by whom was it given and what has become of it?

To an Officer. Did you ever receive a commission, and if so, by whom was it signed, and what has become of it?

7. State the names of persons to whom you are known in your present neighborhood, and who can testify as to your character for veracity, and their belief of your services as a soldier of the Revolution.

The Court will see that the answers to these questions are embodied in the declaration, and they are requested to annex their opinions of the truth of the statement of the applicant.

The applicant will further produce in Court, if the same can be done, in the opinion of the Court, without too much expense and inconvenience to him, two respectable persons - one of whom should be the nearest clergyman, if one lives in the immediate vicinity of such applicant, who can testify, from their acquaintance with him, that they believe he is of the age he represents, and that he is reputed and believed in the neighborhood to have been a Revolutionary soldier, and that they concur in that opinion. If one of these persons is a Clergyman, the Court will so certify, and they will also certify, to the character and standing of other persons, giving such certificates.

The traditionary evidence of service is deemed very important, in the absence of any direct proof, except the declaration of the party. And the Courts are requested to be very particular in the enquiry whether the belief is general, and whether any doubts have ever existed upon the subject.

Applicants unable to appear in Court by reason of bodily infirmity, may make the declaration before required, and submit to the examination, before a Judge or Justice of a Court of Record of the proper county, and the Judge or Justice will execute the duties, which the Court is herein requested to perform, and will also certify that the applicant cannot, from bodily infirmity, attend the Court.

Whenever any official act is required to be done by a Judge or Justice of a Court of Record, or by a Justice of the peace, the certificate of the Secretary of State or Territory, or of the proper Clerk of the Court or County, under his seal of office, will be annexed, stating that such person is a Judge or Justice of a Court of Record, or a Justice of the Peace, and that the signature annexed is his genuine signature.

No payments can be made on account of the services of any person, who may have died before the taking effect of the act of June 7, 1832; and in case of death subsequent thereto, and before the declaration herein required is made, the parties interested will transmit such evidence as they can procure, taken and authenticated before a Court of Record, showing the services of the deceased, the period of his death, the opinion of the neighborhood respecting such services, the title of the claimant, and the opinion of the Court upon the whole matter.

[a] The declarant must appear in open Court, unless prevented from doing so by reason of bodily infirmity; in which case the declarant will follow the rule laid down for his guidance.

[b] The declarant must make his declaration in the county where he resides. If he should fail to do so, he must assign a sufficient reason for not conforming to the rule.

[c] The age of the claimant must invariably be mentioned.

[d] The declarant must mention the period or periods of the war when he served.

[e] Every continental officer or soldier must give the name of the Colonel under whom he served; otherwise a satisfactory examination of the claim cannot be had. Every claimant must state, with precision, the length of his service, and the different grades in which he served, in language so definite as to enable the Department to determine to what amount of pension he is entitled. In a case where the applicant cannot, by reason of the loss of memory, state precisely how long he served, he should amend his declaration by making an affidavit in the following words:

"Personally appeared before me, the undersigned, a Justice of the Peace, &c. A. B. who, being duly sworn, deposeth and saith, that, by reason of old age, and the consequent loss of memory, he cannot swear positively as to the precise length of his service; but, according to the best of his recollection, he served not less than the periods mentioned below, and in the following grades: - For year months, and days, I served as a . For months and days, I served as a ; and for such service I claim a pension."

[It is important, in all cases, to determine with precision the period for which each applicant served, and the particular rank he held, as the law directs the pension to be paid according to the grade of the pensioner and the length of his service. The use of the phrase about three or four months, is too indefinite, and all such qualifying expressions are objectionable. Some persons who apply for pensions merely state that they served two years in the militia, &c. without specifying the tours, the names of the officers, and other particulars respecting their service. This form of a declaration is highly objectionable. It must, in every case, be clearly shown under what officers the applicant served; the duration of each term of engagement; the particular place or places where the service was performed; that the applicant served with an embodied corps called into service by competent authority; that he was either in the field or in garrison; and for the time during which the service was performed, he was not employed in any civil pursuit.]

[f] The law makes the relinquishment indispensable.

[g] The opinion of the Court is always required.

[h] The Clerk must give his certificate in every case.

[i] The Clerk must affix his seal, and if it has no device or inscription by which it can be distinguished from any other seal, or if he has no public seal of office, the certificate of a Member of Congress, proving the official character and signature of the certifying officer, should accompany the papers.

Mode of authenticating papers.

[In every instance where the certificate of the certifying officer who authenticates the papers is not written on the same sheet of paper which contains the affidavit, or other papers authenticated, the certificate must be attached thereto by a piece of tape or narrow ribbon, the ends of which must pass under the seal of office of the certifying officer, so as to prevent any paper from being improperly attached to the certificate.]

Proof of Service.

[j] In a case where the name of the applicant is not found on the records of the Department, he must prove his service by two credible witnesses who are required to set forth in their affidavits the time of the claimant's entering the service, and the time and manner of his leaving the same, as well as the regiment, company, and line to which he belonged. The magistrate who may administer the oaths must certify to the credibility of the witnesses, and the official character and

signature of the magistrate must be certified by the proper officer, under his seal of office.

* The notes from [a] to [i] are all equally applicable to the cases of Militia men, Volunteers and State Troops. The proof required by rule in note [j] applies to continental troops only.

[k] This traditionary evidence is indispensable in militia cases.

[l] If a witness cannot be found, the declarant must state the fact.

[m] The answers to the interrogatories must all be written, and sent to the War Department, with the declaration.

His service as a Spy must be particularly specified.

The Battle of Minisink Comes To Life

As one begins to read the applications of former soldiers and their families, the stories of not just the Battle of Minisink, but of so many other Revolutionary War events begin to come to life. Instead of the usual accounts of the number of homes destroyed and lives lost during the raid on Minisink, for example, you now see what the individual person experienced at the time of the event.

Much of the information found in these applications is detailed, and has never before been mentioned in the local history books. As you go through the pension applications found in this work, it is easy to picture an aged man standing before a judge, recounting the experiences of an old soldier who had served his country long ago, or even an old widow describing the dangers the family faced while living on the frontier.

In some instances, the soldier's memory of events several decades earlier could be rather remarkable, but more often than not, he would confuse dates or the unit he belonged to during a certain event. Needless to say, one has to read the wording in the applications very carefully. We had easy access to the pension records, and were able to go through them very methodically over several years. It is estimated that we read thousands of them. Every person who was known to have a link to the area or who it was thought might have traveled through the area during the war was checked.

Working with such a large number of pension applications requires time and patience. Some of the files can be difficult to read depending upon the penmanship or education of the people writing, and some of the pages can be very faded or the wording written so small that it requires extra effort to follow what the authors of the document are trying to convey. Add to this a person's confusion over the order of events, locations, names, and dates, and it gets to be quite challenging. Some of the stories detailing a man's service could be quite an interesting read while others could be matter-of-fact. The information that one can uncover, however, can be well worth the effort.

The depth of a soldier's conviction to the cause of liberty can be ascertained. When Levi Totten, a veteran of the Battle of Minisink, applied for a pension, he stated that "when he was yet a Youth, the Acts of the British Parliament for blocking up, or shutting the Harbor of Boston, and for seizing and carrying Persons accused or suspected of Crimes, to Places beyond Sea for Trial, not by a Jury of their Peers, or the Law of the Land, together with other arbitrary Decrees of the British Government, raised in the mind of your Petitioner, a Spirit of Opposition which never forsook him. That on Reading an Account of the Commencement of Hostilities at Lexington, by a Detachment from the British Army at Boston. Your Petitioner, with Intent to assist, if necessary, in defending the Rights and Liberties of his Injured Country, diligently learned the Use of the Firelock and Bayonet. And on or about the twenty second Day of February AD 1776, He left a Peaceful and comfortable Home, against the Consent of his Parents, and inlisted." His parents, no doubt, were relieved that he survived the war.

One may also learn what service the soldier performed before the battle that raged near the mouth of the Lackawaxen River on July 22, 1779, or what became of him after the war. We were surprised to learn that some of the men who took part in this battle were refugees from Long Island, fleeing the British occupation of New York City, and staying in Orange County. After the cessation of hostilities they would return home again.

Joel Bower, also a veteran of the battle, was another experienced soldier. He was in the thick of the fighting at Monmouth in June, 1778 and gave a detailed account of his experiences in that battle. He claimed that he "was on the right of the left wing in Clintons Brigade," and reports "I stood in the front rank where we sustained a tremendous cannonade as well as a heavy fire of musketry for a considerable time we then were impetuously charged by the Dragoons, and in their second charge a Dragoon aimed a deadly blow at my head which would have instantly killed me, but I raised my musket to shield my head from the meditated stroke, which caused his sword to glance, & cut off my two middle fingers & the little finger about half off, this was my right hand which was holding on the small of my gun, my hand has ever since been verry much disabled from this wound, as he was then passing me he made an underhanded blow at my head, and the point of his sword passed through my left eyebrow which made a deep incission & considerably stunded me, at this instant his horse was shot through the head & fell dead at my feet, as I recovered from the last blow I received, I drew my bayonet & would have terminated his existance had not Adjutant Johnson catched me by the shoulder & bade me desist saying 'no murder here you young rogue the man cries for quarter.'"

Joel Bower went into as much detail describing his other services during the war. When giving sworn testimony for his pension application, a judge remarked that "it seemed more like the narration of one who had risen from the dead, than one who had survived to this late period."

Frontier Life During the Revolution

The men bearing arms in support of their country were not the only people exposed to danger. With the husbands away performing military duty, the wives and children were left alone to manage the household the best that they could. Throughout the war, the frontier was "infested with Indians," as contemporary accounts state. They could strike anywhere, and the families left behind on the frontier were constantly exposed to the threat of a visit by Indians.

Testimony found in several applications that are not included in this work, nevertheless give good examples of the danger families faced. Anna Post, in the application of her father, John Wisner, stated that she "distinctly recollects that while her father and Major Lucky (who was a near neighbor to her father) were out in Service. Two Indians came to the house of Major Lucky, while his wife was alone and called for something to eat which was given them when the Indians presented a knife and frightened Mrs. Lucky very much but the Indians made some signs by which she understand that they intended to give it to her as pay for the food she gave them. Deponent frequently saw the said knife. The general supposition was at the time that the Indians intended to kill Major Lucky."

Lena Elmendorf, who lived in present-day Pike County on the Pennsylvania side of the Delaware, gave testimony in the case of Martinus Decker's application and stated that "she has a distinct recollection of border troubles of the Revolution. She reccollects well that during those troubles she often had to stand and watch in fields while her father and brothers were at work to give the alarm when the Indians were seen lurking in the bushes and at several different times the whole family with others left their homes and crossed the Delaware into New York & New Jersey for safety."

Margaret Terry, the remarried widow of Wilhelmus Westfall, testified that during the raid on Minisink in 1779, she had to run for her life and traveled for nearly five miles before reaching the safety of a fort. In April, 1781, when Montague, New Jersey was attacked, she again had to flee. This time she required the help of others, having just begun to recover from a sickness.

Some of the testimony in the applications is rather sad, as in the case of Adonijah Middaugh, whose father was killed in the Battle of Minisink. Middaugh's application is not included among the files found in this work, but it shows that he was serving in the same company as his father, and claimed that the militia assembled and marched in consequence of the Minisink alarm, and that when they reached Major Decker's Fort on the Horn Road, he and others were sent back to Fort Van Tuyle while the rest of the militia, his father included, pursued the enemy. This was the last time he ever saw his father. One can easily imagine the scene when father and son parted, neither knowing that this would be the last time they would ever set eyes on each other.

New Details On The Battle

Interesting details of the battle that raged on that hill opposite the mouth of the Lackawaxen River so many years ago also began to emerge as our research progressed. Jason Horton was one of the men at Minisink Ford and he gave an interesting account of the battle many years later.

He testified that in "July 1779 the Indians made an attack on the inhabitants of Minisink & made great destruction. A detachment the Goshen militia Commanded by Col B Tustin & the Warwick Regiment Commanded by Col J. Hartshorn were under marching orders in a few hours, arrived at the place of destruction in the after noon of the day in which the Indians had left. We pursued them, travelling all night some on foot & some on horseback. About break of day our Com Officer was informed that the Indians had Encamped about 4 miles in advance Soon after sun rise, we left our horses & baggage under a guard. We then hastened under the pilotage of Cap Duncan, who was well acquainted with the country. I think it was the intention of the Officers to take them by supprise. But the Enemy became apprised of our meditated attack and were prepared for us. Cap Duncan was well Acquainted with Indian warfare & much reliance was placed on him. He was the first man killed. We in a few minutes perceived we were surrounded. Their leader was Brandt. Our men fought until our amunishion failed. When the Officers gave orders for us to retreat in which many were killed. Col B Tustin was supposed to have fallen. & Cap Jones & Cap Little both of Col Tustins regiment. I think not more than one fourth of our men Ever returned home. This battle was near the mouth of the Lackawack. Capt John Wood was taken prisoner by them & conveyed to the British Garison at or near Detroit. On the retreat I fell in with two men, one a Capt Harker from Sussex County was badly wounded we assisted him to Minisink. We arrived at a fort two days after the battle much fatigued, having been nearly three days without provision. There was with us a small detachment from Sussex County Commanded by Maj Meeker, he was wounded in the sholder. There was also with us one Company of artillery Commanded by Cap Brown of the standing army."

His cousin Jeremiah claimed that Jason "made a safe retreat by taking a different course from the way the militia entered the wilderness he did not git back to the encampment untill the second day after the action he was out two nights after the battel." Nathan Stewart claimed to have heard Benjamin Tusten say that he would never leave this ground when the men were driven from their final position. Jeremiah Kinner, one of the refugees from Long Island, stood next to Tusten when he was killed, and Benjamin Gunsolley claimed that he saw Tusten lying dead on the ground. If true, this shows that Lieutenant Colonel Tusten fell before the fleeing militia were out of sight because these survivors of the battle saw him fall. It also places these three men among the handful that safely made their escape from that hill. There is also evidence that some of the men may have been tortured.

Other pensions included in this work detail the manner in which some of the survivors made their way back down the Delaware River after the final shots were fired and other applications mention in passing detail how different columns of men marched to the assistance of Hathorn's command, and where and why they turned around. By taking things said in different applications we were able to piece together a pretty accurate picture of what occurred during those days before, during, and after the battle. We were also able to prove the involvement of an additional unit that has never before been linked to the battle.

As the beaten militiamen made their individual ways home, Colonel John Seward led a party of men up the Pennsylvania side of the Delaware, and reached the mouth of the Lackawaxen before dark. Barnardus Swartwood, who was with him, stated that "on arriving at the Latter place in the Evening the signs of Indians were discovered and after night the Indian Fires were discovered about one mile from the last mentioned Place, It being the same Party of Indians which was ascertained on our return home that had defeated and killed nearly all the New York Militia that same day oposite the mouth of the Lackawaxen River in the State of New York, a counsel was held by the commanding officers of his company as to the propriety of attacking the Indians that night considering the weakness of our Forces and not knowing the strength of the Indians it was decided to be impracticable to make an attack that night, when the day dawned the next morning the Indians had marched off after reconoitering the country along the Delaware River deponent returned home."

As word of the defeat begin to spread, many wives must have been anxious to learn if they were now widows. Unfortunately for Keziah Finch, news came that her husband Nathan had indeed been killed. In the application of John Wisner, not included in this work, his daughter gave testimony in support of his claim, mentioning that Mrs. Finch "went into fits and it was thought she would not live for some days, and that deponents mother was sent for to stay with her during her sickness."

An Old American Soldier

The war eventually ended, but hardships still remained for some of the veterans of the Battle of Minisink. When the War of 1812 broke out, Joel Bower was then living at Niagara, where, according to his pension application, "it was well known that I was an old American Soldier, & not apprehending any trouble & it being difficuilt for me to leave my little effects, & being then an old man I concluded by taking no part in the war, I might remain unmolested, but in this I was mistaken, I was apprehended by one Major Kirby of the incorporated mounted malitia as I was returning from my labour with a hoe on my shoulder & they forced me to run before them 1 1/4 mile to the guard house threatning to blow out my brains. I was then sent across the lake to York where I was locked up in a cold filthy prison where I suffered incredible hardships for more than two years."

In the years after the Revolutionary War, old soldiers would often gather and relive their experiences during that conflict. In the pension application of John Wisner which seems to have held a number of interesting stories, Jasper Armstrong, son of Robert Armstrong of Warwick, recalled seeing his father, Colonel Hathorn, and John Wisner discussing the Battle of Minisink. He told how Colonel Hathorn used to say to the two men that "If you had been with me I never would have followed the Indians so far into the woods." Hathorn said that he was pressured or "overpersuaded" by the other Officers. Apparently, based upon this statement, Hathorn may have had misgivings about following Brant also.

The stories found within the pension applications were not restricted to the Battle of Minisink itself. Often, interesting anecdotes were given when detailing a former soldier's service. Joseph Vail, son of Gilbert Vail who fell at the Battle of Minisink, served on the frontiers himself in the years after the battle, and told of the soldiers entertaining themselves one day. He reported that "the soldiers did not draw their rations of liquor for several days and when they did draw the liquor the blacks of the surrounding Country came into the Camp and helped them to drink what spirits they had drawn Amongst the soldiers there was kept a pewter tankard measuring about a quart Some of the soldiers offered to the blacks the tankard full of spirits if any one would drink it off, one only accepted the offer drank the tankard of spirits and died before night." Though the story is tragic in that the man died, it does go to show the length young men under arms might go to be entertained, even back then.

In another story a former soldier stated that one day while a number of the men were bathing in the Neversink River, one of the men was challenged to swim from one side to the other with the narrator's brother on his back. The challenge was accepted, and the men began their crossing. To the horror of the man recounting the event, he watched as his brother slid from the swimmer's back and drowned, not knowing how to swim.

Shin Plasters

Joel Bower, who gave us the interesting account of his participation in the Battle of Monmouth, also gives us a look at one of the problems the newly formed country faced; a shortage of currency, especially in the frontier regions. In Bower's story, he mentions "shin plasters." A shin plaster was a privately-issued note or form of currency from a business given to customers as change when they made a purchase. These notes were only redeemable at the same business that issued them, thereby generating repeat business. Some dishonest storekeepers knew that few frontiersmen carried currency in wallets and began to toast the notes in ovens, knowing that after a few days in a frontiersman's pocket, the notes would disintegrate. The frontiersman, not to be outdone, found a way to out fox the merchants and by attaching the notes to their shins, they were able to keep them from disintegrating, and this gave rise to the term "shin plasters."

Bower tells how "All the compensation I received for the 6 years I was in the army was not available for more than $25 when pay-day came I with many others paid no attention to it, the paper which we drawed was of but little consequence as for instance when I was on the Frontier in a ceratin Barrack room there was 19 of us, we collected all the 'shin plasters' we could, amounting according to the face of them to $1008, which was all put in a hat & I carried it to a certain old Scotch Lady who kept a kind of grocery, & all I could get for this amount was a half pint of Brandy which I carried to my quarters & then we rationed it out in a teaspoon so that all might get his share."

Location of Project Archives

To create this book the authors collected and organized all of the pensions of the men who fought, or were associated with the battle, in alphabetical order and then by home state where they were from at the time of the battle. Additional research was undertaken and it is now believed that the pensions included in this book are all of the pensions that are known to have survived.

Photocopies of the original pension materials located in the National Archives in Washington, D.C., as they relate to the battle are now located in the archives of the Minisink Valley Historical Society in Port Jervis, New York, and available to the researching public. It is our belief that more discoveries will be made as these pensions are made available to researchers in the future.

The preceding is an excerpt of a presentation made by Mark Hendrickson at the Battle at Minisink commemoration that took place at the battleground park in 2009 along with material written by Peter Osborne.

Bobbett sc

Chapter 5
The Yorkers

But to return to the western frontiers against the Indians I have performd many hard marches and dangerous fatiegues . . . I was out when so many brave men were lost att Lackaway

Jacob Aldrich
Orange County Militiaman at Battle at Minisink

The following pensions are from those enlisted men and officers who served in New York units and are listed alphabetically. The notation at the end of the pension identifies the source of the material and if the application was rejected.

Jacob Aldrich
New York (R.80)

Suffolk Court House N.Y.
Jany 7 1833

Sir,

Enclosed I return the declaration of Jacob Aldrich for a pension with additional evidence of his service.

I am Sir Very Respectfully
Your Obt Servt.
George Miller

Hon. Lewis Cass

State of New York

Orange County ss. On the 30th Day of March in the year 1833 Personally appeared before me Capt David Swezey and being Duly Sworn Deposeth and saith that the within named Jacob Aldrich served under him as a private in the Militia in the Revolutionary War at dif

firent times and places but this Deponent Cannot state the duration of the time of service of the said Jacob Aldrich & this Deponent is a pensioner under the act of the 7th of June 1832

Sworn & subscribed before me

John Hallock Junr A Judge of Orang Com. Pleas

David Swezy

State of New York

Orange County ss. I Asa Dunning Clerk of the said County Do hereby Certify that John Hallock Junr before whom the above affidavit was Sworn was at the time a Judge of the Court of Common Pleas in and for said County and that his name subscribed to the said affidavit is his genuine signature Witness my hand and Seal of Office this 30th day of March 1833

Asa Dunning Clerk

Dear And Respected friend. Not hearing from you since my Grandsones were there I take this opportunity to ask your assistance in a Reccommendation for my services in the Revolution War performed in Orange County from the time that Independance was declard I took an active part under the command of Captain David Sweezy although I did services under Many different Officers. I was att the Building of fort Montgomery after the Evacuation of that I went on frequent towers and fatiegues in Building Fort West Point. But to Return to the western frontiers against the Indians I have performd many hard marches and dangerous fatiegues. I was att the building of the fort att Pean Pack I was there att the time the Indians invaded the Fort and killed Col Neucarcks horse hanging att the gate killed William Steward I was out when so many brave men were lost att Lackaway I there saw James Reeve with his arm broke with a ball and Many others that might be mentioned in a deploreable situation. These duties wer performed from the time above mentioned untill it was considered unnessary to guard the Frontiers which must have been a number of years. Capt David Sweezy Elisha Reeve Daniel Baily James Reeve was my mess mates on many of these towers to this I am ready to give my affidavit which I think will entitle me to the whole pension. Having a recomendation from that county will enable me to appear bold in my petition to our Court. Remember my love to you wife and Father and not forget yourself Harriet has a _____ a little boy and wish you and your wife would come and see us I would wish you to write to me an answer as soon as possible

I remain your friend and uncle &c

Jacob Aldrich

Dear Uncle

I have recvd your note on the subject of your pension for your revolutionary services. As soon as I could conveniently I got the Affidavit of Capt. Swezey which I enclose to you

Capt Swezey is Old Infirm and his recollection and memory verry much Impaired, but I done the best I could with him he well remembered of your Service under him but could not recollect the different times or the Exact Periods.

I write to you from Newburgh where I am attending Court. I left all our friends well yesterday Father is very smart for a man of his age. My respects to all friends

Newburgh Apl. 8th 1833

Yours Respectfully &c
John Hallock Jr

Declaration of Jacob Aldrich for Revolutionary Pension

State of New York

County of Suffolk ss. On this 28th day of May 1833 personally appeared in open Court in the Court of Common Pleas in and for the said County of Suffolk and State of New York Jacob Aldrich aged seventy eight years a resident of the Town of Southold in the said County of Suffolk who being first duly sworn according to law doth on his oath make the following declaration in order to obtain the benefit of the provision made by the act of Congress passed June 7th 1832, that he was drafted in the Militia of the State of New York in the month of July in the year one thousand seven hundred and seventy six under the following named Officers Captain David Swezey, Major Moses Hatfield and Colonel Benjamin Tustin, the Lieutenant's name was Eliud Tryon, in the service of the United States during the Revolutionary War, that at the time of his entering the said service he was a resident of the Town of Goshen in Orange County in the State of New York, that after being drafted as aforesaid he continued in the said service in the said Militia over five years but by reason of old age and the consequent loss of memory he cannot state positively the precise length of his service but according to the best of his recollection he served five years and three months he having been discharged from the said service in the fall of the year one thousand seven hundred and eighty one as he thinks and believes, that he helped to build the Fort at West Point that he went on many tours on the Western frontiers against the Indians, that he helped build the fort at Pean Pack that he was there at the time the fort was invested by the Indians, when Colonel Newkirk's horse was killed, and William Stewart a private was also killed, that he was out at the time Captain Little and many men were killed near Lackawana on the Delaware River, that he was not present at the battle because he was sent back with a file of men to get provisions, that he was an Orderly Sergeant during the whole time of his said service, that he was born in the Town of Southold in the County of Suffolk in the year 1755 as he has always understood and believes but he has no record of his age, that when he was called into the said service he lived at Goshen in Orange County having removed there from Long Island a little before it was surrendered to the British, that soon after the Revolutionary war he removed back the Town of Southold in Suffolk County where he has lived ever since and where he still lives, that he is known to Rev Abraham Luse Jr & Judge David Warner who live in his neighbourhood and who will testify to his veracity and their belief of his services. That he never received any written discharge from his said service. That he hereby relinquishes every claim whatever to a pension or annuity except the present and he declares that his name is not on the pension roll of any agency in any State

Sworn and subscribed the day and year aforesaid

Attest Joseph R Huntting Clk

Jacob Aldrich

We Abraham Luse Jr a clergyman and David Warner one of the Judges of the above Court of Common Pleas both living in the neighborhood of Jacob Aldrich do hereby certify that we are well acquainted with Jacob Aldrich who has subscribed and sworn to the above declaration that we believe him to be seventy eight years of age; that he is reputed and believed in the neighbourhood where he resides to have been a Soldier of the Revolution, and that we concur in that opinion

Sworn and subscribed the day and year aforesaid

Abraham Luse Junr
David Warner

Attest Joseph R Huntting Clk

And the said Court do hereby declare their opinion that after the investigation of the matter and after putting the interrogatories prescribed by the War department that the above named applicant was a Revolutionary Soldier, and served as he states And the Court further certifies that it appears to them that Abraham Luse Jr who has subscribed the preceding certificate is a clergyman, resident in the Town of Riverhead in said County of Suffolk and in the immediate neighbourhood and congregation of the said Jacob Aldrich and that David Warner who has also subscribed the same is a Judge of this Court of Common Pleas, and that their statements are worthy of credit.

I Joseph R. Huntting Clerk of the Court of Common Pleas in and for the County of Suffolk do hereby certify that the foregoing contains the original proceedings of the said Court, in the matter of the application of Jacob Aldrich for a pension In Testimony whereof I have hereunto set my hand and seal of Office this Twenty Eighth day of May one thousand eight hundred and thirty three.

Joseph R Huntting Clk

War Department,
Pension Office,

Augt. 2nd 1833

Sir:

The evidence in support of your claim, under the act of June 7, 1832, has been examined, and the papers are herewith returned. The following is a statement of your case in a tabular form. On comparing these papers with the following rules, and the subjoined notes, you will readily perceive that objections exist, which must be removed, before a pension can be allowed. The notes and the regulations will shew what is necessary to be done. Those points to which your attention is more particularly directed, you will find marked in the margin with a brace, (thus: }). You will, when you return your papers to this Department, send this printed letter with them; and you will, by complying with this request, greatly facilitate the investigation of your claim.

A Statement, shewing the Service of Jacob Aldrich Suffolk County New York

Period when service rendered	Duration of claimant's service			Rank of the Claimant	Names and Ranks of the Company Officers under whom he served	Names and Ranks of the Company Officers under whom he served
	Years	Months	Days			
Draft July 1777	5	3		Orderly Sergt	Major Hatfield, Cap Swezey	Col Tusten

Place of abode when he entered service, and age at the present period	Evidence by which the declaration is supported
78 years, Orange County New York	Traditionary evidence - papers correctly attached

Proof by which the declaration is supported.
I am, respectfully,

Your obedient servant,
J. L. Edwards, Commissioner of Pensions.

After the statement of service a copy of the regulations under the Act of June 7, 1832 is included but not reproduced here. An identical copy is included in the chapter entitled ***A Declaration of Service****.*

State of New York
County of Suffolk ss. Daniel Hallock of the Town of Southold in the County of Suffolk being duly sworn doth depose and say that to the knowledge of this deponent Jacob Aldrich of the said Town of Southold, was in the summer of the year 1776 a soldier in the Revolutionary War, in the company ~~commanded by~~ of Militia commanded by Captain David Swezey in the Regiment commanded by Colonel Benjamin Tustin, in Orange County in the State of New York, that this deponent was at the same time a soldier in the company of Militia commanded by Captain Seth Marvin in the same Regiment where this deponent served eight months and twenty days, that this deponent does not know when the said Jacob Aldrich entered the said service nor how long here served, except that he knows he served, at least, a large part of the season above mentioned, this deponent further saith that he left Orange County soon after his discharge from service, that he at and about the time understood and has always understood and believes that the said Jacob Aldrich continued in Orange County several years after this deponent left, and that he continued to serve in the Militia until near the close of the Revolutionary War. This deponent further saith that he has known the said Jacob Aldrich ever since he was a child, and that they have lived in the same town ever since the close of the Revolutionary War, that this deponent is aged seventy eight years.

Sworn the 16th day of October 1833 before me
George Miller Commissioner of Deeds

Daniel Hallock

State of New York

County of Suffolk ss. John Woodhull of the Town of Riverhead in the said county being duly sworn doth depose and say that he has been acquainted with, and lived in the same parish with Jacob Aldrich for forty years, that this deponent has always understood and believes that the said Jacob Aldrich was a Soldier in the Revolutionary War, as a Militia man, and that he served in Orange County in the State of New York and that vicinity, and that the said Jacob Aldrich is a man of veracity as this deponent has always considered and believes

Subscribed & sworn the 24th day of October 1833 before me

George Miller Commissioner of Deeds

John Woodhull

State of New York

County of Suffolk ss. I George Miller before whom the two foregoing depositions were taken do hereby certify that I am well acquainted with the above named John Woodhull and Daniel Hallock that they are both persons veracity and worthy of belief on oath that the said John Woodhull has been for several years a Judge of the Court of Common Pleas in and for the County of Suffolk.

George Miller
Commissioner of Deeds

Suffolk County ss. Jacob Aldrich the applicant being duly sworn further saith, that he knows of no person living who has any personal knowledge of his service except Captain David Swezey and Daniel Hallock whose depositions are hereto annexed, that the said Captain Swezey lives one hundred and fifty miles from this deponent, and this deponent has not seen him in twenty years, that this deponent has no doubt but what if he could now see the said Captain Swezey he could bring to his recollection facts which would enable him to make oath that this deponent did serve full five years as stated in his declaration, that during the said term of five years and three months service this deponent pursued or attended to no other business or employment.

Sworn and subscribed this 4th day of January 1834 before me

George Miller Commissioner of Deeds

Jacob Aldrich

State of New York
County of Suffolk ss. I Joseph R Huntting Clerk of the said county do hereby certify that George Miller before whom the three foregoing affidavits were taken on the days of the several dates thereof was and still is a Commissioner of Deeds in and for said county duly authorized to take affidavits, that I believe his name subscribed to the jurats of the said affidavits and to the above certificate to be in his own proper hand writing I being acquainted with the same. In Witness whereof I have hereto set my hand seal of Office the 9th day of January 1834.

Joseph R Huntting Clk

Suffolk Court House N.Y.
June 11 1834

Sir,

Enclosed I forward an additional affidavit of Jacob Aldrich which is all the further information he is able to give at present.

Very Respectfully
Your Obt Servt
George Miller

J. L. Edwards Esqr Commissioner of Pensions

State of New York
County of Suffolk ss. Jacob Aldrich of the Town of Southold in the said county, an applicant for a pension under the act of June 7 1832 being duly sworn saith that the whole time which he served or was subject to military orders was as is stated in his declaration now at the War department, about five years and three months, that during that time the Militia were divided into two companies or divisions each of which was in the actual service or on duty, alternately with the other, that the tours of duty, were from two to four weeks in length when one company or division was relieved by the other, that for the last forty five years he has been entirely removed from the place of his service and from all his comrades in arms, and consequently he is unable to state with precision the particular tours of duty in which he served, that he helped to build Fort Montgomery, the Fort at West Point, and the Fort at Pean Pack, and went many tours to the West against the indians and to guard the frontiers, that he with a number of Soldiers viz six men was stationed for the period of three months at Goshen to guard the Jail to

prevent the escape or rescue of three prisoners (named Claudius Smith, Daniel Dulmar and James Keath as he thinks), all of whom were executed in the month of January, but he cannot state in what year, that during all that time he as orderly serjeant had the command and charge of the said six men, that he was an orderly serjeant throughout the whole of his said service, and Captain David Swezey, in his deposition, could only have meant that this deponent was not a commissioned officer when he says that he this deponent served as a private, that the said David Swezey was Captain of the company to which this deponent belonged, but he was not under the immediate command of the said David Swezey near all the time that he was in the actual service, that he has not seen the said David Swezey but once in forty five years, and that was twenty five years ago, that if this deponent could have seen him, he has no doubt he could have enabled him to testify more fully to his service, that the accompanying letter is from a late member of Congress to whom this deponent wrote to procure the deposition of Captain Swezey, and this deponent further saith that when he was not out on actual service in the camp he was employed nearly all his time by his duties as orderly sergeant in keeping up the discipline and organization of the militia in serving citations to Courts marshal, on delinquents and in collecting fines and the like

Jacob Aldrich

I do hereby certify that I have been acquainted with the above named Jacob Aldrich eight years, he having lived within eight miles of me during that time, that he is a man of fair character and worthy of belief on oath and that he subscribed and made oath to the foregoing deposition before this 9th day of June 1834.

George Miller Commissioner of Deeds

State of New York
Suffolk County ss. I Joseph R Huntting Clerk of the said county do hereby certify that George Miller above named is a Commissioner of deeds in and for said county duly authorized to take affidavits and that, his name subscribed to the above certificate I believe to be in his proper hand writing I being acquainted with the same. In Witness whereof I have hereunto set my hand and seal of Office this Twelfth day of June 1834.

Joseph R Huntting Clk

Jacob Aldrich
Claim of his Executor for pension
Power of Atty to me and Vespasian Ellis Esq. sent about 7th March 1852.
___ _______ Jany 17 1853
Please send same to me.

J. M. Baldwin
Atty for Exr.
Nassau
New York

To Comnr. of Pensions.

Rejected

Revolutionary War Pension and Bounty Land Warrant Application Files
Series M-804; Roll # 26

Nehemiah and Rebecca Baylis or Balys or Bayles Continental/New York (W.4,878)

Whereas the Bairer Nehemiah Balys in July 1779 was a Resedenter in Goshen in Orange County and State of New York a Refugee from Jamaca on long Island at the same time I the subscriber commanded the company of Exzemts in Goshen aforesaid & on allarm of the Indians attacking Minisank in the said county the said Nehemiah Balys as a Voluntere joined my company and went with me on the allarm and was with me at the Battle with the Indians on Delaware River on the 22 July 1779 Where I was taken prisoner by Captain Brant & carried to Niagara and from that to Canada where I remained a prisoner upwards of four years and on my Return was Informed the said Nehemiah Balys was wounded in the said action and was for a long time under the care of Docter John Gale and Docter Jonathan Swezy as Witness my hand this Second Day of November 1792

John Wood

State of New-York,
Rensselaer County ss. On this twenty fourth day of April 1818, before

me the subscriber, ~~one of the~~ first Judges of the County of Renssalaer personally appears, Nehemiah Baylis aged sixty three years, resident in Town of Pittstown in the County of Rensselaer in the said State who being by me first duly sworn, according to law, doth, on his oath, make the following declaration, in order to obtain the provision made by the late act of Congress, entitled "An act to provide for certain persons engaged in the land and naval service of the United States in the revolutionary war." That he the said Nehemiah ~~enlisted~~ volunteered in the year 1775 & served with & under Gen. Putman one year & three months in the state of New York in the companys commanded by Captains Skiddmore, Ludlum & Platt of the Continental Troops that he continued to serve in the said corps, or in the service of the United States until the last of August 1776 when he was discharged from service in Westchester state of New York in the year 1777 I enlisted in the Corps of Engineers Capt. Matchings company Col Lambs Artillery & worked on the Chevause de frise in Hudsons River & other works for one year & six months & was honourably discharged, and in 1779 volunteered & was wounded & much disenabled that he was in the battles ~~of~~ with the Hessians on Long Island & at Westchester against Rogers Rangers, & the Deleware against Brandt & the Savages when I was a Volunteer & wounded in the head, thigh & arm & have not received any pension and that he is in reduced circumstances, and stands in need of the assistance of his country for support; and that he has no other evidence now in his power of his said services.

Sworn to and declared before me, the day and year aforesaid.
Josiah Masters

Nehemiah Baylis

I Josiah Masters first Judge, &c. as aforesaid, do certify, that it appears to my satisfaction, that the said Nehemiah Baylis did serve in the revolutionary war, as stated in the preceding declaration, against the common enemy; and I now transmit the proceedings and testimony taken and had before me, to the Secretary for the Department of War, pursuant to the directions of the aforementioned act of Congress.

Josiah Masters

State of New-York,

Rensselaer County ss. I Certify that Josiah Masters Esq. whose signature is affixed to the foregoing declaration, was on the Twenty fourth day of April last ~~one of the~~ First Judges of the court of common pleas of Rensselaer County In

Witness whereof, I have hereunto set my name, and affixed the Seal of the said Court, this Eleventh day of May A.D. 1818

Joseph D Seldon Clerk

State of New York

Rensselaer County ss. I Josiah Masters first Judge of Rensselaer Common Pleas do certify that Nehemiah Baylis within named a resident near my neighbourhood owns a small farm & by his ________ of his family does not suffer for the necessaries of live from his age & injury of three wounds he is unable to perform very little labor, is a meritorious man. I do not know to what exstent the war department construes the word needs assistance from his country for support I have therefore certified the facts. In the case heretofore transmitted it was my rule not to certify, unless the applicant was in extreme indigence.

Josiah Masters

Orange County ss. Personally appeared before me John Homan of the Town of New Windsor County of Orange and State of New York, who being duly Sworn according to Law, deposeth and Saith, Nehemiah Baylis (as he well recollects) did serve as an artificer in Captain Thomas Machins Company in Colonel Lambs Regiment of Artillery for and during one whole year in the Continental Service, in the War of the Revolution, he thinks he Served in the years of 1777 and 1778.

Sworn before me the 4 day of Dec. 1818
Joseph Morrell Commissioner &c

John Homan

I Certify that I am acquainted with the within named John Homan and that he is a man of truth whose affidavit may be Believed.

Dec. 4th 1818

Joseph Morrell Commissr.

State of New-York

Orange County I David M. Westcott Clerk of the said County of Orange do hereby Certify that Joseph Morrell is one of the Commissioners in and for the said County ~~of~~ duly Commissioned and Sworn and by Law authorized to take any affidavit or other Oath to be read or used in any Court of Record within this State, and that the signature "Joseph Morrell" above written is the proper hand writing of the said Joseph Morrell. In witness whereof I have hereunto affixed the Seal of the said County of Orange the fourth day of December 1818.

D M Westcott

Pittstown Sept. 8th 1819.

Having been personally acquainted with Mr. Nehemiah Baylis upwards of twenty one years and having been likewise his principal family physician for a considerable part of that time, I do hereby certify that said Baylis has ever been from my earliest knowledge of him a person of a feeble slender constitution and who in consequence of which has during that time very frequenly been unable to attend to his ordinary occupation, his disorder is a nervous one and was occasioned by overheating and inflaming the system and by credible report was done during several engagements while in the revolutionry service he has also scars of wounds received while in said service which together with his former in injury ought to entitle him to a pension which by an act of Congress was granted for the relief the wounded and disabled revolutionary soldiers of his description.

David Doolittle

Hon. J. C. Calhoun,

Sir,

Enclosed I send you a copy of the Declaration & Oath of Nehemiah Baylis for the purpose of a pension. Mr Baylis has applied for a pension before this in the year 1819. If he could draw his pension the time of his first application I should be much pleased. He is an honest old man, & quite feeble and is in want of his pension to live. His wife is old & infirm & not able to do any thing toward a living Should his application be granted, and I presume it will, you will please to inclose his certificate to me at this place.

Sep 13, 1824.
Lansingburgh
Rensselaer County
New York

Respectfully Yours &c
E. W. Walbridge

State of New York
Rensselaer County

Mayors Court of the City of Troy in said County ss. On this Seventh day of September 1824 personally appeared in open Court being a Court of Record for said City of Troy in the County aforesaid Nehemiah Baylis resident of the Town of Pittstown in the County of Rensselaer aforesaid aged sixty eight years who being first duly Sworn according to law doth on his oath declare that he served in the revolutionary war as follows - to wit that in the year Seventeen hundred and Seventy five he entered the revolutionary army as a Soldier and Served under General Putnam one year and three months in the State of New York in the Companies commanded by Captains Schidmore Ludnum an Platt of the Continental Troops and that in the year Seventeen hundred and Seventy seven he enlisted into the Corp of Engineerss in Capt. Matchins Company in Col. Lambs Regiment of Artillery and was detached to work on the Chevaux-di-frieze in the Hudson River and other works and served one year and Six months and was then Honorable discharged that in the year Seventeen hundred and Seventy nine he this deponent volunteered in a Company Commanded by Capt Wood against the Indians under Col Brandt up the Delaware by some called the Minisink battle in which battle he this deponent was wounded in the head thigh and arm, that this deponent while he was in the Continental service was wounded and otherwise disabled in the battle on Long Island against the Hessians that he was in the battle against Rogers Rangers in West Chester County that while in the service as aforesaid he was surfeited with heat which Together with his wounds renders him wholly Incompetent and unable to provide a livelyhood for himself and wife by manual Labour that his wife is now sixty four years old and quite infirm and that he is absolutely in nead of the assistance of his country for support and I do solemnly swear that I was a resident Citisen of the United States on the 18th day of March 1818 and that I have not since that time by gift sale or in any manner disposed of my property or any part thereof with Intent thereby so to diminish it as to bring myself within the provisions of an act of Congress entitled an act to provide for certain persons engaged in the land and naval service of the United States in the Revolutionary war passed on the 18th March 1818 and that I have not nor has any person in trust for me any property or securities contracts or debts due to me nor have I any income other than what is contained in the Schedule hereto annexed and by me subscribed

Sworn to and declared on the seventh day of September 1824
Ben. Smith Clerk

Nehemiah Baylis

I have no real property whatsoever	
Personal property	
1 mare ten years old worth	$25.00
1 Two year ~~old~~ Colt	$25.00
1 Cow $10, 1 Two year old Steer $5	$15.00
1 Yearling Heifer $5, 1 Calf	$27.00
Two hogs $2.00, 10 Sheep $10	$12.00
1 Old waggon been used Sixteen years	$8.00
1 Waggon Chair $1, 1 plough old $1	$2.00
1 Set drag teeth worth	$1.00
1 Fanning Mill 24 years old	$5.00
1 Old axe 1 ________ 1 Short Chain	$2.00
6 Knives and forks which have been used 20 years 6 pewter Spoons, 12 old plates and one platter and a few other small articles of kitchen furniture	$5.00
5 Chairs been in use about 40 years	$2.00
Pittstown July 27 1824	$109.00

I am a farmer by profession but So old and weak that I am not able to do but very little work.

Nehemiah Baylis

State of New York

City of Troy Clerks Office I Benjamin Smith Clerk of the Court of common pleas called the Mayors Court, held in and for the City of Troy, do hereby certify that it appears to the satisfaction of the Court that the said Nehemiah Baylis did serve in the revolutionary war, as stated in the preceeding declaration, against the common enemy, for the term of nine months under one engagement, on the continental establishment. I also certify that the foregoing oath and schedule thereto annexed are truly copied from the records of the said Court, and I do further certify that it is the opinion of the said court, that the total amount of the value of the property exhibited in the aforsaid schedule does not exceed the sum of one hundred and nine dollars. In testimony whereof I have hereto subscribed my name and affixed the seal of said Court this seventh day of September 1824.

Ben. Smith Clerk

War Department,
Revolutionary Claim.

I Certify that, in conformity with the Laws of the United States, of the 18th of March, 1818, 1st of May 1820, & 1st March 1823, Nehemiah Baylis late a Private in the Army of the Revolution, is inscribed on the Pension List, Roll of the New York Agency, at the rate of eight dollars per month, to commence on the Seventh day of September one thousand eight hundred and twenty four Given at the War Office of the United States, this Sixth day of January one thousand eight hundred and twenty five

J C Calhoun Secretary of War.

★

New York

Nehemiah Baylis of Pittstown in the state of New York who was a Private in the company commanded by Captain Matchin of the regiment commanded by Colonel Lamb in the New York line, for the term of 1777 - 1 year & 6 months Inscribed on the Roll of New York at the rate of 8 Dollars per month, to commence on the 7th of September 1824. Certificate of Pension issued the 6 of Jany 1825 and sent to the Hon. J L Hogeboom House of Representatives

portion of Semi-anl. all'ce ending 4 Mar 1825

48-23

48-23

{Revolutionary claim, Acts March 18, 1818, ~~and~~
-May 1, 1820 & 1 March 1823}

Payable semi-annually. The first payment on this Certificate will be made on the 4th of March in the year 1825 at New York by the Prest U.S. Br Bank Recorded by ___ Beall Clerk, in Book B, Vol. 5 page 168 Exd. J. L. Edwards

★

Saratoga County ss. Jonathan Rouse of Saratoga Springs in Saratoga County in the State of New York, being duly sworn before me the undersigned ~~a Judge~~ Chancellor of the State of New York, deposeth & saith that he is aged seventy six years. That he is the brother of Rebeca Baylys ~~wife~~ widow of Nehemiah Baylis deceased. That he was acquainted with the said Nehemiah Baylys during his lifetime. That the said Rebecca was married to the said Nehemiah Baylis at Pittstown Rensalaer County in the State of New York about the eleventh day of November AD ~~1872 1872~~ 1782. That this deponent was present at the sai marriage of the said Rebeca and the said Nehemiah Baylis, that they were married by Samuel Rowland Esq a Justice of the peace in & for said County of Rensalaer, and this deponent further saith that said Nehemiah Baylis died on the 16 day of September 1825 at ~~Pittsfield M~~ Pittstown aforesaid. That he was present at his death. That it was supposed & this deponent supposes that the said Nehemiah Baylis died of the consumption, ~~and it w~~ and this deponent further saith that the said Nehemiah Baylis, from the time of his service ~~never~~ in the revolutionary war never was a healthy man and this deponent believes that the disease of which he died had its origin in the privations and hardships & suffering which he endured during the said war.

And this deponent further saith that the said Rebecca & Nehemiah lived together as husband & wife ~~until~~ from their marriage until the death of the said Nehemiah, and the said Rebecca has always from the time of his death ~~until~~ remained & still is a widow & unmarried. And this deponent further saith that he is old & infirm in body & unable without great inconvenience to appear in Court.

Jonathan Rouse

I certify that before me came Jonathan Rouse to me known & ~~took~~ & made & subscribed the foregoing affidavit this 18 day of September 1837. And I further certify that he is a man of reputable character & that his affidavit is entitled to credit & that I have long been acquainted with him many years

R Hyde Walworth Chancellor of the State of New York

State of New York

Steuben County ss. On this eleventh day of October 1837, personally appeared before Robert Morrell Esquire circuit Judge of the sixth circuit at a circuit court held in and for the County of Steuben being a court of record Rebeca Bayles a resident of Howard in said County of Steuben and State of New York and who being first sworn according to law, doth on her oath make the following declaration in order to obtain the benefit of the provision made by the act of congress passed July 4th 1836, and the act explanatory of said act passed March 3d 1837, that she was married to Nehemiah Bayles who was a soldier of the Revolution on the eleventh day of November 1782, at Pittstown, Rensselaer County in the State of New York by Samuel Rowland Esquire a Justice of the peace: that the said Nehemiah Bayles died on the 16th day of September 1825 at Pittstown aforesaid of consumption as this deponent believes occasioned by the hardships privations and sufferings which he endured during the Revolutionary war. That this deponent and the said Nehemiah Bayles, from the time of their marriage as above stated, until the death of the said Nehemiah Bayles, lived together as man and wife: that this deponent has not been married since the death of the said Nehemiah Bayles, but still remains his widow.

and as to the rank which the said Nehemiah Bayles held in the army of the Revolution, and the services which he performed during the Revolutionary war, this claimant refers to the evidence and vouchers thereof to be found on file at the war department of the United States.

her
Rebecca X Bayles
mark

Sworn to and subscribed on the day and year above written, before the above named Circuit Judge in open court and read over to applicant. In witness whereof, I have hereunto set my hand and Seal of the Court of Common Pleas of said county.

W. H. Bull Clerk

Bath 24 February 1838

Hon Jno T Andrews
Dear Sir

Enclosed I hand you papers to be presented at the Pension Office to procure a pension for Mrs Balis a widow of a Revolutionary soldier

Please examine these papers & if possible obtain a pension

Yrs truly
R Campbell J P

Bath NY 17 Aug 1838

J L Edwards Esq
Dear Sir

Will you please forward to me a copy of the pension Law passed at last Session of Congress in behalf of the widows of deced pensioners and also the set of forms addopted you under such Law.

Will you inform me that decision has been made in reference to the case of Rebecca Baylis handed to you by the Hon J T Andrews

Yrs respectfully
R Campbell J P

Schaghticoke November 20 ____

William Ray of the Town of Pittstown in the County of Rensselaer & State of New York, do solemnly swear & declare in the presence of Almighty Good, That in the revolutionary war I was a Soldier & while stationed at Newburgh & Newindsor, I was acquainted with Nehemiah Baylis who had enlisted in the Corps of Engineers & I know he worked for more than year in that ________ & Regiment on the Cheveaux de fries

Sworn before me the 20th day of November ____

Josiah Masters first Judge

his
William X Ray
mark

I Josiah Masters certify that the above named William Ray is a credible witness

Josiah Masters first Judge

Bath Steuben Co New York
February 11, 1839

Mr J. L. Edwards

Sir

Mrs. Rebecca Baylis Widow of the late Nehemiah Baylis a pensioner of the revolution has made an application for a pension under the act granting pensions to widows of revolutionary soldiers and her claim has been rejected My object in writing is to have you inform me on what ground her claim was rejected and whether she is not entiled to a pension under the act of last session granting pensions to widows for five years. Yours respectfully

Joel C. Rouse

PS forward me the ncesry papers for making application the different acts

J C Rouse

Whitesboro Oneida Co New York
March 13, 1839

Sir

I have the honor to inform you that I wrote you requesting information in the case of Mrs Rebecca Baylis who has made application for a pension as the widow of Nehemiah Baylis late a pensioner and have recd. no Answer I likewise wrote to the Hon Mr Hall a member of congress from Vermont and he forward me you Answer to him from which I learn that her case was provided for by the act of 1838 If married previous to 1794 and that documentary evidence of her marriage should be procured if practicable I would inform that at ________ procuring the affidavits of Jonathan Rouse who was present at her marriage that I made diligent search for documentary evidence but could find none His affidavit accompanies application of Mrs Baylis and if it should not be sufficient you will please inform me what more would be necessary If you should further evidence as to the credibility of Mr Rouse I could forward you a commission from every Republican Governor of this State from the from Geo Clinton down to Dewitt Clinton

I would request your immediate attention to this and direct your communication to me at Bennington Vermont

Yours Respectfully
Joel C Rouse

James L. Edwards
Com of pensions

Bennington May 24, A.D. 1839.

Dear Sir,

By request of a near relative & friend of the within named applicant I write you on behalf of Rebecca Baylis, and enclose her declaration, with the testimony of Joel C. Rouse. I am informed by her friend my employer as well as by her enclosed declaration, that further and sufficient evidence to establish her rights under the law of July 1838, are now in possession of the department.

Will you have the goodness to examine the enclosed papers as well as the further testimony in your possession, and referred to by the declarant in her last declaration, and as soon as convenient after due consideration, make known to me your decision in her case. Should you decide she is entitled to a pension certificate under the law aforesaid, her relative & friend now here, wishes it may be sent to my care, as he will leave this place for the residence of Mrs. Baylis, about the 20th of June ensuing, or sooner, if the certificate should be granted & received previous to that time. I have no personal interest in this last request, and can only say it will be a safe medium of conveyance to the old lady, & through the hands of her approved agent will be delivered to her.

It would appear by a letter now before me from the the Pension Office, signed by "Geo: Wm. Crump" that previous to the law of July 1838, Mrs. Baylis had preferred her claim for pension as widow of Nehemiah Baylis, and that this application was rejected in March 1838. Among the papers accompanying this her application, I am informed these are ~~certified~~ authenticated papers bearing testimony, important & necessary in the present application, and which it is requested may be examined. This letter was addressed to Hon J. T. Andrews and further states that (Mrs. Baylis) in her declaration it is asserted that he (her late husband) died 10th Sept. 1825, whereas it appears that his stipend was paid to 4 Sept. 1826. It now appears that Nehemiah Baylis died 16 Sept. 1826, this is the correct day. How the error as to the death of Mr. Baylis happened cannot now be explained, it was probably a clerical error, in Mrs. Baylis' first declaration. This is probably unimportant as it is now corrected.

As to service of her late husband Nehemiah Baylis it is presumed the Department is in possession of sufficient testimony, as he was placed on the pension roll I believe in Sept. 1824. This evidence she prays may be resorted to, and examined as testimony in her case.

From this evidence of service, from the authenticated evidence accompanying her first application, and the declaration & testimony within forwarded, it is supposed her claim is established very plainly.

Please consider the case immediately.

Yours &c. with esteem & respect
O. C. Merrill

P.S. I enclose Nehemiah Baylis' certificate of Pension.

J. L. Edwards Esq.
Commissioner of Pensions
Washington City

☆

Washington 7th Jany. 1840

Hon. J L Edwards
Dr Sir

I enclose to you the letter of Robert Campbell Esq. on the subject of widow Rebeca Baylis application for pension, will you please give me the necessary information and papers that I may enclose the same to him.

I am Sir your obdt. Servt.
M Mallory

☆

State of New York
County of Steuben ss. On this eighth day of January A.D. 1840 personally appeared before me Ziba A Leland First Judge of said County and presiding Judge of the Court of Common Pleas of the same county being a Court of Record,

Rebecca Baylis a resident of Howard in said county & state aged Eighty years last July, who being first duly sworn according to law, doth on her oath make the following declaration in order to obtain the benefit of the provision by act of Congress passed July7, A.D. 1838 That she is the widow of Nehemiah Baylis late a United States pensioner who was a private in the army of the revolution. That she has no documentary evidence in her possession of the services of her deseased husband as a private in the army of the revolution, but she is advised that the particulars of her deseased husband's revolutionary services, the names and rank of the field and company officers under and with whom he served, the day he entered the service and the time when he left the service, the nature character and particulars of his engagements and the service he performed and when and where performed and all such further particulars as may be useful in the investigation of the claim she prefers are to be found in the Records and memorials of the government. This portion of testimony she believes was furnished and authenticated and forwarded to the War Department by her late husband at the period he was inscribed on the pension list Roll of the New York agency. She prays this documentary and Record evidence may be considered a part of hir case and referred to as ______ thereof.

She further declares that she was married to the said Nehemiah Baylis on the 11th day of November A.D. 1783, by Samuel Rowland Esqr a Justice of the peace of the town of Pittstown Renssellear County and State of New York. That she knows of no documentary evidence in support of her marriage, that her husband the said Nehemiah Baylis died on the 16th day of September A.D. 1826, and that she has remained a widow ever since that period as will more fully appear by reference to the proof hereto annexed. She prays further that the testimony of her marriage and all other testimony heretofore forwarded that may be useful and proper to bring her claim to a fair and favourable result may be taken as a part of her case.

Sworn to and supscribed the 8th day of January 1840 before me
Z A Leland 1st Judge Steuben Co

her
Rebeca X Baylis
mark

★

State of New York

County of Steuben ss. Betsey Lemm of the Town of Howard in said County of Steuben first being duly sworn according to Law deposeth and saith that she is acquainted with Rebecca Baylis above named and was acquainted with her before the death of her husband Nehemiah Baylis and has resided in the same family with the said Rebecca Baylis ever since the death of her said husband excepting the first three years next after the death of her said husband and that the said Rebecca has never married since the death of her said husband and still remains his widow.

Sworn & subscribed the 8th day of January 1840 before me
Z A Leland first Judge Steuben Co

Betsey Lemm

Common Wealth of Pennsylvania
Erie County ss. Casper M Rouse of Wattsburgh in said County being duly sworn deposeth and saith that he is well acquainted with Rebecca Baylis of the Town of Howard in the County of Steuben in the State of New York who is the Deponents Sister, that he was present at the marriage of the said Rebecca to Nehemiah Baylis and saw the cerimony performed by Samuel Rowland a Justice of the peace of the Town of Pittstown Rensselaer County in the State of New York which Town the said cerimony took place and that said Marriage took place in the fall of one thousand seven hundred and eighty two or one thousand seven hundred & eighty three deponent not certain which of those years the marriage took place of the said Nehemiah and Rebecca but is certain that it was one of those years and in the fall of one of those years

Casper M Rouse

Sworn and subscribed this 29th day of January 1840 before me
James C Price J. P.

State of Pennsylvania
Erie County ss. I, William Kelley Prothonotary of the Court of Common Pleas in and for said county do certify that James C Price Esquire before whom the foregoing affidavit was Sworn to, was at the time and still is an acting Justice of the Peace in and for said county duly commissioned and sworn, to all whose official acts as such full faith and credit is due, and this his signature above is genuine In witness whereof I have hereunto set my hand and the Seal of said court this 30th day of January 1840

William Kelley Pro.

State of New York
Steuben County ss. I Ziba A Leland first Judge of said county as aforesaid do hereby certify that that the above named Rebecca Baylis by reason of bodily in-

firmity and age is unable to appear in court, and that I am personally acquainted with her the said Rebecca and also, the above named Betsey Lemm, and that their moral characters are good & of good standing in ______ and that their statements are entitled to full credit.

Z A Leland

State of New York

County of Steuben ss. I, William Hamilton Clerk of the said County, do Certify, that Ziba A. Leland Esquire before whom the preceeding affidavits, purport to have been sworn, was at the date of the same, First Judge of the County Courts for the said County. And that I am acquainted with his hand writing and verily believe his name subscribed to the said affidavits and Certificate to be his proper hand writing and signature. Witness my hand seal of Office this 10th day of February A.D. 1840

Wm Hamilton Clerk

House of Reps.
Feby 18th 1840

Hon J L Edwards
Com. of Pensions
Sir

send you inclosed the Pension papers of Mrs. Baylis the widow of Nehemiah Baylis a Pensioner which I think you will find present a clear case

Yours very Respy
Amasa Dana

Albany

Rebecca Baylis widow of Nehemiah Baylis who was a pensioner under the Act of March 1818 and who died on the 16 Sep. 1826 of Steuben Co. in the State of N. York who was an Artificer in the company commanded by Captain Machin of the regt. commanded by Col. Lamb in the N. York line for 1 year Inscribed on the Roll of Albany at the rate of 72 Dollars per annum, to commence on the 4th day of March, 1836. Certificate of Pension issued the 20th day of February 1840 and sent to Hon. Amasa Dana, H.R.

Arrears to the 4th of Sep. '39	252.00
Semi-annual allowance ending 4 Mar. '40	36.00
	$288.00

Recorded by D. Brown Clerk,
Book A Vol. 2 Page 69

(Act July 7, 1838.}

Treasury Department,
Second Comptroller's Office,
June 25th 1841.

Sir:

Under the act of the 6th of April, 1838, entitled "An act directing the transfer of money remaining unclaimed by certain Pensioners, and authorizing the payment of the same at the Treasury of the United States," Rebecca Baylis, a Pensioner on the Roll of the Albany, N.Y. Agency, at the rate of Seventy two Dollars per annum, under the law of the 7th July 1838, has been paid at this Department, from the 4th of March, to the 4th Sept. 1840.

Respectfully yours,
Albion K. Parris
Comptroller.

To the Commissioner of Pensions
Present.

To the Hon Jas M Porter Secretary of War

Howard Steuben Co N.Y.
August 9th 1843

Sir

inclosed you will receive the affidavits of Rebecca Baylis & __________ ________ with a certificate thereunto ______ the Clerk of Steuben County for the purpose of obtaining their certificates to enable them to draw their pensions under the act of the 3rd March 1843

if all correct send their certificates to Howard Steuben County New York & if their should be any informality in the papers already inclosed please in form me in what shape to proced to obtain their certificates to enable them to draw their pensions

Respectfully Yours
James Alley

State of New York

Steuben County ss. On the fifth day of August 1843 personally appeared before me Rebecca Baylis a resident of the town of Howard in the County of Steuben and State aforesaid aged Eighty four years who being duly sworn according to law doth on her Oath make the following Oath in order to obtain the benefit of the provisions made by the act of Congress passed on the 3rd day of March 1843 granting pensions to widows of persons who served during the Revolutionary war.

That she is the widow of Nehemiah Baylis who was ~~the~~ an Artificer in the war of the Revolution, but cannot recollect who he served under, and that she is the Identical person named who drew a pension for five years under the act of the 7th July 1838 and that she received seventy two dollars per annum and that the certificate which she drawed her money on was signed by J R Poinsett Dated 20th day of February _____ and that she has not intermarried but continues the widow of the above named Nehemiah Baylis

Sworn and subscribed the 5th day of August 1843 Before me

A. A. Olmstead Justice

hir
Rebecca X Baylis
mark

I certify that the above named Rebecca Baylis is personally known to me and that she is the Identical person which has heretofore drawn a pension for five years under the act of the 7 July _____

Howard August 5th 1843

A. A. Olmstead Justice

State of New York

Steuben County ss. I William Hamilton, Clerk of the said County, do Certify, That Augustus A. Olmstead Esquire before whom the preceeding affidavit purports to have been Sworn, was at the date of the same, One of the Justices of the Peace in and for the said County of Steuben, and that I am well acquainted with his hand writing and verily believe his Signature Subscribed to the said affidavit, and also to the preceeding Certificate to be his proper hand writing and Signatures. Witness my hand and Seal of Office this ninth day of August A.D. 1843.

Wm Hamilton Clerk

Michigan

Rebecca Baylis widow of Nehemiah Baylis who was an Artificer N.Y. in the Revolution Inscribed on the Roll at the rate of 72 Dollars per annum, to commence on the 4th day of March, 1843. Certificate of Pension issued the 4 day of Dec. 1843 and sent to James Alley Howard, Steuben Co. N.Y. Recorded in Book A Vol. 1 Page 102 [Act of March 3, 1843.]

House of Reps.
Washington City
12 Decr. 1843

J. L. Edwards Esq.

Dr. Sir

I am requested to Enquire why Mrs. Rebecca Bayles' Pension Certificate does not come to hand.

Also to ascertain ~~whether~~ if she would be entitled to back pay on her Husbands account, he not worth $500 when the bill passed and during that time.

Would she not be entitled to a pension from the fact her Husband was wounded and disabled while in service, she shewing these facts

With Great Respect yr. mo. obt. Sr.
Wm. S. Hubbell

Application for a Transfer

State of Michigan

Jackson County ss. On this 22nd day of March 1845 Before me the subscriber a Justice of the Peace in and for said County of Jackson personally appeared Rebecca Balys who on oath declares that she is the same person who formerly drew a pension under the acts of July 7, 1838 and March 4, 1843 for services rendered by her deceased husband Nehemiah Balys. That her name was placed on the pension Roll of the State of New York from whence she has lately removed. That she now resides in the State of Michigan where she intends to remain and wishes her Pension to be there payable in future. The following are her reasons for removing from the State of New York to the State of Michigan, viz. That her Niece Betsey Lemm with whom she had lived for a number of years removed to the State of Michigan County of Jackson for which cause it became necessary for her to remove also

Sworn & subscribed to before me the day and year aforesaid
Wm. L. Palmer Justice of the Peace
March 22nd AD 1845

her
Rebecca X Balys
mark

Oath of Witness

State of Michigan
Jackson County ss. On this 22nd day of March 1845 Before me the subscriber a Justice of the Peace in and for said county personally appeared Michael Lemm who on his oath declares that Rebecca Balys is the identical person who drew a Pension as stated in the foregoing affidavit, and is the same person who has sworn to and subscribed the above affidavit.

Sworn to and subscribed before me the day and year aforesaid
Wm. L. Palmer Justice of the Peace

Michael Lemm

Magistrates of Witness

State of Michigan
Jackson County ss. I do hereby certify that Michael Lemm the Witness aforesaid is a person of truth and veracity
March 22nd AD 1845

Wm. L. Palmer Justice of the Peace

County Clerks Certificate

State of Michigan
County of Jackson ss. I Czar Jones Clerk of the Circuit Court of the Jackson County certify that William L. Palmer is a Magistrate as above and that the foregoing signature purporting to be his is genuine In testimony whereof I have hereunto affixed my seal of Office and subscribed my name this this 25 day of March one thousand eight hundred and forty five

Czar Jones Clerk

Mechanics' & Farmers' Bank,
Albany Pension Agency,

James L. Edwards, Esquire,
Commissioner of Pensions.

April 14th 1845

Sir,

In reply to your letter of the 12th inst. I have to state, that Rebecca Baylis a Pensioner under the Act of 3d March 1843 at 72 Dollars per annum was last paid to the 4 day of March 1844

I am, Sir, Very Respectfully, Your ob't serv't,

Thos. W. Olcott
Pres't and Pension Agent,
___ Jenkins

Pension Agency
Detroit 29 March 1845

J. L. Edwards Esq
Commissioner of Pensions
Sir

I enclose Rebecca Balys application for a transfer to this agency Her Post Office address is Norville Jackson Co.

I am Sir respectfully your Obt. Servant
E. P. Hastings
Pension Agent

Widow, &c. File No. 4878
Rebecca Baylis Widow of Nehemiah Baylis
Artfr. Rev War Act: Mch. 3rd 1843
Index: - Vol. A, Page 30
[Arrangement of 1870.]
Revolutionary War Pension and Bounty Land Warrant Application Files
Series M-804; Roll # 182

Henry and Anna Bertholf or Bartholp
New York (R.801)

Affidavit of Abm Van Valen

Passaic County

State of New Jersey ss. Personally appeared before the Subscriber Abraham Van Valen aged 80 years who being duly sworn on his oath deposeth and saith that during the Revolutionary war he resided at Closter in the County and State aforesaid That about the year 1777 Col Hathorn with his command (of which Henry Bertholf was an Officer) was stationed near that place I recollect Henry Bertholf particularly I do not know precisely how long they continued at that station but think about three months and further this Deponent saith not

Sworn and subscribed this 30th March 1846 before

B. W. Vandeoort

his
Abraham X Van Valen
mark

Commissioner to take affidavits in Supreme & Circuit Court I certify that the above named Deponent is a credible person B. W. Vandervoort

Commissioner to take affidavits in Supreme & Circuit Court

Passaic County

Clerks office ss. I John Keenan clerk of the County of Passaic do hereby certify that Benjamin W. Vandervoort before whom the foregoing affidavit was subscribed and sworn was at the time of the taking the same a Commissioner authorized to take affidavits in this State and further that I am well acquainted with the hand writing of the said Benjamin W Vandervoort and verily believe that the signature to the attest of the said affidavit is genuine In Witness whereof I have hereunto set my hand and affixed the seal of the said County the twenty second day of April AD 1846

John Keenan Clerk

Affidavit of James Burt in case of Anna Bertholf an applicant for Pension

State of New York

County of Orange ss. Personally appeared before the Subscriber a Justice of the Peace in and for the County and State aforesaid James Burt a Revolutionary pensioner aged 86 years who being duly sworn on his oath deposeth and saith That during the Revolutionary war he was well acquainted with Henry Bertholf Said Bertholf was an Ensign in the Regt commanded by Col John Hathorn This Deponent cannot particularly specify the tours of service performed by said Bertholf but he is positive that from the character of the Service performed by that Regiment he ~~must~~ the Said Bertholf must have been out at least three months each year during the years 1777 78 79 and 80 This Deponent further saith that the Said Bertholf was stationed at Ramipo in 1777 or 78. This Deponent further saith that the Said Bertholf could not have served less than one year during the war and probably he served three years. The distance of time and the infirmities of age render it impossible at this late date to specify more particularly

Sworn and subscribed this 31st March 1846 before
Joseph Roe Justice of the peace

James Burt

I certify that the above named Deponent a credible person
Joseph Roe Justice of the peace

Affidavit of Robert Karr

State of New York
County of Orange ss. Personally appeared before the Subscriber a Justice of the Peace in and for said County Robert Karr a Revolutionary Pensioner aged 89 years who being duly sworn on his oath says deposeth and says that during the Revolutionary war he was well acquainted with with Henry Bertholf that said Bertholf was an Officer belonging to the Warwick Regiment under the Command of Colonel Hathorn Deponent was attached to the Company commanded by Captain Colvin Shepherd Deponent says that said Bertholf belonged to a different company reccollects that in 1777 that an alarm was given that the Indians were at Minisink That the companys composing the Regiment were called out. That after this several other alarms were given: Is confident that Bertholf was out three different times The distance which they went from Florida in Orange County were the Companys met to Minisink was about Twenty five miles That the time which they were out upon those three alarms was at least nine days it might have been Twelve days was acquainted with Bertholf after the conclusion of the war when he was the Major of Deponents Regiment

Sworn and subscribed the first day of April AD 1846
Milton Barnes Justice of the Peace
I certify That the above named deponent is a credible Witness
Milton Barnes Justice of the Peace

Robert Karr

Affidavit of Jeffery Wisner

State of New York

County of Orange ss. Personally appeared before the Subscriber a Justice of the Peace in and for the County aforesaid Jeffery Wisner of said County and State aged 77 years who being duly sworn on his oath deposeth and saith that during the Revolutionary war he was well acquainted with Henry Bertholf That Said Bertholf was an Ensign in the Regt of Col Jno Hathorn This deponent further saith that during the war he was frequently at the House of the Said Bertholf and he there learned that Said Bertholf was constantly absent on military duty He has likewise heard the Said Bertholf say that he was constantly out during the revolutionary war ~~and Suffered much and lost much property in consequence thereof~~

Sworn and subscribed this 2nd day of April 1846 before

Joseph Roe Justice of the peace

Jeffery Wisner

I certify that the above named Deponent is a credible person

Joseph Roe Justice of the peace

State of New York.

Orange County,

Clerk's Office, ss. I, John C. Wallace, Clerk of the Court of Common Pleas, in and for the County of Orange, Do Hereby Certify, That Joseph Roe and Milton Barnes before whom the foregoing declarations ~~was~~ were subscribed and sworn, was at the time of taking the same a Justices of the Peace in and for said county; and further, that I am well acquainted with the hand writing of the said Joseph Roe and Milton Barnes and verily believe that the signature to the ~~said~~ attest of the said Declaration is genuine. In Witness Whereof, I have hereunto set my Hand and affixed the Seal of the said County, the Second day of April 1846

John C. Wallace Clerk

Declaration of Anna Bertholf

In order to obtain the benefit of the Act of Congress of the 7th July 1838 "Entitled an Act granting half pay and pensions to certain widows" or any other Act under which the testimony may place it

State of New York

County of Orange ss. On this 2d day of April 1846 personally appeared before the Subscriber Anna Bertholf of the State and County aforesaid aged Eighty nine years, who being duly sworn on her oath ~~deposeth and saith~~ doth make the following declaration in order to obtain the benefit of the provision made by the Act of Congress passed July 7th 1838 "Entitled an Act granting half pay and pensions to certain widows" or any other Act that may apply to her case That she is the widow of Henry Bertholf who was an Ensign during the Revolutionary War She further states that her husband the said Henry Bertholf went out on duty soon after her first child was born which was on May 30th 1776 and continued out until late in the fall of that year He was stationed at that time about Closter This Deponent further saith, that in consequence of his absence during the summer and fall no crops were gathered She further declares that he was out at least 6 months during each year until 1780 This Deponent further saith that she was married to the Said Henry Bertholf on the twenty first day of November 1773 and that he died on the 24th January 1818

Sworn to and subscribed on the day and year above written before John Sutton Judge of the

Court of Common pleas of the County of Orange

her
Anna X Bertholf
mark

I certify that the within named Anna Berhtolf is a credible person, and that she cannot in consequence of bodily infirmity appear in open Court and make the within declaration.

John Sutton Judge of the Court of
Common pleas for said County

State of New York

Orange County ss. Personally appeared before me the Subscriber Henry W. Bertholf who being duly sworn on his oath says that Anna Bertholf widow of Henry Bertholf cannot in consequence of her great bodily infirmity appear in open Court and make the within declaration.

Sworn and subscribed before me April 2, 1846.

John Sutton Judge of the Court of Common pleas
for said County

Henry W. Bertholf

I certify that Henry W. Bertholf the above named Deponent is a credible witness Dated April 2 1846.

John Sutton Judge

★

State of New York

County of Orange ss. I, John C. Wallace, Clerk of the County aforesaid, Do certify that John Sutton before whom the accompanying Declaration was made, was at the time of the taking of the same a Judge of the Court of Common Pleas in and for said County, the same being a Court of Record having a ~~seal~~ Clerk and a seal.

And I do further certify that I am well acquainted with the hand writing of the said John Sutton and verily believe that his signature thereto is genuine. In Witness whereof, I have hereunto set my hand and affixed the seal of the said County this Sixteenth day of April AD 1846.

John C. Wallace Clerk

★

State of New York

Orange County ss. Personally appeared before the Subscriber one of the Judges of the Court of Common Pleas of said County Henry W Bertholf aged fort Eight who being duly sworn on his oath deposeth and says that he is the grandson of Henry Bertholf now deceased who was an officer in the Orange County Militia during the Revolutionary War as deponent has understood. That deponent lived in the same house with his grandfather until the time of his death which took place on the Twenty fourth day of January AD 1818, that Deponent is well acquainted with the handwriting of his said Grandfather having frequently seen him write. That the accompanying Dutch Testament and PSalm Book published as appears therein in 1753 and having upon the Clasp thereof the Initials of the name of said Henry Bertholf is an old family record belonging to the widow of said Bertholf deponent has known said Book upwards of forty years to be in the family: That the entry made in said Book at the end thereof refers to ~~is~~ the Marriage of his Grandfather to his Grandmother and also the birth and christening of his father Jacobus Bertholf now deceased: That from the knowledge which deponent has of his Grandfathers handwriting he is well satisfied that the same is his proper handwriting and were made by him in the Dutch language deponent is acquainted with that language Sworn and Subscribed the 2d day of April AD 1846

John Sutton Judge of the Court ofCommon pleas in Said County
certify that the above named Deponent is a credible witness
John Sutton Judge of the Court of Common pleas for
said County

Henry W. Bertholf

State of New York
County of Orange ss. Personally appeared before the Subscriber a ~~Justice of the Peace~~ Judge of the Court of Common Pleas in and for the County aforesaid John W Bertholf of said County who being duly sworn on his oath deposeth and saith that the accompayning dutch Bible published in "1753" is an old family Bible belonging to the widow of Henry Bertholf and the record of the marriage of the said Bertholf in the duch language is in the hand writing of the said Bertholf This Deponent further saith that the record on this sheet has been in the family 25 years to his knowledge and is a true family record of the events specified This deponent further saith that Anna Bertholf who is an applicant for a pension is the widow of Henry Bertholf ~~who was a Revolutionary Soldier~~
Sworn and subscribed this 2nd day of April 1846 before
John Sutton Judge of the Court of Common ples of said County
I Certify that the ~~above~~ within named Deponent is a
credible Witness April 2, 1846.
John Sutton Judge

John W. Bertholf

State of New York
Orange County,
Clerk's Office, ss. I, John C. Wallace, Clerk of the Court of Common Pleas, in and for the County of Orange, Do Hereby Certify, That John Sutton Esqr before whom the foregoing Affidavits ~~was~~ were subscribed and sworn, was at the time of taking the same, a Judge of the Court of Common Pleas in and for said county the same being a Court of Record &c; and further, that I am well acquainted with the hand writing of the said John Sutton and verily believe that the signature to the ~~said~~ attest of the said Affidavits is genuine. In Witness Whereof, I have hereunto set my Hand and affixed the Seal of the said County, the Second day of April 1846

John C. Wallace Clerk

Births.
Henry Bertholf (Son of Jacobus and Elizabeth Bertholf)
was born the 12th day of August 1750.

Anna Vandervoort (Daughter of John and Catharine Vandervoort)
was born ____ July 1757.
Jacobus Bertholf (Son of Henry and Anna Bertholf)
was born May 30th 1776.
Mary Wisner (Daughter of Henry and Susan Wisner)
was born November 11th 1779.
Henry W. Bertholf (Son of Jacobus & Mary Bertholf)
was born January 1st 1798.
John W. Bertholf (Son of Jacobus & Mary Bertholf)
was born ~~September~~ May 30th 1811.
James Nelson Bertholf (Son of Jacobus & Mary Bertholf)
was born September 5th 1815.

Marriages.

Henry Bertholf was Married to Anna Vandervoort
21st November 1773.
Jacobus Bertholf was Married to Mary Wisner
the 23rd day of June 1796.

Deaths.

Henry Bertholf Departed this life 24th January 1818. Aged 67 Years 5 Months and 12 Days.

New Brunswick
May 11 1846

Hon J L Edwards
Sir

I herewith send the papers in the case of Anna Bertholf widow of Henry Bertholf The papers are not so full as the rules of the Office require I could have procured much more definite testimony from the same witnesses but I am descidedly of the opinion that no witness at this distance of time can specify as the rules require without violating his integrity I think you will concur with me that these witnesses who engaged in the Service of their country rendered in Orange County were compelled to render a large amt of Service Bertholf was one of those and a man of large property and influence he resided between the British posts and the Indian Settlements and must have been subject to constant duty I send two Commissions one dated in 1787 the other in 95 These Commissions prove the estimation in which Bertholf was held at that time by his neighbors evidently the result of his Services during the war The fact that the record of the State Dept (of NY) confirms a portion of the evidence will justify a favorable inference in reference to the residue The record in the Bible is in the dutch language and ______

"In the year 1773 Nov 21 I was married"

Respt Your Obt Sr.
James C Zabriskie

THE PEOPLE of the State of NEW-YORK,
BY THE GRACE of GOD, FREE AND INDEPENDENT;

To Henry Bartholp Esquire Greeting

WE reposing especial Trust and Confidence, as well in your Patriotism, Conduct and Loyalty, as in your Valour and Readiness to do us good and faithful Service; HAVE appointed and constituted, and by these Presents, DO appoint and constitute you the said Henry Bartholp Captain of a Company in the Regiment of Militia in the County of Orange whereof Henry Wisner Esquire is Lieutenant Colonel Commandant. YOU are therefore, to take the said Company into your Charge and Care, as Captain thereof, and duly to exercise the Officers and Soldiers of that Company in Arms, who are hereby commanded to obey you as their Captain and you are also to observe and follow such Orders and Directions, as you shall from Time to Time receive from our General and Commander in Chief of the Militia of our said State, or any other your Superior Officer, according to the Rules and Discipline of War, in Pursuance of the Trust reposed in you; and for so doing, this shall be Your Commission, for and during our good Pleasure, to be signified by our Council of Appointment. IN TESTIMONY whereof, We have caused our Seal for Military Commissions to be hereunto affixed. WITNESS our Trusty and Well-beloved George Clinton, Esquire, our Governor of our State of New-York, General and Commander in Chief of all the Militia, and Admiral of the Navy of the same, by and with the Advice and Consent of our said Council of Appointment, at our City of New-York, the twenty Sixth Day of September in the Year of our LORD, One Thousand Seven Hundred and Eighty Six and in the Eleventh Year of our INDEPENDENCE.

Passed the Secretary's Office, 18th June 1787.
Robt Harpur D. SECRETARY.
Geo. Clinton

Orange County ss. I do hereby certify that the within Named Captain Henry Bartholph appeared before me Thomas Moffat Esquire one of the Commissioners appointed to Qualify all Officers both Civil and Military within said County and took and Subscribed the Oath of Abjuration and Allegiance and the Oath of his Office agreeable to the directions of the Law in that ~~that~~ Case made and provided Sworn and subscribed the 14th day of August 1787

Thomas Moffat

The People of the State of New-York,
BY THE GRACE OF GOD, FREE AND INDEPENDENT:
To Henry Bertholf Esquire GREETING.

WE, reposing especial trust and confidence, as well in your patriotism, conduct and loyalty, as in your valour and readiness to do us good and faithful service, HAVE appointed and constituted, and by these presents, DO appoint and constitute you the said Henry Bertholf Second Major of the Regiment of Militia in the County of Orange whereof David McCamly Esquire is Lieutenant Colonel Commandant. YOU are therefore, to take the said Regiment into your charge and care, as Second Major thereof, and duly to exercise the Officers and Soldiers of that Regiment in Arms, who are hereby commanded to obey you as their Second Major and you are also to observe and follow such orders and directions, as you shall from time to time receive from our General and Commander in Chief of the Militia of our said State, or any other your superior Officer, according to the Rules and Discipline of War, in pursuance of the trust reposed in you; and for so doing, this shall be your commission, for and during our good pleasure, to be signified by our Council of Appointment. IN TESTIMONY whereof, we have caused our seal for military commissions to be hereunto affixed. WITNESS our trusty and well-beloved John Jay Esquire, our Governor of our State of New-York, General and Commander in Chief of all the Militia, and Admiral of the Navy of the same, by and with the advice and consent of our said Council of Appointment, at our City of New York the Seventh day of December in the Year of our Lord, one thousand seven hundred and Ninety five and in the Twentieth Year of our Independence.

Passed the Secretary's Office. the 21st day of December 1795.
Lewis A. Scott SECRETARY.
John Jay

Orange County ss. I do hereby Certify that on the 25th Day of June 1796 the within named Henry Bartholf Esqr. appeared before me and took and subscribed the necessary Oaths prescribed by Law to Quallify him to the Office within mentioned

Reuben Hopkins Clerk of Orange County

In pursuance of a law passed the 27th of April, 1784, entitled "An act for the settlement of the pay of the Levies and Militia, for their services in the late War, and for other purposes therein mentioned," the following certificates have been issued, bearing interest at five percent. per annum, viz:

10th June 1779	Henry Bartolf Ensn.	2	9	4
25th June 1779	Henry Bartolf Ensn.	1	10	10
1st July	Henry Bartolf Ensn.	1	3	10

State of New-York,
Comptroller's Office, Albany, 29th Apl. 1846.

I hereby certify the above to be true extracts from Book No. 1 now remaining in this office, and that the service for which the certificates above named were issued, were performed under the following officers, viz:

First under Capt. Peter Bartolf in Col. John Hathorn Regt.
Second under Capt. Peter Bartolf in Col. John Hathorn Regt.
Third under Capt. Richard Bailey in Col. John Hathorn Regt.

Philip Phelps
Dep. Compt.

Pension Office,
July 11, '46

Sir,

The declaration and evidence furnished by you in the case of Anna Bertholf has been examined and filed.

The evidence as to the date of the marriage is satisfactory, but the service is not established in such manner as our regulations demand, ~~and~~ nor to such an amount as the laws require. While the witnesses who testify to various terms of service fail to specify the length and duration of each tour, the certificate of the Comptroller, in admitting three tours, shews by the amt. which was paid for them that they were of short duration. To obtain a pension at all, at least 6 months service must be proved, yet in this case not half that term is so established as to justify us in admitting it. The case should be set up under the act of 1836 if part of the service was rendered after marriage

James C. Zabriskie, Esq.
New Brunswick, N.J.

State of New York
Sugar Loaf Nov. 17th 1846

Dear Sir

Enclosed Contains Mrs. Anna Bertholf's application for a pension, we have according to instructions received by letter directed to Zabriskie made application under the Act of 1836 which will show by her declarations and aded some more testimony to the same, that is to say some more than we had to the first

I am Respectfully your most Obdt. Servant
Henry W. Bertholf

J. L. Edwards Esqr.
Commissioner of Pensions

Declaration
In order to obtain the benifit of the third Section of the act of
Congress of the 4th July 1836 State of New York

County of Orange ss. On this Sixteenth day of September Eighteen Hundred and Forty Six Personally appeared before the Subscriber one of the Judges of the Court of Common Please in and for the County of Orange and State aforesaid Anna Bertholf aged Eighty nine years who being ~~who being~~ duly sworn according to law doth on her Oath make the following Declaration (to the best of her memory and Recollection) in order to obtain the benifit of the Provision made by the act of Congress passed July 4th 1836 (that She is the Identical Anna Bertholf who made a Declaration before John Sutton one of the Judges of the Court of Common Pleas In the month of March or April) that she is the widow of Henry Bertholf who was Ensign under Capt Peter Bertholf and that the Company belonged to Colonel John Hathorns Regiment

And that her Husband the Said Henry Bertholf served six month in year Seventeen Hundred and Seventy Six to the best of her memory and Recollection, where he was Stationed during that time she cannot State at this late day, by which She knows the services in the year 1776 was by finding a and old day Book of the Said Henry Bertholf her husband. By Reference to the affidavit of Henry W Bertholf will make it more fully appear which will accompany this ~~affidavit~~ Declaration She further saith that there is none now living by which the Services before stated can be proved

She would further State that on account of old age and the loss memmory She cannot State to what Company he belonged nor what Regiment during the before named six month services

She further declares that her husband the Said Henry Bertholf served in Each of the following years that is to say in the year Seventeen Hundred and Seventy Seven, Seventy Eight, Seventy nine, and Eighty in each year at least two month to the best of her memory and Recollection, by which She includes the fifty one days services Rendered by her husband the Said Henry Bertholf as a Substitute for Capt Peter Bertholf and in which the Said Henry Bertholf acted as Captain during that time, the proff of the fifty one days is hereunto annexed

She further declares that her Husband the Said Henry Bertholf was engaged with the Indian in a Battle with them, at a place called Minisink, the year she cannot State

She further declares that he was Stationed during the war of the Revolution at the following Places namely Minisink, Haverstraw, Paramus, Closter, Hoppertown and at the entrance of Smith Clove now Ramapough at which latter place he was Stationed for a considerable time

She further Declares that her husband the Said Henry Bertholf has informed her of all the Places above named, where he was Stationed during the war of the Revolution

She further declares that the Said Henry Bertholf was elected by the Company to which he belonged Ensign and acted as Such during the war of the Revolution under Capt Peter Bertholf & Col. John Hathorn

She further declares that on account of her exstreme old age and the loss of Memory She cannot State the month nor the tours with any degree of certainty or as minutely as the war Department Requires for which She asks the Indulgence of the Honorable Secretary of war and the Commissioner of Pensions and must now Submit her claim

She has However endeavoured to give the years in which her husband the Said Henry Bertholf Tours of duty were performed in the war of the Revolution And She has Stated no more than what She conscientiously believes to be True, to err is Humane to forgive is devine, If However She Should have erred in any of the foregoing Statements it is and error of the head and not of the heart

She further declares that She was married to the Said Henry Bertholf on the twenty first day of November Seventeen Hundred Seventy three that her Husband the aforesaid Henry Bertholf died on the twenty fourth day of January Eighteen Hundred and Eighteen and that She has Remaind a widow ever since that period as will more fully appear by Reference to her Declaration made before John Sutton one of the Judges of the Court of Common Pleas in the month of March or April 1846

Sworn to and Subscribed on the day and year above writen before
John Sutton one of the Judges of the Court of Common Pleas

her
Anna X Bertholf
mark

We Harman D Clark Residing in the Town of Warwick and Nathaniel Knapp of the Town of Chester Both of the County of Orange do hereby certify that we were well acquainted with Anna Bertholf widow of the late Henry Bertholf who has subscribed and sworn to the above Declaration we believe her to be eighty nine years of age that it is Reputed and believed in the Neighborhood where she Resides to be the widow of Henry Bertholf who was a Soldier of the Revolution and that we concur in that opinion

Sworn and subscribed before me this 16th day of September 1846
John Sutton Judge

Harman D Clark
Nathaniel Knapp

I John Sutton one of the Judges of the Court of Common Pleas do hereby declare my oppinion after the investigation of the Matter that the above named Applicant Anna Bertholf who is the the widow of Henry Bertholf who was and officer in the war of the Revolution and I do further certify that Harman D Clark and Nathaniel Knapp Both of the County Orange are credible Persons and that their statement are entitled to credit And I do further testify that Anna Bertholf is not able to attend Court on account of old age and bodily infirmities

John Sutton

State of New York
Orange County ss. Personally appeared before the Subscriber one of the Judges of the Court of Common Pleas in and for the said County and State aforesaid Henry W Bertholf aged forty Eight years who being first duly sworn according to Law Deposeth and saith that he is the grandson of Henry Bertholf of the County of Orange and State aforsaid and the leaf Containing the Record of services Rendered in the war of the Revolution as a Substitute for Peter Bertholf then Capt of a company of which his grandfather was Ensign as the Deponent has always been Informed by his Grandfather the Said Henry Bertholf and which he verily believes to be true

And the deponent further saith that the said leaf accompaning this affidavit was taken from and old accompt Book, which Book was in possession of my grandfather the Said Henry Bertholf

And the deponent further saith that the said Henry Bertholf was a Blacksmith by trade and that it appears the Said Henry Bertholf worked at his Trade for a living (as the Deponent has always been Informed by his grandfather the Said

Henry Bertholf) He kept a day Book in which he charged his Neghbours for work in his line of Business

And the deponent further saith by examining the year Seventy Six, In which year the only Charge is July 29th 1776, and Reads as follows, shod one horse one Shilling by which it appears from the Record of his accompt Book that he must have been from home on duty during that Season

And the Deponent further saith that that his Grandfather the Said Henry Bertholf died on the Twenty fourth day of January Eighteen Hundred and Eighteen, and that his Grandmother the Said Anna Bertholf has Remained a widow ever since that Period and as yet Remains a widow

Sworn to and subscribed before me this 17th day of September 1846

John Sutton

Henry W. Bertholf

I do herby certify to the Honerable Secretary of War and Commissioner of Pensions that Henry W Bertholf who has sworn and subscribed to the foregoing affidavit is a credible Person and that full faith and confidence can be placed in his statement

John Sutton Judge

Note, words, on and erasure which Reads as follows viz, said Book he finds work done every year Except, which was done before subscribed the same

John Sutton Judge

State of New York

Orange County ss. Personally appeared before the Subscriber one of the Judges of the Court of Common Pleas in and for the County of Orange and State aforesaid John W Bertholf who being first duly sworn according to law deposeth and saith that he is the grand son of Henry Bertholf the leaf accompaning this affidavit was taken from and old accompt which Book was in Possession of the Said Henry Bertholf

And deponent further saith that the Said Henry Bertholf was a blacksmith by trade and that it appears the Said Henry Bertholf worked at his trade for a living. He kept a day Book in which he charged his Neighbours for work in his line of Business

And the deponent further saith by examining the said Book, He finds work done at his trade every year, except the year Seventy Six In which year the only charge is July 29th 1776, and Reads as follows, shod one horse one Shilling, by which it appears from the Record of his accompt Book that he must have been from home on duty during that Season

And the Deponent further saith that his Grandfather the Said Henry Bertholf died on the twenty fourth of January Eighteen Hundred and Eighteen and that his grandmother the Said Anna Bertholf has Remained a widow ever since that Period and as yet Remains a widow

Sworn to and subscribed before methis 16th day of September 1846 John Sutton

John W Bertholf

I do hereby certify that John W Bertholf who has sworn and subscribed to the foregoing affidavid is a credible Person and that full faith and confidence can be Placed in his Statement

John Sutton one of the Judges of the Court of
Common pleas in and for said County

State of New York

Orange County ss. I John C. Wallace Clerk of the Court of Common Pleas in and for the County aforesaid do hereby certify that John Sutton who has signed his name to the Declaration of Anna Bertholf and also to the affidavit of Harman Clark and Nathaniel Knapp also to the affidavits and certificates of Henry W Bertholf and John W Bertholf is one of the Judges of the Court of Common Pleas (duly Commissioned and sworn) in and for the County of Orange and State aforesaid and that his signatures to each of them are Genuine In Testimony Whereof I have hereunto set my hand and affixed the seal of the said County this seventeenth day of September in the year of our Lord one thousand Eight hundred & fort six

John C. Wallace Clerk

State of New Jersey

Bergen County to wit Personally appeared before the Subscriber a Justice of the peace in and for said County aforesaid Magdalen Ferdon aged Eighty four years who being first duly sworn according to Law deposeth and saith that she very well recollects that Henry Bertholf from Warwick believes he lived in the County of Orange and State of New York who was an Officer of what grade she cannot state in the war of the Revolution and was stationed with his company at a place called Closter and that the said Henry Bertholf was quartered at her fathers house Mr. Martin Powles as near as she can recollect after the laps of so many years it must have been in the year Seventeen hundred and seventy six believes to the best of her memory he commenced boarding or Quartering at her fathers house the said Martin Powles in the month of October and continued on his station as near as she can recollect untill the month of January following she believes she can conscientiously say that he served at least three months at that time the reasons which she assigns that it must have been in the year seventy six is by her age at that time and by her marriage afterwards

Sworn to and subscribed before me this
25th day of September AD 1846
Albert J Terhune J. P.

her
Magdalen X Ferdon
mark

I do hereby certify that Magdalen Ferdon who has sworn to and subscribed to the above affidavit is entitled to full credit and belief and personally known by me
Albert J Terhune J. P.

State of New Jersey
Bergen County ss. Personally appeared before the Subscriber one of the Judges of the Inferior Court of Common Pleas in and for the County aforesaid Abraham Vanvalen aged Eighty years who being first duly sworn according to law deposeth and saith that he very well Recollects that Henry Bertholf from Warwick County of Orange and State of New York was an Ensign under what Capt. he cannot state but he knows the said Henry Bertholf belonged to Heathons Regiment also from Warwick County & State aforesaid by which circumstance he knows the said Henry Bertholf was and Ensign that the Deponent has often seen the said Bertholf on Parade Performing military duty and acting as Ensign in the war of the Revolution

And the deponent further saith that the said Henry Bertholf was stationed at Closter and Quartered at Martin Rowleson believes he the said Henry Bertholf served three month at least in the Neighborhood of Closter where they were stationed to the best of his Recollection the said Henry Bertholf commenced Boarding or Quartering at the said Martin Rowleson in the month of October and continued Quartering at the above station untill the month of January next which makes the amount of three month as before stated. The Sirvices as above stated was Rendered to the best of his memory and Recollection in year Seventeen Hundred & Seventy Six the Reason He assigns why he knew the said Henry Bertholf was that the said Henry Bertholf a Dutchman and could speak, talk dutch and the deponent then a boy did always speak dutch at that time and the

Deponent further saith that by another circumstance he did know the said Henry Bertholf was that on a certain time he had a dispute with a man about a pair of horse shoes when a scuffel took place and the said Henry Bertholf took the deponents part which made the deponent very intimate with the said Henry Bertholf

Sworn and subscribed before me the 1st day of October 1846

James P Demarest one of the Judges of the Inferior Court of Common pleas

his
Abraham X Van Valen
mark

I do hereby certify that Abraham Vervalen who has sworn and subscribed to the foregoing affidavit is a credible Person and that full faith and confidence can be placed in his statement

James P Demarest

State of New Jersey

Bergen County ss. I Garret G. Ackerson Clerk of the County of Bergen and Clerk of the Inferior Court of Common Pleas in and for said County do hereby certify that Albert J Terhune who has signed his name to affidavit and certificat of Magdalen Ferdon is a Justice of the Peace for said County duly sworn and that his signature is Genuine and that James P Demarest who has signed the affidavid & certificate of Abraham Van Valen is a Judge of the Inferior Court of Common Pleas duly commission and sworn and that his signature is Genuine. In testimony whereof I have hereto set my name and affixed the seal of our said Court this first day of October AD 1846

Garret G Ackerson Clerk

Pen Off
Decr. 7th 46

Sir

The papers of Anna Bertholf under the act of 4 July 36 have been filed. The date of the marriage before the war is duly established but there is neither the specification nor proof of any term of service except the 3 short tours embraced by the Comptrs certificate. The presumption assumed from the silence of his Day book of Blacksmiths work in 76, if well founded, would go to shew that he was in the service during the whole year except a short time in July. This can not be admitted.

The exhibits ought to furnish the data for computing the pension understandingly.

Henry W. Bertholf Esqr
Sugar Loaf Orange Co. NY

Sugar Loaf February 10th 1847

Dr Sir

I received your letter of December 7th (1846) in which you informed me that the papers of Anna Bertholf was ~~placed~~ on file, the date of her marriage is duly established but their is neither the specification nor proof of any term of service except the three Short Tours embraced by the Comptrollers certificate

I wish to state that when her last declaration was forwarded to the Pension Office the affidavit of Mr. Abraham Vanvalen was attached to her declaration who testifies that Henry Bertholf Served three months in Closter in the County of Bergen New Jersey also Mrs. Magdalen Ferdon testifies to the same length of time and to the same Station. And further the leaf out of his account Book Shows that the Said Henry Bertholf Served at least fifty days as a Substitute for his Brother Peter Bertholf on alarms at different places who was at that time a Captain of the company to which he belonged which was also forwarded with her declaration and I would further State that Mr. Robert Carr Testifies that he was with the Said Henry Bertholf then ensign of the company in the Memorable Indian Battle at Minisink whose affidavit is attached to her first declaration (if I am not mistaken) if not it can as yet if necessary be obtained if living when an answer is received

I herewith Send the account Book of my grandfather the Said Henry Bertholf which is all that we can prove or exhibit at present, the applicant cannot State any thing more than already Stated and Specified in her declaration on account of old age and the loss of memory hoping however that you will grant her a pension according to the evidence produced. She needs it in her declining years

I am Respectfully your
most obt. Servant
Henry W. Bertholf

J. L. Edwards Esqr.
Commissioner of Pensions

Pension Office
March 16th, 1847

Sir,

The Old account book which accompanied your letter of the 10th Ulto. in the case of Anne Bertholf widow of Henry Bertholf affords no additional evidence of his service, and of course does not remove the objections to the admission of her claim heretofore communicated to her agents.

Upon referring to the memorandum on the leaf ____ the work now presented said to have been made by Henry Bertholf ~~of his~~ it is beleived that the several short tours therein recorded, is the same as that certified by the Comptroller of New York, which does not amount in the aggragate to one month. This is the only service that has been established to the satisfaction of this Department

Henry W. Bertholf Esqr
Sugar Loaf Orange County NY

Blandinsville, Ill.
May 29, 1905

Commissioner of Pensions
Washington, D.C.
Dear Sir:

Will you please find out for me about the records of Major Henry Bertholf and General Jacobus Bertholf, soldiers in the Revolution? They were in the Orange Co. N.Y., Regiment, I believe. I want to find whether the titles of Major and General were real or acquired in Kentucky style.

Yours Truly,
Frances Bertholf

---

109 N. Third Street
Easton, Penna.
November 18th, 1930

Veterans's Administration
Bureau of Pensions
Washington, D.C.
Rev. & 1812 Wars Sect. M.C.S. M.L.B.

Gentlemen:

I am listing below two of my ancestors who fought in the Revolutionary War. I would like to have a report of the data pertaining to them as shown by your records.

Isaac Otis (Capt.)
B. Sept. 24, 1752

D. Nov. 2, 1838
Resided at enlistment - Bridgewater, Mass.
Resided after service - Cumberland, R.I.
Otisville, N.Y.
Name of widow - first wife: Ruth Brown
second wife: Elizabeth Hopkins
Henry Bertholf (Ensign)
B. Aug. 12, 1750
D. Jan. 24, 1818
Resided at enlistment Warwick, Orange Co., N.Y.
Resided after enlistment Warwick, Orange Co., N.Y.
Name of widow - Anna Vandervoort
Very truly yours,
W. Otis Roe

WOR:BAE

November 25, 1930
Rev. & 1812 Wars Section
MCS:MLB

Mr. W. Otis Roe
109 North Third Street
Easton, Pennsylvania

Dear Sir:

You are advised that it appears from the papers in the Revolutionary War pension claim, R.801, that Henry Bertholf, the son of Jacobus and Elizabeth, was born August 12, 1750.

He married November 21, 1775, Anna Vander voort, who was born in July, 1757, and was the daughter of John and Catharine Vandervoort.

Henry Bertholf died January 24, 1818.

On account of his services in the Revolution, his widow, Anna, applied for pension April 2, 1846, and at that time was living in Orange County, New York.

It was alleged that Henry Bertholf enlisted in 1776 and served at least six months each year until 1780. It was again alleged that he served at least two months each year during the years, 1777, 1778, 1779 and 1780. It was also alleged that he served nine or twelve days in 1777 during alarms at Minisink, as ensign, belonging to the Warwick Regiment under Colonel Hathorn. It was still further alleged that he served at least three months each year during 1777, 1778, 1779 and 1780.

On February 19, 1778 he was commissioned ensign in Captain Peter Bertholf's company in Colonel John Hathorn's regiment of militia in Orange County, New York.

His commission as captain in Lieutenant Colonel Henry Wisner's regiment of Orange County militia was dated September 26, 1786.

His commission as second major in Lieutenant Colonel David McCamly's regiment of Orange County militia was dated December 7, 1795.

The pension law required that the widow furnish proof that the soldier rendered at least six months military service during the Revolution. This requirement was not fulfilled and the pension was not allowed.

Henry and Anna Bertholf had a son, Jacobus, born May 30, 1776 and died prior to April 2, 1846.

Jacobus married June 23, 1796, Mary Wisner, who was born November 11, 1779 and was the daughter of Henry and Susan Wisner.

Jacobus and Mary Bertholf had the following children:

Henry W. born January 1, 1798

John W. born May 30, 1811

James Nelson born September 5, 1815.

Very truly yours
E. W. Morgan
Acting Commissioner

Rejected

Revolutionary War Pension and Bounty Land Warrant Application Files
Series M-804; Roll # 229

Note: Contained within this file are 38 unreadable pages from an account book kept by Henry Bertholf concerning his work as a Blacksmith in 1776.

Joel Bower
New York (S.29,020)

The Right Honourable the Secratary of War of the United States

Dear Sir,

the accompanyins sketch of the services of Joel Bower in the Revolutionary war ~~is for the purpose~~ is for the purpose of obtaining a pension which he supposes to be granted to the soldiers of the Revolution by the government of the United States. I felt a great diffidence knowing my inability & ignorance of the legal course to be pursued, but in the mean time I feel confident that the good sence of the government will excuse my inability, when vindicating the cause and acting in behalf of one whose services has a high claim on the considerations of our country, he is a man who has ever sustained an unimpeachable reputation, of perfect sober steady habits & well respected by all who know him. By misfortune, by sickniss & disease, & being far advanced in age, and reduced to the greatest poverty, with nothing but a bed of straw to lay upon, without food to subsist upon only only as he is supplyed by the hand of charity, and even without clothes to keep him from suffering with cold, a perfect cripple unable to get ½ a mile from his humble cot, now solicits the aid of that government under whose banner it was once his greatest glory to serve, in whose defence he was ready to brave the greatest danger, & in whose cause he freely bled, for that pecuniary assistance which he entertains the firmest hope may be speedily administered. Let the most skeptical person convese with him but a few minuets ~~with him~~ & he will be convinced that he is that person he represents himself to be. In the course of the relations which I have skeched for him, he relates a great number of interesting incidents that transpired during the period of his services, which I deem unnecissary & even to prolix to be communicated to answer the end desired ~~and~~. If on perusal & examination of these documents the government sees proper to grant the pension arrearages &c, they will be pleased if not possible to send it direct to me or to any bank in this Provence, let it if possible be payable to me in the City of Buffalo NY. We will gratefully anticipate an answer with as little delay as possible remember the man is suffering for it. I am unacquainted with the proper person to direct this to, & even with the name of the present Secratary of War, therefore if it does not fall in the proper hands, I trust you will speedily rectify the mistake by placing it where it should be & excuse my ignorance.

Address to me, Pickering P.O. Home District
Canada West. via Lewiston N.Y.

I am with sentiments of the highest
Respect sir your obedient servent
Isaac Hallock

16th Dec. 1845.

From our personal knowledge of the aforesaid Joel Bower, we have no hesitation in stating it to be our firm belief, that he is the person he represents himself to be, and that he is reduced to the lowest degree of Poverty

Alexander Campbell J. P.
P. F. Whitney J. P.

Declaration of Joel Bower for a pension.
A sketch of the military services ~~of~~ of Joel Bower in the
United States army during the war of the Revolution

I was born in South Hampton Old England 20th Feb. 1763. I was about 5 years old when my Father emigrated to America, his name was Phinihas he first settled on Longisland but at the commencement of the Revolution he mooved up the North River & settled in Orange County. I was about this time apprentised to Capt. Samuel Chard Master Carpenter, in the beginning of the year '77 he entered himself 3 sons & 2 apprentices in the govermment Artificials at New Windsor, we drew our arms & accoutrements, we were at the Battle and storming of Fort Montgummary on the 6th day of October of this year, General James Clinton commanded, his Brother George then acting governor of the state was also there with a small body of malitia, the attact commenced from the best of my reccollections between the hours of 12 & 1 oclock which was continued without intermition the remainder part of the day, & under cover of the darkness of the night they succeeded in mounting the wall before we discovered them, the greater part of our soldiers were taken prisoners, the Officers a few soldiers with myself made our escape by a passage out of the fort next to the River which was not known to the British. They kept the fort but 3 days before they evacuated & burned it. A cousin of mine whose name was Samuel Jones lay sick in the fort at this time and not being able to escape, & his comrads was prohibited the privilege of lenging him assistance so he was burned to death, our quarters this winter '77 was at New Windsor, in the spring of '78 we were stationed at West Point where we were engaged in building Barracks, gun carriages, Platforms &c, we were constantly equipped for the common defence, & as liable to be ordered to battle as the Regular Army, some time in the fore part of the month of June of this year we received orders to reinforce Gen. Washingtons army then in New Jersey, as Sir Henry Clinton was about to evacuate Philidelphia and concentrate their forces at New York, we joined a party on the Hackinsack river & thence marched hastily down & joined the main body, in a verry short time the British army made their appearance and a desperate battle ensued which is called the Monmouth Battle. I was on the right of the left wing in Clintons Brigade, I stood in the front rank where we sustained a tremendous cannonade as well as a heavy fire of musketry for a considerable time we then were impetuously charged by the Dragoons, and in their second charge a Dragoon aimed a deadly blow at my head which would have instantly killed me, but I raised my musket to shield my head from the meditated stroke, which caused his sword to glance, & cut off my two middle fingers & the little finger about half off, this was my right hand which was holding on the small of my gun, my hand has ever since been verry much disabled from this wound, as he was then passing me he made an underhanded blow at my head, and the point of his sword passed through my left eyebrow which made a deep incission & considerably stunded me, at this instant his horse was shot through the head & fell dead at my feet, as I recovered from the last blow I received, I drew my bayonet & would have terminated his existance had not Adjutant Johnson catched me by the shoulder & bade me desist saying "no murder here you young rogue the man cries for quarter" I was now sent to the rear of the army, the next day we were conveyed a few miles to a barn which was used as a kind of temporary Hospital, from this place we were sent to Morristown and there I remained untill I had recovered of my wounds so as to join the service again at West Point, where I remained untill the Campaign of '79 we then joined Colonel Haythorns Levies with the intention of attacting New York, thousands of troops were gathered here on this occasion, & a larger quantity of malitia was collected at this time than I see at any other place during the war, but the plan was defeated on the account of Count DeEstaings not being able to get one of his largest ships over the Bar at Sandy Hook to cooperate with the land forces so the enterprise was abandoned, & when this campaign closed we returned again to West Point for winter quarters. At the opening of the Campaign in the spring of '80, I joined Clintons Brigade in the 5th Battallion under the command of Colonel Willet, we crossed the river at West Point & marched down to Kings Ferry & then recrossed the river & marched to Fort Lee where we remained a long time, or untill Major Andre was taken prisoner & executed, we then returned to West Point, where I remained as formerly untill the spring of '82 when I with 39 others of the artificials was sent to the frontiers on the branches of the Deleware to Reinforce Colonel Spauldin for the purpose of keeping the Indians in check who was making great havock amongst the settlers at this time, verry soon after we joined Colonel Spauldings forces the Quarter Master of the regiment was taken sick & left the army, & by the reccommendation of Capt. Dodge to Colonel Spaulding I was appointed to that station in which I acted untill I left the army. In the fall of this campaign a large force of Indians came down upon the settlers & drove off a large number of their cattle, we pursued them & gave them battle but they were to numerous for us we were defeated & obliged to retreat I received a shot in this engagement which passed through my left leg about midway from the foot to the knee I laid under a log the following night, in the morning from the great loss of blood I was scarcely able to raise myself, but with the greatest exertions I succeeded in crawling to the bank of the river where I fell in company with 5 others of the wounded, where we fixed a kind of raft of old logs &c, and floated down the river untill we came to the settlements where we were taken care of, and as soon as I got able to walk I got a furlow, subject to return to the army at shortest notice, I then went home to my parents in Orange County where I remained untill the next season, when we got the news that the army was to be disbanded I then went to New Windsor & received my discharge from the army.

My discharge with many other valuable things that I possessed was burned at the house of Capt. Stephen Hopkins an officer of the Revolution in the year 1809 at Lewiston, N.Y. whose house was burned when he & his wife was absent visiting a certain sick person about ½ a mile distant. All the compensation I received for the 6 years I was in the army was not available for more than $25 when pay-day came I with many others paid no attention to it, the paper which we drawed was of but little consequence as for instance when I was on the Frontier in a certain Barrack room there was 19 of us, we collected all the "shin plasters" we could, amounting according to the face of them to $1008, which was all put in a hat & I carried it to a certain old Scotch Lady who kept a kind of grocery, & all I could get for this amount was a half pint of Brandy which I carried to my quarters & then we rationed it out in a teaspoon so that all might get his share.

I knew nearly all the principle Officers of the Revolution Washington Lay-Fayette & many other distinguished Persons of the Army, was in the habit of freequently dining at my Father's house during the war.

After various misfortunes in life I commenced a kind of trade in salt to Canada & by this means I got considerably acquainted with many people of this Province & a short time before the breaking out of the war of 1812 I had mooved to Niagara where it was well known that I was an old American Soldier, & not apprehending any trouble & it being difficuilt for me to leave my little effects, & being then an old man I concluded by taking no part in the war, I might remain unmolested, but in this I was mistaken, I was apprehended by one Major Kirby of the incorporated mounted malitia as I was returning from my labour with a hoe on my shoulder & they forced me to run before them 1 1/4 mile to the guard house threatning to blow out my brains. I was then sent across the lake to York where I was locked up in a cold filthy prison where I suffered incredible hardships for more than two years, during my imprisonment I endeavoured to acquaint the government of the U.S. of it but whether they ever received the inteligence or not I am unable to state, many years ago while I was in the State of Indianna I received a letter from a friend of mine stating that Congress had granted a pention to the soldiers of the Revolution & that I could draw mine on application, I immediately came to this Provence for the purpose of disposing of some little property I owned here, calculating then to proceed to Washington for the purpose of obtaining this pension, but before I had got my little business accomplished I was siezed with an inflamation which settled in my left knee which has made a perfect cripple of me ever since, almost entirely loosing the use of this leg, so that it is impossible for me to walk a single step without the assistance of a crutch combing with the infirmaties which has accumulated during the various privations hunger, cold, & fatigues of the Revolution, & the hard & cruel imprisonment in the late war all combining to render me a feeble helpless old man, & at this late period soliciting that assistance from the government which I am sure they will not refuse to one who never shrunk from danger when the liberties of the country were ~~in danger~~ at stake, who fought, who bled, & who endured all the privations of the war for 6 long years, ever willing to sacrifice ease & enjoyment with all pecuniary considerations for my countrys good. I should have applyed sooner for my pension but this place is mostly settled by old country people who feel a kind of delicacy in communicating with the government of the United States, untill now I came across Isaac Hallock a citizen of that country who has undertaken the business for me, who & on my behalf I have legally authorized to receive my pension and arrearages or any communication the government may think proper to make, again for me to bring forward any living evidence to substantiate my claim at this late period would be impossible, as all that I know any thing about has passed away. I can now only refer you to the rolls & doccuments of the army in which I acted for proof of my assertions, humbly trusting that the government will be pleased to overlook what is impossible for me to ______, remembering that this appeal is from an old Revolutionary Soldier.

Joel Bower

Home District, C.W. Joel Bower being duly sworn, before us Alexander Campbell Esquire, and P. F. Whitney Esquire two of her Majesties Justices of the Peace, in and for the District aforesaid, say that he is the person described in the above sketch, that the relations contained therein are perfectly true, to the best of his recollection, and that he has authorised Isaac Hallock to act in his behalf. Sworn before us at Pickering in the District aforesaid this 17th day of December 1845

Alexander Campbell J. P.
P. F. Whitney J. P.

To the Honourable the Secratary of War of the U.S.

Dear Sir,

The accompanying sketch of the services of Joel Bower is for the purpose of obtaining a pension which he supposes to be granted to the soldiers of the revolution by the government of the U.S. I felt a great diffidence knowing my inability & ignorance of the legal course to be pursued but in the mean time I feel confident that the good sence of the government will excuse my inability, when vindicating the cause or acting in behalf of one whose services has a high claim on the considerations of our country, he is a person of steady & sober habits, but by misfortunes, by sickness & disease, & being far advanced in age, & reduced to the lowest degree of poverty, with nothing but a bed of straw to lay upon, without food to subsist upon only as he is supplyed by the hand of charity, & even without clothes to keep him from suffering with cold a perfect cripple, unable to get half a mile from his humble cot, now solicits the aid of that government under whose banner, & in whose defence he was ready to brave the greatest dangers, & in whose cause he freely bled for that pecuniary assistance which he entertains the firmest hope may be speedily administered. Let the most skeptical person converse but a few minutes with him & he will be convinced that he is that person that he represents himself to be, in the courese of his relations which I have skeched for him he relates a great number of interesting incidents that transpired during the period of his services which I deem unnecissary & even to prolix to be communicated to answer the end desired. If on perusal & examination of his case the government sees proper to grant the pension arrearages &c they will be pleased if not possible to send direct to me, or let it be payable in the City of Buffalo we will greatefully anticipate an answer with as little delay as possible, remember the man is suffering for the want of it.

I am unacquainted with the proper person to direct this to & even with the name of the present Secratary of War therefore if it does not fall in the proper hands I trust you will speedily rectify the mistake by placing it where it should be. Communications addressed to me

Pickering P.O.
Home District Canada West.
Pickering 16th Dec. 1845

I am sir with sentiments of the highest
respect your obedient servent
Isaac Hallock

From our personal knowledge of Joel Bower we have no hesitation in stating it to be our firm belief that he is the person he represents himself to be & that he is reduced to the lowest degree of poverty

signed P F Whitney J. P.
A Campbell J. P.

A copy of a sketch of the military services of Joel Bower in the U.S. Army, during the war of the Revolution, as transmitted to the Secratary of War the 17th December 1845.

I was born in South Hampton Old England 20th Feb. 1763. I was about 5 yrs. old when my Father whose name was Phineas imigrated to America, he first settled on Long Island but at the commencement of the Revolution he moved up the North River & settled in Orange County. I was about this time apprenticed to Capt. Samuel Chard Master Carpenter, in the beginning of the year '77 he entered himself 3 sons & 2 apperentices in the Government Artificials at New Windsor, we drew our arms & accoutrements, & were at the battle & storming of Fort Montgummary on the 6th day of October of this year, which was taken by the British General James Clinton commanded His Brother George then acting governer of the state was also there with a small body of malitia, the Battle commenced from the best of my reccollections between the hours of 12 & 1 oclock which was kept up during the remainder part of the day, & under cover of the darkness of the night they succeeded in mounting the wall before we discovered them, all the principle Officers & a few privates including myself made our escape by a passage out of the Fort next to the River this was a place unknown to the British, the rest were made prisoners The British kept the fort but 3 days before they evacuated & burned it. A Cousin of mine who at this time lay sick in the Fort & not being able to make his escape & his comrads was prohibited the privilege of lending him assistance, he was burned to death, his name was Samuel Jones Our quarters this winter '77 was at New Windsor, in the spring of '78 we were stationed at West Point where we were engagued in building Barracks Gun Carriages, Platforms &c and were constantly equipped for the common defence, & as liable to be ordered to battle as the Regular Army. Sometime in the fore part of the month of June of this year we received orders to reinforce Gen. Washingtons army then in New Jersey as Sir Henry Clinton was about to evacuate Philidelphia for the purpose of concentrating their forces in New York. We joined a party on the Hackinsack River and then marched hastily down & joined the main body, in a verry short time the British army made there appearance & a desperate battle ensued called the Monmoth battle. I was on the right of the left wing in Clintons Brigade. I stood in the front rank, where we sustained a tremendous canonade as well as a heavy fire of musketry for a considerable time, we then were impetuously charged by the Dragoons, & in their second charge a dragoon aimed a deadly blow at my head which would have instantly killed me, but I raised my musket in front to shield my head which caused his sword to glance, and cut off my two middle fingers & my little finger about half off, this was my right hand which was holding on the small of my gun, & it has been verry much disabled ever since, as he was then passing me he struck an underhanded blow at my head & the point of his sword passed through my left eyebrow which made a deep incission & considerably stunded me, at this instant his horse was shot through the head & fell dead at my feet, as I recovered from the last blow I received, I drew my bayonet & would have terminated his existance had not Adjutant Johnson catched me by the shoulder & bade me desist saying "no murder here you young rogue the man cries for quarter" I was now ordered to the rear of the regiment the next day we were conveyed a few miles to a barn which was used as a kind of temporary hospital, from this place we were sent to Morristown & there I remained untill I had recovered of my wounds so as to join the service again at West Point where I remained untill the campaign of '79 We then joined Colonel Haythorns Levies with the intention of attacting New York thousands of troops were here at this time & a larger number of malitia was collected on this occasion than I see at any one place during the war, but the plan was defeated on the account of Count De Estaings not being able to get one of his largest ships over the bar at Sandy Hook to cooperate with the land forces so the enterprise was abandoned, and when this campaign closed we returned again to West Point. At the opening of the campaign in the spring of '80 I joined Clintons Brigade in the 5th Battallion under command of Colonel Willet, we crossed the river at West Point & marched down to Kings ferry then recrossed the river & marched to Fort Lee where we remained a long time or untill Major Andre was taken prisoner & executed, we then returned to West Point, where I remained employed as formerly untill the spring of '82 when I with 39 others of the Artificials was sent to the frontiers on the branches of the Deleware to Reinforce Colonel Spaulding for the purpose of keeping the Indians in check who was making great havoc amongst the settlers, verry soon after we joined Colonel Spauldings forces the Quarter Master of the Regiment was taken sick & left the army & by the reccommendation of Capt. Dodge to Colonel Spaulding I was appointed to this station in which I acted untill I left the Army. In the Fall of this campaign a large force of Indians came down upon the settlers & drove off a large drove of their cattle we pursued them & gave them battle but they were to numerous for us we were defeated & obliged to retreat. I received a shot in this engagement which passed through my left leg about midway from the foot to the knee I laid under a log the following night, in the morning from the great loss of blood I was scarcely able to raise myself, but with the greatest exertions I succeeded in crawling to the bank of the river, where I fell in company with five others of the wounded where we fixed a kind of raft of old logs &c. and floated down the river untill we came to the settlements where we were taken care of, & as soon as I got able to walk I got a furlow subject to return to the army at the shortest notice. I then went home to my parents in Orrange County, where I remained untill the next season, when we got the news that the army was to be disbanded, I then went to New Windsor and received my discharge from the army.

My Discharges with many other valuable thing that I possessed was burned at the House of Capt. Stephen Hopkins an officer of the Revolution in the year 1809 at Lewiston N.Y. whose house was burned when he & his wife was absent visiting a certain sick person about ½ a mile distant.

I received as a compensation for the 6 yrs. I was in the army an amount not available for more than $25, when pay-day came I with many others paid no attention to it the paper which we drawed was of but little consequence, for instance when I was on the Frontier in a ceratin barrack room there was 19 of us, we collected all the "shin plasters" we could, amounting according to the face of them to $1008, it was all put in a hat, & I carried it to a certain old Scotch Lady who kept a kind of grocery, & all I could get for this amount was a half pint of brandy which I carried to my quarters, & then we rationed it out in a teaspoon so that all might have a share.

I know nearly all the head Officers of the Revolution. Washington, LayFayett, & many other distinguished officers of the army was in the habit of freequently dining at my Father's house during the war.

After various misfortunes through life I commenced a kind of trade in salt to Canada, & by this means I got considerably acquainted with many people of this provence, & prior to the breaking out of the war of 1812 I had moved to Niagara where it was generally known that I was an old american soldier, & not apprehending any trouble, & it being difficult for me to leave my little effects, & being then an old man I concluded by taking no part in the war I might remain unmolested, but in this I was mistaken, I was apprehended by one Major Kirby of the incorporated mounted malitia as I was returning from my labour with a hoe on my shoulder & they forced me to run before them 1 1/4 mile to the guard house threatning to blow my brains out. I was then sent across the Lake to York where I was Locked up in a cold filthy prison where I suffered incredible hardships for 25 months, during my confinement I endeavoured to acquaint the government of the United States of my imprisonment, but whether they ever received the inteligence or not I am unable to say.

Many years ago while I was in the State of Indiana I received a letter from a friend of mine stating that Congress had granted a pension to the soldiers of the Revolution, & that I could draw mine on application, I immediately came to this Provence for the purpose of disposing of some little property I owned here, calculating then to proceed to Washington for the purpose of obtaining this pension, but before I had got my little business accomplished I was siesed with an Inflamation which settled in my left knee which has made me a perfect cripple ever since, almost entirely loosing the use of this leg, so that it is impossible for me to walk a single step without the assistance of a crutch, combining this with the infirmaties which has accumulated during the various privations, hunger, cold, & fatigues of the Revolution, & the hard & cruel imprisonment in the late war, all combining to render me a feeble helpless old man, and at this late period soliciting that assistance from the government which I am sure they will not refuse to one who never shrunk from danger when the liberties of the country were at stake, who fought who bled, & who endured all the privations of the war for 6 long years, ever willing to sacrifice ease & enjoyment with all pecuniary considerations for my countrys good. I should have applyed for my pension several years ago, but this place is mostly settled with old country people who feel a kind of delicacy in communiting with the government of the United States, especially on a subject of this kind untill now I came across Isaac Hallock a citisen of that country, who I have persuaded to under take the business for me, who I have Legally authorised to receive my pension and arearages or any communication the government may think proper to make. For me to bring forward any living evidence to substanciate my claim at this late period would be impossible as all that I now any thing about has passed away, and I can now only refer you to the roll & documents of the different armys in which I acted for proof of my assertions, humbly trusting that the government will be pleased to overlook what is out of my power to ______ remembering that this appeal is from an old revolutionary soldier.

signed Joel Bower

Home District ______ Joel Bower being duly sworn before us P. F. Whitney and A Campbell two of her Majesties Justices of the Peace for said District says that he is the person described in the above sketch & that the relations contained therein are perfectly true according to the best of his reccollections & that he has authorised Isaac Hallock to act in his behalf.

Signed at Pickering this 17th Dec. 1845

Signed P F Whitney J. P.
A Campbell J. P.

Mr L. Cook Esqr.

Pickering 17th Jan. 1846

Dear Sir

By the reccommendation of Mr Joseph Tomlinson Esqr. of Markham I have been induced, to transmit to you the within skeches being a copy of what I sent to Washington, or at least mailed for that purpose but apprehending that the peculiar Loyalty of some of the Provencial Postmasters may have led them to intercept the letter being as it was directed to the secratary of war, & at a time when there was a great excitement in the country ~~from~~ occasioned by the Presidents message and get your advice in the business, as Mr Tomlinson informs me that your Father while living was a government Agent for the settlement of such claims, & that of course you can tell us what steps are necissary to take it has been intimated by some that he will have to make his personal appearance, although there is a possibility of getting him there it would be a considerable of an undertaking in his feeble weak & decripped state, as well as attended with a considerable expence because he must undoubtedly have clothes which he certainly is verry destitute of at present besides other necessaries he seems verry solicitous if the thing can possibly be arranged any other way that it should be, it has been a great number of years since he was able to get out of the immediate neighbourhood in which he lives & during this period has not been able to get the information of what steps were necissary to be taken to obtain his pension therefore if you will be pleased to write and explain what steps should be taken in the case you will not only confer a great favour on an old Revolutionary soldier, but bring me under the greatest obligations

I am sir your obedient servent
Isaac Hallock

J. L. Edwards Esqr
Commissioner of Pensions

Lewiston 20 January 1846

Dear Sir

At the request of a poor old Revolutionary Soldier I herewith enclose you his statement of services rendered during the Revolution, he resides in the interior of Canada and from his helpless situation I suppose it almost impossible for him to appear before a court of Record and if he could he has no evidence except what the Rolls of the army in the war office will furnish. Will you have the kindness to inform me if anything can be done for him.

I am very Respectfully
Lathrop Cooke

P.S. I was acquainted with Bower in the year 1798 and have seen him once or twice since. I believe he is a man of veracity

---

Honourable J. D. Edwards

Pickering Jan. 24th 1846

Sir

In executing the requisite proof to the Declaration of Joel Bower according to the Regulations received from the Pension department we have followed the course of these regulations as near as the forms of this goverment will admit of, and the present state of his health, there is no officers of any court of Reccord nearer than the city of Toronto which is full 26 miles distant and since I transmitted the informal statement under date of 16 or 17 of Dec. last he has been in a verry poor state of health so much so that I think it would endanger his life to take him in his present situation that distance, and his memmory even in this short period of time seems materially changed.

Before I concluded to act as agent for him I verry particularly examined him respecting his services in the war, so as not to be found vindicating the cause of an imposter or one whose claim might be considered doubtful, although I discovered that from age and bodily infirmaties he had utterly forgotten many things that would be of great importance in assisting his claim, but at the same time I conceived that the plausibility of his relations which he distinctly remembered were such as to entitle him to the considerations of a grateful country.

There is an error which we have discovered in the statement made to the department dated 16 Dec last occurring in the campaign of '82 where it was stated a "large force of Indians" & a "large drove of cattle" when the words 'large' in both cases should have been small, which when attested to was not discovered by Mr Bower there are several incidents related in this former statement which we have now omited for brevity considering them of but little use, but in the mean time if on reexamination they are such as may prove servicible in assisting his claim you will see that they have been attested to before the Legal authorities of the country.

If the thing had been possible to have had the affadavit of Mr Bower taken taken before some Judge or higher officers of the Country I should have done it but he is not able to sit up ten minutes at a time and the Justices have been obliged to go to his rustic mansion to accomplish the object, therefore taking the whole situation of the affair into consideration we trust the government will not consider the course we have taken objectionable therefore we now respectfully submit the case to your consideration humbly soliciting a speedy action on the subject.

I am sir with sentiments of the highest
Respect your obedient servent,
Isaac Hallock

P.S. I beg pardon for ~~my~~ the awkward ______ given, as at the time of writing the above I apprehended Mr Edwards Esqr to be the Sect. of War, which now I have Learned to be incorrect

Respectfully &c
Isaac Hallock

Declaration
In order to obtain the benefit of the
Act of Congress of the 7th June 1832

Home District
Provence of Canada ss. On this twenty fourth day of January 1846 person-

ally appeared before us A. Campbell and P.F. Whitney two of her Majesties Justices of the Peace for Said District Joel Bower a resident of Pickering in the Home District and Provence of Canada aged Eighty three years the 20th of February next who being duly sworn deposeth and saith that by reason of old age and the consequent loss of memmory he cannot swear positively as to the precise time he entered the army, neither can he positively state whether he enlisted, volunteered or under what particular form he entered the service, but according to the best of his reccollections he entered the government Artificers in the month of March in the year 1777 at New Windsor which was under the command of Capt. Samuel Chard Master Carpenter, but thinks he had no commission only acting as foreman, but thinks that Capt. Machin who was termed by us the "engineer" had a Commission, states he see one Capt. Moody frequently in company with Capt. Machin but thinks he had nothing to with the Artificers, & that he Capt. Moody afterwards had the command of a company in the Artillery & remained in this station to the end of the war, he drew his arms & equipment from the government stores & is sure that the artificers was at this time a legally authorised branch of the army but has no knowledge of their being at this time organised into companies with their officers as entimated by the pension department, he was first engaged in assisting to put a chevauxdefrize across the river at New Windsor. Nearly all the artificers with whom he was associated was at the battle of Fort Montgummary on the 6th day of Oct. 1777, and the only one of his party who escaped with ~~us~~ him was Jeremiah Smith. James Thompson, Thomas Horton & several others of their party was taken prisoners, and but one ever lived to return his name was Scott Travice during the engagement he was under the command of General James Clinton, whose head was nearly white as snow, his Brother George was there with a small body of malitia he does not reccollect the names of any other officers at the fort at this time, this battle commenced he believes between the hours of 12 & 1 o'clk P.M. which was kept up the remainder part of the day, the night being dark & a piece of wood standing but a short distance from the fort prevented their discovering the British troops untill they had mounted the wall, all the principle Officers a few privats with himself escaped on the side of the fort next the River, the rest were all made prisoners, the British kept possession of the fort but three days before they evacuated & burned it. A cousin of his Saml. Jones perished in the flames as stated in the sketch forwarded to the department under date of 16 or 17 of Dec last, their quarters this winter '77 was at New Windsor. At the opening of the Campaign of 1778 they were stationed at West Point where they were engaged in building Barracks gun carriages platforms &c. untill they received orders to reinforce Gen. Washington's army in New Jersey as Sir Henry Clinton was about to evacuate Philidelphia for the purpose of concentrating their forces in New York, they left West Point & marched to Cacaat a Duch settlement in N.Y. thence they marched to Morristown in NJ, and then to the Hackinsack river, where they joined another party & then marched to the main army, he was in the Monmouth battle but did not know the does not reccollect the names of the Officers, he was stationed on the right of the left wing of Clintons brigade, in this battle he was wounded by a dragoon, by having his two middle finger cut off & the little finger nearly half off, ~~the~~ from the same person he received another wound by an underhanded blow ~~and~~ the point of his sword passed through his left eye brow, at this instant the dragoons horse was shot through the head & fell dead at his feet, and as soon as he had recovered a little he drew his bayonet at his antagonist but Adjutant Johnson catched him by the shoulder saying "no murder here you young rogue the man cries for quarter" he was now ordered to the rear of the regiment, he see a great number fall this day from the excessive heat, want of water &c. he is positive that General Washington and General Lee, highly disagreed this day that Gen. Lee had badly arranged the left wing so that their plan was defeated and the British army gained the brige. Gen. Washington this day was mounted on a ~~milk white~~ dappled gray horse & he see him ride from the extremity of one wing to the other with great velocity, he also states that great indignation was felt by the army towards Gen Lee for his bad management, in this battle. After deponent ______ the next day he was conveyed a few miles with the rest of the wounded to a barn from there they were sent to Morristown where he remained untill he had recovered of his wounds & then he returned again to the service at West Point where he remained until the campaign of '79 he then joined Colonel Haythorns Levies with the intention of attacting N. York, he states that a great number from the service at West Point joined these Levies, that he was in the company commanded by Capt. Seth Marvin one Johnson was their Lieutenant, does not recollect the names of any other officers, this company was stationed at Fishkill, so when this plan was abandoned he returned again to west point for winter quarters, at the opening of the campaign of '80 He joined Clintons Brigade in the fifth battalion under command of Colonel Willet do not recollect the names of any other officers they crossed the river at West Point marched down to Kings Ferry then recrossed the river & marched to Fort Lee where they remained a long time or untill the execution of Major Andre They then returned to West Point, where he remained in the Artificers employed building wood boat repairing vessels that were engaged in the service &c. as there was no particular call in the campaign of 81 ~~nor any particular~~ he did not leave West Point. At the opening of the campaign of '82 he with 39 others was sent to the frontiers on the branches of the Deleware to reinforce Colonel Spauldings forces for the purpose of keeping the Indians in check who were making great distruction amongst the settlers in this section of the country, they marched to Goshen the county town of Orrange County from there to the lower forks of the Deleware where they joined Capt. Abram Westfalls Company Collonel Spaulding on the account of poor health left the service, and Colonel Weisenfelt took the command verry soon after he joined this regiment the quarter master was taken sick & left the army and by the reccommendation of Capt. Doge to Colonel Spaulding he was appointed to this station in which he acted untill he left the army Capt. Dodge did not belong to this regiment, but was acquainted with deponent & was at that time on the frontier, cannot reccollect to what company he belonged, this regiment was divided in companies & extended for 60 miles in length, in the fall of the campaign a ~~large~~ force of Indians came down and drove off a ~~large~~ drove of cattle belonging to the inhabitants, they pursued them & gave them battle but they were to numerous they were defeated & obliged to retreat the company to which Deponent belonged was the only one out of the regiment which was in the battle, the greater part were malitia, and they were commanded by one Capt. Wood of the malitia who was taken prisoner by the Indians, he does not remember the names of any other officers who were in this engagement, deponent saith also that he was wounded in the left leg in this engagement that he laid under a log the following night, that from the great loss of blood he was so weak in the morning that he could scarcely raise himself, but by the greatest exertions he succeeded in crawling to the bank of the River where he fell in company with five other of the wounded, the name of one was Patterson, they constructed a kind of raft of logs &c and floated down the River untill they came to the settlements, and as soon as he got able to walk he got a furlow subject to return to the service at the shortest notice he then went home to his parents in Orange County, and that in the following spring they got information that the army was to be disbanded, he then went to New Windsor and received his discharge from the army, does not reccollect by who his discharge was signed he thinks the time he received his discharge from the best of his reccollections was either the last of May or the first of June in the year 1783, deponent further saith that he was not employed in any civil pursuit from the time he first entered the service in March 1777 untill he had received his discharge as above, deponent further saith that he cannot from the falier of his memmory state how it was or under what particular form that he entered the different branches of the army, only as he supposes that he was detached from one company to another where his services were most needed, neither can he state for what period of time he first entered the service, but that he never was free from service ~~untill~~ from the time he first entered untill he received his discharge, he further states that his Discharges from the army was burned in the House of one Capt. Stephen Hopkins of Lewiston N.Y. whose house was burned in the year 1809. That shortly after the war he settled in Ontario Co N.Y. where he remained many years, & then being engaged in bringing salt to Canada, he lived in various places and for the past 14 yrs. in the township of Pickering Home District Canada. Deponent further says that he never before has applied for his pension and his reasons for not doing so, he states to be, that he never heard that there was a pension for him untill about seven years ago while he was in the State of Indianna, that he then returned to this Provence for the purpose of arranging his business so as to apply for his pension, but was taken verry sick before he had accomplished his design, and that reduced to the lowest degree of poverty and made a perfect cripple by his illness he never found a person to vindicate his cause neither could he ever learn what steps were necissary to take of the purpose of obtaining it, and that from age weakness and debility it was impossible for him to act for himself and that now from his late increased bodily infirmities, he apprehends a possibility that he may not be in a situation of rendering further instructions in his case therefore he has authorized Isaac Hallock of this Township of Pickering, his agent in the settlement of this claim, trusting that he said Isaac Hallock will consummate his designs in this matter, deponent further says that at the time he entered the service at New Windsor he was a resident of Orange Co. N.Y. he further says that he knows of no living evidence that he can bring forward to substantiate his services

He hereby relinquishes every claim whatever to a pension or an annuity, except the present, and he declairs that his name is not on the Pension Roll of any agency in any state. Again deponent says as a further reason for not applying for his pension that he had understood that it would be required by him to bring forward two living evidences to prove his services by, that he had embraced every opportunity in making enqury for such, but was always informed that they were not living and finally he had come to the conclusion that it would be useless to make application, unill Isaac Hallock suggested the idea that it was not requred or believed it would not be in his case, Since taking the above, deponent says that he has called to his reccollection the names of the following officers that commanded in the 5th battalion of Clintons brgade at the period that he deponent was in the battalion. Capt. Ham Trammock a French gentleman Capt. Steward a Scotch gentleman Capt. De Bois, a Duchman, Capt. Vanderbergh, also a Duchman, Capt. Dodge, an Englishman, and Adjutant Johnson, a Scotch man, that after he left this battalion the five battalions were reduced to three.

Sworn and subscribed to before us the day and year above written

Joel Bower

P F Whitney J. P.
Alexander Campbell J. P.

We the aforesaid Justices, do hereby declare our opinion, that the above named applicant was a revolutionary soldier, and served as he states, and we further state, that it is the opinion generally entertained by the Inhabitants of this part of the country, who have been acquainted with him for a number of years that he is what he here declares himself to be

P. F. Whitney J. P.
Alexander Campbell J. P.

Canada
Home District
to wit I, George Gurnett of the City of Toronto Esquire Clerk of the Peace for the said District, do hereby certify that Paul F Whitney of the Township of Whitby and Alexander Campbell of the Township of Pickering Esquires are two of Her Majesty's Justices of the Peace duly qualified, according to Law to act within the said Home District, and I do verily believe that the names "P. F. Whitney J P" and "Alexander Campbell J P" to the annexed Instrument written, are the proper and genuine signatures of the said Paul F Whitney and Alexander Campbell

Given under my hand and Seal of Office at Toronto this Twenty sixth day of January in the Year of Our Lord One thousand Eight hundred and forty six.

G Gurnett
Clerk of the Peace, Home District

Copy of the Declaration of Joel Bower as transmitted to
the Secratary of War. Jan. 1846
mailed in Toronto the 27 Jan.

Copy of the Declaration In order to obtain the benefit of the Act of Congress passed June 7, 1832

Home District
Provence of Canada ss. On this 24th day of January 1846 personally appeared before us P. F. Whitney and A Campbell two of her majesties Justices of the peace for said District, Joel Bower a resident of Pickering Home District and Provence of Canada aged eighty three years the 20th of Feb. next, who being duly sworn, deposeth and saith that by reason of old age & the consequent loss of memmory, that he cannot swear positively as to the precise time that he entered the service, neither can he reccollect under what particular form he entered, whether drafted volunteered or enlisted but according to the best of his reccollections he entered the Government Artificers at New Windsor in the month of March 1777, which was under command of Capt Saml. Chard Master Carpenter but thinks he had no commission, but thinks that one Capt. Machin who was termed but them the "Engineer" had a Commission, states he see one Capt. Moody freequently in company with Capt. Machin but thinks he had nothing to do with the Artificers, & that after this Capt. Moody had the command of a company in the artillery where he remained to the end of the war. Our first work after joining the artificers was putting a Chevauxdefrise across the river at New Wind-

sor, deponent has no knowledge of the artificers being organised into companys with their officers as intimated by the pension department, but is sure that they were a legally authorised branch of the army, he drew his arms & equipments from the government stores and was at the Battle of Fort Montgummary on the 6th day of Oct. before the battle there was a universal alarm, & nearly all with whom he was associated was sent to the fort, the engagement commenced he thinks between the hours of 12 & I oclock P.M. & was kept up the remainder part of the day, the night being dark & a piece of woods standing near the fort prevented them from discovering the British troops untill they had mounted the wall, all the principle Officers a few privates with myself escaped by a passage on the side of the Fort next to the river the rest were all taken prisoners, during this engagement, he was under command of Gen. James Clinton whose head was nearly white as snow his Brother George was also there with a small body of malitia, the British kept the fort but 3 days before they evacuated & burned it. Saml. Jones a cousin of his who at this time lay sick in the fort & his comrads being prohibeted the privilege of carrying him out he perished in the flames, the only one of deponents party who escaped with them was Jeremiah Smith. James Thompson, Thomas Horton and many others of his party who were taken prisoners and but one ever lived to return his name was Scott Travice. Their quarters this winter '77 was at New Windsor at the opening of the campaign of '78 he was stationed at West Point where they were engaged in building Barracks, Gun Carriages, Platforms &c. and were constantly armed & equiped for the common defence & as liable to be ordered to battle as the Regular army. Some time in the fore part of the month of June of this year they received orders to reinforce General Washingtons army in New Jersey, as Sir Henry Clinton was about to evacuate Philidelphia for the purpose of concentrating their forces in New York We left West Point & marched through Caceat a Duch settlement in N. York thence to Morristown in New Jersey thence to River where they joined an other party and marched down to the main army. In a verry short time the British army made their appearance & a desperate battle ensued called the Monmouth battle. He was on the right of the Left wing in Clintons Brigade, he stood in the front rank where they sustained a tremendous canonade as well as a heavy fire of musketry for a considerable time, we then were impetuously charged by the dragoons and in their second charge a dragoon aimed a deadly blow at his head which would have instantly terminated his existance had he not raised his musket so as to elude the meditated blow which caused his sword to glance & cut off the 2 middle fingers & the little finger about ½ off, this was his right hand which was holding on the small of his gun & his hand has ever since been verry much disabled, the Dragoon then struck an underhanded blow at his head & the point of his sword passed through his left eye brow which made a deep incissoon & considerably stunded him, at this instant the Dragoons horse was shot through the head & fell dead at his feet, as soon as deponent had recovered a little he drew his bayonet at his antagonist & would have terminated his existance had not Adjutant Johnson catched him by the shoulder saying "no murder here you young rogue the man cries for quarter" he was now ordered to the rear of the regiment. Deponent further that General Washington & Gen. Lee highly disagreed this day, that Gen. Lee had badly conducted the Left wing of the army so that their plan was defeated, & the british army succeeded in gaining the brige, he states that great indignation was felt by the army towards Gen. Lee for his bad management. Gen. Washington was this day mounted on a Dappled gray horse and he see him ride from the extremity of one wing to the other with great velocity, & that great numbers perished this day from the excesive heat want of water &c does not reccollect the names of any other officers, the next day they were conveyed a few miles to a barn which was used as a kind of temporary hospital from this place he was sent to Morristown where he remained untill he had recovered of his wounds so as to join the service again at West Point, where he remained untill the campaign of '79, he then joined Colonel Haythorns Levies in the company of Capt. Seth Marvin one Johnson was our Lieutenant, for the purpose of attacting N. York this company was stationed at Fishkill, but the plan was defeated on the account of Count De Estaings not being able to get his largest ships over the bar at Sandy Hook to coopperate with the land forces so when this enterprise was abandoned he returned again to West Point for winter Quarters, at the opening of the campaign in the spring of '80, he Joined the 5th Battallion of Clintons Brigade commanded by Colonel Willet, in this Battallion was Capt. Ham Trammock a Frenchman Capt. Steward a Scotchman, Capt. Dodge an Englishman, Capt. De Bois and Capt. Vanderbergh both Dutchman, & Adjutant Johnson a Scotchman, these 5 Battallions were afterwards reduced to three. They crossed the river at West Point & marched down to kings ferry, then recrossed the river & marched to Fort Lee where they remained a long time or untill the execution of Major Andre he then returned to West Point, and the campaign of '81 they had no call, so he remained in the artificers Assisting to build wood boats repairing vessels that were engaged in the service &c. untill the opening of the campaign of '82 when he with 39 others were sent to the frontiers on the branches of the Deleware to reinforce Colonel Spaulding for the purpose of gurding the settlements from the depradation of the Indians they left West Point & marched to Goshen the county town of Orange County, thence to the lower forks of the Deleware, & there joined the company of Capt. Abram Westfall at which time Colonel Spaulding commanded, but on account of poor health he left the regiment and Colonel Wisenfelt took the command, about the time of ~~my~~ his joining the regiment the Quartermaster was taken sick & left the army, and by the reccommendation of Capt. Dodge to Colonel Spaulding, he deponent was appointed to this station in which he acted untill he left the service, this Regiment was divided into companys and extended for 60 miles in length, in the fall of this campaign a small party of Indians came down upon the settlers and drove off a number of their cattle, they pursued them & gave them battle but they were defeated and obiged to retreat the company to which Deponent benonged was the only part of the regiment that was in this engagement, the most of their number was malitia which was commanded by Capt Wood of the malitia & he was taken prisoner by the Indians, deponent was shot through the left leg in this engagement & laid under a log the following night and in the morning from the great loss of blood he was so weak he could scarcely raise himself but by the greatest exertions he succeeded in crawling to the bank of the river where he fell in company with 5 others of the wounded the name of one was Patterson, where they fixed a kind of raft of old logs &c and floated down the river untill they came to the settlements where they were taken care off and as soon as deponent got able to walk he got a furlow subject to return to the service at the shortest notice he then went home to his parents in Orange County where he remained untill the next season when they ~~learned~~ received word that the army was to be disbanded he then went to New Windsor and received his discharge from the army, does not reccollect by who his discharge was signed the time he received his discharg from the best of his reccollection was either the last of May or the first of June 1783, deponent says his discharges was burned at the House of Capt. Stephen Hopkins of Lewiston N.Y. whose house was burned in the year 1809, deponent says he cannot tell how it was or under what particular form he entered the different branches of the army, only as he thinks he was detached from one part to another where his services were most needed neither can he tell whether he first entered the service for any particular period of time but that he never was free from the service from the time he first entered in March 1777 untill he received his discharge in May or June 1783, and during the whole period of his services he was not engaged in any civil pursuit, he says he never before applied for his pension and his reasons for the delay he says was that it was only about 7 years ago that he first learned that Congress had granted a pension to the soldiers of the Revolution & that was ~~while~~ at a time when he was in Indiana that he returned to this Provence to arrange his business so as to apply but was taken verry sick, which made him a perfect cripple and reduced to the lowest degree of poverty, and perfectly unable to apply for himself and finding no one who was willing to act in his behalf & also that at the time he first learned there was a pension for him he understood that he would have to substanciate his claim by two living evidences, that he had embraced every opportunity in makin the enquiry, but always received for answer they were not living, finally he had come to the conclusion that it would be useless to apply & he had given the matter up as useless untill Isaac Hallock sugested the Idea that it was not required & at least he believed it would not be in his case, therefore deponent says he has appointed Isaac Hallock of this Township of Pickering his agent in the settlement of this his claim trusting that the said Isaac Hallock will consumate his design in this matter. He hereby relinquishes every claim whatever to a pension or an annuity except the present and he declairs that his name is not on the pension Roll of any State Agency in any State.

Sworn to, and subscribed the day and year first above written

signed P. F. Whitney J. P
A. Campbell J. P.

Signed Joel Bower

Pickering Home District Canada West
April 3d 1846

Hon. Wm. L. Marcy
Sir

The Declaration of Joel Bower was made out on the 24th day of Jan. last & mailed two or three days afterwards and as yet we have received no inteligence respecting it.

Mr Bower has been urging me to write, & I have put him off for some time, untill I begin to apprehend that either the Declaration or the answer has got misscarried.

I enclosed the Declaration to you. I intended to have sent it to Mr Edwards, but a merchant from Rochester informed me that ~~that~~ he was nothing more than a Clerk in the office & that you was the propper one to direct to, therefore if I have erred it has been for being missinformed. The health of Mr Bower has considerably improved since we sent the Declaration, and he feels verry anxious to hear from there

In the letter to the Declaration I mentioned an error in the 1st statement, there is yet an other in the name of the River called the "Hackinsack" it is not the Hackinsack but a River south of Morristown, at the time I first took it down he was at quite ~~at~~ a loss to reccollet the the name of the River, & I mentioned the name Hackinsack not knowing in what part of the state it was, & the name being fresh in his memmory he thought that to be it, but on looking on an atlas I see he was mistaken I enquired of him he says the River he means was south of Morristown.

His reasons for the delay or for not applying sooner for his pension, I find on enquiry can be thoroughly substanciated, ~~for~~ and I believe if I had not assisted him he never would have got any other person too, the impression in the minds of the people was (as with him) that he would have to proove his claim by surviving evidence and there also appears to be a kind of lack of confidence in the minds of many people here as regards the payment of such claims even if well substanciated, all these causes have had a tendency to keep the thing back.

He wished me to make a further inquiry whether the law did not grant him payment for his services independent of the pension.

I am perfectly ignorant of the correct proceeedings to be taken, therefore I hope you will direct the course to be taken, & how, if any thing further is requisite.

If the papers have come to hand and you are not prepared to give a decission immediately be pleased to drop a line stating the same as it would give him great satisfaction to know that the affair was not laid on side.

I am ~~sir~~ with sentiments of the highest
Respect, sir your obedient servant
Isaac Hallock
Pickering P.O. Home District Canada W.
via Lewiston

(The cover page is dated June 1846)

Pickering Home District Canada W.J
June 15th 1845.

Sir

The Declaration of Joel Bower was executed on the 24th day of Jan. last & mailed 2 or 3 days afterwards, and just at the moment I was enclosing it to you (as we all supposed that you was the Sect.) a merchant from Rochester told me what the Sect.'s name was & informed me that I must direct it to him, he appeared to be a man of good understanding, & I done as he advised without giving the idea a moments consideration, but on reflection I believe that each Office in the Department must have its principle manager, and if so I certainly have committed an error by not forwarding it to you, but I trust that your superior good sense will pardon an error when committed so unintentonally as I done that. I also mailed a letter explaning the thing on the 3d of April but receiving no inteligence from either the Declaration or the inquiry. I am induced again to solicit attention to the case, if the Declaration is not correctly done we will do it over again, if more proof is required, or if his personal appearance is nesessary (although scarcely possible for him to endure the journey) be pleased to inform us, or if he is not entitled to the benefit of the Law I would take it verry kind to be informed as I have lent him the assistance from charitable motives or if an action on his case requires time a moments writing would convey the information to us, therefore I beg you would be pleased to drop us a line giving us some information in the case immediately

I am sir with sentiments of

Respect your obedient servant
Isaac Hallock

J. D. Edwards Esqr

* * *

Mr J. D. Edwards Esqr

Pickering 16th July 1846

Sir

The decision of the Pension Office in the case of Joel Bower was duely received, and being confident that the Government requires only to be thouroughly convinced that the applicant is no impostor, and in order to effect this I would solicit on his behalf that you would be pleased to admit him to a personal examination at the department, where he is willing to undergo a most scrutanising examination on the merits of his ~~claim~~ services

I deem myself somewhat in the fault of his case being so perfectly ignorant of military logick that I have conveyed his statements in rather an uninteligable form and from my indefinitely understanding what was required by the department

The decision of his case has called forth the sympathy of several of the highly respectable inhabitants of this country who thought it best to have him immediately conveyed to Washington that he might answer for himself but I conceived it prudent before such a step was taken to enquire of you whether you would be pleased to admit him to a personal examination, & to solicit your answer as soon as possible as it will be necissary that he should preform the journey before the commencement of the cold weather.

I am sir with sentiments of
respect your obedient servant
Isaac Hallock

On the opposite page I have sketched a little addition to his statement that may throw a little more light on the case

I. H.

* * *

Pension Office,
July 24th 1846

Sir.

In reply to yours of the 16th inst. in the case of Joel Bower, you are informed that all claimants are entitled to personal interviewing upon the subject matter of their claims respectively with the Officers of the Department, who are bound by duty and impelled by inclination, to afford them every opportunity and facility to establish them. The alleged service in Haythorns, Willetts and Weisenfelts Regiments of the New York Troops, should be sustained by the certificate of the Comptroller of New York. The service in the Corps of Artificers is discredited by the rolls.

I am, Sir, Very respectfully
Your Obt. Servt.
J. L. Edwards

Isaac Hallock, Esq.
Pickering, Upper Canada
via Lewiston, New York

In pursuance of a law passed the 27th of April, 1784, entitled "An act for the settlement of the pay of the Levies and Militia, for their services in the late War, and for other purposes therein mentioned," the following certificates have been issued, bearing interest at five percent. per annum, viz:

1782 Dec 26	3979 Joel Bowers	20	19	6
1779 Nov 21	1022 Joel Bowers	2	5	

State of New-York,
Comptroller's Office, Albany, August 12 1846.

I hereby certify the above to be true extracts from Book No. 1 now remaining in this office, and that the service for which the certificates above named were issued, were performed under the following officers, viz:

~~First~~ - Second under Capt. Abraham Westfall in Lt Col. Weissenfels Regt.
~~Second~~ First under Capt. Seth Marvin in Col. John Hathorn Regt.

A. C. Flagg, Compt.

New York - Albany

Joel Bower of in the State of New York who was a Private in the commanded by Captain of the Regt. commanded by Col. Weisenfeldt in the line for 9 months 2 days

7 11

Inscribed on the Roll of Albany, N.Y. at the rate of 30 Dollars 22 Cents per annum, to commence on the 4th day of March, 1831. Certificate of Pension issued the 14th day of Augt. 1846 and sent to Isaac Hallock, of Pickering, U. Canada

---

Arrears to the 4th of March, 1846 $453._
Semi-anl. allowance ending 4 Sept. '46 15.__
468.__

Recorded by E W Robinson Clerk,
{Revolutionary Claim, Act June 7, 1832.}
Book E2 Vol. 4 Page 8

Payments to be made at Albany, N.Y. by Ths. W. Olcott Agent for paying Pensioners in the Agency of Albany, N.Y. Recorded in the Pension Office, in Book E.2 Vol. 4 Page 8 by E. W. Robinson Clerk.

War Department.
Revolutionary Claim.

I certify that in conformity with the law of the United States, of the 7th June, 1832, Joel Bower of New York who was a Private in Weisenfeldt's Regiment is entitled to receive Thirty dollars and Twenty-two cents per annum, during his natural life, commencing on the 4th of March, 1831, and payable semi-annually, on the 4th of March and 4th of September, in every year. Given at the War Office of the United States, this 14th day of August one thousand eight hundred and forty-six.

W. L. Marcy
Secretary of War.

Examined and Countersigned
J L Edwards Commissioner of Pensions.

Pickering Sept. 14th 1846.

Sir

Enclosed is the Comptrollers Certificate for the services of Joel Bower in Colonel Willetts Regiment. A Pension Certificate was issued on the 14th Aug. last, for services in Weisenfeldts Regiment also you informed me it included services in Colonel Haythorns, but from the best of my reccollections the amount of credit mentioned in the other certificate differs from an amount mentioned in a letter from the Comptroller that accompanied the enclosed certificate to me, he say "we find his name under Capt. Seth Marvin, Colonel John Haythorn in 1779 £3.1.5. other services may appear on examination" whether this be the same contained in the former certificate you will know on comparison, & if not & it is likely to make any material difference I will procure the Comptrollers certificate for it and send to you or the above remark might have been made on account of his the Dept Comptroller being absent when the certificate in Aug. was procured & his having no knowledge of it.

I beg the indulgence of once more calling your attention to his services in the Artificers, although the claim has been examined, still I am confirmed in the idea that a reexamination may place the thing in quite a different light. I probably should not have ~~again~~ renewed this subject again if I had not had the privilege of examining the Books at Albany where I found his whole statement so perfectly correct, the names of all the Officers as well as Privates just as he had stated with many other corresponding circumstances. Now in the first place he states that his service was performed in four different Corps of the Revolutionary Army three out of the four he fully sustains according to his declaration in the most definite particulars. And the question verry naturally arises how is it possible for him to be so mistaken about his services in the Artificers. While in this part of the army he was in the Battle of Fort Montgummary and also the Monmoth Battle where he was severely wounded the marks of which time will never erase although he might not even at the time have known that they were organised into companies yet let it be remembered that he was but a boy of 14 yrs old that a degree of needlessnes might have been the cause of his not understanding in reality the true position of this part of the army, or is there not a possibility that he belonged to a party that was not organised into the Regiment, the Comptroller Mr Flagg gave me the name of a Colonel who he said had the command of a Regiment of Artificers, but Bowers says he is quite sure he had nothing to do with the company to which he belonged. While on our return in the Cars between Trenton and Jersey City he remarkably pointed ~~to~~ out the verry place where he marched when going to Monmoth, he showed in what direction it was where the battle was fought and gave many other particulars, which was asserted to be correct by those who were acquainted with that part of the country. Although he had never before been in that part of the State, he also pointed out the particular places along the Hudson River where most of the important events of the revolution had transpired, he showed us the verry spot where he assisted to put the Chevauxdefrise across the River where he was engaged in repairing vessels and finally where the whole of his services had been preformed (except the last campain in the Weisenfeldt Regiment) It was remarked by a certain Judge who had long listened to his relations, "that it seemed more like the narration of one who had risen from the dead, than one who had survived to this late period & now rehearsing to us the events of that important era as correctly ~~as correctly~~ as we now heard them." Again at the time I first see him & commenced a correspondence in his behalf, he had laid sick for a number of years & reduced to the lowest degree of Poverty, we were perfectly ignorant of every proceeding necissary to be taken, and when he mentioned his services in the Artificers I had no knowledge of such a company being associated with the army. I used his own provincialism & called them "artificials" no minor consideration could ever have induced me to have exposed my ignorance so effectually as I have done & apprehended I should do from the verry commencement but we owe them a debt of gratitude and one that should be cherished by every citizen of the union and it was from this consideration that caused me at first to vindicate his claim and it is from the same impulse that induces me again to solicit a reexamination of this part of his services, that from the sincerity of his statements combined with his character for truth & veracity there never has been a doubt left in the mind of any person who has conversed with him as to the truth of his declarations. I anticipated that he would have been admitted to a personal examination at the department as I had understood from former advice, where I was sure he would have satisfactorily sustained this part of his services, but as this was not the case, & the amount already obtained is not adeqquate to his necessity, and the only alternative now left is to solicit a reexamination of his case which under the more plausible pretext of his having satisfactorily proved three fourths of his declaration to be true, will place the fourth in quite a different light, he often asserts that he wishes for nothing but what the law of the country has granted and as far as that has made appropriations to his benefit he wishes to share its indulgence.

I am with sentiments of the highest
respect sir your obedient servant
Isaac Hallock
Pickering Canada
via Lewiston N.Y.

J. D. Edwards Esqr.
Commissioner of Pensions

P.S. J. Bowers wishes me to solicit an early attention to his case as he thinks of removing to some of the western states if this affair can be concluded before the season is to far spent.

yours &c
I. H.

In pursuance of a law passed the 27th of April, 1784, entitled "An act for the settlement of the pay of the Levies and Militia, for their services in the late War, and for other purposes therein mentioned," the following certificates have been issued, bearing interest at five percent. per annum, viz:

15th Decr. 1780 Joel Bower priv 19 18 2

State of New-York,

Comptroller's Office, Albany, 7th Sept. 1846.

I hereby certify the above to be true extracts from Book No. 7 now remaining in this office, and that the service for which the certificates above named were issued, were performed under the following officers, viz:

First - under ~~Capt.~~ Levies that joined the 5th N Y Regt.

Second - under ~~Capt.~~ under Col. Willett

Philip Phelps
Dep. Comptr.

Pickering Oct 14th 1846

Sir

Enclosed is the Pension certificate of Joel Bower returned according to instruction for the purpose of receiving the increase or pay for 7m. 14d. in Colonel Willetts Regiment.

I am sir your most obedient servant
Isaac Hallock
Pickering Canada via Lewiston N.Y.

J. D. Edwards Esqr
Com. of Pensions

New York

Joel Bower of in the State of N.Y. who was a Pri in the commanded by Captain of the Reg commanded by Col. Wesenfeldt in the Revolution for 16 months, 13 days NY Increased from $30.22 Inscribed on the Roll of Albany N.Y. at the rate of 54 Dollars 77 Cents per annum, to commence on the 4th day of March, 1831. Certificate of Pension issued the 30 day of Octo. 1846 and Isaac Hallock, Pickering U. Canada

Arrears to the 4th of Semi-anl. allowance ending

Increased from $30.22

Recorded by D. H. Wood Clerk,

{Revolutionary Claim, Act June 7, 1832.}

Book E2 Vol. 4 Page 8

Payment to be made at Albany by T. W. Olcott Agent for paying Pensioners in the Agency of New York Recorded in the Pension Office, in Book E2 Vol.4 Page 8 by D. H. Wood Clerk. Duplicate.

War Department.
Revolutionary Claim.

I certify that in conformity with the law of the United States, of the 7th June, 1832, the pension of Joel Bower of New York who was a Private in the Revolution and pensioned at $30.22 per annum is increased to $54.77 dollars and cents per annum, during his natural life, commencing on the 4th of March, 1831, and payable semi-annually, on the 4th of March and 4th of September, in every year, deducting all former payments. Given at the War Office of the United States, this 30th day of October one thousand eight hundred and forty-Six.

Examined and Countersigned

Secretary of War.

Commissioner of Pensions.

Duplicate.

House of Representatives

Sir

Dec 19, 1846

I must ask you to peruse the enclosed letter from Mr. Hallock, long as it is: and that you will return it to me at your convenience. I know nothing of the case beyond his statements, but I presume you intended to allow Mr. Bower as much as the facts and the law would justify. I would enquire if he was not entitled to back pay and whether any was allowed to him.

I remain Sir
respectfully & truly yours
W. Hunt

Col. Edwards
Comr. of Pensions

I would be glad to have you suggest whether any thing further can be done for the old soldier.

Lewiston N.Y. September 10th 1847

J. L. Edwards Esquire
Dear Sir;

Joel Bower A poor old decriped Revolutionary pensioner of the United States called on me and says the Agent who drew his pension (Isaac Hallock) has refused to in form him what amount he received, or any thing concerning the amount drawn. Will you do the old man the kindness to write me and if not too much trouble give me a Statement of the amount he received on Bower's Power of Attorney in order that he may be enabled to oblige him to settle with him.

I am Very Respectfully your obt Servt
Lathrop Cooke

1777 Entered the Artificers in the month of March at New Windsor, Capt. Machin was the only one who he can reccollect that he believes had a commission, although he was termed "engineer," was at the Battle of Fort Montgummary, the artificers were drilled by drill sargents from the Regulars the name of one was Charles McKinney

1778 was stationed at West Point, was with a detachment under Lieutenant Rose that was sent to reinforce Gen. Washingtons army in N.J. was at the battle of Monmoth where he received two wounds from a dragoon

1779 N.Y. He entered the levy Regiment of Colonel Haythorn for this Campain in the company of Capt. Seth Marvin

1780 N.Y. He joined the 5th battalion of Clintons Brigade for this campaign commanded by Colonel Willett in the company of Capt. De Bois who was absent from duty on account of poor health. Was present at the execution of Major Andre

1781 was in the Artificers

1782 N.Y. Joined the levy Regiment of Colonel Spaulding afterwards commanded by Colonel Wisenfelt in the company Capt. Abram Westfall, was in an engagement with the Indians & was wounded by a shot, received his discharge in the spring following

1777 Joined the artificers he thinks during the war believes that Machin was a commissioned Capt. although they termed him "engineer" but states he always had the command of the artificers to the end of the war, they were drilled by drill sergants from the Regulars the name of one was Charles McKinney, he thinks the name of the Capt. who commanded the party that was sent to Fort Montgummary was Thomas Horton who was taken prisoner & did not live to return

1778 When they were sent to reinforce Washingtons army in New Jersey they were commanded by Lieutenant Rose

1779 He entered the company of Capt. Seth Marvin in Colonel Haythorns levies for this campaign

1780 He entered for this campaign in the 5th battallion of Clintons brigade, commanded by Colonel Willet in the company of Capt. De Bois who was not on duty on account of poor health

1781 Was in the artificers

1782 He joined for this campaign the levy Regiment of Colonel Spaulding, & afterwards commanded by Colonel Wisenfeldt, in the company of Capt Abram Westfall, Sergant George Hall commanded the party when they were sent to the frontier to join the regiment

He says that the 3 campaigns of '79, '80 & 82 he only entered for those particular campaigns & whether designated by a certain no. of months or the active opperations of the army he is unable to state but is sure the term of service expired with the close of the campaigns, the campaign of '82 he thinks was 9 mo. but is not distinct in his reccollection

Joel Bowers name was found in book 7, page 77, which stood under Colonel Drakes Regiment but at the side of the page was written in pencil "Levies that joined the 5th N.Y. Brig. under Colenl. Willett" the person who made said note was absent.

35163 Joel Bowers 19.18.2
Dec. 15, 1780.

Invalid. File No. 29,020
Joel Bower Pri Rev War
Act: June 7 32
Index: - Vol. 2, Page 45
[Arrangement of 1870.]
Revolutionary War Pension and Bounty Land Warrant Application Files
Series M-804; Roll # 301

John and Diana Brink
New York (W.15,759)

Sir,

Washington D. C. Jany 13, '52

Enclosed herewith please find evidence in support of the claim of Diana, widow of John Brink for pension under the act of 4th July 1836. The papers enclosed which are marked "A" were presented in 1839 to the Pension Office, and the claim was then suspended for an alledged insufficiency of evidence of service, the testimony as to the date of the marriage was admitted then to be sufficient.

Those enclosed marked "B" though got up for the purpose of presenting them directly to Congress were first filed in the Pension Office, when the claim was again suspended for the same reason. All the papers were then with drawn from the Pension Office by Hon Mr Carroll and having been presented to the House of Representatives were referred to the Committee on Revolutionary Pensions, before being acted on by the Committee, it appears, they were withdrawn by Mr Carroll and sent to James G. Austin of Phelps New York for the purpose of enabling him to take copies of them. Austin, instead of returning them to the House files, transmitted them to one G. F. Yates of Schenectedy N.Y. who asserted that he could procure record evidence of the service and would then be able to get an allowance of the claim by the Pension Office, he has been endeavoring to get the papers from Yates ever since and has at last succeeded. Yates had misplaced them and supposed and asserted that they were lost.

In addition to the evidence contained in the papers above spoken of, I herewith enclose an original certificate of Cornelius Dubois, Muster Master in the war of the revolution, which conclusively establishes the fact of the claimants husband having enlisted for a period of eight months in that war, you will perceive that he was mustered in the "Ulster County Militia" and I understand that Minisink, the place of Brinks residence, was during the war within the limits of Ulster County. It seems to me, that on fair consideration and examination of the evidence in support of this claim, there can be no question as to it justness nor as to the sufficiency of the evidence presented in support of it. Three witnesses who were with Brink in the service, and whose credibility under oath is strongly certified, give evidence to a service of several years. John Vanauken who was a drummer boy in the war was a pensioner under act of 1832, Henry Van Etten is also a pensioner under the same act, and though James Van Vliets name is not upon the roll which was published in 1835 I presume he is also a pensioner.

Very Respectfully
Your Obdt servt
Tho. Ewing Jr.

James E. Heath Esq
Comr of Pensions

Diana Brink widow of John
For Pension act of 1836 original Papers bearing date in 1839
"A"

Phelps Ontario Co NY
June 26, 1839

Hon. J L Edwards
Sir
Enclosed is Widow Diana Brink's application for a Pension, please give it your attention at as early a day as possible and oblig

Respectfully Yours
J G Austin

State of New York
Cayuga County ss. Henry Van Etten being duly sworn according To Law says, that he resides in the town of Niles in the County and State aforesaid, that he is seventy Eight years of age, that he was born in the town of Minisink in the County of Orrange, in this State, Where he resided allways, till his removal to the town and county first above mentioned, about 19 years since that he was well acquainted with John Brink from his infancy to his death, which took place at Phelps in the County of Ontario and state aforesaid on the 7th day of February 1821, being at his death, sixty years of age, that the said John Brink served as a private in the Revolutionary War, and during the war from 1777 to the close of the same, that the said John Brink was a native of said town of Minisink, and allways lived there untill his removal to Phelps aforesaid, that in or about the month of June in the year 1777, said Brink was draughted and went out in a company commanded some times by Capt Jones, and other times under Capt Moses Cortright, in the Regiment Commanded by Col Thurston, and Major John Decker of What was called the Goshen & Minisink Regiment, at the time he was draughted he went out as a releif of a former Draught in the same Regiment and was out a number of days, he thinks about 20 days, some point on the North River, was discharged and returned home and soon after was ordered out as well as the whole company under the command of Capt Cortright, said Brink was marched in said Company under the command of Col Thurston aforesaid and Major Decker on to the North River, immediately after the storming of Fort Montgomery, was marched up the River towards Kingston before arriving there news came that the Brittish had burnt Kingston and had retired, was discharged with orders to Major Decker, I think, to keep that part of the Regiment belonging to Minisink, upon the Frontiers in the neighbourhood of the Deleware and Neversink Rivers, that in the spring of 1778, said Brink together with the company he belonged, I think under the command of Lieut Vantyle, was stationed at Carpenters point, and other points on the Deleware and Neversink Rivers, to watch the movements of the Indians, then in or about that part of the country under Brant, Indian Chief, and was allmost the whole of that season under arms, some times stationed at said points, or marching against small parties of Indians and the Torys, who were con-

tinually harrassing & plundering the inhabitants or burning their dwellings and other property, at one time the said Brink was in a Skirmish with the Indians at a place called Van Aukens Fort or Picket, that in the spring of the year 1779, the said Brink together with myself and all that belonged to the aforesaid company were ordered out against the Indians and the Torys (who made their appearance in that neighbourhood), and were out allmost the whole of that season in actual duty, that the said John Brink, as well as myself and others, were on the 21st day of July 1779, ordered out to fight the Indians & Torys under Brant after they had burnt Minisink he was marched as aforesaid after the Indians, and came up with them at Beaver Brook or at the mouth of the Lackawaxen River, said John Brink was in that Battle, and served under Col Thurston who was killed, Major Wood taken prisoner, and a great many killed, that from that time during the year 1780 and 1781, the said Brink together with myself and many others under the command of Major Decker, Lieut John Vantyle Capt Cortright, were continually on Gaurd on the Frontiers, or fighting small parties of Indians who continued to Harrass that place that the said Brink served in that part of the Regiment belonging to Minisink, under Major Decker, the Col in the neighbourhood of Goshen, that the said Brink, when not in actual duty, marching he was stationed at different points above stated, and served as a good soldier during the war from 1777, to the close 1783, that the said John Brink at the time he entered the service above stated, he lived in Minisink aforesaid, that from inquiries which this deponent has made, the Officers above stated, under whom said Brink served are all dead, and none of them left, a part of the time the said John Brink served in the above mentioned company under Capt Cortright, Jones, and Lieut Vantyle, & Major Decker the Regiment was commanded by Col Hatfield, the whole time of the service of the said John Brink was five years.

Subscribed and sworn to before me this 11th day of May 1839

C. Cuykendall Justice of the Peace

Henry Van Etten

Cayuga County ss. I, Cornelius Cuykendall a Justice of the Peace in and for the county aforesaid do certify that I am well acquainted with Henry Van Etten, who took and subscribed his name to the above affidavit to be of good character and is Entitled to the Highest Credibility, and that the facts therein stated by him I have no doubts are substantially true.

C. Cuykendall Justice of the Peace

Cayuga County ss. We Martin Cuykendall, and Simeon Swartwout, residents of the town of Owasco in said County, do certify that we are well accquainted with Henry Van Etten, who took and subscribed his name to the above affidavit, to be of good character, and entitled to the highest credibility, and that we have been accquainted with him ever since our infancy, and that his statements in said affidavit, we have not the least doubts are substantially true as therein stated by him

Owasco May 11th 1839

Martin Cuykendall
Simeon Swartwout

State of New York

Cayuga County ss. Henry Van Etten being duly sworn according to Law says, that he resides in the town of Niles in the County and State aforesaid, that he is Seventy Eight years of age, that he was born at Minisink in the County of Orrange and state aforesaid, that he was well accquainted with John Brink, whose service in the Revolutionary War, is proved in another affidavit hereto attached, was well acquainted Dianna Brink, said John Brink's wife, that the said John Brink and Dianna Brink his wife, were both born in Minisink aforesaid and lived in the same place of myself, that he was present at the marriage ceremony of the said John & Dianna Brink said marriage took place at Minisink aforesaid on the seventeenth day of February 1778, that they were positively maried on that day as husband and wife by Solomon Cuykendall Esquire a Majistrate duly qualified to perform such ceremonies, that the said John and Dianna, have lived together ever since, till the death of John Brink, which took place at Phelps on the 7th day of February 1821, that he never heard their marriage disputed by their friends who were numerous, that they have raised a large family of children, that this deponent has made inquiries and cannot find that the Majistrate who married said John & Dianna Brink aforesaid, ever left any reccord of the said marriage either among his reccords or with any member of his family, that it was not customary in those days to keep Reccords of Marriages, that the maiden name of Mrs Dianna Brink was Westfall.

Subscribed and sworn to before me this 11th day of May 1839

C. Cuykendall Justice of the Peace

Henry Van Etten

I, Cornelius Cuykendall a Justice of the Peace in and for the County of Cayuga above mentioned do certify that I am well acquainted with Henry Van Etten who took and subscribed his name to the above affidavit, to be of good character and entitled to the Highest credibility.

C. Cuykendall Justice of the Peace

New York State,

Cayuga County, ss. I, William Richardson, Clerk of said County, certify that Cornelius Cuykendall Esq. was, on the 11th day of May 1839 a Justice of the Peace in and for said County, duly qualified: And that I know his hand writing, and believe the signature on the annexed instrument, purporting to be his, is genuine. In Testimony Whereof, I have hereunto affixed the seal of said County, and set my hand this 11th day of May 1839

W Richardson Clerk

Declaration of Mrs Dianna Brink

State of New York

Ontario County ss. On this 23rd day of May one thousand Eight hundred and thirty nine, personally appeared before the Judges of the court of Common Pleas, in and for said County, said court being a Court of reccord in said County, and now in session, Mrs. Dianna Brink, a resident of the town of Phelps, in the County of Ontario, aged seventy Eight years, who first being duly sworn according to Law, doth on her oath make the following declaration, in order to obtain the benefit of the provisions of the Act of Congress, passed July the 4th 1836, Entitled " an Act granting half pay and pensions to Certain Widows or Orphans where their husbands and fathers have died of wounds received in the Military service of United States in certain Cases, and for other purposes" - that she is the widow of John Brink (whose service is proved by Henry Van Etten and others, and their affidavits of such service is herewith transmitted) who was a private in the Revolutionary army, and served in a Company Commanded at diferent times By Captains Jones, Moses Cortright, and Lieut. Vantyle in the Regiment Commanded by Col. Thurston, Major John Decker and others she does not recolect, said Regiment was called the Goshen and Minisink Regiment of the New York State Militia, that her husband the said John Brink lived at the time he Entered the service above mentioned, in Minisink, in the County of Orrange in the State of New York, that in or about the month of June 1777, the said John Brink was draughted into and went out in the aforesaid company under the command of Capt Moses Cortright, and marched on to the North River in the State of New

York at some point she thinks New Windsor or New Burgh, and was on duty as a soldier about 20 or 30 days, whas discharged and returned home to Minisink and soon after he was ordered out as Well as the whole company to which he belonged, and marched under the command of Capt Moses Cortright, in the Regiment commanded by Col Thurston aforesaid and Major Decker, on to the North River soon after the Storming of Fort Montgomery, from some point on said River he marched in said Regiment up the River towards Kingston on said River but before arriving there, news came that the Brittish had burnt that place and retired, he was then discharged and marched back to Minisink under Major Decker in said Regiment - said Major Decker, then had orders to keep that part of said Regiment belonging to Minisink aforesaid, upon the Frontiers as it was then called, and in the neighbourhood of the Deleware and Neversink Rivers in said County of Orrange, to watch the movements of the Indians who began to commit depredations on said Rivers, that in the spring of the year 1778, her husband John Brink together with a part or all of the company to which he belonged, under the Command of Lieut Vantyle, was stationed at Carpenters point, and other points on the Deleware River and allso upon the Neversink River, to gaurd the Indians who then infested that part of the country, said Regiment was then commanded (or the Minisink part of said Regiment) by Major Decker, Col Thurston then being on the North River with the Rest of said Regiment, was the greatest part of that year viz (1778) under arms, stationed at points on said Deleware River, or marching against the Indians and Torys who where continually harrassing & plundering the inhabitants burning their Houses, stealing their cattle, and otherwise destroying their property, at one time said Brink (and at that time her husband) was in a skirmish with the Indians & Torys at what was called Van Aukens Picket Fort, on the Deleware River, that in the spring of the year 1779, her husband the said John Brink, was called out as well as the whole company and commanded by Lieut Vantyle and Major Decker and marched against the Indians who made made their appearance in Large numbers upon the Deleware and Neversink Rivers, that her husband aforesaid, was in the action at Beaver Brook (as it was called) on the 21st day of July 1779, that he served on that day under Col Thurston, Major Decker, Wood & others that her husband was after that Battle, dury the remainder of that year, stationed at diferent points, or fighting small parties of Indians and Torys, allso during the year 1780 her husband was the greatest part of the year in actual duty as a soldier in the remnant of the company before mentioned, and in fact untill the close of the War her husband, was continually Engaged in fighting and driving the Indians and Torys, or gaurding the Frontiers inhabitants, their property &c from depredation by the Indians, that she has made Enquiries if any of the Officers under whom her husband Served is living and can find none, and that she has no Documentary Evidence in proof of her Claim, and no other Evidence that she at present able to find Except the affidavits herewith transmitted. She further declares positively that she was married to the said John Brink on the Seventeenth day of February seventeen hundred and seventy Eight, By Solomon Cuykendall a Justice of the Peace Legally authorized to perform marriage ceremonys, that her husband the aforesaid John Brink died at Phelps in the County of Ontario and State of New York on the seventh day of February Eighteen hundred and twenty one that the whole time of the service of her husband in said war was about four years, that she was married to him previous to the last period of his service viz at the time above stated, that she is the Widow of the said Brink and has remained his Widow Ever since the death of her husband the said John Brink

Sworn to and subscribed on the day and year first above written before John _____ Judge of Ontario County Courts

her
Dianna X Brink
mark

State of New York

Ontario County I Thomas Hall Clerk of Ontario County Courts do hereby Certify that the foregoing contains the Original proceedings of the said Court in the matter of the application of Dianna Brink for her pension. In Testimony Whereof I have hereunto set my hand & seal of Office this 23rdday of May A.D. 1839 at Canandaigua

Thomas Hall Clerk
pr. _____ Dep. Clerk

James Van Vliet's affidavits

State of New York

Aleghany County ss. James Van Vliet being duly sworn according to Law, says that he resides in the town of New Hudson in the County of Aleghany, and state aforesaid, that he is Eighty three years of age that he was born in the town of Minisink in the County of Orrange in the State of New York, that he was well acquainted with John Brink, from his infancy to the death of said Brink, which took place at Phelps in the County of Ontario and state aforesaid, on the 7th day February 1821, being at his death sixty years of age, that the said John Brink served as a private in the Revolutionary army and during the war from the year 1777, to the Close of the same, that the said John Brink, was a native of the aforesaid town of Minisink was born and allways lived in said town of Minisink, untill his removal to Phelps in the County of Ontario, that in or about the month of June in the year 1777, said John Brink, was draughted into and went out in a Company Commanded by Captain Moses Cortright (and others), in the Regiment Commanded by Col Thurston, Major John Decker in the Goshen and Minisink Regiment of New York State Militia, that he served in said regiment of New York State Militia from the time above stated viz (1777) to the close of the revolutionary War, that at the time he was first draughted, he marched out as the relief of a former Draugt in said Regiment, that he was out on duty a private in said Regiment at some point on the North River about 30 days, was discharged and returned home to Minisink, and was soon after ordered out as well as the whole of the company to which he belonged, under the Command of Captain Moses Cortright, and Col. Thurston & Major Decker, was marched on to the north River to some point thinks New Burgh or New Windsor soon after the storming of Fort Montgomery, from that point was marched up the said River, towards Kingston, but before ariving there, News came that the Brittish had burnt that place and had retired, was then discharged, with orders to Major Decker of said Regiment, to keep that part of said Regiment belonging to Minisink, upon the Frontiers, as it was called, in the neighbourhood of the Deleware and Neversink Rivers, that the said Brink was during the whole of that year viz (1777) in actual duty as a soldier, and when not in actual duty, held himself as a minute man as ordered, when not under marching orders, that in the Spring of the year 1778 thinks in April, that the said John Brink, together with the whole Company to which he belonged, and then under the Command of Lieut Vantyle, was ordered to march out and ocupy Certain points on the Deleware & Neversink Rivers, to gaurd the Frontiers, and to watch the movements of the Indians, who began to make their appearance in large numbers in that neighbourhood under the Command of Brant, Indian Chief, that the said Brink assisted in building, Vanaukens Picket Fort, was during that year or during that Campaign the whole time on duty as a soldier, under the Command of Lieut. Vantyle, Major Decker & others, the principal point where the said Brink was stationed that season was Carpenters point on the Deleware River, at one time that year the said John Brink as well as myself, was in an action or skirmish with the Indians at Vanakens Picket Fort, where the women and children had fled for protection from the Indians, who, had made and were making great destruction among the inhabitants such as burning their Houses, stealing their property &c, that in the Spring of the year 1779, the said Brink was called out as well as the whole company to which he belonged, to gaurd again the Frontiers, as the Indians made terrible havoc among the inhabitants upon the Deleware & Neversink Rivers, sometimes said Brink was stationed at Carpenters point, and at Vanaukens Picket, and on the 21st day of July that year viz (1779) the said Brink as well as myself and all the company, was ordered to embodey themselves, and march under the Command of I think Col Thurston, Major Decker, Major Wood & others, who had come on from Goshen and other places, against the Indians & Torys, who had assembled in large numbers, in the neighbourhood of the Deleware River, that on the 22d day of July of that year viz (1779) the said Brink was in the action or Battle of Beaver Brook or at the mouth of the Lackawaxen River, together with myself and many others, were in that hard fought battle with Indians & Torys under Brant Indian Chief, where Col Thurston was killed, and a great many others, Major Wood taken prisoner and others, that after that action, the remnant of said Regiment and the company to which the said Brink belonged, got home as well as they could, and during the rest of that year, the said Brink, together with the remnant of the company, was ordered to keep gaurd at certain points to protect the inhabitants from the Tomahawk of the Indians, and that during the years 1780, was on duty a great part of the time on duty either skirmishing with small parties of Indians or stationed at points to gaurd the Frontiers, or as a minute man that when at home or relieved was obliged to work, with our muskets in hand, that during the year 1781, was in actual duty under the Command of Lieut. Vantyle, Major Decker, either to keeping Garrison at Carpenters point, Vanauken Picket, or held himself as a minute man in said Regiment, that this deponent was with the said John Brink the greatest part of his service above stated, and that the said Brink served as a good soldier, that the whole time of the service of the said Brink in said war was about four years or more, that at the time the said Brink entered said service, he lived in the Town of Minisink in the County of Orrange and State first above mentioned, that this deponent has made inquiries, if any of the officers under whom the said John Brink served is still living, and cannot find that any of them are now living, But are long since dead that one of the Captains under whom said Brink served was Jones and the Regiment I think was after the war or about the close commanded by Col Hatfield do not reccolect particularly about this.

Subscribed and sworn to before me this 15 day of June 1839
A. K. Allen Justice of the Peace

James Vanfleet

Aleghany County ss. I A K Allen a Justice of the Peace in and for the county aforesaid do hereby certify that I am well acquainted with James Van Vliet who took and subscribed his name to the above affidavit to be of good character & sound Judgment, and his statements therein are entitled to the highest credibility.

A. K. Allen Justice of the Peace

State of New York

Aleghany County ss. James Van Vliet being duly sworn according to Law says, that he resides in the Town of New Hudson in the County and State aforesaid that he is Eighty three years of age, that he was born in the Town of Minisink in the County of Orrange and State aforesaid, that he was well acquainted with John Brink, and Dianna Brink his wife, that he was born and brought up in the same place with them, viz. Minisink aforesaid, that the maiden name of Mrs Dianna Brink, was Westfall, that he was present at a marriage ceremony at Minisink aforesaid on the 17th day of February Seventeen Hundred and Seventy Eight, that the said John and Dianna, were on that day, married as husband and wife in my presence by Solomon Cuykendall Esq. a Magistrate duly qualified to perform such ceremonies, that he has known them ever since that day and from their infancy, to the death of said Brink, that he has never heard the Legality of their marriage questioned by any one, the said Brink died at Phelps Ontario County on the 7th day of February 1821, that the said Magistrate who married said John & Dianna, has long since died and has left no reccord of marriages by him performed either among his reccords or with any member of his family, that it was not customary in those days to keep reccords of marriages

Subscribed and sworn to before me this 15 day of June 1839

James Vanfleet

A. K. Allen Justice of the Peace

Aleghany County ss. I, A. K. Allen a Justice of the Peace in and for the County aforesaid do certify that I am well acquainted with James Van Vliet who took and subscribed his name to the above affidavit to be of good character of sound Judgment and entitled to the highest credibility and that the statements made in the foregoing affidavit are entitled to credibility

A. K. Allen Justice of the Peace

Aleghany County We Jonas Eastwood a Clergyman in the _____ church in the Town of New Hudson in the aforesaid county and I Leonard Nichols of the same town and County aforesaid do hereby Certify that we are well acquainted with James Van Vliet, who took and subscribed his name to the foregoing affidavits to be of good character and his statements, we have no doubt are substantially true as therein stated by him.

June 15th 1839

Jonas Eastwood
Leonard Nichols

State of New York

Aleghany County Clerks Office ss. I Saml. Van Wickle Clerk of said County, do Certify that A. K. Allen was on the 15th day of June 1839 a Justice of the Peace in and for said County duly qualified: and that I know his hand writing, and believe the signatures on the anexed instruments, purporting to be his, is genuine. In testimony whereof, I have hereunto affixed the seal of said county and set my hand this 15th day of June 1839

Saml. Van Wickle Clerk

H. W. Brinks affidavit &c

State of New York

Ontario County ss. Henry W Brink being duly sworn says that he resides in the town of Phelps in the County and State aforesaid, that he is thirty one years of age, that he was well acquainted with John Brink and Dianna Brink his wife, that he was present at the death of the said John Brink which event took place at Phelps in the County and State aforesaid on the 7th day of February 1821, that he has been acquainted with Mrs Dianna Brink, ever since the death of her husband John Brink, that ever since the death of said John Brink Dianna Brink, his wife has remained a widow, that the said John and Dianna Brink, kept a family reccord, and the said reccord was destroyed by the Burning of said Brinks House about 18 years since, and that from diligent inqiry which this deponent has made, he cannot find that Solomon Cuykendal Esquire, the Majistrate who married said John and Dianna Brink, ever left any reccord of such marriage, among His reccords or with any member of his family

Subscribed and sworn before me this 24th day of June 1839

James G. Austin Justice of the Peace

Henry W Brink

I, James G. Austin one of the Justices of the Peace of the County of Ontario and State of New York, do certify that I am well acquainted with Henry W Brink, who took and subscribed his name to the above affidavit to be of good character and entitled to the Highest credibility.

James G. Austin

State of New York

Ontario County ss. Mrs Phebe Brink being duly sworn according to Law says that she resides in the town of Phelps in the County and State aforesaid, that she is fifty years of age, that she was well acquainted with John Brink and Dianna Brink, his wife, that she is the daughter of the said John and Dianna Brink, that she was present at the death of her father John Brink, which took at Phelps in the county and state aforesaid on the 7th day of February 1821, that she has resided near her mother, Mrs Dianna Brink, ever since her fathers Death, that her mother, the said Dianna Brink is now a widow and has remained a widow ever since the death of her husband John Brink aforesaid, that the said John and Dianna Brink kept a family reccord, that said reccord was destroyed By the burning of said John Brinks house about 18 years since, that from diligent inquiry which she has made, she cannot find that Solomon Cuykendal Esquire the Majistrate who married her father and mother, ever left any reccord of such marriage among His records or with any member of his family, that the said Majistate has been dead a great number of years.

Subscribed and sworn to before me this 24th day of June 1839

James G. Austin Justice of the Peace

Phebe Brink

I, James G. Austin one of the Justices of the Peace in and for the County of Ontario and above mentioned, do certify that I am well acquainted with Phebe Brink, who took and subscribed her name to the above affidavit to be of good character and entitled to the highest credibility

James G. Austin

State of New York

Ontario County ss. I Thomas Hall Clerk do hereby Certify that James G. Austin whose name is subscribed to the within Affidavits and Certificates was on the day of the date of the same a Justice of the Peace in and for said County duly elected and qualified and authorized to take the same and that his several signatures within written are genuine In Witness Whereof I have hereunto set my hand and seal of said County this 25th day of June 1839 at Canandaigua

Thomas Hall Clerk

War Department
Pension Office
Oct. 4, 1839

Sir,

The papers of Mrs. Diana Brink, widow of John Brink, have been examined and filed. Of the service alleged to have been rendered by the claimant's husband, some evidence may be had by application to the Comptroller of the State of New York, Bates Cook, Esqr. The statements of the witnesses are indefinite, except as to one tour of twenty, and one of about thirty days.

I am respectfully
Your Obt. Servt.
J L Edwards

J. G. Austin, Esqr.
Phelps N. York

Comptrollers Office
Albany Nov 20, 1839

James G Austin Esq
Sir

The books and papers relating to the revolutionary period having been indexed and put in order for reference, I have caused diligent search to be made for the name of John Brink, but without success, his name cannot be found in such index

I am very Respectfully
Your Obt Servant
Philip Phelps
Dep Compt

Sir

Albany Decr 3d 1839

I reenclose the affids. and letter referred to in your favour of the 28 ulto. I regret that the name of the individual sought for could not be found, and also that in apprising you of that circumstance these papers were inadvertently omitted to be sent.

I am very respectfully
Yr Obdt Servt
Philip Phelps
Dep Comptroller

James G Austin Esqr
Phelps Ontario Co

Additional testimony of service of John Brink
late husband of Widow Dianna Brink on file at War Department
State of New York
Ontario County ss. Capt. John Vanauken being duly sworn according to Law says that he resides in the Town of Junius in the County of Seneca, and State aforesaid, that he is seventy two years of age, that he formerly resided in Minisink Northhampton County & State of Pensylvania (it being his native place) that he was well accquainted with John Brink the husband of Widow Dianna Brink, whose application for a pension is now on file at the War Department, that he was accquainted with said Brink from his infancy to his death, the he knows the said John Brink, to have served as a private in the Revolutionary War, that the said John Brink aforesaid served in a company commanded by Capt. Moses Cortright in the Regiment commanded by Col. Thurston & Major John Decker in the Goshen and Minisink Regiment of New York State Militia, that the said John Brink, was draughted in June 1777 and went out up the Delaware River, and across the country on to the North River in the County of Orrange & State of New York, in the Regiment commanded by Col. Thurston aforesaid and company commanded by Moses Cortright Capt, was out about 40 days in that march, was discharged and returned Home, and immediately was called out on to the then frontiers to gaurd the inhabitants their dwellings & property from the depredations of the indians, and remained on duty as a soldier under arms in the neighbourhood of Carpenters point in State of New York from 1st of August to Nov that year, in the winter of 1778 in February, the said Brink was called out to pursue the Indians, who were making their appearance on and about the Delaware & Neversink River, and who were plundering the Inhabitants, said John Brink served during February & March of that year on duty as a soldier under Capt Cortright aforesaid, that in the spring of 1778 in May he thinks, he was called out, marched under Capt Cortright & Lieutenant Vantyle, on to the Neversink River, and Deleware River, was on duty as a soldier all that summer about five Months, had his house and all his goods destroyed by the Indians was Engaged in a skirmish with Indians under Brant at What was called Vanaukens Picket Fort, in Nov 1778 was called out and marched under the aforesaid Capt Cortright down the Deleware River in

pursuit of the Indians was out 20 days, and when not on actual duty held himself as a minute man, that during the Winter of 1779 the said John Brink together with the company he belonged and then commanded by Lieutenant Vantyle, for three Months was stationed at the fort, to protect the inhabitants who assembled there for protection from the Indians, in May 1779 the said Brink was Draughted to serve as a private and did march out in the same company under the aforementioned officers, was marched on to the North River, at or near New Windsor, after the storming of Fort Montgomery, was marched up the North River to near Kington, through the county back to the Deleware River and remained in actual duty under arms, some part of that Campain was stationed at Carpenters point on the Deleware River, that the said Brink was ordered out and did march against the Indians in the month of July that year and was in the Battle of Beaver Brook with Indians under Brant Indian Chief, where almost all the Officers were killed and but few soldiers escaped, the whole service as performed by the said John Brink that year was six months, that during the Winter following viz 1780 the said Brink acted as a minute in the Militia, and frequently turned out to fight small parties of Indians who infested that nieghbourhood (viz Minisink Carpenters point & Neversink), was on duty that winter 2 months, in the spring of 1780 the said Brink Entered the service under Lieutenant Vantyle, Capt Willhelmus Westfall & Major Decker and was a private in said company and continued in the service that summer six months, sometimes marching against the Indians, and at other times stationed at Carpenters point, Vanaukens Picket Fort and at other points on the Deleware & Neversink Rivers was discharged he thinks in Nov that year, that during the Winter of 1781, the said Brink acted as a minute man, and was frequently called out against the Indians and Torys, that in Campaign of 1781 or summer of 1781, he served under the same officers before mentioned and continued in the service as a private under Capt. Westfall & Lieutenant Vantyle & Major Decker and served four months in actual duty as a soldier, and the remainder of that season three months was relieved from actual duty but under order to hold himself as a minute man, which he did, in 1782 he served six months under the same Officers, & was on duty on the then frontiers, (Deleware & Neversink Rivers) in the County of Orrange, some times was on gaurd at the Picket Fort called Vanaukens Fort Carpenters point & other points on said Rivers, that the said John Brink, he this deponent knew verry well was frequently with him in service, that the said John Brink at the time he Entered the service of the Revolutionary War resided in Minisink, Orrange County & State of New York that this deponent Lived about two miles distant from the residence of said Brink, that he knew the said Brink to have been a good soldier, the whole service performed by the said Brink in the Revolutionary as before stated was in all about 5 years and performed as aforesaid and under the aforesaid officers in the New York State Militia, was accquainted with the said Brink from infancy to his death, that when the said Brink was not on actual duty, he kept himself as a minute man, is and was well accquainted with Henry Vanetten & James Van fleet whose affidavit respecting the service of the said John Brink are now on file at the Warr department in application of Dianna Brink for a pension.

Subscribed and sworn before me this 31st day of January 1840

James G. Austin Justice of the Peace

John Vanauken

Ontario County ss. I Certify that I am well accquainted with Capt. John Vanauken who subscribed his name to the above affidavit, and that his character is good and his statements Entitled to credit.

James G. Austin
Justice of the Peace

State of New York

Ontario County Clerks Office ss. I, Thomas Hall Clerk of said County do hereby Certify that James G. Austin Esqr whose name is subscribed to the within affedavit and certificate was at the time of taking said affedavit and certificate a Justice of the Peace in and for said County dwelling in the said County duly elected and qualified and authorized to take the same and that his name within written is his genuine hand writing In witness whereof I have hereunto set my hand and affixed the seal of said county this Eleventh day of February AD 1840

Thomas Hall Clerk

Henry V. Etten - June '77 under Capt Jones & Cortwright (alternates)

Col Thurston 20 d - after Ft. Montgomery.

Spring'78 Lt V. Tyle Capt Cortwright almost whole season

'79

80 & 81

served from spring of '79 to close of the war - whole time 5 years - Part of the time under Col Hatfield.

James Van Vliet - June 77 30 d - then during the whole year.

Apl '78 for the whole year or campaign.

spring 79 21st July ordered to embody - out the remainder of the year

'80 a great part of the year _____
81 the same - served 4 yrs or more.

Jno Van Auken - June 77 40 d - then from 1 Aug to Nov.
Feby 78 & Mch - then in May for 5 mos - in Nov 20 ds.
Winter '79 3 mos. - at the Fort. 6 mos this year
80 winter 2 mos - spring 6 mos
81 4 mos
82 6 mos
5 yrs in all.

War Department
Pension Office
Feb. 22, 1840

Sir,

In compliance with your request I have the honor to report more specifically the objections to the claim of Mrs. Diana Brink. It is alleged in the claimant's declaration that her husband was in the service in 1777, 1778, 1779 and 1780 in Capt. Moses Cortright's company. On the first examination of the papers reference was made to the Comptroller of New York for further proof. According to his certificate no evidence could be found there of the service in question. From the statement of John Van Auken, offered as additional evidence, it would appear that John Brink served as follows, viz. about forty days on the North River, commencing in the month of June 1777, and from the 1st of August till the 1st of November on the frontiers. In 1778 during the months of February and March, and from and after the month of May about five months. In the winter of 1779 three months at "the Fort." In May 1779, after the storming of Fort Montgomery, six months; In the winter of 1780 two months; in the Spring of that year in Capt. Wilhelmus Westfall's company, six months; in the same company in 1781, six months, and seven months in 1782. But the witness is evidently mistaken in the service he has undertaken to describe. If Capt. Moses Cortright's company was in actual service in 1777, 1778 & 1779 for the terms stated, it was in pursuance of a draft authorized by law. From the best evidence that has been collected of the services of Capt. Wilhelmus Westfall, it appears that that officer was not in actual service to exceed one month during the war; and it may be added that Fort Montgomery was taken in October 1777, and not in 1779. In June 1777 a draft of 4 months men was authorized; in April 1778 a draft of 8 months men; in 1779, March or April, a draft of 9 months men, and in 1781 a draft of 4 months men.

If the claim to the alleged service in 1777, 1778, 1779 and 1780 is sustainable at all, the different tours may be so described by some witness, that his account thereof will be such as to inspire confidence in his statement.

I have the honor to be
Very very respectfully
Your Obdt. Servt.
J L Edwards

Hon. R. Strong
House of Representatives

Dr Sir

Otis Ville August 6, 1843

Soon after the passage of the Pension law of 1832 Thomas Dunn of the town of Lyons Wayne County New York made a declaration before the court of said County setting forth his services a commander of a team of baggage wagons in the Revolutionary War which was as I am informed transmitted to the Pension Office in Washington City for the purpose of obtaining a Pension but which has never been granted. Will you do me the favour if it is not incompatible with the rules of your office to inform me of the reason why a Pension has not been granted to the said Thomas Dunn

Will you also inform me of the reason why a Pension has not been granted to Dian Brink widow of John Brink who resides in the town of Phelps Ontario County NY who made a declaration in 1838 or 39 and forwarded the same to the Pension Office through a Mr Orson who acted as her agent

In June last I forwarded to your office the declaration of the widow Van Gorder widow of James Van Gorder with other affidavits which had been returned to me they not being duly authenticated and not having heard from them since I ask the favour of being informed whether they have been received or not

Yours truly
John W Martin
Otis Ville Orange County New York

Hon J L Edwards
Comr of Pensions

Dianna Brink widow of John
For Pension act of 1836

Papers bearing date in 1846 except Letter of Comptroller of N.Y. bearing date in 1839

"B"

State of New York

Ontario County ss. John Van Auken being duly Sworn according to Law, doth on his oath say, that on the 31st day of January ~~1840~~ he was Called on by James G Austin Esq. in behalf of Mrs Dianna Brink, then making application for a Pension as the Widow of John Brink late of the town of Phelps in said County of Ontario, for a statement of the facts within this deponents Knowledge in relation to the services rendered by the said John Brink in the War of the Revolution, that this deponent did on the said 31st day of January 1840, make oath to a statement so far as his memory at that time served him of the services of the said John Brink in the said Revolution, that his said statement was reduced to writing by the said James G Austin Esq. and Read to this deponent, that subsequently the said Austin has Read to this deponent a Copy of the said affidavit or statement by him made as aforesaid which this deponent recognizes to be true Excepting in a statement which is in the following words & figures viz "In May 1779, the said Brink, was drafted in said Company and marched out under Capt Westfall, Lieut Van Tuyle and Major Decker was marched on to the North River at or near New Windsor soon after the Storming of Fort Montgomery" this deponent further says that when said statement was Read to him he did not discover the Error in the year which should have been in the year 1777, instead of the year 1779, as in his former affidavit a Copy of which has been read to him. And this deponent say that he feels Confident in saying that he gave Mr Austin the true time viz 1777, when the said service so refered to in his former affidavit was rendered, and that the mistake in the year must have been made by Mr Austin in taking down his said statement so made as aforesaid. And this deponent further says, that all the dates given in his former affidavit as the same has been read to him are true excepting as aforesaid, and are in accordance with his present Recollections. And this deponent feels Confident in saying that the services rendered by the said John Brink, was Rendered as stated in his former affidavit, excepting the eror above mentioned And this deponent further says, that he well Recollects the name of Col Johnson, Col Pawling & Col Weissenfelt & particularly Capts. Henry Pawling and Derick Westbrook, who were on the east side of the mountains & on the North River, that he Recollects of the said Brink being at one time during the Revolution with Cols Pawling & also with Col Weissenfelt, who were principally latter part of the War upon and along the North River, & deponent further says that he has no doubt but that the services named in the annexed Comptrollers Certificate relates to a portion of the services Rendered in the Revolution by the said John Brink late husband of the said Dianna Brink, that he knew the said John Brink to have been a great part of the time in the service either on the Deleware, over the mountains, and upon the North River, that the Officers named by him in his former affidavit, under whom the said John Brink served, were Militia Officers, and several times during the Revolution, the said Brink was over the mountains on the North River, up into Ulster County, & among the Highlands, that when the said Brink was on duty upon the north River, deponent Cannot say under what Officers he served, but Remembers well of hearing the names of the aforesaid Officers to Wit Capt Pawling, Capt Westbrook, Col Pawling & Col Weissenfelt, deponent Remembers of the said Brink being absent nearly the whole of the year immediately before the Close of the war, that he knew the said John Brink & the said Dianna his widow at the time they were married in 1778, & that he has known as husband & wife & to live together as such from 1778 to the death of the said John Brink in the year 1821.

Subscribed and sworn the 31st day of January 1846 before me

Henry C Swift J.P.

Henry Vandemark Witness

John Vanauken

State of New York

Ontario County ss. James G Austin being duly sworn according to Law, doth on his oath say that on Reading to the said John Van Auken a Copy of an affidavit made by him on the 31st day of January 1840, in behalf of Dianna Brink then making application for a Pension for, and on account of services by her late husband John Brink in the War of the Revolution, that the mistake refered to by the said John Van Auken in the foregoing affidavit, was made by this deponent in reducing his former statement to writing & that when the said former affidavit was Read to him the said John Van Auken the mistake was not by him then noticed, that this deponent feels Convinced that the mistake was his, from the Circumstance of the said Van Auken noticing the eror on hearing the Copy of his former affidavit Read, and immediately Correcting the same, & this deponent further says that in the fall of the year 1839, he Requested the Comptroller of the State of New York, to search the Military Records, Rolls & c. in his Office for the services of the said John Brink, late husband of the said Dianna Brink, & that the annexed Letter is his Reply, that deponent has never from that time to the present (from the general information he has been able to Collect in Relation to the services of the said Brink) doubted Evidence of his said Brink service was to be found upon such Records& that he has Recently caused another search to be made & the annexed Certificate of the Comptroller of the State of New York is the Result of such second search & Deponent has no Doubt, but the services therein detailed relates to the service of the husband of the said Dianna Brink now his widow.

Subscribed and sworn this 31st day of January 1846 before me
Henry C Swift J.P.
Henry Vandemark Witness

James G. Austin

State of New York
Ontario County ss. I Henry C Swift a Justice of the peace in & for said County do hereby Certify that I am well accquainted with John Van Auken and James G Austin who severally took and subscribed their names to the two foregoing affidavits, that they a persons of good moral character and that their several statements therein contained are entitled to Credit.

Henry C Swift J.P.

State of New York
Ontario County Clerks Office ss. I Alexander H. Howell Clerk of the County of Ontario in said State do hereby certify that Henry C. Swift before whom the three preceding affidavits were taken and by whom the preceding certificate was made was at the time of the taking such affidavits and making the said Certificate a Justice of the Peace in and for said County duly elected qualified and sworn to act as such Justice and duly empowered by the laws of said State to take such affidavits and make such certificate And that being acquainted with his hand writing I know his signature subscribed to said affidavits and certificate is genuine. In Witness whereof I have hereunto subscribed my name and affixed the seal of the said County this second day of February A.D. 1846.

Alex. H. Howell Clerk

Dianna Brink widow of John Brink
Petition to Congress
(June 24, 1846 Referred to the Committee
on Revolutionary Pensions.)

To the Congress of the United States

Your Petitioner humbly represents that her late husband John Brink who died in the town of Phelps in the County of Ontario, in the State of New York, on the

7th day of February 1821, was a soldier in the War of the Revolution, and served his Country faithfully to the end thereof, and she further represents that she was married to him the said John Brink in the year 1778, to wit on or about the 17th day of February 1778, and which was prior to his last period of service, and about one year after he first volunteered or went into the service, that her said husband was in the Service as a soldier and private most of the time from the early part of the year 1777, to the close of the War, under Officers who are now dead so far as her information extends, that she and her husband resided during the Revolution on the Deleware River in the town of Minisink in the now County of Orange in the State of New York, and which place during the Revolution, was called the Frontiers, and which Frontiers was almost continually anoyed by the Indians who infested that Region and west of the Deleware River. That her late husband was on duty as a soldier at various points on said Deleware River the Neversink & upon the North River, most of the time from 1777 to the close of the War. Evidence of the service of her said husband so far as she was then able to Collect the same, was filed in the Office of the Commissioner of Pensions, that Recently she has caused search to be made in the Comptrollers Office of State of New York, for service of her said husband, and the accompanying Certificate is the Result of such search, & she confidently believes that the service named in the said annexed certificate Relates to the service of her said husband & that the Evidence of her marriage to her said late husband is also on file in the Pension Office & she most Respectfully refers your honorable body, in Relation thereto & that she further Represents, that some time since she made application to the Commissioner of Pensions & filed her proofs of Service, Marriage & Death of her said husband, which said proofs of Service of her said husband, was all that she was then able to Collect, that she did in the fall of 1839, make application to the Comptroller of this State of N.Y. for a search, for service of her said husband & that the accompanying annexed letter of the Comptroller informed her agent that no service could be found, & in Consequence of the absence of such Record at that time & the decision of the Commissioner, that she has delayed any further prosecution of her Claim till the present, that she was not able to give the names of all the Officers under whom her said Husband served she is confident that her husband did serve in the Revolution as her proofs show & that she further Represent that she is now an old widow, that she has not intermarried, but continues the widow of the said John Brink & that she now resides in the town of Phelps in the County State of New York. And finally trusting to the good sense and liberality of Congress towards the Patriots of the Revolution, she now makes her appeal, confidently believing that a careful examination of her proofs &c that you will place her upon the Pension Roll under the act of July 4th 1836, in behalf of her late husband who she knows was a good soldier in the Revolution & Served as is alleged.

Subscribed and sworn this 31s day of January 1846 before me

Henry C Swift J. P.
Benjamin Shepperd Witness

her
Dianna X Brink
mark

To the Congress of the United States

The undersigned residents of the town of Phelps in the County in the State of New York do hereby Certify that the above named petitioner Mrs. Dianna Brink, Widow of John Brink who was a soldier in the Revolution, but now dead, is of good Character, and her statements are entitled to Credit, and we do further Certify that her late husband John Brink was always reputed to have been a soldier in the Revolution and served his Country faithfully, and we do hereby Cordially unite with her, in her above petition for a pension, believing as we do, that her Claim, is an honest and Just one And in placing her upon the pension Roll would but be an act of Justice to an old deserving Patriot of the Revolution.

Phelps Ontario County N.Y. January 1846.

Benjamin Shepperd
John Watson
A J Edmonston
Jonas Beall
Richard _____
Joseph _____
Geo Hubbell
Caleb Bannister
George Cobb
Cyrus Jackson
Abram Cuddeback
H McLouth
Owen Edmonston
Cornelius Horton
W. W. Redfield
M D Morrison
Wilbur P. Horton
Saml. E. Norton
Jacob D Westfall
C. A. Coplin

Charles E. Hobby
Geo. W. Breck
J. B. Breck
Wm. Dillon
Moses B. Whitmore
Sam J Symons
Wm. J. Hooker
W. T. Hotchkiss
Edwin Gilbert
James S McBride
John Q Howe
Wm. _____
John M Horton
Isaac Decke Junior
John C. Price
B. F. Hawks
Elias Cost
Simeon D Shuler
_____ Shear

Comptrollers Office
Albany January 28th 1846

Sir

I transmit enclosed a certificate, which contains all the evidence remaining in this office, relative to the services of Jno. Brinck in the war of the Revolution

Yr. Obdt Servt
Philip Phelps
Dep Compt

J. G. Austin Esq
Phelps NY

In pursuance of a law passed the 27th of April, 1784, entitled "An act for the settlement of the pay of the Levies and Militia, for their services in the late War, and for other purposes therein mentioned," the following certificates have been issued, bearing interest at five percent. per annum, viz:

10th May 1780	John Brinck	priv	1	4	10__
25th June 1780	John Brinck	priv		15	10__
28th Sept 1780	John Brink	priv		13	____
22nd Oct 1780	John Brink	priv		5	2__
25th Apl 1782	John Brink	priv	1	0	10
25th Decr 1779	John Brink	priv	4	17	9
20th Oct 1780	John Brink	priv	7	16	5
31st Decr 1782	John Brink	priv	21	10	2
			38	4	2

State of New-York,
Comptroller's Office, Albany, 28th Jany 1846.

I hereby certify the above to be true extracts from Book No. 1 & 2 now remaining in this office, and that the service for which the certificates above named were issued, were performed under the following officers, viz:

First -under Capt. Petrus Roosa in Col. Johannes Johnsons Regt.
Second under Capt. Mattw. Jansen in Col. Johannes Johnsons Regt.
Third under Capt. Mattw. Jansen in Col. Johannes Johnsons Regt.
Fourth under Capt. Mattw. Jansen in Col. Johannes Johnsons Regt.
Fifth under Lt. Joseph Crawford in Col. Johannes Johnsons Regt.
Sixth under Capt. Robert Hunter in Col. Albert Pawling Regt.
Seventh under Capt. Dirck Westbrook in Col. Albert Pawling Regt.
Eighth under Capt. Henry Pawling in Col. Fredk. Weissenfels Regt.

Philip Phelps
Dep. Compt.

J. L. Edwards Esq
Commissioner &c

Washington March 6th 1846

Will you do me the favor to examine the Letter and certificate of the Comptroller of the State of New York dated Jany 28th 1846 In relation to the services of John Brink, a Revolutionary Soldier (which letter is herewith inclosed attached to the papers of Diana Brink) and see if said letter together with the proofs now on file in your office in regard to the claim of Diana Brink, to be placed on the Pension List, does not effectually remove the objections heretofore made by you.

If there are still objections, The inclosed papers being forwarded to me to present to Congress, I have to request the favor of you, to enable me to do so, to return the inclosed papers to me together with all the proofs connected with the case on file in the Pension Office

I have the honor to be Resplly

C. H. Carroll

Sir

Albany N.Y. July 24th 1848

I was engaged sometime since to aid in the prosecution of the claim of Diana Brink widow of John Brinck for a pension under the act of 4 July 1836.

At the time I was engaged I failed to discover any documentary proof of John Brinck's services in the revolution. I have recently found the enclosed, which I think cannot fail to prove available. It is a species of proof which the N.Y. Comptroller's Records only occasionally ~~afford any widow~~ furnish. John Brinck hailed from the Regiment & County described in the certificate of the Muster Master.

Respectfully Your Obt Servt
G. F. Yates

To Col. J L Edwards
Commr of Pensions.

I Certify that John Brinck is mustered to serve untill the first January 1783 in the New York State Levies Sergt. Cornelius Brinck's Class in Colo. Johans. Snyders Regt. of Militia in Ulster County this 29th April 1782

Corns. DuBois Lt. M. M.

Washington DC Oct 2, 1851

Sir,

I respectfully request a reconsideration of the claim of Diana Brink for rev. pen. under the act of 1836, now in the New York suspended file.

By the time the case comes up I hope to be able to present such further evidence in the claim as may be necessary to establish it.

Very Respectfully
Your Obd Svt
Tho. Ewing Jr.

James E. Heath Esq
Comr of Pensions

Brief in the case of Diana Brink decd County and State of New York act 4 July 1836

Claim, ("original," ~~or "for increase."~~)

Proof exhibited, (if original.)

Is it documentary, traditionary, or supported by rolls? If either, state the substance.

Parole

(If for increase.)

Has additional evidence been filed since the admission of the claim? If so, what?

Name and residence of Agent
T. Ewing Jr. Present

Examining Clerk.

State of New York

Ontario County ss. Simon Westfall aged 74 years and Eli Middaugh aged 52 years both residents of Phelps Ontario County & State of New York being duly Sworn according to Law Severally Depose and say that they were well acquainted with Old Mrs Dianna Brink (widow of John Brink) deceased that the said Dianna Brink was (at the time of her Death hereinafter mentioned) an applicant for a Pension on account of the Revolutionary services of her said late Husband the aforesaid John Brink that the said Dianna Brink died at the town of Galen Wayne County & State of New York on the 24th day of August 1848 that they were present at her funeral & burial, that they have known the said Dianna & her husband & the children of their family ever since 1810 that at the time of the death of the said Dianna aforesaid she left three children only, she and her said late husband surviving and whose place of residence & ages are as follows, Cornelius Brink of the town of South Nankin in the County of Wayne and State of Michigan aged about fifty Eight years and Abigal Westfall aged about sixty years and who resides in the town of Plymouth in the County of Wayne and State of Michigan and Jane Shepperd of the town of Phelps Ontario County and State of New York who was One of the Administrators of the Estate of the said Dianna Brink with Benjamin Shepperd of the said town of Phelps and that the said Jane Shepperd Administrator as aforesaid died at the said town of Phelps on the 2d day of January 1852 and that they attended and was present at her burial, and that the affidavit of Benjamin Shepperd hereto annexed is the son of the aforesaid Jane Shepperd and the surviving Administrator of the Estate of the said Dianna Brink

Subscribed and Sworn before me this 28th day of January 1852 by theabove named Simon Westfall & Eli Middaugh who are personally known to me & that they are creditable persons & not interested in the claim

Henry C Swift a Justice of the Peace of Ontario County

Simon Westfall
Eli Middaugh

State of New York

County of Ontario ss. Benjamin Shepperd being duly sworn according to law, deposes and says that he was the duly appointed administrator of the Estate of Dianna Brink late of Phelps in said county deceased with Jane Shepperd and that as such administrator he makes the declaration & application (in behalf of the children of the said Dianna) for a Pension under the act of Congress of the 4th July 1836, which was due the said Dianna at her death, on account of the Revolutionary services of her late Husband (John Brink in the War of the Revolution) and for the proofs & facts, as well as all papers connected with this application respectfuly Refers to the papers hereto annexed, and also to the original

Declaration & proofs heretofore filed at the Pension Office by the said Dianna Brink previous to her death.

And that any certificate for such Pension which may be issued upon this application I Request that the same may be transmitted To James G. Austin of the City of Washington D.C. & the said Austin is hereby authorized to Do & Receive all, every thing, in relation to the premises, as I might or Could do were I personally present.

The words with Jane Shepperd interlined before Execution
H C Swift

Benjamin Shepperd administrator
of the estate of Diana Brink

Subscribed, sworn and acknowledged before me this 28th day of January 1852 by the above named Benjamin Shepperd who is personally known to me to be the individual who signed and swore to & acknowledged the foregoing & that he is a Credible person.

Henry C Swift a Justice of the
Peace of Ontario County

State of New York,

Ontario County Clerk's Office ss. I, Reuben Murray, Jr., Clerk of said County, Do hereby Certify, That Henry C Swift Esquire, before whom the annexed instruments were subscribed acknowledged and sworn, was at the time the same purports to have been so subscribed acknowledged and sworn, a Justice of the Peace in and for said County, duly elected, sworn and authorized to act as such. And further, that I am well acquainted with the hand-writing of such Justice and verily believe his signatures subscribed to the jurat of said affidavits & certificates to be genuine. In Testimony Whereof, I have hereunto subscribed my name, and affixed the Seal of the said County, this 29th day of January 1852.

Reuben Murray Jr Clerk

State of New York

Ontario County ss. I George Willson 2d Surrogate of in and for the County aforesaid do hereby certify that a Surrogates Court held at Canandaigua in said County on the 9th day of February 1852 said Court being a Court of record. It was proven to my satisfaction by the testimony of Simon Westfall and Eli Middaugh whose affidavits are hereto annexed that Diania Brink (widow of John Brink late of the town of Phelps in said county deceased) died at the town of Galen in the County of Wayne & State of New York on the 24th day of August 1848 and at the time of the Death of the aforesaid Diania Brink she was the widow of the aforesaid John Brink and that she was an applicant for a pension under the act of July 4th 1836 on acct of the Revolutionary services of her late husband, and further that at the time of her death aforesaid she left three children only _____ her late Husband Surviving, and whose ages and places of residence are as follows viz Cornelius Brink of the town of South Nankin in the County of Wayne & State of Michigan aged about 58 years, and Abigail Westfall of the town of Plymouth County of Wayne & State of Michigan aged about sixty years and Jane Shepperd of the town of Phelps County of Ontario & State of New York aged about Sixty Two years all of whom are above the age of Twenty One years

And I do further certify that the above named Jane Shepperd one of the children above named and Benjamin Shepperd of the town of Phelps aforesaid were duly appointed administrator and administratrix of the Estate of the aforesaid Diania Brink that they were duly sworn & authorized by the laws of this state to act as such administrator & administratrix

And I do further certify that the above named Jane Shepperd administratrix is dead and that the above named Benjamin Shepperd is the sole surviving administrator of the Estate of the said above named Diania Brink deceased In testimony whereof I have hereunto subscribed my name and affixed my Official Seal at Canandaigua this 9th day of February 1852

George Willson 2d Surrogate of Ontario Co

Henry Van Etten May 11, 1839 was present at the marriage ceremony of John & Diana Brink, Feby 17, 1778.

John Brink died Feby 7, 1821, being then sixty years old.

James Van Vliet died Feby 7, 1821 aged 60 years

1777, Capt Moses Cortright, served from 1777 to close of war. Was in service with John Brink. Was present at marriage ceremony Feby. 17, 1778 between John & Diana Brink

Dianna Brink May 23, 1839. 78 yrs _____
John Van Auken Pensioner _____ N.Y. $88.00 _____
Henry Van Etten Pensioner _____ N.Y. act 1832 $80

Dianna Brink died August 24, 1848
Cornelius Brink of Michigan Feby 9, 1852
Jane Shepperd N.Y. decd
Abagail Westfall Michigan

New York

Diana Brink widow of John Brink deceased who died on the of in the State of who was a private in the Revolutionary War Inscribed on the Roll of N. York at the rate of 40 Dollars Cents per annum, to commence on the 4th day of March, 1831 & Ending 24th August 1848. Certificate of Pension issued the 16th day of Feb. 1852 and sent to _____ Jr

Arrears to thc 4th of

Semi-annual allowance ending

Recorded by R. McRae Clerk, {Revolutionary Claim, Act July 4, 1836, Section the }

Book D Vol. 2 Page 362

Sir:

Washington, April 23d 1852

Enclosed I have the honor to present the declaration of the Admr. of Diana Brinck, decd. widow of John Brink, asking for a re-examination of her case, and an increase of her pension, under the act of 4th July 1836, together with a power of Atty. to me to prosecute the claim. Also, the deposition of Simon Westfall, in support of said claim

A pension at the rate of $40 per annum has been recently allowed in this case. I think the evidence most fully and satisfactorily proves much more than two years service, and consequently that a full pension should have been allowed. Three or four old soldiers, who served with said Brink, who are themselves pensioners, and men of good character, testify fully and clarly to more than two years services by said Brinck, within their own knowledge

I am, very respectfully
Your obt servt
J. J. Coombs

J. E. Heath Esq
Comr of Pensions

Diana Brink widow of John Brink Decd
New York Roll act 1836
to J. J. Combs Power of Atty
Additional Evidence and Power of Atty

State of New York
County of Ontario ss. Benjamin Shepperd a Resident of the town of Phelps in the county and state aforesaid being duly sworn according to law, deposes and says that he is the administrator of the Estate of Diana Brink (widow of John Brink) deceased, and that as such administrator, he has lately received the amount of Pension which has been allowed, the said Diana Brink at her death (as the widow of said John for his services in the Revolution) and that in as much as the said Diana Brink was Entitled to a larger pension than he has Received in her behalf, as the proofs on file will show he makes this deposition to obtain the increase to which she is Entitled under the act of 4th July 1836 as by Refference to her application & proofs on file will show, and for this purpose he hereby constitutes and appoints J. J. Coombs Esq of the City of Washington D.C. his true and lawful attorney for me & in my name place & stead to take such steps & measures in the premises as will be advisable in the premises & such, attorney or attorneys as he may appoint under him for the purposes aforesaid hereby Ratifying all my said attorney or his substitutes may lawfully do or cause to be done in the premises.

Sworn & subscribed before me this 13th day of March 1852
Henry C Swift Justice of the Peace of Ontario County

Benjamin Shepperd

State of New York
County of Ontario ss. Be it known that before me the subscriber a Justice of the peace in & for said county, duly authorized by law to administer oaths personally appeared Benjamin Shepperd (the administrator of Diana Brink deceased) and to me known to be the individual described in & who Executed the within instrument, and acknowledged that he Executed the same for the uses and purposes therein mentioned

Dated Phelps this 13th day of March 1852

Henry C Swift Justice of the Peace of Ontario County

State of New York,
Ontario County Clerk's Office ss. I, Reuben Murray, Jr., Clerk of said

County, Do Hereby Certify, That Henry C Swift Esquire, before whom the annexed instrument was subscribed and sworn and acknowledged was at the time the same purports to have been so subscribed and sworn and acknowledged, a Justice of the Peace in and for said County, duly elected, sworn and authorized to act as such. And further, that I am well acquainted with the hand-writing of such Justice and verily believe his signature subscribed to the jurat of said affidavit & acknowledgment to be genuine. In Testimony Whereof, I have hereunto subscribed my name, and affixed the Seal of the said County, this 15th day of March 1852.

Reuben Murray Jr Clerk

Additional Evidence
Dianna Brink deceased
Pensioner act of 4th July 1836
New York Roll
Claim for increase

State of New York

County of Ontario ss. Simon Westfall a resident of the town of Phelps in said County, being duly sworn according to law deposes and says, that he is Seventy Four years of age, that he knew John Brink and Dianna Brink in their life time, and has known them from his infancy, having lived & was brought up near them & near the Deleware River, that this deponent from his Early infancy & boyhood always heard it talked and it was universally believed by all the old acquaintances of the said John Brink, that he was a soldier in the Revolutionary war & ~~to his knowledge~~ the fact was never disputed to his knowledge that he always understood & it was universally believed that the said Brink while he was in the service as aforesaid, had his house burned with all his house hold goods, by the Indians, and further that John Van Auken, and Henry Van Etten, two old Revolutionary soldiers & pensioners & who testified to said Brinks services in the application of Dianna Brink his widow for a pension & whose papers are on file in her said application, are men of truth & veracity & their statements are Entitled to Credit, & also that Vanfleet who also testified in her behalf is a man of Credit & also his statements are Entitled to Credit, that he knew ~~He knew~~ the said Brink at his death having lived near him in Phelps aforesaid for about Fifteen years, and prior to that time lived in the neighborhood of the Early Residence of said Brink & his wife, both now deceased, that he has no interest whatever in the claim for a pension in behalf of the heirs & verily believes that the claim is a Just and meritorious one.

Subscribed & sworn this 9th day of April 1852 before me by the above named

Simon Westfall who I am personally acquainted with & that he is a man of truth & veracity &

his statements are Entitled to Credit.

H. C. Swift Justice of the peace of the said County

Simon Westfall

State of New-York,

County of Ontario ss. I, Reuben Murray Jr., Clerk of the County Court, in and for the County and State above mentioned, do hereby certify, that H. C. Swift whose genuine signature appears above, is, and was at the time of signing the same a Justice of the Peace in and for the County and State aforesaid, duly commissioned and sworn; that all his official acts as such are entitled to full faith and credit; and that the aforesaid County Court is a Court of Record, having general jurisdiction. Given under my Hand and Seal of Office, at Canandaigua this 10th day of April 1852

Reuben Murray Jr Clerk

L P Waldo Esq
Comr of Pensions

Washington D.C.
July 5, 1853

Sir

I desire to Examine the papers in the case of Diana Brink (widow of John) who was admitted under act of 1836.

Respectfully yours
Jas. G. Austin
Atty in the Case

Additional Evidence & asks a Reexamination
Rev Pension of Dianna Brink widow of John Brink

L. P. Waldo Esq
Comr of Pensions

Washington D.C.
July 22d 1853

Sir

In the Case of Dianna Brink (widow of John Brink decd.) of New York who was placed on the Pension Roll under act of July 4th 1836. I have to Respectfully ask a Reconsideration of her claim and increase to a full Pension.

There are several witnesses, themselves revolutionary pensioners, who each swear to more than 2 years Service on the part Brink, the Office resisted the Evidence on the ground, it would appear by the Letter of the 22d July 1840 that Van Auken placed the Storming of Fort Montgomery in '79, and not in 77, when it actually took place but bothe the Witness and the agent who drew up the papers swear this was a mere Clerical error, and we find that Van Etten who gave his deposition about the same time correctly places this Event in 1777, this is proof positive of the Character of the error

If any reliance is to be placed in human testimony, Brink served more than 2 years under Capt Jon Capt Cortright, Capt Westfall & Lieut Van Tyle, the records Show a service amounting to $95 or 19 ½ months in which the names of those Officers do not occur, and one of the Witnesses swear that Brink was a portion of the time serving under the Officers named in the Comptrollers Certificate. That the Comptrollers Certificate Cannot be taken as a final guide is Evidenced by the facts in this Case, he at first certified that he could not find the name of John Brink at all - he afterwards certified to service of John Brink amounting to $95 & upwards - now, we also afterwards furnish from the Board of Muster Master Du Bois, a Cotemporary certificate preserved in the family Showing a service of 8 months & over under Col Snyder whose name is not Contained in that of the Comptroller - what other guide can be taken, than the Oaths of men pronounced Credible and trust worthy, the witness are clearly disinterested, they are certified to be credible - they Show that they were in a position to know the Truth of what they State, it is further to be observed that they do not particularly state the Service Shown by Du Bois' certificate, which service is to be added to that Established by them.

It is Respectfully submited that the Evidence in this case is Conclusive of the service of Brink more than 2 years & that the pension be increased to a full pension in lieu of the amount allowed otherwise must we not totally discredit the Testimony of Van Auken & Van Etten 2 old soldiers & pensioners & who were born & Raised in the Same neighborhood with Brink. Saying nothing of the other Evidence presented in this Case I have not been able to discover the point upon which the pension was allowed for one half the amount Established by the Evidence in the Case.

Your long and valuable Experience in claims of this nature, as well as your good Judgment in applying Evidence in these Claims warrants me in the belief that you will See the Justice of increasing this pension to a full pension. In connection with the forgoing I have to Say that the Claimant & Witnesses Van

Auken & Van Etten were Long Known to me & I do not Hesitate to Say that they were Honest respectable and trust worthy people.

Very Respectfully Yours
James G. Austin Present.

Dianna Brink wid of John Brink
act 4th July 1836
Letter to Comr asking a ReviewL.

P. Waldo Esq
Comr of Pensions

Washington D.C.
Dec 8th 1853

Sir

In February 1852 Dianna Brink of N.Y. widow of John Brink decd was placed on the Pension list under the act of July 4th 1836 at the Rate of $40 per annum, and paid to her Children, who have ever since felt disatisfied as to the amount allowed, believing their father served a sufficient length of time to have entitled their mother to a full pension, the Record Evidence, together with a mass of Parol Evidence of Soldiers who served with Brink & of others who Resided in the same neighborhood show, Conclusively & Satisfactorily (If any Credit can be placed in human testimony) that Brink served over 2 years & instead of the allowance of $40 one years service, the Claim Should have been allowed for 2 years service which is fully proved if any Service is proved at all. At the Earnest Solicitation of the surviving heir, I did on the 22d July last write your Honor a letter Reciting the Evidence in the Case, and pointed out the grounds upon which it was believed that the Claim for 2 years service should have been allowed instead of for one year, and Respectfully asked that you would personally review the whole case, and grant the parties that increase the Case merited, in which review I had no doubt you be satisfied from the proofs that the claim should be increased. In answer to my Letter & arguments, I have Recd a Report from your office, with simply stating whenever I filed any additional material Evidence the Claim would be Reexamined & Reported to me, without I believe, the Case ever having been reviewed by you personally, and believing my letter and argument were not duly Examined, in connection with the proofs filed in the Case, & having filed some additional evidence subsequent to the admission of the Claim, which I think has been overlooked. I am induced to again Call your attention to this Claim, and Respectfully ask you to reexamine this claim & would Respectfully call your attention to my arguments filed with the papers in July last.

The application for increase has been made & on file, & some additional Evidence also filed since the admission of the Claim.

I would remark here that John Vanauken a Pensioner & who testifies in the Case, although he appears to have been of Pa. it will be borne in mind that he was a Resident of the same neighborhood with Brink & but an imaginary line between them a place adjoining N.Y. & Pa lines.

Trusting you will be able to give this case your personal review & grant the Claimant the increase, the Claim was Examined and adjusted under your predecessor & has not been reexamined to my knowledge although twice heretofore asked.

Very Respectfully Yours
James G. Austin Present
my authority will be found with the papers

☆

Brief in the case of Diana Brink, decd. widow of John, Ontario County and State of New York, act 4th July, 1836.
Claim, (~~"original,"~~ "for increase.") from $40. to $80. per annum

Proof exhibited, (if original.)

Is it documentary, traditionary, or supported by rolls? If either, state the substance.

A Certificate was issued for the benefit of the surviving children at $40 per annum, from 4th March, 1831, to August 24th 1848 - the date of the claimant's death.

(If for increase.)

Has additional evidence been filed since the admission of the claim? If so, what?

No additional evidence has been filed since the above mentioned Certificate was issued, but upon an examination of the proofs, which are on file, I am fully satisfied that, John Brink, performed much more than two years actual military service, & that this Office was in error in its former computation of the period of said service.

James G. Austin Esqr. Present.
Name and residence of Agent

S. C. Boyntone
Examining Clerk.

☆

Diana Brink, decd. widow of John, New York. Act of July 4th 1836, for increase. Admitted - Private, two years, at $80 per Annum, deducting former payments - from March 4th 1831 to Aug 24th 1848, when she died. Certificate payable to Benjamin Shepperd, Administrator for the benefit of, Cornelius Brink, & Abigail Westfall, her only surviving children. Certificate to James G. Austin, Present.

⋆

New York Albany

Diana Brink widow of John decd who died on the of in the State of New York who was a private in the commanded by Captain of the commanded by in the Revolution for 2 years N.Y. Inscribed on the Roll of Albany at the rate of 80 Dollars 00 Cents per annum, to commence on the 4th day of March, 1831 & Endg 24th August 1848. Certificate of Pension issued the 5th day of January 1854 and sent to James G. Austin

Arrears to the 4th of Present
Semi-annual allowance ending
Increased from $40
{Revolutionary Claim, Act July 4, 1836, Section the }
Recorded by H R Weightman Clerk,
Book D Vol. 2 Page 362

⋆

Widow, &c. File No. 15,759
Dianna Brink widow of
John Brink Pri Rev War
Act: July 4th 36
Index: - Vol. A, Page 88
[Arrangement of 1870.]
Revolutionary War Pension and Bounty Land Warrant Application Files
Series M-804; Roll # 341

⋆

John and Orilla Brooks
New York (W.10,451)

Application for a new Certificate

State of Ohio

Huron County ss. On the 20th day of Jany AD 1825 before me the subscriber a justice of the peace for said county of Huron personally appeared John Brooks who on his Oath declared that he is the same person who formerly belonged to the company commanded by Capt. Cornelius Hartenburgh in the Regiment commanded by Colonel Rodolphus Ritzma and that he is the same John Brooks who formerly belonged to the company commanded by Capt. Abraham Cuddeback in the Regiment commanded by Colonel James McClaughry in the servis of the United States. That his name was placed on the pension Roll of the State of N York. That he Received a Certificate under the signature of Richard Varic & Richard Platt on which he has drawn his pension till within 18 months last passed and which is hereunto attached

Sworn and Subscribed before me the day and year above said

John Brooks

Enos Gilbert Justice of the Peace

Norwalk Jany 20th 1825

State of Ohio

Huron County ss. We the subscribers practising physicians do hereby certify that after careful examination in the case of John Brooks who is now on the pension Roll of the state of New York we are of opinion that his disability does still continue which arose from various wounds Received in the Revolutionary war and further that the degrees of Disability under which he labours at present is the same being the fourth and no less than the original degree of disability for which he was placed on the Pension Roll

Sworn to and acknowledged before me this day and year above said Justice of the Peace of the State of Ohio Huron County (seal)

Enos Gilbert

Daniel Tilden _____

Elis E Ford Physician & Surgeon

State of Ohio

Norwalk Jany 20, 1825

Huron County Clerk's Office ss. I David Gibbs clerk of the Court of Common Pleas in & for said County do certify that Enos Gilbert before whom the within affidavits appear to have been made was at the time of making the same a lawfull Justice of the Peace in & for said County duly commissioned & sworn & that full faith is due to all his official Acts also that David Tilden, & Elis E. Ford are to me well known & that they are respectable practising Physicians in sd. County & that their names on the opposite side hereof are in their own proper hand writing. Given under my hand & official seal the day & year above written.

D. Gibbs Clerk

By Richard Varick & Richard Platt of the City of New York Esquires

This is to certify that in Pursuance of the Law of the State of New York Entitled "An Act making provision for Officers Soldiers & Seamen who have been disabled in the Service of the United States passed the 22nd of April 1786 We have examined John Brooks of Peenpack in the County of Ulster Labourer and find that in the Year one thousand seven hundred and Seventy six he enlisted as a private in Captain Cornelius Hardenbergh's Company in the third New York Regiment in the Service of the United States commanded by Colonel Rudolphus Ritzema; That while in the Service aforesaid in the Company & Regiment aforesaid to wit on the twenty eighth Day of October in the same year at the Battle of White Plains he received a Wound by a Musket Ball which has in great Measure deprived him of the Use thereof That afterwards to wit in the Month of July in the Year one thousand seven hundred and Seventy nine he was a Private Soldier in Captain Abraham Cuddeback's Company in the Regiment of Ulster County Militia commanded by Colonel James McClaughry and that upon an Invasion of the Frontiers of Ulster County by a party of Savages & others under the Command of Joseph Brandt he was ordered to march with a party of Militia under the Command of Colonel Thurston to repel the Invasion of the Enemy: That the Party of Militia then being under the Command of Colonel Thurston or of Colonel Hathorn were defeated by the Enemy and he received three other wounds in the same arm & _____ and one other wound in his Left arm above the Elbow by means of which _____ Wounds he is in a great Measure rendered incapable of obtaining his Livelihood by Labour; That on the fourth Day of December last he was thirty one Years of Age: Therefore in pursuance of the Law aforesaid We do further certify that upon the Principles of the Act of Congress of the seventh Day of June 1785 the said John Brooks is entitled to from the Treasurer of this State a yearly Pension of thirty six Dollars from the first Day of January in the Year one thousand seven hundred & seventy seven, if not before paid. Given under our

Hands this eighteenth Day of April 1787.

Richd. Varick
Richd. Platt

We the subscribers practicing Physicians of the Township of Milan Huron county State of Ohio do hereby certify that after a careful examination in the case of John Brooks who is now on the Pension Roll of the Ohio Agency We are of opinion that his disability does still continue viz one musket ball wound through the right wrist two bullet wounds in the right arm one saber wound in the left arm at the elbow and one bullet wound through the right groin which wounds occasion a partial loss of the use of those limbs when they were received, and further that the degree of disability under which he labours at present is ~~total~~ 3/5 being not less than the original degree of disability for which he was placed on the Pension Roll.

Sworn and subscribed before me this 4th day of September 1832
Frederick W Fowler Justice Peace

Lyman Fay
Amos B Harris

I certify that I am acquainted with the Physicians who have signed the above certificate and that they are reputable in this profession I also certify that the words (of Florence) were interlined below the first line on the second page of this instrument before the signing and sealing of it.

Frederick W Fowler Justice Peace

The State of Ohio
Huron County Clerks Office ss. I David Gibbs Clerk of the Court of Common Pleas in & for said County do certify that Frederick W. Fowler is a Magistrate as within & that the signatures within purporting to be his are genuine. Given under my hand & Official Seal the day & year above written

D. Gibbs Clerk

Ohio

John Brooks an Invalid, originally placed on the New York roll, transferred to Ohio, Feb. 11, 1825, and notification thereof sent to Hon. M. Bartley H.R. New certificate issued (conformably to the order requiring formal certificates from the War Office) allowing him $4.80 per month, commencing at the time at which the transfer took effect. viz. 4 Sept. 1824.

Norwalk

State of Ohio

Huron County ss. On this 6th day of March AD 1833 personally appeared in open Court before the Court of Common pleas within & for said county now sitting John Brooks a resident of Florence in said County aged Seventy seven years who being first duly sworn according to Law, doth on his Oath make the following declaration, in order to obtain the benefit of the act of Congress passed June 7th 1832. That he entered the service of the United States under the following named officers and served as herein stated. That on the 3d day of March AD 1776 the declarant, then living in Peen beck Ulster County State of New York, enlisted into the Army of the United States into a Company commanded by Captain Cornelius Hardenburgh & in the third New York Regiment commanded by Colonel Rudolphus Ritzmas in the Continental Line. He enlisted for one year. Edward Lumsbury was first Lieutenant James Newkark was second Lieutenant & Jeremiah Clark was Ensign and that Alexander McDougal was Brigadeer General of the Brigade in which he served that they went from Peen Beck after he enlisted & marched by Land to Esopus in the State of New York and from thence by Water to New York that he was one of the American Army who retreated from Long Island, was wounded at Flat Bush that he staid on York Island untill sometime in the following October evacuated York Island about the time Fort Washington was taken, that he then marched to the White plains, was there in a Battle & wounded in his right arm, after he was wounded he obtained a Furlough, he went home to Peen Beck was absent about four weeks when he returned to the Army about the 4th of December 1776. He then went to Trenton & assisted in taking the Hessians & was there dismissed. That about the middle of April AD 1777 he enlisted into the Militia in a Company commanded by Captain James Tillford in a Regiment commanded by Colonel John Snyder. William Faulkner was Major and James Clinton was Brigadeer General of the Brigade He enlisted for Eight months then living at said Peen Beck. that he went from thence to Fort Clinton about Eighteen miles below West point, that he was orderly Seargant and assisted in Building Redoubts below Fort Montgomery, that he kept Garrison & staid there untill Fort Clinton and Fort Montgomery were taken, that he escaped in the Dark from Fort Clinton and joined his company at New Burgh, that he continued untill the 7th of December of the same year, when he was dismissed That he in fact served nine months under Said last Enlistment. There were in the same Regiment Captain John Felter a Captain Rose Lieutenant John Stocker &c &c The declarant returned to Peen Beck, and then served on the Frontier in Fort Deupuy, that he went from this Fort to Beaver Creek where he found Brant and his Indians under his command, and he engaged in a Battle with them and there received three different wounds by Musket Balls Buckshot & Swords, one of which broke his right thumb which has always since been useless to him. After the last mentioned time, viz in the year 1779, he continued in service untill April AD 1780, when he was discharged. which discharge was burnt in his house at Neversink which was accidentally burnt. That he has no Documentary Evidence & that he knows of no person whose testimony he can now procure who can testify to his service. He also kept Garrison in Dewitts Fort from about the year 1780 to the time of the peace. When he served at Beaver Creek against Brant Col. Hathorn was Colonel of the Regiment, James Thurston was Lieutenant Colonel who was killed in the Battle. Thomas Dewitt was Major, & Brazilla Tyler was Captain who was also killed. He was appointed Ensign under Captain Tyler before said Battle & received ~~a Lieuten~~ an Ensign's Commission, & after said Tyler's death, he received a Lieutenants Commission under which he served as Lieutenant untill the close of the War which two Commissions were burnt in his house in the Township of said Peenpeck. That he was born in Ridgfield Fairfield County State of Connecticut in the year AD 1755 he has no record of his age. That when he served at Fort Montgomery the officers, besides which he has mentioned. There were Major Lewis Dubois, Capt Black of the Artillery, Capt Chambers, Capt Rose & a Lieutenant Tom Rose, &c &c. The Declarant says that the Revd Joshua Philips & Elijah Bills can testify respecting their beleif of his revolutionary services & his Character for veracity. In 1776 he served six months in taking St John under General Montgomery from 1776 to 1777, he served one year at New York, White plains & Trenton, after this he served at Fort Montgomery Eight months, in 1778 he served Eight months from Fort Stanwix to Orange County Line on Frontiers under Capt. Felter in 1779 he served at Peenbeck & on the Frontiers & including his other Services, he is certain that he served in the revolutionary war not less than five years as a private Soldier Ensign & Lieutenant. That since the Revolutionary he lived in said Peenpeck seven years He then lived ten years in Thomson Town Sullivan County State N York. He after this lived in Pompey Onandaga County N York, from thence removed to said Florence his present residence. He hereby relinquishes every claim whatever to a pension or annuity except the present & declares that his name is not on the Pension roll of the agency of any State only on that of agency of the State of Pennsylvania at Pittsburgh Branch bank of the U States

Sworn in open Court March 6, AD 1833
D. Gibbs Clerk H. C. P.

John Brooks

☆

We Joshua Philips a Clergyman residing in the Town of Berlin in said County of Huron, and Elijah Bills residing in Hartland in sd. County, hereby certify that we are well acquainted with John Brooks who has subscribed and sworn ~~& subscribed~~ to the above declaration, that we believe him to be Seventy seven years of age: that he is reputed & beleived in the Neighborhood where he resides to have been a Soldier of the Revolution & that we concur in that opinion and that we also consider him as a man of veracity Sworn & subscribed the day & year aforesaid

D. Gibbs Clerk H. C. P.

Joshua Philips
Elijah Bills

☆

And the said Court do hereby declare their opinion after the investigation of the matter and after putting the interrogatories prescribed by the War Department, that the above Applicant was a Revolutionary Soldier & served as he states. And the Court further certify; that it appears to them that Joshua Philips who has signed the preceding Certificate is a Clergyman resident in the town of Berlin and that Elijah Bills who has also signed the same is a Resident in the town of Hartland and is a credible person and that their statement is entitled to credit

D. Higgins Presd. Judge

☆

I D. Gibbs Clerk of the Court of common pleas within & for the county of Huron State of Ohio do certify that the foregoing contain the original proceedings of the said Court in the matter of the application of John Brooks for a pension. In Testimony whereof I have hereunto set my hand & Seal of Office this 15th day of March AD 1833

D. Gibbs Clerk H. C. P.

☆

Norwalk

State of Ohio

Huron County ss. On the 31st day of July AD 1833 personally appeared before me Timothy Baker one of the Associate Judges of the Court of Common Pleas for said county John Brooks a Resident in Florence in said county who being firstly duly sworn according to the Law, doth on his Oath make the following supplementary declaration in addition to his former declaration made before the Court of Common Pleas for said county then sitting in Norwalk in said county on the 6th day of March AD 1833, in order to obtain the benefit of an Act of Congress passed 7th June 1832. That he has seen a Letter from J L Edwards dated at the Pension office of the War Department dated 27th June 1833 addressed to D Gibbs Esq. at Norwalk Huron County Ohio upon the subject of his the Declarants Pension application. The Declarant supposes that must have been a mistake in addressing said letter to said Gibbs instead of Lewis B Sturges who has been his only agent in making his said application. That said Edwards states in said Letter there is a discrepancy in his said former declaration of the 6th of March 1833 of his age with that upon which his claim for an invalid Pension was granted The Declarant does not know or conjecture what is the discrepancy mentioned but the fact that he was born on the 4th Dec 1755 as mentioned in his declaration of the 6th of March 1833, he as fully believes, as he does any other fact of which he has not absolute personal knowledge, he always through Life to this time so understood & heard it so said by his Parents & Family connections & if the time of his Birth & age are mentioned a differently in his papers when he was allowed his Invalid Pension, it must have been owing to a mistake of himself or his agent who transacted his former application. He can not conceive of any interest which he coud have at any time to state the time of his Birth or his age untruly. The Declarant says; when he served under Captain Tyler as Ensign as mentioned in his former declaration he served for six months & after said Tyler's death when he served as Lieutenant as mentioned in his said former declaration he served as much as Two years both times he served in the York Troops called Rangers raised by the authority of the State of New York & his commissions were signed in both instance by George Clinton Governor & his appointments were made by said Governor Clinton & he believes by the Council of Appointment. He began to serve as Lieutenant sometime in May 1780 and served as such untill the Army was disbanded. He was regularly & honorably discharged & his discharge & commissions was burnt in Peenpeck formerly Ulster County now Orange County State of New York. He has frequently made Enquiries respecting the officers under whom he served in both Declarations mentioned and he cannot find any or any others living upon whom he can call to testify to his said Service

Sworn & Subscribed before me at Norwalk this 31st day of July 1833

Timothy Baker Judge of Huron Com Pleas

John Brooks

I hereby certify that the above named John Brooks is a man of truth & veracity so far as I am acquainted with him, which has been an acquaintance of twelve years

Timothy Baker

⭑

I D. Gibbs Clerk of the court of common Pleas within and for the county of Huron State of Ohio; do certify that Timothy Baker Esq is an Associate Judge of said Court and that the signature on the preceding side of this sheet purporting to be his is genuine In Testimony whereof I have hereunto set my hand and affixed my Seal of Office in Norwalk in said county this 31st day of July AD 1833

D. Gibbs

⭑

I concur in opinion with Hon Timothy Baker as to the character of the aforesaid John Brooks for truth & veracity, having been acquainted with said Brooks for Ten or Twelve years past.

D. Gibbs

⭑

War Department,
Feby. 15, 1834.

Sir:

In answer to your letter of the 31 ulto. in regard to John Brook's application for a Pension, I have the honor to transmit, herewith, a report from the Pension Office, which contains the views of this Department in relation to it.

Very respectfully yr. mo. obt. svt.
Lew Cass

Lewis B. Sturges, Esq.
Norwalk, Huron County, Ohio.

⭑

War Department
Pension Office
14th Feb 1834

Sir

In obedience to your order, I have the honour to inform you, the claim of John

Brooks has been examined and the papers are herewith enclosed.

The claimant alleges to have served as an Ensign and Lieutenant, and for such service claims a pension.

His declaration is unsupported by any other than traditionary evidence.

As it has been the rule in this office, generally, to require of applicants who claim as Officers, either direct proof of service, or a Commission, such evidence is required in this case.

I have the honour to be Very Respectfully

Your Obt Servt
J. L. Edwards

Hon Lewis Cass
Secretary of War

War Department,
Pension Office,
14th Feby 1834

Sir:

The evidence in support of your claim, under the act of June 7, 1832, has been examined, and the papers are herewith returned. The following is a statement of your case in a tabular form. On comparing these papers with the following rules, and the subjoined notes, you will readily perceive that objections exist, which must be removed, before a pension can be allowed. The notes and the regulations will shew what is necessary to be done. Those points to which your attention is more particularly directed, you will find marked in the margin with a brace, (thus: }). You will, when you return your papers to this Department, send this printed letter with them; and you will, by complying with this request, greatly facilitate the investigation of your claim.

A Statement, shewing the Service of John Brooks

Period when service rendered	Duration of claimant's service			Rank of the Claimant	Names and Ranks of the Field Officers under whom he served
	Years	Months	Days		
1776				priv	Cap. Hardenburg Col. Ritzmas
1777		9		priv	Cap. Tilyard Col. Snyder

Place of abode when he entered service, and age at the present period	Proof by which the declaration is supported.
aged 77 N. York	Traditionary

I am, respectfully,
Your obedient servant,
J. L. Edwards,
Commissioner of Pensions.

After the statement of service a copy of the regulations under the Act of June 7, 1832 is included but not reproduced here. An identical copy is included in the chapter entitled ***A Declaration of Service****.*

War Department,
Pension Office,

Sir:

The evidence in support of your claim, under the act of June 7, 1832, has been examined, and the papers are herewith returned. The following is a statement of your case in a tabular form. On comparing these papers with the following rules, and the subjoined notes, you will readily perceive that objections exist, which must be removed, before a pension can be allowed. The notes and the regulations will shew what is necessary to be done. Those points to which your attention is more particularly directed, you will find marked in the margin with a brace, (thus: }). You will, when you return your papers to this Department, send this printed letter with them; and you will, by complying with this request, greatly facilitate the investigation of your claim.

A Statement, shewing the Service of John Brooks Huron Co. Ohio

Period when service rendered	Duration of claimant's service			Rank of the Claimant	Names and Ranks of the Field Officers under whom he served
	Years	Months	Days		
Enlisted 1776	1			Private	Cap Hardenberg 3 N. York Regt Col Ritzmas Continental Line Brig Gen McDougal
Enlisted 1777 Militia		9		Private	Cap Tillford Col Snyder Maj Faulkner Brig Gen Clinton

continued in service from 1778 till Apl 1780 also kept Garrison from 1780 till peace is certain he served not less than 5 years	Col Hathorn & Col Thruston Maj Dewitt

Place of abode when he entered service, and age at the present period	Proof by which the declaration is supported.
aged 77 Peenbeck Ulster Co N. York	Traditionary

I am, respectfully,
Your obedient servant,
J. L. Edwards,
Commissioner of Pensions.

*After the statement of service a copy of the regulations under the Act of June 7, 1832 is included but not reproduced here. An identical copy is included in the chapter entitled **A Declaration of Service**.*

Norwalk Ohio 27 Feby 1834

J L Edwards Esq.
Sir,

I have received the Letter from the Honl. Lew Cass Secretary of War of the 15th instant inclosing yours to him of the 14th.

If I understand your communication the only remaining difficulty is that as he cannot produce his Commission as an Ensign & Lieutenant it is required that he shoud bring positive proof, that he served as a Commissioned Officer. It is unfortunate that Capt. Brooks the applicant, after a lapse of time of 50 years, knows of no person living that he can get to swear to the fact. I cannot doubt of the truth of his statement; judging of his Character & the support of it from the Court Judge Baker David Gibbs &c yet if the rules of the War Dept. cannot dispense with positive proof he will acquiese in your not considering him as a Commissioned Officer and woud then respectfully claim for his services in the Revolutionary war for his services and for these he supposes that under the Law of 7th June 1832 independent of his pension as an Invalid on account of his wounds he is entitled to: His services therein as appears by his first & supplementary declaration and papers are as follows; that after his services in the Continental Army in which he enlisted in the year 1776, he in April 1777 enlisted in a Militia Company commanded by Captain Tillford in a regiment commanded by Colonel Snyder & served for 9 months that after his services in 1777 that he continued in the Militia service in a Company commanded by Capt Brazilla Tyler Colo Hathorns

Regiment to 1780, & from 1780 to the time of peace that he served against Brant at Beaver Creek & at other places mentioned that Capt Tyler was killed in Battle that he served until the close of the War that his Commissions were burnt in his house at Peenbeck. He swears in his supplementary declaration he served in Tylers Company before Tyler was killed for 6 months and after his death he served as Lieutenant for 2 years that he both times he served in the New York troops called rangers raised by the authority of the State of New York The result then is, that he firstly in 1777 served in the militia under Capt Tillford & Colo Snyder for 9 months then under Tyler for 6 months & after Tylers death he, as he says, serving as Lieut, for 2 years making in the whole at least 9

6

24 months

39 months 3 years & 3 months

which is more than sufficient to entitle him to a full pension as a Private. If I am mistaken in my views upon the subject please to inform me as I have before observed I suppose his claim under the Law of 7 June 1832 is independent of, and in addition to his Pension as an Invalid Pensioner. I respectfully refer you particularly to his Supplemental declaration of 31st July 1833. I have the authority from Capt Brooks to say to you that if you don't find sufficient proof of his having served as a Commissioned Officer that you woud allow him for his services as a Private and you will oblige him & me to come to as early a decision as possible and will you also be pleased to decide on the Cases of Philip _____ & Josiah Ward.

Yours very respectfully
Lewis B Sturges

Pittsburgh

John Brooks of Huron Co in the State of Ohio who was a Pr in the company commanded by Captain Hardenberg of the Regt. commanded by Col in the New York line for _____ Inscribed on the Roll of Ohio _____ at the rate of 80 Dollars per annum, to commence on the 4th day of March, 1831. Certificate of Pension issued the 12th day of March 1834 and sent to Lewis B Sturges, Norwalk

Arrears to the 4th of March 1834	240.00
Semi-anl. allowance ending 4 Sep 1834	40.00
	$280.00

Recorded by J. T. Sprague Clerk,
{Revolutionary Claim, Act June 7, 1832.}
Book E Vol. 8 Page 107

Hon. Joseph H. Barrett
Coms of Pensions.

Norwalk Ohio
March 10th 1868

Mrs. Orilla Brooks, widow of John Brooks deceased, who was a Revolutionary Pensioner, and drew his pension at the Pittsburgh Pa. Agency, is desirous of being placed on the pension list, John Brooks died Sept 22d 1846. The widow has not since marrid. His Pension certificate is lost, and I would ask for a copy if consistent with your rules. I am not able to find the regiment or character of his service, the papers having been lost.

I am very respectfully &c
Frank Sawyer

Norwalk Ohio
June 15th 1868

Sir:

Enclosed please find Declaration of Orrilla Brooks, widow of John Brooks, a Revolutionary soldier, decd. asking for Pension, also proof of marriage &c.

I have no form for this claim of pension, but trust the proof is sufficient. Please advise with me in the matter.

Very respectfully &c
Franklin Sawyer
atty for clamt.

Hon. Jos. H Burrett
Coms of Pensions.

State of Missouri
Adair County, ss: On this 4 day of June AD 1868, personally appeared before me David Wells a Justice of the Peace in and for said county Mrs. Orrilla Brooks widow of John Brooks deceased, who makes oath and says that she is 71 years of age that she was married to the said John Brooks at Huron County Ohio on the 5th day of February 1838 and that her said husband John Brooks died on the 22d day of Sept AD 1846 That they were married by John W. Brooks a Justice Peace of the county of Huron state of Ohio.

She further says that her said husband John Brooks was a private in the war of the Revolution, and received a pension at the rate of Eighty dollars per year under the act of June 7th 1832 and was paid at Pittsburgh Pa.

She further says that she has never again married since the death of her said husband John Brooks, and that she is still his widow

She further says that she makes this declaration for the purpose of being placed on the pension roll and receiving the pension due her by the law of Congress, since the death of her said husband.

She hereby appoints Franklin Sawyer of Norwalk Huron County Ohio attorney with full power to do and perform all acts and make any and all necessary receipts in and about procuring the said pension, and hereby declares said power irrevocable.

Her post office address is Kirksville county of Adair and State of Missouri.

Witnessed by
John G. Ely
Adelbert Lowe

Orilla Brooks

Sworn to and subscribed before me this 4 day of June AD 1868, and also at the same time & place personally appeared John G. Ely of the county of Adair State of Mo. and Adelbert Lowe of the county of Adair and state of Missouri persons whom I certify to to respectable and entitled to credit, and who being by me duly sworn, say, that they were present and saw Orilla Brooks sign her name to the foregoing declaration; and they further swear that they have every reason to believe from her appearance, their knowledge of, and acquaintance with her that she is the identical person she represents herself to be; that they have no interest in the prosecution of this claim

John G. Ely
Adelbert Lowe

State of Missouri
County of Adair ss. Sworn to and subscribed before me the day and year above written; and I certify that I am not interested in this claim or concerned in its prosecution. That I believe the affiants to be credible persons, and the declarant to be the person She represents herself to be; and the said applicant acknowledged the signing and sealing of the foregoing power of attorney to be her free act & deed and that the foregoing declaration and affidavit were read and explained

to applicants before signing Witness my hand the date last above written

David Wells Justice of the Peace

State of Missouri,

I County of Adair ss. I, John Richey, Clerk of the County Court aforesaid, certify that David Wells whose name appears on the within certificate, was at the time of signing the same an acting Justice of the Peace within and for said County, duly appointed and commissioned as such. That I am well acquainted with the handwriting of the said David Wells and that I believe his signature to the within to be genuine. Witness my hand and the Seal of said Court hereunto affixed at office in Kirksville this 6th day of June A.D. 1868

John Richey Clerk.

The State of Ohio

Huron County ss. I DH Fox Judge of the Probate Court within and for the County of Huron do hereby certify that the following is a true and correct Transcript taken from the Marriage Records in this Office where the same is by law required to be kept to wit:

John Brooks To Orrilla Hancock

John Brooks & Miss Orilla Hancock were by me joined in marriage within three months last past Given under my hand this 16th day of March 1838

J W Brooks Justice Peace

In testimony whereof I have hereunto set my hand & affixed the seal of Said court at Norwalk this 15th day of June AD 1868

D H Fox Judge

Brief in the Case of Orilla Brooks Widow of John Brooks, Who was a private in Rev War. Kerksville Adair County and State of Missouri

Act 2d Sec Feb 3 1853.

Claim For New Pension Certificate,

Proof Exhibited,

She claims pension as the Widow of John Brooks who was a pensioner on the roll of Ohio paid at Pittsburgh agency at $80.00 pr year & died in 1846. Identity and continued Widowhood properly shown Marriage Copy of record as well as affidavits Admitted Oct 14, 1869 to pension at $96. Pr ann. to commence ~~Feb 27, 1869~~ Aug 3, 1868 the date of filing last important evidence of identity.

Name and residence of Agent.

Frank Sawyer
Norwalk Ohio

W. E. _____
Examining Clerk.

Brief in the Case of Orilla Brooks Widow of John Brooks Who was a private Rev War. Kerksville Adair County and State of Missouri.

Act 2d Sec Feb 3, 1853.

Claim For New Pension Certificate,

Proof Exhibited,

She claims a pension as the Widow John Brooks who was pensioner under the Act of June 7, 1832 at $80.00 pr. ann. Roll of Pittsburg & paid to Sep 4, 1846 & died Sep 22, 1846. She alleges that she is 71 years of age.

State of Missouri

Adair County, ss. Ransom F Hancock of said county being first duly Sworn says that his age is thirty seven & a half 37 ½ years, that he is a son of Orilla Brooks, widow of John Brooks deceased. That his father was the former husband of said Orilla Brooks.

He further states that he knew John Brooks in his lifetime for (7) seven years, and that he was a Revolutionary pensioner, and drew such pension up to the date of his death Sept 22d 1846. That the said Orilla Brooks is his widow, and is now an applicant for pension as his widow. That during the life of the said John Brooks they lived in ~~Berlin~~ Florence Erie County Ohio. That his mother the said Orrilla removed with affiant to the state of Missouri in the year 1861 and has since remained a member of his family

That the reson she has not applied for the pension at an earlier date is because she was ignorant of her right thereto. That Judge Zahunna Philips of Berlin Erie Co. Ohio had formerly acted as agent in drawing the pension for said John Brooks, and this affiant believing that his mother probably was entitled to it wrote said Philips in 1867 in regard to it, and that said Philips ascertained the fact of her right to the same & advised his mother to apply & sent her the name of Franklin Sawyer of Norwalk as an agent, who has prepared the application as affiant believes as soon as he could procure the testimony.

Affiant further says he has no interest in the claim.

~~R. F.~~ Ransom F. Hancock

Subscribed & sworn to before me this 24th day of July 1868, and I further certify that the said affiant is the person he represents himself to be and that he is a credible person & further that I have no interest in this claim.

John Richey Clerk of Adair County Court Mo

★

Treasury Department,
Third Auditor's Office,

Sir:

December 15, 1868

In reply to your letter of the 9th inst, I have to state that the pension of John Brooks under Act 7 June 1832, appears to have been last paid to September 4 1846.

His pension for the few remaining days of his life (eighteen), does not appear to have been applied for by any person.

Very respectfully
John Wilson
auditor

Hon. C. C. Cox
Commissioner of Pensions

★

NorwalkOhio

Sir:

February 24th 1869

Herewith please find enclosed the affidavits of Almira Henderson & Ransom F. Hancock, and of Zahunna Phillips as to the identity of Mrs. Orilla Brooks, called for in your letter of 18th December, which I trust will be sufficient.

I am very respectfully Your obedient servant

Franklin Sawyer
Atty for claimt.

Hon. C. C. Cox
Commissioner of Pensions
Washington DC.

★

State of Missouri

Bates County ss. Personally appeared before me, a County Clerk within and for said County and State Almira Henderson and Ransom F. Hancock of said county of Bates who being first duly sworn, by me on oath, State; That they have been acquainted with Orilla Brooks, widow of John Brooks deceasd, who was a revolutionary Pensioner: that they have known her for thirty years last past, and know her to be the Identical Orilla Brooks widow of said Pensioner, and that they were acquainted with the said John Brooks deceased for six years or mor during his lifetime while he resided with his wife the said Orilla Brooks and further that they have no interest in this claim whatsoever. Almira Henderson further says that her age is fifty-eight years, and Ransom F. Hancock states that his age is thirty eight years.

Witness
A. Henry
Fayett Hancock

Almira Henderson
Ransom F Hancock

Subscribed and sworn to before me this 9th day of February AD 1869.
John D Myers County Clerk.

The State of Ohio

Huron County, ss. Zahunna Phillips of Berlin Erie county Ohio being first duly sworn maks oath and says that he was well acquainted with John Brooks, a revolutionary Pensioner, in his lifetime, and that he made out the necessary papers for drawing the pension of said Brooks for several years, and as he belivs from 1838 to 1846, and forwarded them regularly to Thomas Jackson, a pension agent at Pittsburgh, who would remit the mony to affiant, by whom it was paid to the said John Brooks. He further further says that he is acquainted with Orilla Brooks widow of the said John Brooks decd. & the present applicant for pension. That he was acquainted with her before her marriage with the said John Brooks deceased, her name then being Orilla Hancock, and well acquainted with her during her marrige life with the said John Brooks deceased, as they were his neighbors. After the death of said John Brooks, the said Orilla Brooks left the neighborhood to live with her children by her former husband - Hancock - & went west, & finally to the State of Missouri. Affiant further says that he saw the said Orilla Brooks at her home with her son in Bates County Missouri on the 8th or 9th days of February, instant, and recognized & knows her to be the identical Orilla Brooks, widow of the said John Brooks deceased, for whom he formerly made out the pension papers, and who is now an applicant for said pension. Affiant is for the reasons aforsaid and the circumstances stated able to state positively the identity of the applicant Orilla Brooks, as the widow of the said John Brooks deceased.

Affiant further says that the said Orilla Brooks applied to affiant some two years ago to know about her right to said pension and to assist her in getting it, if entitled thereto. Affiant placed the matter in the hands of Franklin Sawyer the present attorney in the case who was then doing business as pension agent, and that he has assisted Said Sawyer in preparing the case, and went to Missouri at his request for the purpose of being able to identify the said Orilla Brooks, for this service he expects to be compensated so far as expenses, but does not regard himself as having any interest in the claim or it collection.

Zahunna Phillips

Subscribed & sworn to before me this 24th day of February AD 1869 and I further certify that the said Zahunna Phillip is the person he represents himself to be and that he is a credible witness, and further that I have no interest in this claim or its collection.

D H Fox Probate Judge

Norwalk Ohio
June 12th 1869

Sir:

I have the honor to enquire as to the present condition of the claim for Pension of Orilla Brooks, widow of John Brooks, deceased, a Revolutionary Soldier.

Additional evidence was forwarded by me Feb. 24, last, in reference to which I have not heard.

I am very respectfully &c
Franklin Sawyer
Atty for claimt.

Hon. H. Van Aernam
Commissioner of Pensions
Washington D.C.

---

19 July 1869

In the case of Orrilla Brooks, who claims a pension as the widow of John Brooks, a Soldier of the Revolution and how the proof of identity is not satisfactory. The _____ addl witness, disinterested and highly credible, ~~and disinterested~~ are reqd, who knew the husband ~~during his life~~ and wife during the life of the former, and who have known the widow since, giving the facts ~~as~~ in the case, and showing distinctly that she is the person she represents herself to be

Mr Franklin Sawyer
Norwalk Ohio

Norwalk Ohio
Aug 4th 1869

Hon. John Sherman
Mansfield O.
Dr Sir:

Something over a year ago I made an application for pension for one Orilla Brooks, widow of John Brooks, a Revolution Soldier who died in 1846. Soon after the widow removed to Missouri, ignorant of any right to a pension, she was a second wife.

Three years ago when Judge Zahunna Phillips and I wer in the Ohio Legislater he receid a letter from her, asking as to her right to a pension. She knowing that he had formerly attended to the business for her husband.

So soon as we could get at it the application was made, but the commission was not satisfied with the identity, Judge Phillips went to Mo. saw the widow, and knew the widow of John Brooks, as she had represented, and his affidavit of the facts was forwarded to the Department. The commission is still dissatisfied with the identity. Now I cannot make the identity any more certain unless some one who knew her in this State should again go to Mo. This is imposible.

I wish you would be kind enough to inclose on this letter, what you know of Judge Phillips character, standing and worth as a man, also _____ as to myself, and forward it to _____, and ask him to hand it to the Commissioner of Pensions with such comments as he may make, as I suppose he knows Judge Phillips as wel as myself. I would hope this might be sufficient, as Mrs. Brooks is very old, and very poor, and certainly entitled to the pension

I am very truly yours
Franklin Sawyer

Department of the Interior.
Received Oct 6th 1869. Dated Oct 4th 1869. From Hon. Jno. Sherman Mansfield, O.
Subject.

Refers letter from Franklin Sawyer, Norwalk, O., rel. to pension claim of Mrs.

Orvilla Brooks, widow of Jno., a revolutionary soldier.

Action.

Oct 6th 1869.

Respy referred to the Com of Pensions.

Geo. T. Metcalf Chief Clerk, Pensions.

At the request of Gen Sawyer of Norwalk Ohio I write to assure you that Z. Phillips Esq of Erie Co Ohio is a citizen of the highest credit & standing when under oath may be relied on with entire confidence

John Sherman

Transfer Letter.

For Finance Division, Transfer Section.

Pension Agency,

Sir:

St. Joe Mo. Mar. 15th, 1876.

In compliance with yours of the 29th Jany instant, notifying me of an application for a transfer from this agency by Orilla Brooks, Widow of John, act Feb 3rd 1853 - 2d Sect. a Pensioner under Certificate No. 7,298, the following report is made:

1. Date to which last paid: March 4th 1876
2. Rate at which last paid: $96 per annum
3. Rate at which to be paid: Same
4. Are semi-annual or annual examinations required?
5. Is he exempt from biennial examination?
6. How many children entitled to the additional $2 per month?
7. Has any child died or been abandoned?
8. How many children, or sisters and brothers; and are all the children entitled to the $2 per month additional?
9. Names of minors who have died before 16 years of age:
10. Date when minor's pension terminates:
11. Name of the active guardian:
12. If overpaid, suspended, re-enlisted, re-married, or other objection to a transfer, report it: 13.Date When last payment was made: March 15th 1876

P.O. Amherst, Lorain Co. Ohio

Respectfully,

Jno. J. Clement

To

Commissioner of Pensions, Pension Agent.

Washington, D.C.

Application for a Transfer.

State of Ohio,

County of Lorain, ss. On this 28 day of June, 1876, before me, a Notary Public personally appeared Orilla Brooks late a widow of John Brooks private in Revolutionary War of Co. Regiment , now a Permanent resident of the County of Lorain, in the State of Ohio, who being duly sworn, declareth that she is the same person whose name was placed, on the 3rd day of August, 1868, upon the list of army pensions, at the rate of $96 dollars Cents per month, from the 3rd day of August, 1868.

That said Orilla Brooks was last paid at the Pension Agency of Macon Mo to the 4 day of March, 1876 but now desires and applies for the payment of said pension at, and the transfer of her name to the roll of, the Pension Agency at Cleveland Ohio. I wish the transfer made because I now reside in _____ Ohio, and will find it more to my convenience to draw my money through the agency at Cleveland Ohio My Post Office address is North Amherst Ohio

Sworn and subscribed to before me the day and year aforesaid.

E H Hinman Notary Public

witnesses

_____ Johnson

E H Hinman

Her
Orilla X Brooks
Mark

Evidence of Identity

State of Ohio,

County of Lorain, ss. I, E H Hinman a Notary Public, in the County above named, do hereby certify that I have the most satisfactory evidence, viz: from personal knowledge that Orilla Brooks, who has this day appeared before me to take the oath of identity, is the identical person named in the pension certificate, which she has exhibited before me, numbered 7298, and bearing date at the Department of the Interior the 19 day of October, 1869, and signed by W. T. Otto acting Secretary of Interior. Given under my hand at North Amherst O. this 28 day of June, 1876

E H Hinman Notary Public

State of Ohio,

County of Lorain, ss. I, Wm. N Briggs, Clerk of the Common Pleas Court of the County and State aforesaid, do hereby certify that E H Hinman is Notary Public, duly commissioned and qualified; that his commission was dated on the 29th day of May, 1876 and will expire on the 29th day of May, 1879; and that his signature above written is genuine. Given under my hand and the seal of said County this 1st day of July, 1876.

Wm. N Briggs, Clerk.

Orilla Brooks

Under an order of the Commissioner of Pensions dated January 29, 1876, you will hereafter be paid at the Agency within the geographical limits of which you are now a resident. Your name will, therefore, be transferred to the Roll of the Agency at in your district & state In order that there may be no delay in the payment of your pension you should At Once forward your certificate to the Commissioner of Pensions for indorsement of transfer. Inclose this Notice with your certificate.

Jno. J. Clement
U.S. Pension Agent.
(Stamped 5 Jul 1876 Department of the Interior Pension Office)

Columbus O

Dec 14 1906

To the Commissioner of Pensions:

Dear Sir: I desire to secure the record of the military service of John Brooks, an ancestor, who was a soldier in the war of the revolution. All I know of him is that he was born in Connecticutt in 1755, and died at Florence O. in 1846, and that he drew a pension of about $96 per annum from the U.S. and one of $37.50 from the State New York, and that he probably entered the service of the Colonies from New York.

Respectfully
J.H. Galbraith
263 15th Ave Columbus O.

Jan. 3, 1907.

Mr. J. H. Galbraith,
263 15th Ave.,
Columbus, Ohio.
Sir:

In reply to your letter dated the 14th ultimo and received in this Bureau on the 15th, you are advised that from the papers in claim, Widow's File #10451, it appears that John Brooks was born December 4, 1755, at Ridgefield, Fairfield county, Connecticut, and resided at Peenpack, Ulster county, New York when he served in the New York troops as follows: enlisted in 1775, served six months, and was at the taking of St. Johns, Canada from March 3, 1776, as Private in Captain Cornelius Hardenbergh's company, Colonel Rudolph Ritzema's 3d New York regiment, he was wounded in action at Flat Bush, Long Island, sustained gunshot wound of right arm in battle of White Plains, and was at the battle of Trenton; served one year: enlisted about the middle of April, 1777 as Orderly Sergeant in Captain James Tillford's company, Colonel John Snyder's regiment; nine months: in 1778 he served under Captain John Felter on the frontier; eight months: enlisted in July, 1779, Ensign in Captain Brazilla Tyler's company, Colonel Hathorn's regiment, was in engagement with Colonel Brandt and Indians at Beaver Creek in which he sustained three wounds of right arm, wound of right hand and left arm, Captain Tyler was killed, and soldier served as a Lieutenant from May, 1780 until the army was disbanded.

He also served under Captain Abraham Cuddeback and Colonel James McClaghry, dates and length of service not stated.

Soldier was allowed a service pension on an application executed March 3, 1833, at which time he was a resident of Florence, Huron county, Ohio.

The list of invalid pensioners of Huron county, Ohio, printed in 1835, shows John Brooks pensioned (for wounds) from January 1, 1777, under the Act of June 7, 1785.

Soldier married in Huron county, Ohio, February 5, 1838, Mrs. Orilla Hancock for his second wife. He died September 22, 1846, and she was allowed pension on an application executed June 4, 1868, at which time she was seventy-one years of age and her post-office address was Kirksville, Adair county, Missouri; in 1876 she resided in Lorain county, Ohio.

Very respectfully,
Commissioner.

Radke & McGowan Attorneys at Law

Mills Building San Francisco
Phone Sutter 3279

Record and Pension Office,
War Department, Washington, D.C.
December 19th 1928
Gentlemen:

I am seeking information for the compilation of ancestral data, but if you can assist me, please do not throw this in the stove!

John Brooks, by family tradition, is said to have been a pensioner of the War of the Revolution. I know nothing of his birth, death, or residence, etc., save that he married Rachel _____, and had a son George Brooks born Aug. 29, 1780, (place also unknown) Family tradition says that John Brooks's family was of Connecticut. This may be faulty because his later descendants married into old Connecticut and Rhode Island lines which I already have. The son George Brooks married Mary Clark on Nov. 9, 1809, probably in Ohio, and perhaps in Huron County. I mention this because it may give a clue to a possible later residence of John Brooks mentioned above. Mary Clark was a daughter of Elisha Clark and Elizabeth _____, but that probably is of no assistance.

I once found a John Brooks (Revolutionary soldier) married a Rachel Taylor. I discarded this one because I could not find that he had a son George Brooks. Perhaps you have information otherwise.

The most suspicious pair thus far are John Brooks and Rachel Blizzard (Blizard or Blysard), of whom I know nothing more than that they had a son George Brooks baptised Oct. 29th, 1780 (two months to the day after the birth of "my" George) The baptismal entry is in the Old Dutch Church at Deerpark, N.Y. The only other entry there is of the baptism of a daughter Sarah born a year or so later. There are no other birth or baptismal records, or the records of any marriages or deaths. Where they came from and where they went are unknown to me.

If you can locate my John Brooks and his wife Rachel with dates and residences, and the same for their parents, I shall be most grateful.

Yours very truly,
Delbert W. Radke

January 4, 1928

M Delbert W. Radke
Mills Building
San Fransisco, California
Sir:

I advise you from the papers in the Revolutionary War pension claim, W.10451, it appears that John Brooks was born, December 4, 1755, at Ridgefield, Fairfield County, Connecticut.

While a resident of Peenpack, Ulster County, New York, he enlisted and served with the New York Troops as follows:

In 1775, six months, was at taking of St. John's, Canada.

From March 3, 1776, one year as a private in Captain Cornelius Hardenbergh's Company in Colonel Rudolphus Ritzema's Regiment; he was wounded in action at Flatbush, Long Island, sustained gun shot wound of right arm in the battle of White Plains and was in the battle of Trenton.

From about the middle of April, 1777, he served nine months as orderly sergeant in Captain James Tillford's Company, in Colonel John Snyder's Regiment.

In 1778 he served eight months under Captain John Felter.

From July, 1779, he served as ensign in Captain Brazilla Tyler's Company, Colonel Hathorn's Regiment, was in an engagement with Colonel Brandt at Beaver Creek in which he sustained three wounds of the right arm, a wound of the left hand and left arm; he served as a lieutenant from May, 1780 until the army was disbanded,

He also served under Captain Abraham Cuddeback and Colonel James McClaghry, dates and length of service not stated.

On account of disability resulting from the above mentioned wounds, he was pensioned from January 1, 1777, under the Act of June 7, 1785.

In March, 1833 he was a resident of Florence, Huron County, Ohio.

He died, September 22, 1846.

The soldier married in Huron County, Ohio, February 5, 1838, Mrs. Orrillia or Orrilla Hancock, who was his second wife. She was allowed pension on her application executed June 4, 1868, at which time she was seventy-one years of age and her postoffice address was Kirksville, Adair County, Missouri.

In 1876 she resided in Lorain County, Ohio.

This is the only John Brooks who lived in Connecticut, New York and Huron County, Ohio, that is found on the Revolutionary War records of this bureau.

The names of his first wife and his parents are not given.

Respectfully,
Winfield Scott
Commissioner.

Widow, &c. File No. 10,451
Orilla Brooks wid. of
John Brooks Priv. Rev War
Act: Feb. 3d 1853

Index: - Vol. A, Page 85
[Arrangement of 1870.]
Revolutionary War Pension and Bounty Land Warrant Application Files
Series M-804; Roll # 353

Benjamin and Margaret Carpenter
New York (W.25,392)

State of New York

Orange County ss. Thomas V Wood of the town of Minisink in the said County being duly Sworn deposeth and saith that he is in the Eighty Eight year of his age and that in the war of the Revolution he belonged to the militia of the County of Orange in the State of New York and done duty through the whole of the war in the militia and volunteers of the said State He further saith that in the latter part of July 1776 he volunteered into a company of volunteers under Captain John Wood and in the Regiment commanded by Isaac Nichols that the said company rendezvoused at Goshen about the first of August of the said year and from thence they went to New Windsor from thence they went by Sloops to Fort Montgomery from thence to tarrytown when a head wind prevented them from sailing they then went on shore and went by land to Kingsbridge where the company lay for some time when they got Marching orders and the commenced a retreat towards the white plains where they had an engagement with the enemy from the white plains they retreated to Peekskill where the Regiment untill the first of January 1777 when the men were discharged after a service of five months. In this service he was well acquainted with Benjamin Carpenter who was a private in Captain John Woods Company and served in the said company the full term of Five months and was discharged with the rest of the said company

That in the latter part of July 1777 there was a draft made of a company of men from the Goshen Regiment to serve for three months the said company to be commanded by Captain John Wood that on or about the first of August of the said year the said company rendezvoused at Goshen and went to New Windsor and from thence by watter to Fort Montgomery and Fort Clinton that after the said company had been two month at those forts Captain Wood was detached with a company of selected men to go for one month into the neighbourhood of Rammapoo He further saith that Benjamin Carpenter was drafted into the said company and went with the said company to Fort Montgomery and was left at fort Clinton by Captain Wood when he left for Ramapoo and was in fort Clinton when it was attacked and remained in the said fort untill he saw that resistance was useless when he made his escape out of the backside of the fort crossed the bridge toward fort Montgomery and made his escape through a deep ravine in this service Benjamin Carpenter was out two months under Col Allison who was taken prisoner

That in the spring of 1779 Captain Wood raised a company of Levies for the term of nine months in this company Benjamin Carpenter enlisted for nine months and rendezvoused with the said company at Goshen about the first of May of the said year and from thence the said company marched to Major Deckers Fort on the west side of the Shawangunk Mountain and along the valley of the neversink River and remained there from the first of May untill the twentieth of July of the same year being at least Eleven week when an alarm was made and the company turned out and Joined the militia belonging to the Goshen Regiment under Colonels Hathorn and Tusten and followed the Indians to a place called Beaver brook on the bank of the Delaware River where a most dreadfull battle ensued and the whites were defeated Captain Wood was taken prisoner and that portion of his company that escaped the dreadfull massacre returned to their former headquarters and served out the nine month under Lieutenant Westfall and were discharged at the expiration of the said term of nine months and he further saith that the said Benjamin Carpenter served the full term of nine months under Capt Wood and Lieutenant Westfall and in Col Pawlings Regiment and further this deponent saith not

Sworn and subscribed this 21st day of September 1839 before me
Simeon M Stoddard Commissioner of Deeds

Thomas V Wood

I certify that Thomas V Wood who made and subscribed the foregoing affidavit is a creditable witness

Simeon M Stoddard Commissioner of Deeds

State of New York

Orange County ss. I Albert S Benton Clerk of the Court of Common Pleas in and for said county, do certify that Simeon M Stoddard is a Commissioner of Deeds as above, and that the foregoing signatures purporting to be his are genuine In testimony whereof I have hereunto subscribed my name and affixed the seal of the said county this 7th day of January 1840

A. S. Benton Clerk

State of New York

Orange County ss. On this fourteenth day of December Eighteen Hundred and thirty nine personally appeared before the undersignede a Judge of the court of Common please in and for Orange County Margaret Carpenter a resident of the town of Deerpark in the County of Orange aged Seventy One years who being first duly sworn according to law doth on her oath make the following declaration in order to obtain the benefit of the provision made by the act of Congress passed July 7th 1838 entitled An act granting half pay and pensions to certain widows. That she is the Widow of Benjamin Carpenter who was a private in the Militia of the state of New York and belonged to a company under Captain Thompson and in the Regiment commanded by William Allison as Col Benjamin Tusten Lieut Col and John Decker Major in the town of Goshen County of Orange and State aforesaid, And as she has been informed and believes her said husband Benjamin Carpenter sometime in July or the beginning of August 1776 volunteered into a company of five months men under the command of Capt John Wood and in the Regiment Commanded by Col Nichols the said company on or about the first of August Rendezvoused at Goshen and went from thence to New Windsor from New Windsor they went on the River to Fort Montgomery where they remained a short time from that place they went by watter to a place called Tarytown where they landed from the boat and went by land to Kingsbridge where the company was stationed until after the battle on long Island when they were ordered to retreat towards the White Plains and was in sight when Gen McDougall was attacked from the White Plains they retreated to Peekskill where they remained untill the first of January 1777 when her said husband Benjamin Carpenter was discharged after having faithfully discharged his duty for the term of five months

That as she has been informed and believes her said husband Benjamin Carpenter in July 1777 was drafted to go to the forts along the Hudson River for the term of three months and that sometime in July of the said year her said husband Rendezvoused at Goshen with the rest of the drafted men under Capt John Wood and went by land to New Windsor from thence by watter to Fort Montgomery where they were placed under the officer then in command of Forts Montgomery & Clinton and served untill Fort Clinton was taken and that her said husband remained in the Fort as long as resistance was in their power when he made his escape out of the back side of the fort and made for fort Montgomery but coming to the creek that run between the two forts and being unable to ford the creek he made up the creek through a deep ravine and made his escape and returned home and reported himself to Capt Swezy in whose company he belonged and went immediately with the said company to Murders creek and followed the british shipping up the river towards Kingston and back to New Windsor under she thinks Col Tusten and Major Decker in this service he was out one month and in the service previous at Forts Montgomery and Clinton he was out two months.

That her said Husband Benjamin Carpenter was as she has been informed and believes was out to Fort Dewitt on the west side of the Shawangunk Mountain on the frontier of the County of Ulster in 1778 under Captain Swezy and served in the said company seven months in the Regiment Commanded by Col Newkirk

That in the spring of 1779 in the beginning of May Capt John Wood raised a company of Levies for nine months to serve on the western frontiers against the Indians in the regiment commanded by Col Pawling that the said company rendezvoused at Goshen in the beginning of May of the said year and Marched a cross the Shawangunk Mountain to her this deponents fathers fort called Major Deckers fort who was then Major in Col Allisons Regiment and the said company remained at Her fathers fort untill the twentieth day of July ~~of~~ when the Indians made a decent on the frontiers and burnt Minisink an alarm was made and the Goshen Regiment was called out and Capt Wood went with the said Regiment under Col Tusten in pursuit of the Indians and followed them to a place called Beaver Brook where they were overtaken and a most dreadfull battle ensued and the whites were defeated Col Tusten was killed and Capt Wood was taken prisoner That at this time her fathers fort and house was burnt and all the property therein contained was destroyed she further declares that her said husband Benjamin Carpenter from her own personal knowledge belonged to Capt Woods company at her fathers fort from the first of May 1779 untill the 20th of July of the said year and started with Capt Woods company in pursuit of the Indians after the company had gone about four Miles her said husband with two others were sent in pursuit of Powder they obtained the Powder and started in pursuit of the Regiment under Col Tusten after they had gone some distance they lost the track in the woods and was obliged to return after Capt Wood was taken prisoner the men who belonged to his company was not killed returned and encamped at Van Ettens Fort and done duty their untill her father Major Decker had another Fort built when the company went to the said Fort and remained in the said Fort and in the neighbourhood of the said Fort and along the Neversink River doing duty under officers whose names she does not reccollect for the term of their enlistment and was discharged at her fathers fort about the last of January 1780 She further declares that she thinks that the said company after the battle was commanded by a Lieutenant and by her father Major John Decker who had command of the Fort That her said husband was at least Eleven weeks under Capt Wood before he was taken prisoner that her said Husband served from time to time along the frontier during the years 1780, 1781, 1782 and the beginning of 1783 as a classman and on alarms but from old age and loss of memory she is unable to state the exact periods or the officer that he served under but she has no hesitation in saying that it is her full belief that her said husband Benjamin Carpenter served more than two years during the said war She further declares that she was married to the said Benjamin Carpenter on the ninth day of April Seventeen Hundred and Eighty seven by the Rev Mr Van Benschoeten then Pastor of the Dutch reformed Church in Minisink that her name before marriage was Margaret Decker and that her said husband Benjamin Carpenter Died on the 26th day of February 1820 That she was not married to him prior to his leaving the service but the marriage took place previous to the first day of January Seventeen Hundred and ninety four Viz, at the time above stated That from old age and the infirmities attendant therein and the great distance she lives form the court house she thinks she ought to be excused from attending the sitting of the court of reccord in the said County

Sworn and subscribed before me The day and Year above written
Hulet Clark a Judge of Orange County Com Pleas

Margaret Carpenter

State of New York
Orange County ss. I certify that Margaret Carpenter who made and subscribed the foregoing Declaration is a woman of Respectability and her statements made are entitled to full credit and belief

Hulet Clark

State of New York
Orange County ss. Benjamin Carpenter of the town of Deerpark being duly sworn deposeth and saith that he is thirty Six years of age and that he is the son of Benjamin and Margaret Carpenter and that his father Benjamin Carpenter Died on the twenty Sixth day of February Eighteen Hundred and Twenty and that his mother Margaret Carpenter has ever since remained a widow

Sworn and subscribed this fourteenth day of December 1839
before me and I certify that he is a creditable witness
Hulet Clark a Judge of Orange County

Benjamin Carpenter

I Cornelius C Elting pastor of the Reformed Dutch Church of Deerpark do certify that the following is a true copy from the marriage record of the church of Deerpark viz

"Benjamin Carpenter and Margaret Decker _____ Apr 9. 1787"

said date carried out in full in letters is married April the ninth seventeen hundred and eighty seven I Cornelius C Elting above named, depose and say, that I hold the office of pastor of the Reformed Dutch Church of Deerpark in the town of Deerpark Orange County and state of New York and that the above is a true extract from the records of said church with the exception above named as certified by me

Sworn and Subscribed before me the 14 day of December 1839

____ I certify that he is a creditable man

Hulet Clark a Judge of Orange County

Cornelius C Elting
pastor of sd Church

★

State of New York

Orange County ss. I Albert S Benton Clerk of the Court of Common Pleas in and for the County of Orange Do certify that Hulet Clark before whom the foregoing declaration of Margaret Carpenter and the within affidavit of Cornelius C Elting were subscribed and sworn was at the time of taking the same a Judge of the Court of Common Pleas in and for said county, and further that the foregoing signatures purporting to be his are genuine In witness whereof I have hereunto set my hand and affixed the seal of the said court the 7th day of January 1840

A. S. Benton Clerk

★

H R

To the Commissioner of Pensions

Jany. 13, 1840

Sir,

I send for your favorable consideration the enclosed papers of Mrs. Margaret Carpenter, widow of Benjamin Carpenter, who claims under act of July 1838. Be pleased to give to the papers as early attention as your business may permit, and oblige

Your friend and servt.
Nathaniel Jones

★

War Department
Pension Office

January 16, 1840

Sir,

The papers of Mrs. Margaret Carpenter, widow of Benjamin Carpenter, have been examined and filed. It is alleged that amongst other tours of service the claimant's husband served seven months in 1778 under Capt. Sweezy, and a number of tours after 1779; but there is no satisfactory proof except of 5 months in 1776, three months in 1777 and two months and 28 days in 1779. If it be decided to accept a pension for that term, a certificate will be issued accordingly.

I have the honor to be
Very respectfully
Your Obt. Servt.
J. L. Edwards

Hon. Nathaniel Jones
House of Representatives
Please furnish a certificate for the above ten mo. & 28 days

H Reps Jan. 17th '40
N. Jones.

New York

Margaret Carpenter widow of Benjamin Carpenter who died on the 26 Feb. 1820 in Orange Co. in the State of N. York who was a Private in the company commanded by Captain Wood of the regt. commanded by Col. Nichols in the N. York line for 10 months & 28 days Inscribed on the Roll of New York at the rate of 36 Dollars 44 Cents per annum, to commence on the 4th day of March, 1836. Certificate of Pension issued the 20th day of January 1840 and sent to Hon. Nathl. Jones, H. R.

Arrears to the 4th of Sep. 39	127.54
Semi-annual allowance ending 4 Mar '40	18.22
	$145.76

Recorded by D. Brown Clerk,
{Act June 7, 1832.}

Book A Vol. 2 Page 73

State of New York

Orange County ss. On this 13th day of July 1843 personally appeared before the undersigned a Justice of the Peace in and for the said County Margaret Carpenter a resident of the town of Deerpark in the County of Orange and State of New York aged seventy two years who being first duly sworn according to law doth on her oath make the following declaration in order to obtain the benefits of the provision made by the act of Congress passed on the 3d of March 1843 granting pensions to widows of persons who served during the Revolutionary war That she is the widow of Benjamin Carpenter who was a private in the Militia of Orange County and belonged to the Regiment commanded by Col William Allison and served in the Militia under Col Allison and in the Levies under Capt John Wood and Col Pawling untill the close of the war That her annual amount of Pension which she drew from the Government was Thirty Six Dollars and forty cents That she still remains the widow of the above named Benjamin Carpenter

Sworn to and subscribed the day and year above written before me
Abraham J. Cuddeback Justice of the peace

Margaret Carpenter

Note the words under Col Allison in the first Entirlined

⁂

Orange County ss. I Abraham J Cuddeback do certify that I am well acquainted with Margaret Carpenter who made and subscribed the above affidavit and believe her to be the person who was pensioned in the town of Deerpark under the act of July 7th 1838 at the rate above mentioned

July 13th 1843

Abraham J Cuddeback

⁂

State of New York,

Orange County, ss. I, Lebbeus L. Vail, Clerk of the Court of Common Pleas, in and for the County of Orange, Do Hereby Certify, That Abraham J. Cuddeback before whom the foregoing affidavit was subscribed and sworn, was at the time of taking the same, a Justice of the Peace in, and for said county: and further, that I am well acquainted with the hand writing of the said Justice and verily believe that the signature to the said affidavit & Certificate is genuine. In Witness Whereof, I have hereunto set my Hand and affixed the Seal of the said County, the 10th day of October 1843

Lebbeus L. Vail, Clerk

Otis Ville New York
Oct 9 1843

Dr Sir

With this letter I send to your care the declaration of Margaret Carpenter made under the act of Congress passed March 3 1843 Please forward certificate to me

Your Obt St
John W Martin

Hon James L Edwards Comm of Pensions

New York

Margaret Carpenter widow of Benjamin Carpenter who was a Private in the New York line for 10 mo 28 days Inscribed on the Roll at the rate of 36 Dollars 44 Cents per annum, to commence on the 4th day of March, 1843. Certificate of Pension issued the 3rd day of January 1844 and sent to John W Martin Otisville New York

Recorded in Book A Vol. 1 Page 104

[Act of March 3, 1843.]

State of New York

Orange County ss. On this 2nd day of July 1849 personally appeared before the undersigned a Justice of the Peace in and for the County aforesaid Margaret Carpenter a resident of Deerpark in the County of Orange and State of New York aged 79 years who being first duly sworn according to law doth on her oath make the following declaration in order to obtain the benefits of the provision made by the act of Congress passed on the 2d day of February 1848 granting ~~half pay and~~ pensions to widows of persons who served during the Revolutionary war That she is the widow of Benjamin Carpenter who was a private in the militia of Orange County and State of New York and belonged to Col Allisons Regiment That she received annually under the act of 17th June 1844 Thirty six dollars and forty four cents She further declares that she is still a widow

Sworn to and subscribed the day and year above written before me
Eli Van Inwegen Justice of the peace

Margaret Carpenter

I certify that Margaret Carpenter above named is known to me personally and I know her to be the same person that she represents herself to be in the above declaration

Eli Van Inwegen Justice
Note The words half pay erased and the word widow interlined before signing
Eli Van Inwegen Justice

State of New York,

Orange County ss. I, ~~Albert S. Benton~~ Joseph W. Gatt Clerk of the County of Orange, Do Hereby Certify that Eli Van Inwegen whose name is subscribed to the certificate of the proof or acknowledgment of the annexed instrument, and thereon written, was, at the time of taking such proof or acknowledgment, a Justice of the Peace for said county, dwelling in said county commissioned and sworn and duly authorized to take the same. And further, that I am well acquainted with the hand writing of such Justice and verily believe that the signature to the said Certificate of Proof or Acknowledgment is genuine. In Testimony Whereof, I have hereunto set my hand, and affixed the Seal of the said county the eleventh day of September 1849

J. W. Gatt Clerk.

New York City

Margaret Carpenter, widow of Benj. Carpenter, who served in the Revolutionary war, as a Private Inscribed on the Roll at the rate of 36 Dollars 44 Cents per annum, to commence on the 4th day of March, 1848. Certificate of Pension issued the 19th day of Sept 1849 and sent to John W. Martin Mount Hope N.Y. Recorded on Roll of Pensioners under act February 2, 1848, Page 42 Vol. 4

R. W. Cushman

H. G. Carpenter
Ithaca, N.Y.
2/22/08

Chief of Bureau of Pensions
War Dept. Washington D.C.
Dear Sir

I am desirous of securing a copy of whatever the Department may furnish relative to a Pension or Land Bounty Right of Benjamin Carpenter a soldier of the Revolution from Albany County & other counties in New York State.

He was my great grandfather & I wish this information to enable me to become a member of the Society of the "Sons of the American Revolution"

I am, Sir
Respectfully yours
H. G. Carpenter

E. H. Addison C. A. Smedal

Addison & Smedal
Attorneys at Law
Nevada, Iowa

War Department,
Bureau of Pensions,

June 30, 1926

Washington, D.C.
Gentlemen:

Re: Benjamin Carpenter,
4th Regiment, Orange County militia, New York.

For the purpose of joining the D.A.R. I am desirous of obtaining the date of enlistment, date of discharge, rank and any other information your records may disclose concerning the above named. Enclosed is copy of letter received from the State Historian, Albany, New York.

Tradition is that Benjamin Carpenter married Margaret Decker, daughter of a Major Decker, New Jersey, and I ask if you can give me any record of Major Decker.

Upon receipt of statement remittance for this service will be promptly mailed you.

Yours very truly,
Imo Davisson
Imo Davisson, nee Carpenter

Rev. & 1812 Wars Section

August 23, 1926.

Mrs. Imo Davisson,
Box 736, Nevada, Iowa.
Madam:

I have to advise you, that from the papers in the pension claim, W.25392, it appears that Benjamin Carpenter enlisted in July, or August 1776 and served until 1783, at various times, amounting in all to ten months and twenty-eight days, as a private, under Captains Thompson, John Wood and Swazy and Colonels William Allison, Isaac Nichols, Tuston, Newkirk, Pawling and Hathorn, in the New York Troops.

He married April 9, 1787, Margaret Decker and he died February 26, 1820. She was allowed pension on her application executed December 14, 1839, while living in Deer Park, Orange County, New York. aged seventy-one years. Their son Benjamin was thirty-six years old in 1839.

The names of the parents of soldier and his widow are not stated.

If you wish a search of the records for any particular officer of the Revolution named Decker, in addition to his full name, some of the following data should be furnished: approximate dates of his birth and death, places of residence at enlistment and after service, given name of widow.

Respectfully,
E. W. Morgan
Acting Commissioner.

Box 94
Massena N.Y.

Veteran's Administration
Washington D.C.

Jan. 28, 1937

Dear sirs -

I should like to receive information as to the military record of a certain Nathaniel Hitchcock born at Wilbraham Mass. May 13, 1732. He married in 1755 Sarah Severance both of Springfield Mass.

The military record of a Jerrah Hitchcock of Dekalb who fought in the war of 1812 He was born Oct 7, 1794 and married Abigail Wilson born Sept. 21, 1804 at Depeyster, N.Y.

Would also like the name of the Revolutionary soldier upon whose war service Margaret Carpenter who was living with her son Benjamin Carpenter at Deer Park, Orange Co. N.Y. in 1840 received a pension.

very truly yours
Laural W. Clark.

February 10, 1937
Benjamin Carpenter, W.25392

Jerrah Hitchcock, W. O. 32388
BA-J/MCS

Laural W. Clark
Box 94
Massena
New York
Dear Sir:

Reference is made to your letter of January 28th, requesting information relative to Revolutionary War soldiers, Nathaniel Hitchcock who was born at Wilbraham, Massachusetts, May 13, 1732, and ---- Carpenter whose widow, Margaret Carpenter, was pensioned while living at Deer Park, New York, and War of 1812 soldier, Jerrah Hitchcock who married Abigail Wilson.

The Revolutionary War records of this office fail to afford any information in regard to a Nathaniel Hitchcock who was born May 13, 1732, at Wilbraham, Massachusetts.

You are furnished herewith, the records of the Revolutionary War soldier, Benjamin Carpenter, W.25392, and the War of 1812 soldier, Jerrah Hitchcock, W.O. 32388.

Very truly yours
A. D. Hiller
Executive Assistant to the Administrato

February 10, 1937
BA-J/MCS
Benjamin Carpenter
W.25392

Laural W. Clark
Box 94
Massena
New York
Dear Sir:

The data which follow were obtained from papers on file in the pension claim, W.25392, based upon the military service of Benjamin Carpenter.

He enlisted in July or August, 1776, and served at various times until sometime in 1783, amounting in all to at least two years as private with the New York troops under Captains Thompson, John Wood and Sweazy, Major John Decker and Colonels William Allison, Isaac Nichols, Tusten, Newkirk, Pawling and Hathorn. The troops he served with rendezvoused at Goshen. He was at Fort

Clinton when it was captured, was in a severe battle with the Indians at Beaver Brook on the Delaware River and was at Major John Decker's Fort on the west side of Shawangunk Mountain along the valley of Neversink River when it and the house of Major Decker were burned by the Indians.

He married April 9, 1787, Margaret Decker who was the daughter of Major John Decker noted above.

Benjamin Carpenter died February 26, 1820.

His widow, Margaret Carpenter, was allowed pension on her application executed December 14, 1839, at which time she was living at Deer Park, Orange County, New York, and she then gave her age as seventy-one years. In 1849 she was living there and gave her age as seventy-nine years.

In 1839 their son, Benjamin Carpenter, was living at Deer Park, Orange County, New York, aged thirty-six years. There are no further data relative to the family of the soldier.

Very truly yours
A. D. Hiller
Executive Assistant to the Administrator

Revolutionary War Records Section.
Department of the Interior,
Bureau of Pensions.
Washington, D.C., , 19

In reply to your request of , received for a statement of the military history of Benjamin Carpenter a soldier of the Revolutionary War, you will find below the desired information as contained in his (or his widow's) application for pension on file in this Bureau.

Dates of Enlistment or Appointment	Length of Service	Rank.	Officers under whom Service was Rendered.	
			Captain	Colonel
July or Aug 1776 to 1783	served at various times in N.Y. militia 10 mos. And 28 days	Pvt.	Thompson John Wood Swazy	William Allison Isaac Nichols Tuston Newkirk Pawling Hathorn

Battles engaged in, Taking of Fort Clinton.

Residence of soldier at enlistment,
Date of application for pension, Dec 14 1839. Her cl. was al..
Residence at date of application, Deer Park, Orange Co. N.Y.
Age at date of application, She was 71 years old.
Remarks: Soldier married Apr. 9, 1787 Margaret Decker Son Benjamin was 36 years old in 1839 Soldier died Feby 26, 1820 The above noted is the only soldier

Very respectfully,
Commissioner.

Widow, &c. File No. 25,392
Margaret Carpenter Widow of
Benj. Carpenter Priv. Rev. War
Act: Feb. 2nd 1848
Index: - Vol. A, Page 122
[Arrangement of 1870.]
Revolutionary War Pension and Bounty Land Warrant Application Files
Series M-804; Roll # 474

Uriah and Rebecca Chapman
New York (W.6,653)

Bureau of Pensions.
Uriah Chapman

His widow states in ____ that ~~he served~~ his first service was for 3 month in NJ in the winter of 1776-77 but does not give the names of officers In the spring of 1777 enlisted in a co of rangers raised in Orange Co N.Y. comd by Capt Elijah Winter and marched to Penna. During the tour he was in an engagement with the Indians and was shot in the left arm the ball passing throgh the arm into his side after his recovery again enlist in a co of Rang com by Capt Abraham Westfall & servd for about one yr. She also states that she was married to the soldier Aug 13, 1781, and that he died July 24, 1827 Pen was granted Date of birth not given

I Samuel Wood of La Porte Indiana of lawful age being duly sworn do depose

& say that I was acquainted with Mrs. Rebecca Chapman for the term of about eight years at Chili, Monroe County, New York, who now lives at Conneaut, Ashtabula County, Ohio widow of Uriah Chapman and that she is entitled to credit in whatever she may say and that she bore during my acquaintance with her an unimpeachable character for truth & veracity

Sworn to & subscribed before me at Conneaut, Ashtabula
County, Ohio on this 5 day of December A.D. 1840
Sebastian F. Taylor J. Peace

Samuel Wood

The State of Ohio

Ashtabula County ss. I, A. C. Hubbard Clerk P. T. of the Court of Common Pleas within said County certify that S. F. Taylor, on the fifth day of December 1840, was a Justice of the Peace within said County, and that the within signature, purporting to be his, is genuine. In testimony whereof I have hereunto affixed my official seal & subscribed my name this 13th day of December AD 1841.

A. C. Hubbard Clerk P. T. of said Court

Isaac Van Gorden's Statement

I Isaac Van Gorden aged Eighty years, residing at Conneaut, County of Ashtabula & State of Ohio, being duly sworn, do depose & say, that I was acquainted with Uriah Chapman during the Revolutionary war. That I was stationed at Capt. Chambers Fort at Minisink Orange County New York when the said Chapman was brought in, wounded by the Indians, & his wound it was said was in his arm & body, the ball passing through his arm & lodging in his body, near his back bone. I did not see the wound as I was soon called on to march towards Trenton. I knew him until long after the close of the war and knew him to be the same man, for whose services Rebecca Chapman his widow now seeks to draw a pension. He always during my acquaintance with him, complained of the wound at times.

In presence of S. F. Taylor

his
Isaac X Van Gorden
mark

State of Ohio
Conneaut Oct 30, 1841
Ashtabula County ss. I Sebastian F. Taylor a justice of the Peace within & for said County do hereby certify, that the above affidavit was drawn up by me as directed by Isaac Van Gorden & that he signed it as above & made solemn oath to its truth.

Sebastion F. Taylor J. Peace

★

The State of Ohio
Ashtabula County ss. On this first day of December in the year of our Lord one thousand eight hundred and forty one personally appeared before me, James M Bloss, Associate Judge within and for the County of Ashtabula and State of Ohio, Rebecca Chapman a resident of Conneaut in the County of Ashtabula & State of Ohio aged Seventy eight years, who being first duly sworn according to Law, doth on her Oath make the following declaration in order to obtain the benefit of the provission made by the act of Congress passed July 7th 1838, entitled "an act granting half pay and pensions to certain widows" That She is the widow of Uriah Chapman who was a private in the army of the revolution. That he served in the army under General Washington three months in New Jersey in the winter of 1776 '7 as a substitute for his brother Asa Chapman that he afterwards in the spring of 1777 enlisted into a company of Rangers at (Minisink Orange County New York) under Captain Elijah Winter marched to Lackawaxen in Pennssylvania and was in that vicinity for a considerable time. While there was attacked by the Indians and wounded severely, was taken to Wallkill Ulster County New York where he remained Sick in consequence of his wounds for about twelve weeks That after his recovery he again enlisted as a Ranger under Captain Abraham Westfall and served for five months; was in a battle with the Indians called the "Delaware fight" between the Delaware and Neversink Rivers: That afterwards he served as a ranger under the same Captain Westfall above named for the Term of Seven months
Sworn to and subscribed on the day and year above written before me
Jas. M Bloss Associate Judge

Rebeca Chapman

★

The State of Ohio
Ashtabula County ss. I, A. C. Hubbard Clerk Protempore of the Court

of Common Pleas within the County aforesaid, do hereby certify, that James M. Bloss is an Associate Judge of said Court in said County, and that within signature, purporting to be his, is genuine. In testimony whereof, I have hereunto affixed my seal of Office, and subscribed my name, this 13th day of December in the year of our Lord one thousand eight hundred and forty one.

A. C. Hubbard Clerk of the Court of Com. Pleas
Ashtabula County

I Harvey Guthrie of Conneaut Ashtabula County Ohio of lawful age being duly sworn do depose and say that I have been acquainted with Mrs. Rebecca Chapman for the Term of about Thirty years that she now lives at Conneaut Ashtabula County Ohio and that she is the widow of Uriah Chapman and is entitled to credit in whatever she may say and that she has during my acquaintance with her, sustained an unimpeachable character for truth and veracity

Sworn to and subscribed before me at Conneaut Ashtabula
County Ohio on this first day of December AD 1841
J M Bloss Associate Judge

Harvey Guthrie

The State of Ohio
Ashtabula County ss. I, A. C. Hubbard Clerk P. T. of the Court of Common Pleas within the said County of Ashtabula, certify that James M Bloss is an Associate Judge of said Court within said County, and that the within signature, purporting to be his, is genuine. In testimony whereof, I have hereunto affixed my seal of Office, and subscribed my name, this 13th day of December in the year of our Lord one thousand eight hundred and forty one.

A. C. Hubbard Clerk P. T. of the Court of Common Pleas
Ashtabula County

Sir

Washington City Jany 28, 1842

I herewith transmit the papers of Mrs Chapman for a Pension

With great respect your Obt Svt
J L Edwards
J R Giddings

State of New York

County of Wayne ss. Be it known, that on the 15th day of July, 1842, before me, the subscriber, a commissioner of deeds in and for said county personally appeared Phebe Van Auken, of the town of Sodus, in said County of Wayne, and made oath in due form of law, that she is well acquainted with Rebecca Chapman, widow of Uriah Chapman, that this deponent was present at the time of the marriage of the said Rebecca and the said Uriah, that they were married on the 13th day of August, in the year 1781, in the town of Goshen (now Minisink) in the county of Orange and State of New York, by Rev. Mr. Benedict, that this deponent has no record proof of the time and place of said marriage, nevertheless, she distinctly recollects the time and place as above stated.

Subscribed and sworn this 15th day of July in the year 1842, before me Alanson M Knapp Comr. of deeds of Wayne County

Phebe Vanauken

I hereby certify that I am well acquainted with the above named Phebe Van Auken, that she is a person of irreproachable character, and her statements entitled to full credit on oath or otherwise

Alanson M Knapp Comr. of deeds &c

State of New York

Wayne County Clerks Office I James Hawley clerk of said County do hereby Certify that Alanson M. Knapp, before whom the above affidavit was taken was at the taking of said affidavit a Commissioner of deeds in and for said County duly authorised to act as such Commissioner and his name as subscribed to the above Oath is genuine In testimony whereof I have hereunto set my hand and affixed the Seal of said County this 19th day of July A.D. 1842

James Hawley Clerk

Dr Sir

Jefferson August 31, 1842

The claim of Rebecca Chapman widow of Uriah Chapman a Revolutionary Soldier was forwarded ~~to you~~ Mr Giddings last spring with instructions to procure a pension for said widow. I suppose the papers are in your Office on file. We now forward you the testimony of her marriage I do not know how the matters stand but am informed that this testimony is all that is wanting. Please communicate to B. Carpenter Esq Conneaut Ohio

Yours,
A Bayley

State of Ohio
Ashtabula County ss. Jacob Schoonover of North Hampton Summit County and State of Ohio being duly sworn deposes and says that this Deponent was acquainted with Uriah Chapman who was reputed to have been a Soldier of the Revolution This Deponent was born in the town of Minisink Orange County and State of New York in which place he resided until he was eight years old when deponent removed to the town of Riga now Chili in the County of Monroe and State aforesaid that at that time Uriah Chapman resided in the town of Riga (now Chili) aforesaid and Deponent knew said Chapman from that time until his death both Uriah Chapman and this deponent continued to reside in the said town of Riga (now Chili) with the exception of about two years until said Chapmans death Deponent further says that from his earliest recollections of said Chapman he was reputed a soldier of the Revolution which this deponent has never heard questioned Deponent further says that he was ~~present~~ in the neighbourhood at the death of the said Uriah Chapman ~~which~~ and saw his body after death & was present at his funeral his death took place on the twenty fourth day of July A.D. eighteen hundred and twenty seven at Chili in the County of Monroe aforesaid Deponent saw the ball extracted from the body of said Chapman after his death and always understood that he had been wounded in the war of the Revolution Deponent has often seen the scar of the wound which was through the left arm and into the left side Deponent has been acquainted with Rebecca Chapman ever since he was eight years old and from his earliest knowledge of the said Rebecca and Uriah until the death of the said Uriah they have lived together and been reputed as husband & wife

Deponent knows the said Rebecca Chapman to be the same person who has made application for a pension as the widow of the said Uriah Chapman. This Deponent was forty five years old on the twenty sixth day of December last, and further Deponent saith not.

N.B. in the twenty third line of the first page the word "present" erased and the words "in the neighbourhood" interlined also in the twenty fith line first page the word "which" erased and the words "and saw his body after death and was present at his funeral his death" interlined all before subscribing ________ & making oath thereto.

Jacob Schoonover

The said Jacob Schoonover further on his said oath deposes and says that the said Rebecca Chapman has remained unmarried and the widow of said Uriah until the present time.

Sworn to and subscribed before me
this 17th day of October A.D. 1842

B Randall J. P.

The State of Ohio
Ashtabula County ss. I, A. C. Hubbard Clerk of the Court for the County and State aforesaid, do hereby certify that B. Randall Esquire is a Justice of the Peace within and for said County of Ashtabula duly commissioned and qualified according to the Laws of the State of Ohio, that his commission is dated December 24th A.D. 1840 and will expire on the twenty fourth day of December A.D. 1843 and that the signature above written purporting to be his, is genuine Given under my hand and the seal of said Court this 26th day of October A.D. 1842

A. C. Hubbard Clerk

State of Ohio
Ashtabula County ss. I, Olive Bratt of Lawful age first being duly sworn depose and say that I was living in the family of Uriah Chapman in the Town of Chili, Monroe County & State of New York at the time said Uriah Chapman died, and was present at the time of his death, which was on the twenty fourth day of July AD one thousand eight hundred and twenty seven said Chapman had been sick and confined to his house for seven or eight months prior to his death which sickness and consequent death was occasioned by a wound said to have been received by the said Chapman during the war of the Revolution while engaged as a soldier therein. I saw the bullet extracted from the side of the said Chapman the day after his decease by his attending Physicians Mr ______ & Mr Thompson, which bullet he had carried in his body from the time he received the wound until the day above alluded to, and this deponent further saith that she is acquainted with Rebecca Chapman who has made application for a pension and that the said Rebecca Chapman is the widow of the said Uriah Chapman and has lived in my family most of the time since the death of her husband Uriah Chapman and that the said Rebecca has remained up to the present time unmarried the widow of the said Uriah Chapman and further this deponent saith not

N.B. The word "unmarried" interlined

Sworn to and subscribed before me this 17th day of October AD 1842 B.

Randall J. P.

Olive Bratt

The State of Ohio
Ashtabula County ss. I, A. C. Hubbard Clerk of the court of Common Pleas in and for said county do hereby certify that B. Randall Esquire is a Justice of the Peace within and for said county of Ashtabula duly commissioned and qualified according to the Laws of the State of Ohio; that his commission is dated December 24th AD 1843 and will Expire on the 24th day of December AD 1843 and that the signature above or within written purporting to be his, is genuine. Given under my hand and the seal of said court this 26th day of October AD 1842.

A. C. Hubbard Clerk

State of Ohio
Ashtabula County ss. On this ninth day of January in the year of our Lord one thousand Eight hundred and forty three, personally appeared before me James M Bloss associate Judge of the Court of Common Pleas, within and for the county of Ashtabula and State of Ohio Rebecca Chapman a resident of Conneaut in the county of Ashtabula and State of Ohio aged seventy nine years who being first duly sworn according to Law, doth on her oath make the following declaration in order to obtain the benefit of the provision made by the act of Congress passed July 7th 1838 entitled "An act granting half pay and pensions to certain widows:" That she is the widow of Uriah Chapman who was a private in the army of the Revolution, and resided at the time he entered the service in the Township of Minisink, Orange County, & State of New York That he served in the army under General Washington three months in New Jersey in the winter of 1776'7 as a Substitute for his Brother Asa Chapman. That he afterwards in the Spring of 1777 enlisted into a company of Rangers under Captain Elijah Winter, marched to Lackawaxen in Pennsylvania and was in that vicinity for a considerable time while there was attacked by the Indians and wounded severely, was taken to Wallkill, Ulster County New York where he remained sick (in consequence of his wounds) for about twelve weeks That after his recovery he again enlisted as a ranger under Captain Abraham Westfall and served for five months in the year 1778, was in a battle with the Indians called the Delaware fight between the Delaware and Neversink Rivers That afterwards he served as a Ranger under the same Captain Westfall above named for the term of seven months The applicant

has not been able as yet to obtain any documentary evidence in support of her claim. Uriah Chapman the deceased soldier was not a pensioner, never having applied therefor. The applicant further declares that she was married to the said Uriah Chapman on the thirteenth day of August in the year seventeen hundred & eighty one at Minisink Orange County New York That her husband the aforesaid Uriah Chapman died on the Twenty fourth day of July Eighteen hundred & Twenty seven and that she was a widow on the fourth day of July 1836 and still remains the widow of the said Uriah Chapman. That she was not married to him prior to his leaving the service, but the marriage took place previous to the first of January seventeen hundred and ninety four viz at the time above stated.

Sworn to and subscribed on the day and year above written before me and I do further certify that the said Rebecca Chapman cannot from bodily infirmity attend the Court of Record as I am fully satisfied James M Bloss Associate Judge

Rebecca Chapman

The State of Ohio

Ashtabula County ss. I, A. C. Hubbard Clerk of the Court of Common Pleas within and for the County and State aforesaid do hereby certify, that James M. Bloss before whom the within Declaration was made, was, at the date thereof and now is, an Associate Judge of said court in said County of Ashtabula, duly commissioned and qualified agreeably to the Laws of the State of Ohio to whose official acts full faith and credit is and ought to be given, and that the signature to the within certificate purporting to be his, is genuine. In witness whereof I hereto set my name and the Seal of said Court at Jefferson in said County, this 14th day of January AD 1843.

A. C. Hubbard Clerk

David Whitney Certificate

I hereby certify that I am, and have been for a number of years past, personally acquainted with Phebe Van Ocken named in a deposition certifying to the marriage of Uriah & Rebecca Chapman, and from my personal knowledge believe her to be a person of good moral carachter and undoubted veracity

Dated at Willoughby this 3rd May 1843

David Whitney

We the undersigned have been acquainted with the above named David Whitney for a series of years and know him to be a man of good moral carachter and of unimpeachable veracity

Willoughby 3rd May 1843

S W Hall
O. H. Sharp
B Brainard
C. C. ________
N Clark
D Christy P M
___ Brainard ___ ___

Certificate of Service of Uriah Chapman

In pursuance of a law passed the 27th of April, 1784, entitled "An act for the settlement of the pay of the Levies and Militia, for their services in the late War, and for other purposes therein mentioned," the following certificates have been issued, bearing interest at five percent per annum, viz:

30th Nov. 1780 Uriah Chapman private
£17 - 4 - 10 pri 6 mo. 9 ds.

State of New-York,
Comptroller's Office, Albany, May 17th 1843.

I hereby certify the above to be true extracts from Book No. 2, now remaining in this office, and that the service for which the certificates above named were issued, were performed under the following officers, viz:

First - under Capt. Abraham Westfall in Col. Albert Pawling Regt.

Philip Phelps
Dep. Comptroller

Declaration of Rebecca Chapman
In order to obtain the Benefit of the act of Congress
of the 7th July 1838, entitled
"An act granting half pay and Pensions to certain Widows"

State of Ohio
Ashtabula County ss. On this 12th day of June AD 1843 personally appeared before me, (at Conneaut Ashtabula County Ohio) Rebecca Chapman a resident of Conneaut in the County of Ashtabula and State of Ohio aged Eighty years, who being first duly sworn according to Law doth on her oath make the following declaration in order to obtain the benefit of the provission made by the

act of Congress passed July 7th 1838 entitled "an act granting half pay and pensions to certain widows" That she is the widow of Uriah Chapman, who was a private in the army of the Revolution and has remained his widow since his death to the present time, that he served in the army under General Washington three months in New Jersey in the winter of 1776 '7 as a substitute for his Brother, Asa Chapman. That he afterwards in the Spring of 1777 enlisted into a company of Rangers at Minnisink, Orange County New York under Captain Elijah Winter, marched to Lackawaxen in Pennsylvania and was in that vicinity for a considerable time, while there was attacked by the indians and wounded severely by a ball passing through his left arm into his side, which remained there untill his death and ever rendered him unable to perform bodily labor as he was wont to do before he received the injury. He was taken to Walkill Ulster County New York where he remained sick in consequence of his wound for about twelve weeks. That after his recovery he again enlisted as a ranger under Captain Abraham Westfall and served for five months, was in a battle with the indians called the Delaware fight between the Delaware and Neversink rivers. That afterwards he served as a ranger under the same Captain Westfall for the Term of seven months. That said Uriah Chapman at the time he first entered the service of his country resided in the town ~~of~~ now called Minisink, Orange County, New York She further declares that she was married to the said Uriah Chapman on the 13th day of August AD 1781 by the Revd Mr Benedict in the town now called Minisink Orange County New York. That her husband, the aforesaid Uriah Chapman died on the 24th of July AD 1827 in the town of Chili Monroe County and State of New York. That she was not married to him prior to his leaving the service, but the marriage took place previous to the first of January AD 1794 viz. at the time above stated The said Uriah Chapman was not a pensioner, never having applied _________

Rebecca Chapman

Sworn to, and subscribed before me on the day and year above written. And I do hereby further certify that the above named Rebecca Chapman the signer of the foregoing declaration is unable, from bodily infirmity to appear before the Court

Conneaut June 12th AD 1843.

Jas M Bloss Associate Judge

The State of Ohio

Ashtabula County ss. I, A. C. Hubbard, Clerk of the Court of Common Pleas within the County and State aforesaid, do hereby certify that James M. Bloss is, and at the time of subscribing the annexed certificate, was an Associate Judge of said Court in said County of Ashtabula, duly commissioned and qualified agreeably to the Laws of the State of Ohio; and that his signature to said certificate is genuine. In witness whereof I hereto set my hand and the seal of said Court at Jefferson this 14th day of June, AD 1843.

A. C. Hubbard Clerk

Conneaut June 12th 1843

J. L. Edwards Esq.
Commissioner of Pensions
Sir

Enclosed you will find another declaration made, by Mrs Rebecca Chapman in order to obtain the benefit of the act of Congress of the 7th July 1838, entitled "An act granting half pay and Pensions to certain widows" we hope the difficulties pointed out by you in the former declaration, are herein obviated & all the requirements of the Law fulfilled. In addition to the declaration of the claimant, you will find the Certificate of Philip Phelps Deputy Comptroller at Albany New York of the service of Uriah Chapman deceased, husband of the claimant, which fully establish claim as far as the service is concerned. We also send the Certificate of David Whitney who certifies as to the character of Phoebe Van Ocker the person whose deposition you have on file proving the marriage of the claimant to Uriah Chapman & also the certificate of sundry gentlemen of unquestionable veracity as to the veracity of the said Whitney. You will also find on file the deposition of Olive Bratt and Jacob Schoonover heretofore forwarded to your office, which fully show the death of the said Uriah Chapman, together with the additional fact that the Ball which entered his arm & side while engaged in the service of his country, and which disabled him through life, was taken from the vicinity of the spine after his death. You also have the deposition of Isaac Van Gorden who was himself a revolutionary soldier who certifies that he saw the said Uriah Chapman while engaged in the war of the revolution & saw him brought into the fort after he received the wound alluded to in the claimants declaration & that he knew him many years afterward & that he was in truth the same Uriah Chapman whose widow the claimant represents herself to be. The Town or family Record is not in existence which shows the marriage of the parties & we have proved the fact by living witnesses we have also forwarded the depositions of James & Harvey Guthrie, gentlemen who reside in this Borough who ____ unimpeachable character for truth & veracity who were acquainted both with Uriah Chapman & his widow the claimant & they fully from the character for truth & veracity of Uriah Chapman & the aplicant ______________ I have as briefly as possible recapitulated the testimony in the case, in order to direct your attention to the proofs and if after all you should still think of us as was thought of the young men in the gospel, that we lack something yet, it will be somewhat difficult to keep up the courage of the old Lady whose claim has so often been postponed however I hope that as the case now stands, it will be fully satisfactory without any further testimony and that you will make the neccessary order in her case, that she may be enabled to receive the small pittance that is due her and thereby relieve the pressing wants of the worthy widow of a soldier of the Revolution, please write immediately.

Yours &c
B. Carpenter Atty for claimant

P.S. There is no further testimony that can be reached by the applicant, & if it you deem a part of the claim of service is not sufficiently proven you can disregard so much, and please send what she is entitled to as proved by the Comptrollers certificate.

Yours &c
B. Carpenter

State of New York
Munre County ss. Ebenezar Skinner this day personally appeared before me being duly sworn saith that he is now seventy six years old and that he was present and saw Uriah Chapman and Rebecca St John Lawfully married by the Rev Mr Benedict in the town of Minasink in the County of Orange State of New York in the month of August in the year seventeen hundred and eighty one also that the above named Uriah and Rebecca have since lived together as husband and wife and that the said Uriah died in the town of Chile County of Monroe State of New York sixteen years ago this summer and the said Rebecca is now the widow of the late Uriah Chapman and further this deponant saith that he knows that Pheba St John and now Pheba Vanocker is a sister of the above named Rebecca and was presant at the time of the marriage of the above named Uriah and Rebecca and that the said Pheba is a woman of good moral character

Sworn and subscribed before me this the 31st day of August 1843
Elisha Harmon J. P.

Ebenezer Skinner

I do certify the above named Ebenezar Skinner is to me personally known and that he is a man of good moral character

Elisha Harmon J P

State of New York

Monroe County ss. I certify that Elisha Harmon Esqr is a Justice of the peace in and for said county, and that I am well acquainted with his hand writing and verily believe that the above signature purporting to be his, is genuine. Witness my hand & seal of Office this first day of September 1843.

James W. Smith Clerk

★

The State of Ohio

Jefferson Township Sept 4, 1843

Ashtabula County ss. I Daniel St John Chapman being duly sworn say that I am a son of Uriah Chapman and Rebecca Chapman and that I am forty seven years of age according to the best dates which I possess. That I have from my earliest recollection lived with my parents until the decease of my father in July AD 1827, and since that period my mother has mostly resided in my family. I further say that I always understood that my parents were married in the State of New York in the Town of Minisink, and that no public records are kept of marriages in that State to my knowledge nor have I at any time heard of such record or documentary evidence of the marriage of my parents I further say that I ~~am~~ have never seen nor heard of any family record of said marriage between my parents, nor was there any record of births deaths or marriages kept in my fathers family since my recollection ~~And further saith not~~ I further know that my parents lived and cohabited as man and wife and were reputed as such since my earliest recollection. And further saith not.

Daniel St John Chapman

★

I Brewster Randal a justice of the Peace for the county aforesaid certify that I am personally acquainted with Daniel St John Chapman and know him to be a man of good moral character and that full faith and credit is and ought to be given to his testimony, and that he personally appeared before me on the the day and year aforesaid and made solem oath to the truth of his foregoing affidavit

B. Randall J. P.

Pleas direct to B. Carpenter Esq Conneaut, O

★

The State of Ohio

Ashtabula County ss. I, A. C. Hubbard, Clerk of the Court of Common Pleas within the County and State aforesaid do hereby certify that Brewster Randal Esquire, is a Justice of the Peace within the County of Ashtabula aforesaid, duly commissioned and qualified ~~and that~~ agreeably to the Laws of the State of Ohio; and that his signature to the within certificate is genuine. Given under my hand and the seal of said Court at Jefferson in said County this fourth day of September AD 1843.

A C Hubbard Clerk

Ohio

Rebecca Chapman widow of Uriah Chapman who was a pensioner under the Act of   and who died on the 24th July 1827 of Ashtabula Co in the State of Ohio who was a Private in the company commanded by Captain A Westfall of the Regiment commanded by Col. Pawling in the New York line for 6 mo 8 days Inscribed on the Roll of Pittsburgh Pa at the rate of Twenty Dollars Eighty Eight Cents per annum, to commence on the 4th day of March, 1836. Certificate of Pension issued the second day of November 1843 and sent to B. Carpenter Conneaut, Ohio Whole amount due $104.40

Recorded by Geo C Ames Clerk,
[Act July 7, 1838.]
Book D   Vol. 1   Page 385

J L Edwards Esqr.
Sir

You have hereto attached the declaration of Mrs Rebecca Chapman widow of Uriah Chapman in order that she may avail herself of the pension allowed by act of Congress of 3rd of March 1843 to certain widows who have received pensions under the act of July 7th 1838. You will please forward the necessary papers to me, directed "Conneaut Ashtabula County Ohio"

B. Carpenter Atty for Rebecca Chapman

Declaration of Mrs Rebecca Chapman

"In order to obtain the benefits of the Act of Congress of the 3rd March 1843 granting pensions for one year to certain widows who have received pensions under the act of July 7th 1838"

State of Ohio
Ashtabula County ss. On this Twenty first day of February AD one thou-

sand eight hundred and forty four personally appeared before me George Morton, Justice of the Peace in and for the county aforesaid, Rebecca Chapman, a resident of Conneaut in the County of Ashtabula and State of Ohio aged Eighty one years who being first duly sworn, according to Law, doth on her oath, make the following declaration in order to obtain the benefits of the provission made by the act of Congress, passed on the 3rd of March 1843, granting pensions to widows of persons who served during the Revolutionary war. That she is the widow of Uriah Chapman who was a private in the war of the Revolution and served under Captains Westfall & Elijah Winters. That her name is inscribed on the pension list at the rate of Twenty Dollars and Eighty Eight cents per annum commencing on the 4th of March 1836 & terminating on the 4th of March 1841 which amount she has received. She further declares that she is still a widow.

Rebecca Chapman

To the truth of the within statement of Mrs Rebecca Chapman I am fully satisfied Sworn to and subscribed before me this 21st day of February 1844 & I certify that the said Rebecca Chapman is to me personally known

Witness

Geo Morton Justice of the Peace for Conneaut township
Ashtabula County, Ohio

Margaret Chapman
B. Carpenter

Michigan

Rebecca Chapman widow of Uriah Chapman who was a Private in the N. York Line Inscribed on the Roll at the rate of 20 Dollars 88 Cents per annum, to commence on the 4th day of March, 1843. Certificate of Pension issued the 21st day of March 1844 and sent to B Carpenter Conneaut Ohio

Recorded in Book A Vol. 1 Page 157

[Act of March 3, 1843.]

Oakland, Oakland Co. Mich.
Jan. 6, 1846

Honored Sir -

You will by your records I suppose see that the subscribers name is enrolled on your pension list, while living in Ohio with my son Daniel Chapman in Conneaut, Ashatbula Co. this the special agency of Esq Carpenter (I think Benjn.) of the same place, I drew as my memory serves (as I have no written minutes with me) as follows - in Sept. 1843 - $104.24 - in May 1844 - $20.00, and none since, being told the Law applicable to my case was suspended in its opperation for 4 yrs. I removed to Oakland Oakland Co. Mich. in Oct. 1844, where I now live with my son in law Joshua Howell, being informed that others in similar continued to draw, I applied to E. P. Hastings Esqr. at Detroit pension agent for Mich. who told me (by letter) I must apply to the department at Washington and have the agency transfered from O. to Mich. which I wish to have done immediately he requested me to inform him under what Law I drew as this is not at present in my power, you would confer a favor on me, when you transfer the agency to him you would tell him under what Law. Esqr. Hastings told me if I had a Certificate it must be sent to him, but I have none, that and all other papers on the subject was left in O. or sent to Pittsburgh Pa. I was told on leaving O. there was no writings there that would be of use to me in Mich. Be so kind dear Sir as to inform me what I must do in this case to obtain a letter (be it more or less) to relieve my pressing wants, that I may go down to my grave in my old age, with my heart full of gratitude to my country.

respectfully yours,
Rebecca Chapman

Witness J. Howell I took hold of her hand in writing the last part of her name

---

Oakland Aug 3rd 1846

Sir,

I herewith enclose the necessary papers according to your instructions for the guidance of Rebecca Chapman in obtaining a transfer from the Ohio Pittsburgh Agency to that of Michigan.

I am respectfully Your Obt Servant
Joshua Howell

P.S. The instructions were directed to Daniel Chapman son of the Widow, but ~~directed~~ to my place of residence and I am doing the the business for her. Chapman still lives in Ohio.

J. H.

Application for a transfer

County of Oakland ss. On this 7th day of July 1846, before me, the subscriber, a Justice of the Peace for the said County on her Oath declares, that she is the widow of Uriah Chapman the same person who formerly belonged to the Company commanded by Capt A. Westfall in the Regt. in which Meeker was

Major, in the service of the U.S. that her name was placed on the pension Roll of the State of Ohio, from whence she has lately removed; that she now resides in the State of Michigan where she intends to remain, and wishes her pension to be there payable in future. The following are her reasons for removing from Coneaut, Ashtabula Co. to Oakland, Oakland Co. Mich. I had two daughters & one son in Ohio. I had lived with them alternately one daughter is now dead, the family broke up, the other daughter & family now in Mich. I had one son and one daughter with families, in Mich. before I came here expecting to spend the remainder of my days with my children here, who live all in the same Town. I am now living with my son in law Joshua Howell

Sworn and subscribed to before me this 7th Day of July A.D. 1846
Jesse Decker Justice of the Peace

Rebecca Chapman

State of Michigan

County of Oakland ss. On this 7th day of July 1846 before me the Subscriber a Justice of the Peace in and for said County personally appeared Joshua Howell who on his oath declares that Rebecca Chapman the individual who signed the foregoing affidavit is the Individual named in an original certificate of Pensions, which is not now in her possession from the War Department, and who drew a pension in the State of Ohio as stated in her said affidavit

Sworn and subscribed this 7th day of July 1846 Before me Jesse Decker Justice of the Peace

Joshua Howell
(Widow to sign here)
Rebeca Chapman

State of Michigan

County of Oakland ss. I do certify that Joshua Howell the witness aforesaid is a person of truth and veracity

Dated July 7th A.D. 1846

Jesse Decker Justice of the Peace

State of Michigan

County of Oakland ss. I Joseph R __________ Clerk of the County of Oakland aforesaid do hereby certify that Jesse Decker Esqr. was on the seventh day of July AD 1846 a Justice of the Peace within & for said County duly Elected & sworn & that his term of Office commenced on the 4th day of July AD 1846 & will Expire on the 4th day of July AD 1850, & that I am well acquainted with his hand writing & that his signature subscribed to the within papers are his own proper hand writing Given under my hand & the Seal of the Circuit Court for said county (my Seal of Office) at Pontiac this 15th day of July AD 1846

Joseph R _______ Clerk

J L Edwards Esqr.
Commissioner of Pensions

Pension Agency
Pittsburgh Pa Augt 15 1846

Dear Sir

Rebecca Chapman pensioner under Acts of 3rd March 1843 &c at 20.88 dollars per annum was last paid at this Agency up to 4th March 1844;

As directed in yours of the 12th inst. I have marked her name transfered to the Michigan Agency

I am most respectfully Your Obt Svt
Thos M Howe
Pension Agent
By __________

J. L. Edwards Esq.
Commissr. of Pensions

Pension Agency
Detroit Apl. 18 1849

Sir,

I enclose the application of Rebecca Chapman for the benefits of the provision of the act of Congress passed on the 2d day of Feby 1848

I am Sir respectfully your obedient servant
E P Hastings Pension Agent

State of Michigan

County of Oakland ss. On this fifth day of April 1849 personally appeared before me Jesse Decker a Justice of the Peace in and for said County of Oakland Rebecca Chapman a resident of Orion in the County of Oakland aged Eighty five years, who being first duly sworn, according to law, doth on her oath, make the following declaration, in order to obtain the benefits of the provision made by the act of Congress, passed on the 2d February 1848, granting pensions to widows of persons who served during the Revolutionary War. That she is the widow of Uriah Chapman who was a Private and served under Capt Westfall of the Rangers in the Regiment commanded by Col Tuston, that she does not know the name or number of the Regiment and that she is the same Individual who drew a pension under the Act of June 17, 1844 at the rate of Twenty dollars and Eighty eight cents pr. annum. She further declares that she is still a widow

Sworn to and subscribed on the day and year above written Before me Jesse Decker Justice of the Peace

Rebecca Chapman

Witnesses Wm Chapman & Martin __________

State of Michigan

Oakland County ss. I Jesse Decker a Justice of the Peace of oakland county Do Certify that the said Rebeca Chapman was not able to sign her name without assistance She being So Infearm _____ with old age that She held the Pen in her hand and her Son Wm. Chapman assisted her in Writing her name She being so fible not able to Write

Dated orion April 5th 1849

Jesse Decker Justice of the Peace

Michigan

Rebecca Chapman widow of Uriah Chapman NY who served in the Revolutionary war, as a private Inscribed on the Roll at the rate of 20 Dollars 88 Cents per annum, to commence on the 4th day of March, 1848. Certificate of Pension issued the 26 day of Apl 1849 and sent to E. P. Hastings Detroit Mich. Recorded on Roll of Pensioners under act February 2, 1848, Page 364 Vol. 2

Attorneys-at-Law,
Bradford, Pa.
W. B. Chapman & Son,

Will B. Chapman,
Notary Public.
Jany 6, 1892

Commissioner of Pensions
Washgtn DC
Dear Sir

I am complig a breaf Histry of my Fathr Family. My Grand Fathr Uriah Chapman was in the ware of the Reveuletin and his widw Rebea Chapman abot 1840 obtain'd a pension. If you wud furnish me with a copy of his aplication I coud obtain facts fer my History, that is imposobul to obtain in any oth wey Hoping that I am not asking too much

W B Chapman

Department of the Interior,
Bureau of Pensions,
Washington, D.C.

April 28, 1892

Sir:

Replying to your request for a statement of the military history of Uriah Chapman, a soldier of the Revolutionary war, you are informed that his widow, Rebecca, states in her application for pension on file in this Bureau that his first service was for 3 months with the New Jersey Troops during the Winter of 1776-77, but does not give the names of his officers. In the Spring of 1777, he enlisted with a company of Rangers raised in Orange County, N.Y., commanded by Capt. Elijah Winter, and marched to Pennsylvania.

During this tour he was in an engagement with the Indians and was shot in the left arm, the ball passing through the arm into his left side. After his recovery he again enlisted in a company of Rangers commanded by Capt. Abraham Westfield, and served for about one year.

The widow also states that she was married to the soldier August 13, 1781, and that he died July 24, 1827. Her application for pension was allowed.

Very respectfully,
Commissioner.
W. B. Chapman, Esq.,
Bradford, Pa.

Widow, &c. File No. 6,653
Rebecca Chapman Widow
Uriah Chapman Pvt. Rev. War

Act: Feby. 2 1848
Index:- Vol. A, Page 115
[Arrangement of 1870.]
Revolutionary War Pension and Bounty Land Warrant Application Files
Series M-804; Roll # 522

William Curry
New York (S.49,282)

BRIEF in the case of William Curry of Elmira in the State of New York (Act 7th June, 1832.)

1. Was the declaration made before a Court or Judge? in open Court

2. If before a Judge, does it appear that the applicant is disabled by bodily infirmity?

3. How old is he? Eighty four years

4. State his service, as directed in the form annexed.

Period.

Duration of Service.

Rank.

Names of General and Field Officers under whom he served.

Years Months Days

States that he entered the service in May 1775 as a Volunteer in a company of Militia under Capt. Wisner, in Colo. Hathorns Regt. and continued to serve in said compy until the close of the war Periods of service not defined, but alleges that he was in active service more than two years

5. In what battles was he engaged? none named

6. Where did he reside when he entered the service? County of Orange - N. York

7. Is his statement supported by living witnesses, by documentary proof, by traditionary evidence, by incidental evidence, or by the rolls? by Traditionary and living testimony

8. Are the papers defective as to form or authentication? and if so, in what respect? the papers are defective in authentication

Certify that the foregoing statement and the answers agree with the evidence in the case above mentioned.

Wm. Gordon Examining Clerk.

★

At a Court of Common pleas held at the Court house in the village of Elmira in and for the County of Tioga in the State of New York commencing on the first Tuesday in September 1832 the same being a court of Record so declared by the Constitution and laws of the said State by which the said Court is constituted, & having a Seal & Clerk

Present. Hon G. B. Baldwin first Judge
Darius Bentley
John R. Drake Esqrs Judges
Elijah Shoemaker
Joseph L Darling

State of New York

Tioga County ss. Personally appeared in open court on the 4th day of September 1832 before the above named Judges now sitting William Curry, aged eighty four years & upwards a resident of the town of Elmira in the said County of Tioga who being duly sworn according to law, doth on his oath make the following declaration in order to obtain the benefit of the act of Congress passed June 7. 1832.

That during the war of the Revolution the deponent resided in the County of Orange in the State of New York, in which there were a great many tories and persons attached to the British government, and that the Inhabitants of the said County were much annoyed during the greater part of the contest by attacks from Tories & occasional detachments of British Troops & refugees from the City of New York where a considerable British force continued during the whole of the war. The inroads of the Indians along the Minisink frontier on the western part of the said County were also very frequent, and companies of militia were consequently required to be constantly in motion, for the defence of the Inhabitants at the various points where they were the most exposed. That the deponent very early in the commencement of the war about the month of May 1775 entered as a volunteer in a company of militia in the town of Warwick then commanded by Capt. Henry Wisner in the Regiment commanded by Col. John Hathorn: Thomas Goldsmith was the Lieutenant and Joseph Todd Ensign; that the said deponent continued to serve in the said company until the close of the war. During his engagement in the said company, the Captain (Wisner) was promoted, and Peter Bertholf was appointed Captain and the Lieut. Goldsmith resigned and Henry Bertholf was appointed in his place. That the said company was engaged in active service, in various parts of the County of Orange, Rockland &c. as occasion required. The particular periods in which they were employed at various positions he can not now state, but recollects that while in the service, he was on one occasion stationed at Ramapo where there were barracks & public stores for a month or more, once at Kakiat about two weeks, at Paramus for a month or more, four or five weeks in the English neighborhood near Tappan, four or five weeks at Suffern's in Rockland County, and at various other places - in the mountains & along the North River in the Tory neighbourhoods, and frequently along the Minisink frontier, when there were Indian alarms.

He was also out when the Minisink Indian battle was fought in August 1777, when Fort Montgomery was taken he was at Ramapo with other troops guarding stores which were kept there. He was not engaged in the Minisink battle having been stationed about five miles distant from the scene, where a small company were kept to take charge of the horses which were left there. He saw Major Meeker the day after, who recieved a wound in the Battle, & was in command of some Jersey militia who were in the engagement.

That his services during the war would amount to more than two years constant service.

In answer to such of the Interrogatories propounded by the War Department as are not answered in the above relation he states that he was born on Middle Patent in the County of Westchester NY on the 13th July (O.S.) 1748. that he removed to the County of Orange town of Warwick about 1773, continued there until 1788 when he removed to Cherry Valley for a year, thence to Tioga Co. NY where he resided two years or nearly, & thence to Ulster in the County of Bradford Penna. where he has resided for more than 30 years. For a few years past he has had no fixed place of residence having children residing in both New York & Pennsylvania, and has spent his time among them as inclination dictated. That he has no documentary evidence of his services, and can prove them by Joseph McConnel and Alexander Miller of the said Town of Elmira. to whom & the Revd. Simeon R. Jones of Southport in the said County of Tioga he would refer. That the Continental officers whom he knew were Genl James Clinton & Col. Livingston & the militia officers with whom he served (other than his own) were Gen Geo Clinton, Cols Seward & Tustin of Orange County.

He hereby relinquishes every claim whatever to a pension or annuity except the present & declares that his name is not on the Pension Roll of the agency of any State.

Sworn & subscribed the day & year aforesaid Green M Tuthill Clk

William Curry

On the 4th day of September 1832 personally appeared in open court before the above named Judges now sitting Joseph McConnel, a resident of the Town of Elmira in the said County of Tioga, aged sixty five years and upwards, who being duly sworn according to law doth on his oath declare, that he is well acquainted with William Curry who has subscribed & sworn to the foregoing declaration; that he knew him during the Revolutionary war, as the deponent's father lived a near neighbour to the said Curry in the town of Warwick in the said County of Orange, & continued so to reside there until the close of the war. That said Curry belonged to Capt. Henry Wisner's company of militia in Col John Hathorn's Regiment in the said County of Orange; after Capt. Wisner was promoted, Peter Bertholf was appointed Captain of the said Company: the deponent well recollects that the said Curry was frequently out in service as a private in the company commanded by the said Captains Wisner & Bertholf in Hathorn's Regiment at different periods during the Revolution, as calls for troops in various places in the said County of Orange & the adjoining country were of almost continual occurrence during the greater part of the war. The deponent has no doubt and verily believes that the said William Curry served during the said war a term not less than two years. The deponent was about nine years of age at the commencement of the Revolutionary Struggle & about sixteen at its close, & his recollection of the material events which transpired during the time is clear and distinct

Subscribed & sworn the day & year aforesaid Joseph McConnel
Green M Tuthill Clk

We Simeon R Jones, a clergyman residing in the Town of Southport, in the said County of Tioga & Lathrop Baldwin residing in the Town of Elmira in the said county do hereby certify that we are well acquainted with William Curry who has subscribed & sworn to the foregoing declaration: that we believe him to be of the age stated in his declaration: that he is reputed & believed in the neighborhood of his residence to have been a soldier of the Revolution, and that we concur in that opinion

Sworn & subscribed the day & year aforesaid In open court
Green M Tuthill Clk

Simeon R. Jones
Lathrop Baldwin

And the said court do hereby declare their opinion after the investigation of the matter and after putting the interrogatories prescribed by the War Department that the above named applicant was a Revolutionary soldier and served as he states:

That it appears to them that Joseph McConnel who has made & subscribed the foregoing affidavit in confirmation of the said applicants declaration is a credible & disinterested witness, in whose statement the said court place implicit reliance

That Simeon R Jones, who has subscribed the foregoing certificate is a clergyman resident in the Town of Southport in the said County of Tioga & that Lathrop Baldwin, who has also signed the same is a resident of Elmira in the said county & that they are credible persons & their statements are entitled to credit

G. B. Baldwin First Judge

I Green M. Tuthill, Clerk of the said Court do hereby certify that the foregoing contains the original proceedings of the said Court in the matter of the application of William Curry for a pension In Testimony whereof I have hereto set my hand & affixed my Seal of Office the 4th day of Septemr. 1832

Ribbon attached 19 Dec 1832 Green M Tuthill

Green M Tuthill Clk

State of New York

Tioga County ss. William Curry being duly Sworn before the undersigned a Justice of the peace in & for said county deposeth & saith That in consequence of old age & consequent loss of memory he cannot state the precise length of his Service in the several years of the Revolutionary war but according to the best of his recollection & after mature reflection he states that he served not less than the periods and in the grades & years mentioned below

In 1775 in the company commanded by Capt. Henry Wisner, in Col John Hathorn's Regt. Orange Co. (N.Y.) militia as a private Three months 3.00

In 1776 as a private in the same company 6 ½ months 6 ½

In 1777 as a private in the same company then commanded by Capt. Bertholf 4 ½ months 4 ½

In 1778 as a private in the same company 3 months 3

In 1779 as a private in the same company (Bertholfs) 2 months 2

In 1780 as a private in the same company 2 months 2

In 1781 was out but a short period, as a private in same comp 2 weeks ½

In 1782 in the same company including 4 days pressed with his Team to carry provisions to New Windsor 2 weeks ½

22 mo

making all his Services as a private Twenty two months as above detailed in addition to his declaration, for which he claims a Pension

Sworn & Subscribed before me this7th day of Jany 1833 W. Maxwell Justice of the peace Tioga County N.Y.

William Curry

State of New York

Tioga County ss. I Green M. Tuthill Clerk of the said county do hereby certify that William Maxwell Esquire before whom the within affidavit of William Curry was made was when the same was taken a Justice of the peace in & for said county duly sworn & qualified; and that from long acquaintance with the signature of the said Justice I verily believe his signature subscribed to the said affidavit is genuine. Witness my hand & seal of office the 7th day of January 1833

Green M. Tuthill

BRIEF objections to the admission of Pension Claims, of which those answered in the negative, in red ink, will apply to the claim of William Curry which require further proof or explanation.

I. Does the declaration show where the applicant resides? yes

II. Did the declarant make the declaration in the county where he resides? yes

III. If the applicant did not make the declaration in the county where he resides, is there any reason assigned for his not doing so?

IV. Has the declarant mentioned the period or periods of the war when he served? no

V. Has he mentioned with precision the length of his service, and the different grades in which he served, in language so definite as to enable the department to determine to what amount of pension he is entitled? no. a detailed account of each tour is required - an allegation of agragate service unstisfactory.

VI. Has he given the names of the officers under whom he served, in conformity with the regulations? yes

VII. Has he made a relinquishment of every claim to any other pension than the one under the Act of June 7, 1832? yes.

VIII. Has the Court given their opinion? yes

IX. Has the Clerk given his certificate? yes

X. Is the Clerk's seal affixed? And if so, has it a device or inscription by which it can be distinguished from any other seal? yes

XI. Has the applicant obtained the evidence of a clergyman and another respectable citizen as to their belief respecting his age and the general belief in his neighborhood relative to his revolutionary service? yes

XII. If the clergyman's affidavit has not been produced, is there any reason assigned for not obtaining it?

XIII. If the applicant has no documentary evidence, and has not obtained the testimony of at least one living witness, has he stated in his declaration that such proof cannot be had? States that he has no documentary evidence - but has adduced the testimony of one living witness

XIV. If the applicant has produced a witness or witnesses as to his service, has the magistrate who administered the oath certified that he, she, or they are persons of credibility? the court has And is the official character and signature of the magistrate certified by the proper officer under his seal of office?

XV. Does it appear from any of the papers that the witness was in such a situation, or of such an age, as to have a personal knowledge of the applicant's service? yes

XVI. Are the papers authenticated as the regulations direct? no

XVII. Are the seven interrogatories prescribed by the War Department answered satisfactorily? And, if not, which of them is not so answered? the court certify that the interrogatories were put to the applicant

EXPLANATORY NOTES.

QUERE V. In a case where the applicant cannot, by reason of the loss of memory, state precisely how long he served, he should amend his declaration by making an affidavit in the following words: "Personally appeared before me, the undersigned, a Justice of the Peace, &c. A. B. who, being duly sworn, deposeth and saith, that, by reason of old age, and the consequent loss of memory, he cannot swear positively as to the precise length of his service, but according to the best of his recollection he served not less than the periods mentioned below, and in the following grades: For year months and days, I served as a . For months and days, I served as a ; and for such service I claim a pension."

It is important in all cases to determine with precision the period for which the applicant served, and the particular rank he held, as the law directs the pension to be paid according to the grade of the pensioner and the length of his service. The use of the phrase about three or four months is too indefinite, and all such qualifying expressions are objectionable.

QUERE XIV. Proof of Service.

In a case where the name of the applicant is not found on the records of the Department, he must prove his service by two credible witnesses, who are required to set forth in their affidavits the time of the claimants entering the service, and the time and manner of his leaving the same, as well as the regiment, company, and line to which he belonged. The magistrate who may administer the oaths must certify to the credibility of the witnesses, and the official character and signature of the magistrate must be certified by the proper officer, under his seal of office.

QUERE XVI. Mode of authenticating papers.

In every instance where the certificate of the certifying officer who authenticated the papers is not written on the same sheet of paper which contains the affidavit, or other paper authenticated, the certificate must be attached thereto by a piece of tape or narrow ribbon, the end of which must pass under the seal of office of the certifying officer, so as to prevent any paper from being improperly attached to the certificate.

New York

William Curry of Elmira in the State of New York who was a Private in the company commanded by Captain Wisner of the Regt commanded by Col Hathorn in the New York line for 21 months and 24 days Inscribed on the Roll of New York at the rate of 72 Dollars 61 Cents per annum, to commence on the 4th day of March, 1831. Certificate of Pension issued the 25th day of January 1833 and sent to Ths Maxwell Elmira NY

Arrears to 4th of Sept 1832	$108.91 ½
Semi-anl. allowance ending 4 March '33	36.30 ½
	$145.22

Recorded by Jno Cromwell Clerk,

{Revolutionary Claim, Act June 7, 1832.}

Book D Vol. 5 Page 158

Newark Valley, N.Y.,
Dec. 15, 1916.

To the Pension Department,
Washington, D.C.

Will you kindly send me such information as you can regarding William Curry, a pensioner of the Revolutionary War. He was in the battles of Minisink, Cowpens, Guilford Court House, Hobkirk's Hill and Eutaw Springs Such information as you can give, or confirmation, will be appreciated, as I hope to use it in becoming a member of the D.A.R.

Yours very truly,
Mrs. C. P. Lyman

September 14, 1930.
BA-J/ady
William Curry
S.49282

Mrs. S. K. Frost
76 Remsen Street
Brooklyn, New York
Dear Madam:

Reference is made to your letter relative to William Curry, a soldier of the Revolutionary War.

The data contained herein were obtained from the papers on file in the Revolutionary War claim, S.49282, based upon the military service of William Curry in that war.

William Curry was born July 13 (old style), 1748, in Middle Patent, Westchester County, New York. The names of his parents are not shown. About 1773 he moved to Warwick, Orange Couny, New York.

While a resident of said Warwick, he enlisted about May 1775 and served three months as private in Captain Henry Wisner's Company, Colonel John Hathorn's New York Regiment. He enlisted in 1776 and served six and one half months as private in Captain Henry Wisner's New York Company. He enlisted in 1777 and served four and one half months as private in Captain Henry Wisner's New York Company. He rendered five subsequent tours of service as private in Captain Peter Bertholf's New York Company; three months in 1778, two months in 1779, two months in 1780, two weeks in 1781, and two weeks in 1782.

In 1788 he moved from Warwick, New York, to Cherry Valley, one year later he moved to Tioga County, New York, and nearly two years later moved to Ulster, Bradford County, Pennsylvania, where he resided more than thirty years.

He was allowed pension on his application executed September 4, 1832, while a resident of Elmira, Tioga County, New York. He died May 6, 1844, place not stated.

In 1832 soldier had children living in New York and Pennsylvania, no names are stated. The name of soldier's wife is not given and there are no further family data.

This is the record of the only William Curry, all spellings of the surname searched, that is found on the Revolutionary War records of this office.

No charge is made for the search of the records and this report.

Very truly yours
A. D. Hiller
Executive Assistant to the Administrator

La Junta Colo.
2/23/35

Mr A. D. Hiller
Asst Administrator
Veterans Administration
Washington D.C.
Dear Sir

Will you kindly give me the military history and service of William Curry a Revolutionary Soldeer in State of N.Y. Also a pensioner He was born 6/29 1748 in Newburgh Orange Co N.Y. died 5/6 1844 in Lodi Seneca Co N.Y. He made his home with his daughter Elizabeth, wife of Aaron Miller. After the death of his wife Charity Lockwood who was born 3/12 1747 in Orange Co NY, & died 10/12 1832 in Athens Pa

Very truly yours
(Mrs H. R.) Sarah Mercer Carpenter
Route #1 Box 72
La Junta Colo

Revolutionary War Records Section,
Department of the Interior,
Bureau of Pensions.
Washington, D.C., , 19

In reply to your request of , received for a statement of the military history of William Curry a soldier of the Revolutionary War, you will find below the desired information as contained in his (or his widow's) application for pension on file in this Bureau.

DATES OF ENLISTMENT OR APPOINTMENT	LENGTH OF SERVICE	RANK.	OFFICERS UNDER WHOM SERVICE WAS RENDERED.		STATE
			CAPTAIN	COLONEL	
1775	3 mos.	Pvt	Henry Wisner	John Hathorn	New York
1776	6 ½ mos.	Pvt			
1777	4 ½ mos.	Pvt			
1778	3 mos.	Pvt	Peter Bertholf		
1779	2 mos.	Pvt			

Dates of Enlistment or Appointment	Length of Service	Rank.	Officers under whom Service was Rendered.		State
1780	2 mos.	Pvt			
1781	2 weeks	Pvt			
1782	2 weeks	Pvt			

Battles engaged in, He stated he was not in Minisink battle having been stationed about five miles distant where a small company were taking charge of horses

Residence of soldier at enlistment, Warwick, Orange Co. N.Y.

Date of application for pension, Sept 4, 1832 His cl. was al.

Residence at date of application, Elmira, Tioga Co. N.Y.

Age at date of application, born July 13 (old style) 1748 at Middle Patent Westchester Co. N.Y. died May 6, 1844

Remarks: In 1832 it was stated that he had children residing in New York and Penn no names are stated

---

April 3, 1935.
BA-J/ady
William Curry
S.49282

Mrs. H. R. Carpenter
Route 1, Box 72
La Junta, Colorado

Dear Madam:

Reference is made to your letter relative to William Curry, a soldier of the Revolutionary War.

The data contained in herein are obtained from the papers on file in the Revolutionary War claim, S.49282, based upon the military service of William Curry in that war.

William Curry was born July 13 (old style), 1748, in Middle Patent, Westchester County, New York. The names of his parents are not shown. About 1773 he moved to Warwick, Orange County, New York.

While a resident of said Warwick, he enlisted about May 1775 and served three months as private in Captain Henry Wisner's Company, Colonel John Hathorn's New York Regiment. He enlisted in 1776 and served six and one half months as private in Captain Henry Wisner's New York Company. He enlisted in

1777 and served four and one half months as private in Captain Henry Wisner's New York Company. He enlisted in 1778 and served three months as private in Captain Peter Bertholf's New York Company. He enlisted in 1779 and served two months as private in Captain Peter Bertholf's New York Company. He enlisted in 1780 and served two months as private in Captain Peter Bertholf's New York Company. He enlisted in 1781 and served two weeks as private in Captain Peter Bertholf's New York Company. He enlisted in 1782 and served two weeks as private in Captain Peter Bertholf's New York Company.

In 1788 he moved from Warwick, New York, to Cherry Valley, one year later he moved to Tioga County, New York, and nearly two years later moved to Ulster, Bradford County, Pennsylvania, where he resided more than thirty years.

He was allowed pension on his application executed September 4, 1832 while a resident of Elmira, Tioga County, New York. He died May 6, 1844, place not stated.

In 1832 soldier had children living in New York and Pennsylvania, no names are stated. The name of soldier's wife is not given and there are no further family data.

This is the record of the only William Curry, all spellings of the surname searched, that is found on the Revolutionary War records of this office.

Very truly yours
A. D. Hiller
Executive Assistant to the Administrator

Mrs. S. K. Frost
76 Remsen Street
Brooklyn, N.Y.

Pension Bureau
Revolutionary War Dept.
Washington, D.C.
Sept. 8, 1938

Dear Sirs: How much do I have to pay for a copy of the pension papers issued about 1832, for service in the Revolutionary War on William Curry - born 1748, died in Lodi, N.Y. 1844. He served in Orange Co. N.Y. under Col. John Hathorn. An early reply will be deeply appreciated by

Yours truly
(Mrs) S. K. Frost

Revolutionary War Pension and Bounty Land Warrant Application Files
Series M-804; Roll # 716

John Daniels
New Jersey/New York (S.16,752)

State of New York

Essex County ss. Joseph Edwards of Lewis in the County of Essex and State aforesaid, being Seventy Seven years of age, & now having his name on the pension list & having Served as a Soldier in the war of the revolution, being duly sworn Saith that he assisted in & was a soldier at the erection of fort Lee & fort Constitution in the year 1776, & served as a Soldier more than three years. That in the same year, a part of the time, he this deponent was stationed on Long Island, & was in a battle there at a place called Flatbush. And this deponent farther Saith that he has conversed with John Daniels, whose declaration for a pension is hereto annexed, & from a variety of like incidents which happened at the building of those forts & also at the said Battle & related by the said Daniels feels fully sensible, the said John Daniels must have been there, at the building of said Forts, & at said Battle. And this deponent farther Saith that ever since he knew the said John Daniels, which has been some years, he has understood & now believes he the said John Daniels was a Soldier of the Revolution, & which he now verily believes & has never heard the same doubted.

Sworn & subscribed this 5th day of Sept 1832 before me

D. Edson Sup Court Commissioner

his
Joseph X Edwards
mark

I certify to the credability of Joseph Edward of Lewis in the State of N York the witness in the above affidavit

Reuben Whallon 24 March 1834

State of New York

Essex County ss. I Leonard Stow Clerk of the Court of Common Pleas of the County of Essex do hereby certify that Dean Edson whose name is subscribed to the Jurat of the within affidavit and before whom the same purports to have been sworn was at the date of the same a Supreme Court Commissioner in and for the said county duly commissioned and sworn and authorised by law to do certain duties of a Judge of the Supreme Court out of court and of a Judge of the Court of Common Pleas out of court And further I am well acquainted with the hand writing of the said ~~Judge~~ Commissioner and verily believe the signature to the said Jurat is the genuine signature of the said Commissioner In testimony whereof I hereunto subscribe my name and affix my seal of office this tenth day of October 1832

L Stow Clerk

State of New York

Essex County ss. On this twenty fifth day of September in the year of our Lord one thousand eight hundred and thirty two personally appeared in open Court before the Court of Common Pleas in & for Said County now sitting John Daniels a resident of the Said town of Essex in the County of Essex and State aforesaid aged Seventy two years, who being first duly sworn according to law, doth on his oath, make the following declaration in order to obtain the benefit of the Act of Congress passed June 7th 1832. That he entered the Service of the United States under the following named Officers and Served as herein stated. That he firstly enlisted in the Militia Service at Westfield in the State of New Jersey in a Company commanded by Capt. --------- Col Heard Comidant of the Regiment that the Company was embodied, & he this deponent joined them at Woodbridge in the State of New Jersey, & went over to New York, & the next day crossed over to Long Island for the purpose of disarming and regulating the Torys there, that he remained there with Said Company over one month - this was in the winter of Seventeen hundred & Seventy six. And the deponent farther Saith that in the month of May following he enlisted in the Regular Service at the same town of Westfield, State of New Jersey, in what was called the five months Service, Lieut. Hodges and Capt Amos Swan belonged to Said Company, & Col. Courtland commanded the Regiment. That this deponent went with Said Company under Said enlistment to Elizabeth Town point in New Jersey, & remained there a few days only, and & then was ordered to, & went to the City of New York, & that very soon thereafter there was a call made of and from each Company to go and build Fort Lee, and he this deponent was one of the hands Selected & went with Genl Lee & remained there about six weeks or two months, & then was ordered to rejoin his Company again, which he did at Long Island & there remained on duty till the Island was given up to the British That he this deponent was engaged in a Severe battle at Flat Bush, & Genl Sterling was taken prisoner there. That this deponent then repaired to New York with Said Company, & from thence was ordered and went to Fort Washington, in a day or two after the British took New York. That this deponent remained in Service at fort Washington for some time, was taken Severely sick there, obtained a furlow & went home & remained about three weeks, & then rejoined his Company again at Fort Lee. That the British drove the American forces from thence to Hakwaknar Bridge about four miles Col Courtlands forces being the rear guard. That from thence he with said Company marched to Fox Hall & from thence to Springfield, & from thence to Trenton and from thence round to Morristown & this deponent was there discharged some days after his time of enlistment was out. And this deponent farther Saith that Soon after his discharge last aforesaid, the British forces came out near Springfield, near to this deponent, & in full view, that he this deponent shouldered his musket, and went before the British to Springfield where a pretty sharp action was fought in which this deponent was engaged, & was attached to Col Fords regiment. And this deponent farther Saith that he soon after this went to reside with his Uncle Stephen Bunnel at the drownded lands so called in Orange County New York. And this deponent farther Saith that he soon thereafter enlisted to drive teams under the Command of Capt. Cummins, he thinks, for nine months, helped move Sullivans Army, from Sussex Court House to Haverstraw & draw the teams for nine months, & was generally employed in transporting Stores, & provisions, from Sussex Court House to New Windsor & was then discharged. And this deponent farther Saith that he soon enlisted again, for one month to guard the frontiers & did that duty between Esopus and Minisink, & that for about five years thereafter he was required to be in constant readiness, & prepared for action at all times when required, in driving back the Indians, which was very frequent & more so during the summer seasons. That this deponent was out, & engaged in the action on the Delaware River, under the Command of Col Weasner, & Maj Meeker, who was shot through the shoulder & that three fourth of the Militia there assembled, & in that action were killed. And this deponent farther Saith that being a young man, & without any family at that time, he held himself in readiness at all times for the call of his Country & engaged in little or no other pursuits during the war, was frequently called to prevent the depredations of Indians & Torys & to guard the dwellings of the people where he resided, & their effects, well remembers receiving a _______ from two Justices ~~Tracy & Morrison~~, for the confinement of Tracy & Morrison two refractory Torys, in the Jail of Orange County N York, & took both of them under his charge, & marched them before him, safely to, & lodged them in Said Jail, with no other assistance but his fire arms, well loaded to be sure. That this deponent has no hesitancy in declaring under his oath that he was employed as a soldier & doing duty, for his Country in the Revolutionary war for a term of time in the whole exceeding four whole years; but that he is not able to produce any documentary evidence, & has preserved no written discharges, & that from the lapse of time, & his removal so far from the scenes of those actions at that day he is not able to prove positively his services from a cotemporary, neither can he produce much testimony, except traditionary, & circumstantial other than his own, relative to his Services, without traveling a great distance, in Search of youthful acquaintance, & at too much expense, for him to bear, & perhaps might then should he undertake it be disappointed in his expectations. The Said John Daniels hereby relinquishes every claim whatever to a pension or annuity except the present, & declares that his name is not on the pension roll of any Agency of any State.

Sworn & subscribed the day & year first above written in open Court
Leonard Stow Clerk of Essex Com. Pleas

his
John X Daniels
mark

And the said Court proposed to the said John Daniels the following Interrogatories (viz)

First Interrogatory - Where & in what year were you born - To the Said first Interrogatory the said deponent Saith that he was born at Scotch plains in the State of New Jersey

Second Interrogatory - Have you any Record of your age; and if so, where is it - To the Second Interrogatory this deponent Saith, he has no record of his age except what is contained on a leaf in the family Bible now at his home in Essex.

Third Interrogatory Where were you living when called into Service; where have you lived since the Revolutionary war, and where do you now live - To the third Interrogatory this deponent Saith that he was living at Westfield in New Jersey when he first entered the service. That for while after the war of the Revolution he lived in Morris County & State of New Jersey & for about eight years lived in Pittstown in the County of Rensellaer & State of New York & since then for about forty years he has lived where he now resides in the town of Essex & County of Essex, New York.

Fourth Interrogatory - How were you called into Service; were you drafted, did you volunteer, or were you a substitute, & if a substitute for whom - To the fourth Interrogatory this deponent saith - That he enlisted into Service, & also volunteered Several times; but was never a substitute for any one.

Fifth Interrogatory State the names of some of the Regular Officers who were with the troops where you Served; such Continental & Militia Regiments as you can recollect, & the general circumstances of your Service - To the fifth Interrogatory this deponent Saith that he was acquainted with or knew Col Courtland, Lieut Hodges & Capt Swan - Genl Lee & Genl Putnam, he knew Col Weasner, a Militia Col, & who was killed in the Service on the Delaware River, at a place called the Cook house, & that this deponent cannot now relate any thing farther, than he has related, respecting the general nature of his Service.

Sixth Interrogatory - Did you ever receive a discharge from the Service, & if so by whom was it given, & what has become of it - To the sixth Interrogatory this deponent saith that he does not now recollect of having received any written discharge; & has not preserved any.

Seventh Interrogatory State the names of persons to whom you are known in your present neighborhood & who can testify as to your character for veracity & their belief of your Services as a soldier of the Revolution - To the Seventh Interrogatory this deponent Saith he is acquainted with Judge Edson, & the Rev Elijah Benedict a Clergyman, Nathaniel Rogers, & Benjamin Stafford, & has been acquainted with all these persons for more than twenty years, and some of them nearly forty years.

Sworn & subscribed the day & year aforesaid in open Court
Leonard Stow Clerk of Essex Com. Pleas

his
John X Daniels
mark

State of New York

Essex County ss. John Hoffnagel of Willsborough in the said County, late a member of the Legislature from Said County, & Dean Edson of Essex in Said County, Counsellor at law, & public prosecutor for said County, Do Certify that we have been well acquainted with John Daniels who has subscribed to & sworn to the above declaration, & that we have known him for more than twenty years, that we believe him to be Seventy two years of age; that he is reputed & believed in the neighborhood where he resides to have been a Soldier of the Revolution & that as well from Said reputation aforesaid & from his zeal often manifested on the part of his Country in public meetings & elections, & from his conversation held with & about those who were against the Country in the war of that Revolution, as from his statement made above & entitled to credit, we concur in the opinion that he must have been and was a Soldier of the Revolution.

Sworn & subscribed the 26th day of September 1832 in open court Leonard Stow Clerk of Essex Com. Pleas

John Hoffnagle
D Edson

⁂

State of New York
Essex County ss. Elijah Benedict a Clergyman residing in the town & County of Essex & State of New York & Nathaniel Rogers residing in the same place hereby Certify that we are well acquainted with John Daniels who has subscribed to & sworn to the above declaration & that we believe him to be a man of truth & veracity, & that we have known him for more than twenty years, & that we believe him to be Seventy two years of age, & that he is reputed & believed in the town where he resides to have been a Soldier of the Revolution & that we concur in that opinion.

Sworn & subscribed this 26th day of September 1832 ~~before~~ in open court
Leonard Stow Clerk of Essex Com Pleas

E Benedict
Nathaniel Rogers

⁂

And the said court do hereby declare their opinion after the investigations of the matter and after putting the interrogatories prescribed by the War Department that the above named or ______ applicant was a revolutionary soldier and served as he states And the court further certifies that it appears to them that Elijah Benedict who has signed the preceding certificate is a clergyman resident in the town & county of Essex and that Nathaniel Rogers who also has signed the same is a resident of the same place and is a credible person and that their statement is entitled to credit

Signed By order of the court.
Leonard Stow Clerk

⁂

I Leonard Stow Clerk of the Court of Common Pleas of the County of Essex do hereby certify that the foregoing contains the original proceedings of the said court in the matter of the application John Daniels for a pension In testimony whereof I have hereunto set my hand and seal of office this 26th day of September in the year 1832

Leonard Stow Clerk

⁂

State of New York
Essex County ss. John Daniels of Essex in the County of Essex aforesaid, in further support, of his claim to a pension, being duly sworn saith, that the particular Services, he rendered, his Country, as a Soldier, in the Revolutionary war, at Minisink, on the Delaware River, & mentioned in his declaration hereto annexed consisted in driving back the Indians, who were frequently led on, and sometimes commanded by torys, that being a frontier settlement, that he was under the immediate command of Capt. Cole, & Maj. Meeker, in the Regiment of Militia commanded by Col Weasner, for the most of the time for five years, tho this deponent believes that a Lieutenant, had the command of said Company a part of the time, in the name of Capt. Cole, that the Indians made frequent incursions, on the frontier at Minisink, & this deponent as a Soldier, attached to said Company, & with Said Company, was engaged in a variety of encounters with them, both by night, & by day, that he was in a severe battle during said time, had with Indians, & Torys, on the Delaware River, & mentioned in his declaration hereto annexed, in which engagement, Col. Weasner, was killed, & also two thirds, at least, or more, of the Militia, who were engaged in that action, were killed or taken prisoners, by the Indians, that the surviving Militia retreated from that battle ground, & this deponent with them, that Maj. Meeker was shot through the shoulder, but not mortally wounded, & that from the appearance of several dead bodies, which were seen the next day on said battle ground, which were fastened to the ground, with stakes, & had the appearance of being burnt, this deponent, has no doubt but they were put to death in that barbarous manner, according to the savage custom sometimes adapted at that day. And this deponent farther saith, that during that portion of his Services, he took two men convicted of toryism, & bad conduct, to prison, as is mentioned in the declaration hereto annexed, And this deponent farther Saith, that during said term of service, at Minisink, he was at a number of times, in said Company, engaged in driving off Indians, who frequently made secret marches, into that part of the Country, for the purpose, of burning buildings, plundering & murdering the inhabitants. That one Van Ashers, Barn was burnt, & his Brother was killed, in the highway, by the Indians. That an old man by the name of Cole was taken prisoner, by the Indians, & lead off into the woods and killed, & that this deponent, together with a party of said Company, made pursuit after said Indians, & found the body of said Cole. And this deponent farther saith, that a variety of plundering, & other depredations, were frequently committed there, by the torys, and Indians, & that the Services above mentioned of this deponent, & the guarding of the dwellings, the property, & persons, of the inhabitants, from those & similar depridations, & destruction, was the particular Service, rendered by the deponent, on Said frontier, at Minisink aforesaid. And this deponent farther Saith, that he cannot, from his recollection, undertake to state, the time or times, he entered, & left the

service, & feels that too many years have passed away, from him to undertake to state with precision those dates but that he could state, with certainty, many incidents, happening at the time, like those above mentioned, & which made such an impression on his mind & memory, that he could not, he supposed, to forget them, when the recollection of times, days, & months, of the commencement, & termination of any particular service, will long since have escaped his recollection, so that he could not, in conscience, undertake to detail them, at this late day. And this deponent farther Saith, that it is extremely difficult, for him at his advanced age, & in reduced circumstances, to take a journey to the place where he remembered those Services, in search of positive proof of the same, should he at last, be required to do so, or suffer his application, for relief to be rejected and farther Saith not.

Sworn & Subscribed this 3rd day of August 1833 before me
D. Edson Sup Court Commissioner &c.

his
John X Daniels
mark

State of New York
Essex County ss. I Dean Edson a Supreme Court Commissioner authorised & empowered to do the duties of a Justice of the Supreme Court at Chambers, certify that John Daniels, who subscribed, & made oath, to the within affidavit, is, & has been known to me, for many years, & that he is a man, of truth & veracity and his declarations & statements under oath are considered entitled to full credit in the Courts of this State. In witness whereof I have hereunto set my name this third day of August in the year of our Lord one thousand eight hundred and thirty three.

D. Edson

State of New York
Essex County ss. I Leonard Stow Clerk of the County of Essex do hereby certify that Dean Edson Esq. whose name is subscribed to the above certificate and before whom the foregoing affidavit purports to have sworn was at the date of the same a Supreme court commissioner of the said state & in & for said county duly commissioned & sworn and subscribed by the Laws of said state to do _____ duties of a Judge of the Supreme court out of court and of a Judge of the court of common Pleas out of court. And further I am well acquainted with the

hand writing of the said commissioner & verily believe the signature to the said certificate and to the Jurat of the preceeding affidavit to be his genuine signature & hand writing And further that the foregoing contains the original proceedings of the said court in the matter of the application of John Daniels for a pension together with the supplemental affidavit here on written. That they are the identical papers and evidence in full as certified by me under my seal of office this twenty sixth day of September and the tenth day of August last past as above contained In testimony whereof I hereunto subscribe my name and affix my seal of office this tenth day of August in the year one thousand Eight hundred and thirty three

Leonard Stow Clerk

State of New York

Essex County ss. John Daniels of Essex in the County & State aforesaid being duly sworn Saith, that in the year one thousand Seven hundred & Seventy Seven at Minisink in the County of Orange in the State of New York the middle of June he was enrolled in a Company of Militia commanded by Capt. Cole & Lieutenant Whiteker & did actual service in Said Company, engaged in the war of the Revolution at the place last aforesaid for two months before he was discharged That he was enrolled at the same place in the same Company, same Officers in the first of June in the year 1778 & did duty in said Company & in actual service in Said war for the term of two months & was then discharged.

That he was again enrolled in said Company at the same place, & same Officers, the middle of May 1779 & & did actual service in that campain for the term of more than three months & was then discharged.

That about the first of May 1780 he was again enrolled in said Company, same place & Officers and did duty in said Company engaged in actual Service in Said war & under the same Officers for the term of three months or more & was then discharged. That in the year 1781 & in the year 1782 & in the year 1783 he was enrolled in said Company, under the same Officers & at the same place & did duty therein as a soldier in Said war (viz) in 1781 May 1st was enrolled & did duty for three months or more & was then discharged, & in 1782 - Middle of May was enrolled & did like duty for more than three months & was then discharged. And that in the year 1783 first of May was enrolled in Said Company and did like duty as a soldier for more than two months & was then discharged. And this deponent farther Saith that Col Wisner & Maj Meeker mentioned in his affidavit heretofore made, & hereto annexed, were Militia Officers of the New York Militia & farther Saith not.

Sworn & subscribed this 10th of March 1834 before me

D. Edson Sup Court Commissioner

his
John X Daniels
mark

State of New York
Essex County
Clerks Office ss. I Edwd S Cuyler clerk of the court of Common Pleas of the County of Essex do hereby Certify that Dean Edson before whome the foregoing affidavit of John Daniels was taken was at the day of the date when the same purports to have been taken a commissioner of the Supreme court of the State of New York duly Appointed & Commissioned for & Residing in the Said County of Essex and qualified to Administer Oaths &c and that I Am well Acquainted with the hand writing of the Said Superior court commissioner and believe his Signature to the Jurat of the said affidavit to be his genuine Signature. In testimony Whereof I have hereunto Set my hand And affixed my official Seal of Office this 10th day of March in the year 1834

Edwd S. Cuyler Clerk

War Department,
Pension Office,
183

Sir:

The evidence in support of your claim, under the act of June 7, 1832, has been examined, and the papers are herewith returned. The following is a statement of your case in a tabular form. On comparing these papers with the following rules, and the subjoined notes, you will readily perceive that objections exist, which must be removed, before a pension can be allowed. The notes and the regulations will shew what is necessary to be done. Those points to which your attention is more particularly directed, you will find marked in the margin with a brace, (thus: }). You will, when you return your papers to this Department, send this printed letter with them; and you will, by complying with this request, greatly facilitate the investigation of your claim.

A Statement, shewing the Service of John Daniels, of Essex county, N. York

Period when service rendered	Duration of claimant's service			Rank of the Claimant	Names and Ranks of the Company Officers under whom he served	Names and Rank of the General and Field officers
	Years	Months	Days			
Enlisted 1776		1		private	Capt Swan Lieut Hodges	Col. Heard Col. Courtland
1777		5				Col. ______
Enlisted		9*				
Enlisted 1778		1				
Employed long as a spy on the frontier					Col. Measner	

Battles in Which the applicant engaged	Country through which he marched	Place of abode when he entered service, and age at the present period	Evidence by which the declaration is supported
Flatbush Fort Lee	Long Island Eliasbeth town Fox Hall	72 Westfield New Jersey	tradition
Springfield	Springfield Morristown		
Haverstraw			
Stillwater	Esopus to Minisink		

* a teamster this tour

I am, respectfully,
Your obedient servant,
J. L. Edwards,
Commissioner of Pensions.

*After the statement of service a copy of the regulations under the Act of June 7, 1832 is included but not reproduced here. An identical copy is included in the chapter entitled **A Declaration of Service**. At the bottom of the regulations is a note that says that "His Service as a spy must be particularly specified."*

New York
John Daniels of Essex Co in the State of N. York who was a Private in the

commanded by Captain of the Regt. commanded by Col. Heard in the N.J. and N.Y. Troops line for 2 years Inscribed on the Roll of New York at the rate of 80 Dollars per annum, to commence on the 4th day of March, 1831. Certificate of Pension issued the 25 day of March 1834 and sent to Hon. R. Whallon H. R.

Arrears to the 4th of March 1834

240.00
40.00
$280.00

Semi-anl. allowance ending 4 Sep 1834

Recorded by D. Brown Clerk, {Revolutionary Claim, Act June 7, 1832.}

Book E Vol. 4 Page 96

Rome N.Y. RFD #5
June 13, 1921

Department of the Interior,
Bureau of Pensions,
Attention Commissioner of Pensions
Dear Sir:

I would like as soon as possible the records of two Revolutionary Pensioners namely Isaac Burnham and John Daniels They were on the rolls for St Lawrence Co. N.Y.

Isaac Burnham was a Private in Connecticut Line. Lived in DeKalb, N.Y. placed on Pension Rolls Sept 30, 1818; Drew pension from April 29, 1818 at the Age of 69.

John Daniels was born in 1760 or 1763. He died in Canton, N.Y. Mar 12, 1847. He was married to Mrs. Thankful (Gaylord) Frisbie who died Aug 4, 1846 aged 71.

Hoping for an early reply:
I am, Sincerely,
Roger F. Williams

Rev. War Section.

June 28, 1921.

Mr. Roger F. Williams,
R. D. #5, Rome, N.Y.
Sir:

In response to your letter dated the 3d instant, you are advised that it appears from the papers in the Revolutionary War pension claim S. File No. 16,752, that John Daniels was born ~~in 1760~~ at Scotch Plains, New Jersey. He was a resident of Westfield, New Jersey, when he enlisted in 1776 as a Private in the New Jersey troops, serving at various times under Captain Amos Swan and Colonels Heard, Courtland and ford, and was in the battles of Flatbush, and Springfield. Soon after the last named battle he removed to Orange County, New York, and drove teams under Captain Cummins for 9 months.

The middle of June 1777, he enlisted at Minisink, New York, under Captain Cole and *Colonel Wisner and was in the battle of Minisink and also guarded the frontier against Indians between Esopus and Minisink until 1783.

He was allowed pension on an application executed September 25, 1832, while a resident of Essex, Essex County, New York, 72 Yrs

There is no data ~~on file~~ as to his family.

Very respectfully,
H. M. Vandervoot
Acting Commissioner.

* See Lossing Vol. 1 page 671.

Independence, MissouriJ
June 18, 1930

Bureau of Revolutionary War Pensions
Washington, D.C.
Gentlemen:
Would you kindly send me the Rev. War record of
(1) Benjamin Cutler (1747 - 1828) of Plainfield, N.Y
(2) Benoni Cutler (1737 - 1807) who died at Guildhall, Vermont
(3) John Daniels (1755 - 1833) a pensioner died at Plainfield
(4) William Cutler, who served in Conn, credited to Plainfield, N.Y.

Also have you a record of Alpheus Cutler, said to have served in War of 1812, born in Plainfield in 1788, died sometime after 1860 in Iowa?

Very sincerely yours,
Inez S. Davis

July 31, 1930
AWF:MLB
Rev. and 1812 Wars Section

Inez S. Davis
Independence, Missouri
Dear Madam:

You are advised that it appears from the papers in the Revolutionary War pension claim, S.16752, that John Daniels was born at Scotch Plains, New Jersey.

He was a resident of Westfield, New Jersey when he enlisted in 1776 as a private in the New Jersey troops, serving at various times under Captain Amos Swan and Colonels Heard, Courtland and Ford, and was in the battles of Flatbush and Springfield.

Soon after the last named battle, he removed to Orange County, New York, and drove teams under Captain Cummins for nine months.

The middle of June 1777, he enlisted at Minisink, New York, under Captain Cole and Colonel Wisner and was in the battle of Minisink and also guarded the frontier against Indians between Esopus and Minisink until 1783.

He was allowed pension on his application executed September 25, 1832, while a resident of Essex, Essex County, New York, aged seventy - two years.

There are no data ~~on file as to his family.~~ rel to wife or chil

Very truly yours
E. W. Morgan
Acting Commissioner

Invalid. File No. 16,752
John Daniels Pri Rev War
Act: June 7th 32
Index: - Vol. 1, Page 428
[Arrangement of 1870.]
Revolutionary War Pension and Bounty Land Warrant Application Files
Series M-804; Roll # 737

David Dean
New York (S.29,118)

BRIEF in the case of David Dean of Hopewell in the
State of N. York
(Act 7th June, 1832.)

1. Was the declaration made before a Court or a Judge? in open Court

2. If before a Judge, does it appear that the applicant is disabled by bodily infirmity?

3. How old is he? Sixty nine years

4. State his service, as directed in the form annexed.

Period when service rendered	Duration of claimant's service			Rank of the Claimant	Names and Ranks of the Field Officers under whom he served
	Years	Months	Days		
March 1778		4		Private	Col. Thurston
Septemr 1778		3		Private	Col. Cantine
April 1779		4		Private	Col. Thurston

States that he was afterwards called out to serve in the Militia at various times, and "verily believes that he thus served more than three months." and claims for 14 months service

5. In what battles was he engaged? None stated

6. Where did he reside when he entered the service? Goshen N. York

7. Is his statement supported by living witnesses, by documentary proof, by traditionary evidence, by incidental evidence, or by the rolls? Traditionary evidence alone

8. Are the papers defective as to form or authentication? and if so, in what respect? the papers are correct in form and authentication

Certify that the foregoing statement and the answers agree with the evidence in the case above mentioned.

Wm. Gordon Examining Clerk.

☆

State of New York

Ontario County ss. On this twenty seventh day of August 1832 personally appeared in open Court before the Judges of the County Court now sitting David Dean a resident of the town of Hopewell County and State aforesaid aged sixty nine years who being first duly sworn according to Law, doth on his Oath make the following declaration in order to obtain the benefit of the act of Congress passed June 7, 1832.

That he entered the service of the United States under the following named Officers and served as herein stated.

1st At Goshen in the County of Orange and State of New York about the first March 1778 he enlisted for four months in Captain Sweeseys Company in Colonel Thurstons Regiment and first marched to Ramappo and there done duty until discharged.

2nd In September following enlisted for three months in Captain Coe's Company, Colonel Cantine's Regiment, was mustered at Goshen and marched to Peen Pack Dewitts Fort, and remained there to the end of the term, and was then discharged.

3rd About the middle of April 1779 enlisted for four months in Lieutenant Webbs Company Colonel Thurstons Regiment was marched to West Point and done duty there until the term of my enlistment expired and then returned home to Goshen.

4th At Goshen as aforesaid, he was after called out in the Militia, and was once sent to Hurley for Provisions for the army and was out sixteen days; and he verily believes that he has thus served more than three months. He always turned out when called upon, and never refused, and during this time done duty at the Prison, keeping guard when Claudius Smith was confined there and four other prisoners. Smith and two others were hung, two of them were pardoned. He was in Captain Sweeseys Company and Colonel Thurstons Regiment. John Reid of Canandaigua was in the service with him at Beaver Brook where Colonel Thurston was killed; this was in 1779. He verily believes that he was in actual service more than eighteen months, but from the infirmities of age, and the privations that he has borne, he is unable to state with precission many services done and performed by him. He only claims from the Country, compensation for fourteen months in all.

He was born in Little Britain in Orange County and State of New York on the 31st day of March 1763. Has no documentary evidence Has no record of his age, and knows no person now living who served with him except John Reid. He lived in Goshen until 1799, when he removed to Canandaigua, and has resided there ever since, all his services was in the Militia Has seen General Putnam at West Point, and did duty there. Has seen General Clinton at Fort Lee and Fort Montgomery, saw General Washington at Peramus. He knows John Reid and Adam Nichols of the town of Canandaigua

He hereby relinquishes every claim whatever to a Pension or annuity except the present and declares that his name is not on the Pension Roll of the agency of any State

Sworn & subscribed the day and
year aforesaid in open Court
David Dean

We John Reid and Adam Nichols residing in the Town of Canandaigua in the county & State aforesaid hereby certify that we are well acquainted with David

Dean who has subscribed & sworn to the above declaration, that we believe him to be sixty nine years of age that he is reputed and believed in the neighborhood where he resides to have been a soldier of the Revolution & that they concur in that opinion. Sworn & subscribed the day & year aforesaid in open Court

John Reid
Adam Nichols

☆

And the said Court do hereby declare their opinion, after the investigation of the matter, and after putting the interrogatories prescribed by the War department, that the above named applicant was a Revolutionary soldier and served as he states. And the court further certifies that it appears to them that John Reid and Adam Nichols, who have signed the preceding certificate, are both credible persons, and that their statement is entitled to credit.

☆

I, Charles Crane, Clerk of the said Court, do hereby certify that the foregoing contains the original proceedings of the said Court in the matter of the application of David Dean for a pension. In testimony whereof, I have hereunto set my hand and seal of office, this first day of September A.D. 1832.

Charles Crane

☆

BRIEF objections to the admission of Pension Claims, of which those answered in the negative, in red ink, will apply to the claim of David Dean which require further proof or explanation.

I. Does the declaration show where the applicant resides? yes

II. Did the declarant make the declaration in the county where he resides? yes

III. If the applicant did not make the declaration in the county where he resides, is there any reason assigned for his not doing so?

IV. Has the declarant mentioned the period or periods of the war when he served? yes

V. Has he mentioned with precision the length of his service, and the different grades in which he served, in language so definite as to enable the department to determine to what amount of pension he is entitled? no Militia service should be set forth in detail

VI. Has he given the names of the officers under whom he served, in conformity with the regulations? yes

VII. Has he made a relinquishment of every claim to any other pension than the one under the Act of June 7, 1832? yes

VIII. Has the Court given their opinion? yes

IX. Has the Clerk given his certificate? yes

X. Is the Clerk's seal affixed? And if so, has it a device or inscription by which it can be distinguished from any other seal? yes

XI. Has the applicant obtained the evidence of a clergyman and another respectable citizen as to their belief respecting his age and the general belief in his neighborhood relative to his revolutionary service? Has not obtained the evidence of a clergyman but has adduced the testimony of two credible persons

XII. If the clergyman's affidavit has not been produced, is there any reason assigned for not obtaining it? none

XIII. If the applicant has no documentary evidence, and has not obtained the testimony of at least one living witness, has he stated in his declaration that such proof cannot be had? yes

XIV. If the applicant has produced a witness or witnesses as to his service, has the magistrate who administered the oath certified that he, she, or they are persons of credibility?

Has the official character and signature of the magistrate been certified by the proper officer under his seal of office?

XV. Does it appear from any of the papers that the witness was in such a situation, or of such an age, as to have a personal knowledge of the applicant's service?

XVI. Are the papers authenticated as the regulations direct? yes

XVII. Are the seven interrogatories prescribed by the War Department answered satisfactorily? And, if not, which of them is not so answered? the court certify that the interrogatories were put to the applicant but not embodied

EXPLANATORY NOTES.

QUERE V. In a case where the applicant cannot, by reason of the loss of memory, state precisely how long he served, he should amend his declaration by making an affidavit in the following words: "Personally appeared before me, the undersigned, a Justice of the Peace, &c. A. B. who, being duly sworn, deposeth and saith, that, by reason of old age, and the consequent loss of memory, he cannot swear positively as to the precise length of his service, but according to the best of his recollection he served not less than the periods mentioned below, and in the following grades: For year months and days, I served as a . For months and days, I served as a ; and for such service I claim a pension."

It is important in all cases to determine with precision the period for which each applicant served, and the particular rank he held, as the law directs the pension to be paid according to the grade of the pensioner and the length of his service. The use of the phrase about three or four months is too indefinite, and all such qualifying expressions are objectionable.

QUERE XIV. Proof of Service.

In a case where the name of the applicant is not found on the records of the Department, he must prove his service by two credible witnesses, who are required to set forth in their affidavits the time of the claimants entering the service, and the time and manner of his leaving the same, as well as the regiment, company, and line to which he belonged. The magistrate who may administer the oaths must certify to the credibility of the witnesses, and the official character and signature of the magistrate must be certified by the proper officer, under his seal of office.

QUERE XVI. Mode of authenticating papers.

In every instance where the certificate of the certifying officer who authenticated the papers is not written on the same sheet of paper which contains the affidavit, or other paper authenticated, the certificate must be attached thereto by a piece of tape or narrow ribbon, the end of which must pass under the seal of office of the certifying officer, so as to prevent any paper from being improperly attached to the certificate.

☆

In the matter of David Dean's application for a Pension

State of New York

Ontario County ss. Personally appeared before me the undersigned a justice of the Peace in and for said county, David Dean, who, being duly sworn deposeth and saith, that the reason he did not have a Clergyman to support his original application before the court of said county on the 27 day of August 1832, was, the society to which he belongs is without a settled minister, and knew of none that he could conveniently procure. He is required to answer the interrogatories prescribed by the War department. The 4 first of them he has answered in his original declaration, to the 5th interrogatory he now says, under the service stated under his first head, that Colonel Hawthorns Regiment was at Ramappo at the same time, but under the 2nd head there was no other Troops on duty besides the company to which he belonged, and under the 3rd head he now says that the company to which he belonged was stationed most of the time at the "Gap" near West Point, and he cannot state the names of any other Regiments then on that station, under the 4th head he says that his service was almost always on fatigue and guarding the Goal at Goshen. He is required to set forth his service in detail, under the first head he served four months, under the Second head, Three months, under the third head, four months, and under the 4th head not less than Three months and sixteen days, and in all not less than fourteen months and sixteen days, as a private for which he claims a pension

Sworn to and subscribed this Ninth day of January 1833 before me Henry Chapin J. P.

David Dean

State of New York

Ontario County ss. I, Charles Crane, Clerk of said county, do hereby certify, that Henry Chapin, whose name is subscribed to the within certificate of affidavit, was, at the date thereof, a Justice of the peace in and for said county; and that I am well acquainted with his hand writing, and verily believe the said signature to be his proper act and deed. In testimony whereof, I have hereunto set my hand and seal of office this tenth day of January A.D. 1833

Charles Crane

Michigan

David Dean of Hopewell in the State of New York who was a Private in the Company commanded by Captain Coe of the Regt. commanded by Col Cantine in the New York line for 11 months Inscribed on the Roll of New York at the rate of 36 Dollars 66 Cents per annum, to commence on the 4th day of March, 1831. Certificate of Pension issued the 1st day of March 1833 and Sent to Hon ____ Ho Reps

Arrears to the 4th of Sept. 1832	$54.99
Semi-anl. allowance ending 4 Mar '33	18.33
	$73.32

Recorded by Jno. Cromwell Clerk, {Revolutionary Claim, Act June 7, 1832.}

Book D Vol. 5 Page 81

State of Michigan

County of Wayne ss. On this twentieth day of October 1836 before me, a Justice of the Peace in and for said County, personally appeared David Dean who, on oath, declares that he is the same person who formerly belonged to the company commanded by Captain David Sweezy, in the Regiment commanded by Colonel Thurston in the service of the United States; that his name was placed on the Pension Roll of the State of New York, from whence he has lately removed; that he now resides in the State of Michigan, where he intends to remain, and wishes his pension to be there payable in future. The following are his reasons for removing from the State of New York. His children removed to Michigan and as he is dependant on them for support he likewise removed to Michigan.

Sworn and subscribed before me, Oct. 20, 1836

H. A. Noyes Justice of the Peace

David Dean

⋆

State of Michigan

County of Wayne ss. On the twentieth day of October 1836 before me, a Justice of the Peace in and for said county, personally appeared, David Dean, who, on oath, declares, that he is the identical person named in an original certificate of pension in his possession from the War Department, of which I certify the following is a true copy:

⋆

War Department

Revolutionary claim:

I hereby certify that in conformity with the Law of the United States, of the 7th June 1832 David Dean of the State of New York who was a Private in the Revolutionary army is entitled to receive Thirty six Dollars and sixty six cents per annum, during his natural life, commencing on the 4th of March, 1831, and payable semi-annually on the 4th of March, and 4th September in every year. Given at the War Office of the United States, this 1st day of March one thousand eight hundred and thirty three

Examined and countersigned

J. L. Edwards

Lew Cass

Secretary of War

and is also the same person who has sworn to, and subscribed the above affidavit

Sworn and subscribed before me, Oct. 20, 1836

H. A. Noyes Justice of the Peace

David Dean

⋆

State of Michigan

County of Wayne ss. I do hereby certify that the witness aforesaid is a person of truth and veracity. Oct. 20, 1836

H. A. Noyes Justice of the Peace

State of Michigan

County of Wayne ss. I, Isaac S. Rowland, Clerk of the court of said County of Wayne, certify, that H. A. Noyes is a magistrate duly elected and qualified and that the foregoing signatures purporting to be his, are true and genuine. In testimony whereof I have hereunto affixed my seal of office and subscribed my name this twentieth day of October in the year 1836

Isaac S Rowland, Clk.
David E Hasbaugh, Dept.

J. L. Edward Comr. of Pensions
City of Washington

Mechs. & Farmer's Bank
Albany Pension Agency
November 14, 1836.

James L. Edwards Esquire,
Commissioner of Pensions.
Sir,

In reply to your letter of the 10th inst. I have to state that David Dean, a Pensioner under the act of June 7, 1832, was last paid to the 4th of March 1836.

I am, Sir, very respectfully
Your Obt. Servt.
Thos W Olcott
President & Pen Ag

War Department.
Pension Office,
Nov 18th 1836

Sir:

In conformity with the regulations of this Department, you have this day been transferred from the Roll of Albany to that of Michigan to take effect from and after the 4th March 1836

I am, Sir, respectfully,
Your obedient servant,

J. L. Edwards
Commissioner of Pensions.

Mr. David Dean
U.S. Pensioner.
Detroit Michigan

Charlevoix, Mich., March 27th, 1924.

Bureau of Pensions,
Dept. of the Interior,
Washington, D.C.
Sirs:

Will you please give me the military record of David Dean (my great grandfather) who enlisted in the Revolutionary War from Goshen, Orange Co., New York, born ~~New B~~ Little Britain, same county. Enlisted March 1778. He applied for and received a pension.

I would also like to know if his son Luther Dean was in War of 1812, as his brother Gabriel was in that war and probably Luther was also. I do not know as he was in the war or that he drew a pension if he was in war, but would be glad to be informed on that point.

Very respectfully,
Herbert D. Dean

May 1, 1924.

Herbert D. Dean,
Charlevoix,
Mich.
Sir;

I have to advise you that from the papers in the Revolutionary War pension claim, S.29118, it appears that David Dean was born March 31, 1763 at Little Britain, Orange County, New York.

While residing at Goshen, Orange County, New York, he enlisted March 1, 1778 and served four months as a private in Captain Sweezey's Company, Colonel Thurston's New York Regiment.

He enlisted in September 1778 and served three months as a private in Captain Coe's Company, Colonel Cantine's New York Regiment.

He enlisted in April 1779 and served four months as a private under Lieutenant Webb in Colonel Thurston's New York Regiment, and he served three months and sixteen days date of enlistment not stated, in Captain Sweezey's Company, Colonel Thurston's New York Regiment.

He was allowed pension on his application executed August 27, 1832, while a resident of Hopewell, Ontario County, New York.

In October 1836, he stated that he had lately moved to Wayne County, Michigan to live with his children, but their names are not stated, and there is no further data on file as to his family.

The 1812 War records of this Bureau fail to afford any information in regard to pension having been claimed on account of the services of any Luther Dean.

Respectfully,
Commissioner.

Northville, Michigan
Feb. 11, 1929.

Bureau of Pensions
Dept. of Interior
Washington, D.C.
Gentlemen:

Will you kindly send me the Revolutionary War record of David Dean, who according to family record was born in Little Britain, Orange Co. New York, March 31, 1763, enlisted from Goshen, Orange Co. and applied for pension Aug. 27, 1832. If there is any record of his coming to Wayne County, Michigan, will you include that. Kindly advise me as to any charges in the matter.

Yours very truly
(Miss) Helen Cunningham
317 Dunlap
Northville, Michigan

May 23, 1929.

Helen Cunningham,
317 Dunlap,
Northville, Michigan.
Madam:

I advise you that from the papers in the Revolutionary War pension claim, S.29118, it appears that David Dean was born March 31, 1763 at Little Britain, Orange County, New York.

While residing at Goshen, Orange County, New York, he enlisted March 1,

1778 and served four months as a private in Captain Sweezey's Company, Colonel Thurston's New York Regiment.

He enlisted in September 1778 and served three months as a private in Captain Coe's Company, Colonel Cantine's New York Regiment.

He enlisted in April 1779 and served four months as a private under Lieutenant Webb in Colonel Thurston's New York Regiment, and he served three months and sixteen days date of enlistment not stated, in Captain Sweezey's Company, Colonel Thurston's New York Regiment.

He was allowed pension on his application executed August 27, 1832, while a resident of Hopewell, Ontario County, New York.

In October 1836, he stated that he had lately moved to Wayne County, Michigan to live with his children, but their names are not stated, and there are no further data on file as to his family.

Respectfully,

E. W. Morgan,
Acting Commissioner.

Revolutionary War Records Section.Department of the Interior,
Bureau of Pensions.

Washington, D.C., , 19

In reply to your request of , received for a statement of the military history of David Dean a soldier of the Revolutionary War, you will find below the desired information as contained in his ~~(or his widow's)~~ application for pension on file in this Bureau.

Dates of Enlistment or Appointment	Length of Service	Rank.	Officers under whom Service was Rendered.		State
			Captain	Colonel	
Mar 1, 1778	4 mos.	Pvt.	Sweezey	Thurston	NY
Sept 1778	3 mos.		Coe	Cantine	
Apr 1779	4 mos.		Lt Webb	Thurston	
	3 mos. 16 day		Sweezey	Thurston	

Battles engaged in,

Residence of soldier at enlistment, Goshen Orange Co NY

Date of application for pension, Aug 27, 1832. His cl. was al.

Residence at date of application, Hopewell, Ontario Co NY

Age at date of application, ~~69~~ b. Little Britain Orange Co NY Mar 31, 1763.

Remarks: In Oct 1836 he stated that he had lately removed to Wayne Co Mich to live with his children, but their names are not stated & there is no further data on file as to his family

Respectfully,
Commissioner.

Revolutionary War Pension and Bounty Land Warrant Application Files
Series M-804; Roll # 777

Francis DeWitt
New York (R.2,919)

State of New York

County of Onondaga ss. On this twenty seventh day of May in the year one thousand eight hundred and thirty four, personally appeared in open Court before the Court of Common Pleas in and for the County of Onondaga now sitting at the Court House in the village of Syracuse in the said County, Francis DeWitt, a free man of color, and a resident of the town of Skeneateles in the said County and State, aged eighty seven years, who being first duly sworn according to law, does on his oath make the following declaration in order to obtain the benefit of the act of Congress passed June 7, 1832, viz.

That in May and he thinks in the third year of the Revolutionary war, he entered the service of the United States as a Ranger under Captain Jacob DeWitt, then of Ulster County in the said State, and served in a company of Rangers for about three years thereafter. That their duty was to scout on the back frontier then harassed by the Indians and tories. That James Bull was Lieutenant and he believes John English the Ensign.

In that year, he believes 1777, a fort was built of pickets around Capt. Jacob DeWitt's house, which was at a place called Peen Pack on Neversink River in the precinct of Mamakating to protect the neighborhood from the incursions of the Indians and tories, and Capt DeWitt's Rangers when not called out were engaged in protecting and defending this fort and other forts of a similar kind on the Neversink and this deponent at such time was stationed at Fort DeWitt.

The first season he enlisted he and his company went up to Cochecton on the Delaware on a scouting expedition and were there two or three weeks. Also the same season encamped at Mongaup on the Delaware.

They were also ordered out and went the same season to Ramapo and Katyack and were gone till winter set in.

The next year this deponent with the Rangers was out in the back country lying on the Delaware and our troops then had a battle with the Indians and tories at the Beaver Brook. Deponent was present but Daniel Van Vleet now of Owasco in the County of Cayuga was there, and the father of James Terwilliger hereinafter named, was killed. Several companies were engaged, among others he remembers Capt. Jones and Wood's, and Capt. Wood was taken prisoner. Col. Thurston commanded our troops, who were defeated and fell back to the Neversink.

In 1778 Capt. afterwards Col. Newkirk of the militia commanded Fort DeWitt, which was attacked by the Indians & tories while this deponent was stationed there. This deponent was wounded in the leg with a bullet at this attack.

This deponent remembers lying a while on Croton River while ordered out with the Rangers, and also of lying between Newburgh and New Windsor while Gen Washington was there

He was on constant duty while at Fort DeWitt as guard or scout. When the Rangers were at the forts, they were ordered out frequently in squads of 10 to 15 as scouts and spies to look out for and follow the tracks of the Indians. And this deponent was sent out very often as one knowing the woods. In 1779, Capt. Wood commanded Fort DeWitt & the troops there

The company of Rangers was broke up and the men turned into other companies in the month of October, He thinks three years after he commenced with them. This deponent then went into a militia company and served in that till the war closed. Capt. Abraham Cuddeback commanded it. King and Harden were Lieutenants and Samuel DePuy was ensign.

This deponent further declares that his name is not on the pension roll of the agency of any State. That he is unable to take any witnesses to the Court now sitting, & can produce none but those who volunteer to attend for him; that he resides 20 miles from the place where the Court is now sitting.

In answer to interrogatories propounded by the Court, pursuant to the instructions of the War Department, the deponent says;

1. That as he has always heard & believes, he was born eighty seven years ago last December on the fifteenth day, old style, at the Napenaugh Kill in the County of Ulster and State of New York.

2. A record of his age was made in the Dutch family bible of Capt. Jacob DeWitt in whose family deponent was born & lived. That bible he is informed & believes is in Marbletown, Ulster County.

3. When he first enlisted in the war he was living at Peen Pack in the then precinct of Mamakating and County of Ulster & State of New York. He also lived there after the war. About 35 years ago he came to Onondaga County, and

then lived in Manlius about four years. He then lived about three years in Genoa, Cayuga County and has ever since lived where he now lives in the now town of Skaneateles.

4. He entered the service as a volunteer.

5. Of the regular officers who were with the troops where he served, he recollects Gen. James Clinton, Col. Van Cortland, Major afterwards Col. Fish, Col. James McCloughry. Recollects McCloughry's regiment well. He has above stated, he says, the general circumstances of his services.

6. He received a Certificate of his faithful service which his second wife burnt up some years ago in a quarrel where he now lives.

7. He says he can name many persons to whom he is know in his present neighborhood, & who can testify as to his character for veracity & their belief of his services as a soldier of the revolution. He names David Van Vleet of Owasco (adjoining Skaneateles.) Widow Van Auken of ditto, (whose life deponent saved when Fort DeWitt was attacked by the Indians in 1778.) Mrs. Jane Cuddeback of Skaneateles, who was the daughter of Capt. Jacob DeWitt. He also names John DePuy, Benjamin DePuy, Cornelius Wallace, James Terwilliger. He says they all know of his services in the Revolutionary War. He also names Solomon Davis, Richard Tallcot, Augustus Kellogg, Lewis H. Sandford of Skaneateles

~~his~~
~~Francis DeWitt~~
~~mark~~

The deponent further says he is unable to procure the attendance of any clergyman who can testify respecting him on account of the distance and his poverty And further says not.

his
Francis X DeWitt
mark

Sworn and subscribed the day year first aforesaid, The names "Solomon Davis, Richard Tallcot, Augustus Kellogg & Lewis H. Sandford" first inserted, and

"his
Francis DeWitt
mark"

crossed out, before signing as Witness Lewis H. Sanderford
R. S. Hess clerk of Onondaga County

State of New York
Onondaga County ss. Benjamin De Puy of the town of Salina in the said County, being duly sworn, deposes and says as a witness in the said Court of

Common Pleas above mentioned, that he is in his seventieth year. That at the beginning of the Revolutionary War, he knew Francis DeWitt, colored man, above named, who then resided at Peen Pack now in the then precinct of Mamakating, in the then County of Ulster, and State aforesaid. That this deponent then lived about a mile from him. That the said Francis DeWitt helped to defend Fort DeWitt, erected as stated by him, and deponent knows that he served in thee Rangers under Captain Jacob DeWitt. The Rangers were gone a considerable time to Ramapo when Francis DeWitt went with them as he heard That he was frequently out on scouting parties, and as deponent then heard and believed and still believes, the said Francis DeWitt from early in 1777 to the end of the war was constantly in the service. Deponent lived at Fort DePuy till October 1778 when it was burnt by the tories & Indians, & our people retreated to Fort DeWitt, & deponent went from there to Fort Gumaer which was above below both the others. In that year the Indians & tories destroyed that country on the Neversink and attacked Fort DeWitt then under Captain Newkirk who had some of the nine months militia with him. The attack was kept up four or five hours and deponent saw the said Francis De Witt there as a soldier and fighting bravely the whole time. Deponent heard at the time and have no doubt of it, that the said Francis DeWitt and another man, while stationed at Fort DeWitt, turned out and with great difficulty and danger rescued from the Indians a soldier who was passing between Forts DePuy and DeWitt.

That the country where those forts were, was the frontier, and thinly peopled and for several years of the war, was much distressed by the incursions of the Indians and tories, and depended principally upon themselves for protection in their forts aided by the Rangers and occasionally the nine months militia.

That this deponent has received a Certificate for a pension the present year.

Sworn to and subscribed the day and year first aforesaid, "in the then precinct of" interlined & "now" struck out. Also "above" struck out and "below both the others inserted. Also "on the Neversink" interlined before signing. Also "as he heard"

R. S. Hess Clerk of Onondaga County

Benjn. DePuy

State of New York

Onondaga County ss. Cornelius Wallace of the town of Pompey in the said County, being duly sworn as a witness in the said Court of Common Pleas above mentioned does depose and say, that he is in his seventieth year. That during the Revolutionary war he was a soldier in the Forts on the Neversink River at a place called Peen Pack in the precinct of Mamakating in the County of Ulster, which was then on the frontier of the settlements West of Hudson River. That during the year 1777, the inhabitants of that neighborhood put up three Forts on that River and went into them for their safety, and all the male inhabitants who were old enough, while there were engaged for several years in defending them against the Indians and tories. The upper fort was called Fort DeWitt. This deponent was brought up in the neighborhood, and has known Francis DeWitt, a colored man, now of Skaneateles in the said County of Onondaga, a colored man, ever since deponent was a boy. That he knows that the said Francis DeWitt served under Capt. Cuddeback in the militia in defending the said Forts for the two or three last years of the revolutionary war and was constantly engaged in the service during that time, being mustered in the company and standing guard like the others. And this deponent who was at Fort Gumaer (the lower fort,) and was a boy in the early part of the war, heard and has no doubt that the said Francis DeWitt previously to his being under Capt. Cuddeback was in the Rangers under Capt. Jacob DeWitt. And that this deponent has received a pension for his services in said war. And further this deponent says not.

Subscribed & sworn the day & year first aforesaid.

R. S. Hess clerk of Onondaga County

his
Cornelius X Wallace
mark

State of New York

County of Onondaga ss. James Terwilliger of the town of Skaneateles in the said County, being duly sworn in the said Court of Common Pleas, deposes and says, that he knew the above named Francis DeWitt, colored man, before the Revolutionary War, and then lived at the same place. Knew him during the whole of the war, and knew that he went out with the scouts and Rangers several times, and in the intervals was standing sentry at the Forts above mentioned on the frontier. Knows that from the year 1777 till near close of war the said Francis DeWitt was nearly all the time engaged at the Forts, in the Rangers and as one of the militia; and heard and believes he served constantly during all that time.

Deponent heard of the above circumstance stated by Benjamin DePuy, of the said Francis DeWitt's rescuing a man from the Indians, and also heard of his saving the life of Esther Van Auken now of Owasco in Cayuga from the Indians and tories with great hazard when Fort DeWitt was attacked in 1778.

Deponent was some of the time at Fort DeWitt during the war.

Deponent's father was killed in a battle between the tories & indians, & the militia and rangers, in the back country on the Delaware; and he then heard and has no doubt the said Francis DeWitt was engaged in that battle ~~and was wounded~~

Sworn to and subscribed the day and year first aforesaid.

R. S. Hess clerk of Onondaga County

"and was wounded" struck out

his
James X Terwilliger
mark

State of New York

Onondaga County ss. Solomon Davis of the town of Skaneateles in the County of Onondaga, being duly sworn, deposes and says, that he is aged sixty years and upwards.

That he knows Francis DeWitt of the said town, a colored man, and has known him ever since this deponent was a boy. That this deponent when he was a boy lived near Captain Jacob DeWitt's with whom the said Francis DeWitt then lived, about a mile and an half distant.

That this deponent has no doubt that the said Francis DeWitt is at least eighty seven years of age; and this deponent further says that the said Francis DeWitt is reputed and believed in the neighborhood where he resides to have been a revolutionary soldier, and this deponent concurs in that opinion. And that such has been the reputation for a great many years past to this deponent's knowledge. And further this deponent says not.

Sworn and subscribed the twenty second day of May in the year 1834 before me.

J. M. Allen Judge of Onondaga County Courts

Solomon Davis

We James Terwilliger, and Lewis H. Sandford residing in the town of Skaneateles in the County of Onondaga, hereby certify, that we are well acquainted with Francis DeWitt who has subscribed and sworn to the above declaration, that we believe him to be eighty seven years of age, that he is reputed and believed in the neighborhood where he resides to have been a soldier of the revolution, and that we concur in that opinion.

Sworn & subscribed the day & year first aforesaid

John Watson First Judge of Onondaga County

his
James X Terwilliger
mark
Lewis H. Sandford

State of New York

Onondaga County ss. Richard Tallcot, of the town of Skaneateles in the County of Onondaga, being duly sworn, deposes and says; that he knows Francis DeWitt of the said town, a colored man, & has known him for many years past. That this deponent believes the said Francis DeWitt is of the age of eighty seven years at least. That the said Francis DeWitt is reputed and believed in the neighborhood where he lives to have been a revolutionary soldier, and that this deponent concurs in that opinion. And that this deponent resides within a mile and an half (or thereabouts,) of the said Francis DeWitt. And further this deponent says not.

Affirmed and subscribed the twenty secondsixth day of May in the year 1834, before me.

Richd. Tallcot

Lewis H. Sandford Supreme Court Commissioner in & for Onondaga County

State of New York

Onondaga County ss. I, Reuben S. Hess, Clerk of the said County & clerk of the Court of Common Pleas in and for the said County, do hereby certify, that at the date of the within deposition of Solomon Davis, James M. Allen Esq was and he still is a Judge of the said Court of Common Pleas, and that the signature "J. M. Allen, Judge of Onondaga County Courts." annexed to the said deposition, is the genuine signature of the said James M. Allen. I also certify that at the date of the above deposition of Richard Tallcot, Lewis H. Sandford was & he still is a Supreme Court Commissioner in the said County appointed & authorized to do the duties of a Judge of the said Supreme Court of said State at chambers, & that the signature "Lewis H. Sandford" thereto annexed is his proper and genuine signature. Witness my hand & the seal of the said County & Court of Common Pleas, the 27th day of May 1834.

R. S. Hess clerk

And the said Court of Common Pleas do hereby declare their opinion after the investigation of the matter, and after putting the interrogatories prescribed by the war department, that the above named applicant Francis DeWitt, was a revolutionary soldier and served as he states.

And the said Court further certifies that it appears to them that Benjamin De Puy, Cornelius Wallace, and James Terwilliger who have signed the preceding depositions, reside as stated in their said depositions, and are credible persons and their statement is entitled to credit. Also that Solomon Davis, James Terwilliger, Lewis H. Sandford and Richard Tallcot who have made and signed the respective depositions next preceding, are residents of the town of Skaneateles in the County of Onondaga, and also credible persons, and their statements are also respectively entitled to credit. And that they reside in the immediate neighborhood of the said applicant.

John Watson
David Munro Judges
J. M. Allen

I, Reuben S. Hess, Clerk of the County of Onondaga & of the Court of Common Pleas of the said County, do hereby certify that the foregoing contains the original proceedings of the said Court in the matter of the application of Francis DeWitt for a pension. In testimony whereof I have hereunto set my hand and the seal of the said Court of Common Pleas, (being the seal of my office) the twenty seventh day of May in the year one thousand eight hundred and thirty four

R. S. Hess clerk

State of New York
Cayuga County ss. Harriet Dewitt of the County of Onondaga being sworn says that she is the widow of Francis Dewitt who made application for a Pension under the act of 1832 his claim was Suspended & he died on the 14th day of December 1837.

her
Harriet X Dewitt
mark

Know all men by these presents that I Harriett Dewitt do hereby constitute

& appoint G. C. Ames Esqr of Washington my true & lawful attorney for me & in my name to ask for an examination of the papers filed by my late husband Francis Dewitt in support of his claim to a Pension & to Establish my claim to the said Pension

Witness my hand & seal this 4th dayof July 1851 In presence of
Aaron Murphy

her
Harriet X Dewitt
mark

State of New York
County of Cayuga ss. On this 4th day of July 1851 before me came Harriet Dewitt above named & acknowledged the foregoing power of attorney to be her act and deed

Harvey Heath Justice Peace

Harriet Shields married Francis Dewitt

Auburn July 24, 1852.

Dr Sir,

I am requested by Harriet Dewitt of Skaneateles County of Onondaga, N.Y. to ascertain whether a Pension has been allowed to her or not. She is the widow of Francis Dewitt, and her application was made something like a year ago, by one Aaron Murfey of Milan Cayuga County N.Y. her husband was a soldier of the Revolution. Mrs Dewitt has heard nothing from Murfey for several months; he will not answer any letters that are written to him on the subject, and it is for the reason that she can learn nothing from him that she requested me to make the inquiry for her of you, Your early attention to the above is earnestly desired.

Yours Respectfully
J H Bostwick

Jas E Heath Esq.

State of New York
Cayuga County Clerk's Office ss. I hereby certify that Harvey Heath was at the time of taking the foregoing affidavit a Justice of the peace in & for Cayuga County and duly authorized to act as such. and that the foregoing is his genu-

ine signature. Given under my hand and the Seal of the County Court of said County this 5th day of October 1852

E. B. Marvin Dep Clerk

Cayuga Co ______ Court

⋆

Rejected

Revolutionary War Pension and Bounty Land Warrant Application Files
Series M-804; Roll # 805

⋆

Nathaniel and Keziah Finch
New York (R.3,542)

To the Com. Of Pensions

Sir

Enclosed please find, the ap. of Solomon Finch, Kesiah Raplee & Hannah Curen for pension and bounty land under acts of Congress granting pensions and bounty land to the widows and Children of decd. Revolutionary Soldiers, they have constituted me their Attorney & I wish you to direct any communications on the subject to me at Tiffin Seneca County Ohio As the applicants are living a great distance apart it is desirable that any certificates or papers which may be issued from the department be so drawn that they can be executed by their Attorney without sending them to the applicants These applications although executed in June and July have only just reached here

Respectfully

Solomon Finch

⋆

The People of the State of New York,
by the Grace of God FREE and INDEPENDANT.
To Nathaniel Finch Gentleman Greeting.

WE reposing especial Trust and Confidence, as well in your Patriotism Conduct and Loyalty, as in your Valour and Readiness to do us good and faithfull Service, Have Appointed and constituted, and by these Presents DO Appoint and Constitute you the said Nathaniel Finch Adjutant of that Regiment of Militia in our County of Orange whereof John Hathorn Esquire is Colonel YOU are therefore, to take the said Regiment into your charge and care as Adjutant thereof and duly to exercise the Officers and Soldiers of that Regiment in Arms, who are hereby commanded to obey you as their Adjutant and you are also to Observe and follow such Orders and Directions, as you shall from time to time receive from our General and Commander in chief of the Militia of our said State, or any other your Superior Officer, According to the Rules and discipline of War, in pursuance of the Trust reposed in you: and for so doing This shall be Your Commission, for and during our good pleasure, to be Signified by our Council of appointment. In Testimony whereof, we have caused our Seal for Military Commissions to be hereunto Affixed. Witness our Trusty and well beloved George Clinton Esquire Governor of our State of New York, General and Commander in chief of all the Militia, and Admiral of the Navy of the same, by and with the Advice and Consent of our said Council of Appointment, at Poughkeepsie the Nineteenth day of February in the Second year of our Independance, and in the year of our Lord, one Thousand Seven Hundred and Seventy Eight

By His Excellency's Command.
Richard Hatfield Secretary

Geo. Clinton

⋆

An act. of the age of the Children of Nathan Finch and his Wife Keziah

1 John Allison Was Born May 12 an 1765
2 Solomon Was Born September ___ an 1766
3 Mary Was Born January 3 an 1769
4 Elizabeth Was Born Novr 9 an 1770
5 Keziah Was Born Aprel 20 an 1773
6 Mathew Was Born Febur 25 an 1775
7 hanah Was Born January 21 an 1777
8 Daniel Was Born January 26 an 1779

⋆

Samuel Finch Affidavit
Exhibit B

The State of Ohio

Seneca County ss. Before me the undersigned a Justice of the Peace within and for said county personally appeared Samuel Finch, of lawfull age, a resident of said county, who being first sworn, makes Oath and says that he is a grand son of Nathaniel Finch & Kesiah Finch refered to in the application of Solomon Finch and others to which this is attached as exhibit B. That he was acquainted with the said Kesiah Finch, widow of the said Solomon Finch, who performed the military service set forth in said application, that she was the identical widow of the said Nathaniel Finch, that she died at Warwick, in Orange County, State of New York some time in the yeare one thousand Eight hundred & thirteen And that Solomon Finch, Kesiah Rapplee, and Hannah Currin are children, and heirs at law of the said Kesiah Finch, widow &c, deceased, named in said application, and that she has no other immediate heirs now liveing. In the above erasures Solomon was written where it should have been Nathaniel

Sworn to and subscribed before me this third day of June AD 1856

Jacob Hossler Justice of the Peace

Samuel Finch

State of Ohio

Seneca County ss. Before me the undersigned a Justice of the Peace within and for said county and state this day came Solomon Finch a resident of Fayett County in the State of Iowa, aged Forty Seven years, who being first duly sworn makes Oath and says that he is the youngest son of John Finch deceased, who was a son of Nathaniel Finch named in the declaration to which this is attached, that the muster Roll and commission hereto attached and made part of said declaration, and also what appeares to be a family record of the family of the said Nathaniel Finch, came to the possession of affiant at the death of his said father, and have remained in his possession ever since & that from conversations had with his said father, and with his mother, and from the general appearance of said papers or documents, he believes them to be genuine, and what they purport to be, and that the ______ names and ages of the children of the said Nathaniel Finch deceased are correctly stated in said record.

Sworn to & subscribed before me this 5th day of June AD 1856

J. W. Wilson J.P.

Solomon Finch

Solomon Finch Affidavit

State of Ohio

Seneca County ss. I Phillip Spielman, Clerk of the Court of Common Pleas in and for said county and state hereby certify that Jacob ______ before whom the affidavit of Samuel Finch appeares to have been taken and sworn to N. P. Caldwell before whom the affidavits of Peter and Sarah Eastman appeare to have been taken and sworn to, and both of which are hereto attached, and Joel

W. Wilson before whom the within affidavit was made and sworn to, were at the time of so doing each acting Justices of the Peace in and for said county and state duly commissioned and sworn, that all their acts as such Justices are entitled to full faith and credit, and that their signatures thereto are genuine. In testimony whereof I have hereto signed my name and affixed the seal of said Court at Tiffin this 5th day of June AD 1856

Philip Spielman Clk
by M Noble Dept Clk

N.B. The affidavit of Peter & Sarah Eastman refered to have been _______ and their testimony taken in open Court; and an entry of the same hereunto filed

---

Solomon Finch declaration

State of New York

Wyoming County ss. On this 28th day of June AD 1856 personally appeared before me a Justice of the Peace within and for the county and state aforesaid, Solomon Finch aged Eighty nine (89) yeares a resident of the town of Castile in said county who being duly sworn according to law declares that Kesiah Finch deceased late of the County of Orange in said state, was at the time of her death the widow of Nathaniel Finch who held the commission of Adjutant in the Regiment of New York Militia commanded by Col. John Hathorn; as will appeare by his Original commission hereto attached and marked; A; and who also served as a Captain in said regiment of Militia as will appeare by reference to the muster rolls at the Office of the War department at Washington; in the revolutionary war declared by the United States against Great Brittain, and was killed at the Battle of Minisink during said war. That he is unable to state the time when the said Nathaniel Finch enlisted, or for what length of time he served in said war. That the said Kesiah Finch was maried to the said Nathaniel Finch in Orange County New York, that the date of said mariage, by whom the ceremony was performed, ~~and the ____ maiden name of the said Kesiah Finch~~ are unknown to declareant That the maiden name of the said Kesiah Finch was Kesiah Allison and that no record either public or private of the same can be found, that the said Kesiah Finch died at the said County of Orange in the yeare 1813 and that no public or private record of the same ~~of the same~~ is in existence to the knowledge of declareant, and that the said Kesiah Finch was at the time of her death the widow of the said Nathaniel Finch, and unmaried, as will appeare by reference to Affidavits hereto attached marked, B.

He further states that he, said declareant, Kesiah Raplee, of the town of Milo in the County of Yates, in said state, and Hannah Currin of the ~~City~~ County of

Bradford and State of ~~New York~~ Pennsylvania, are the only surviveing children of the said Nathaniel and Kesiah Finch, decd, as will appeare by reference to proofs hereto attached marked C.

He makes this declaration for the purpose of Obtaining whatever pension and bounty land, ~~to which~~ the said children of the said Kesiah Finch, may be entitled to under Acts of Congress granting pensions and bounty land to widows and children of deceased revolutionary soldiers.

Solomon Finch

Sworn to and subscribed before me the day and year first above named. And on the same day personally came before me Lewis Finch and Alonzo P Davison residents of said County of Wyoming, to me known as credible witnesses, who being sworn according to law declare that they are personally acquainted with Solomon Finch who has made and subscribed the foregoing declaration, that they were present and saw him sign his name to said declaration, and that from the appearance of the applicant, and their acquaintance with him they believe that he is the identical person he represents himself to be. That they are disinterested in the claim, and reside at the above named place

Lewis Finch
Alonzo P Davison

Sworn to and subscribed before me, and I certify that I am not interested in the claim or concerned in its prosecution, and I believe that the claimant is the person he represents himself to be

Anson Howard Justice of the peace

State of New York
Wyoming County ss. I _________ clerk of the County Court of in and for the county and state aforesaid, do hereby certify that Anson Howard whose name is subscribed to the certifficate of the proof of the annexed declaration, and before whom the foregoing affidavits appeare to have been made, was at the time of makeing the same a Justice of the Peace in and for the county and state above named duly commissioned and sworn, and that all his official acts as such are entitled to full faith and credit. In testimony whereof I have hereunto signed my name and affixed the seal of said court this day of AD 1856
Clerk

Know all men by these presents that I Solomon Finch, a resident of the town of Castile in Wyoming County State of New York, do hereby constitute and appoint Solomon Finch of the County of Fayett and State of Iowa my lawfull attorney to prosecute my claim for whatever pension or bounty land I may be entitled to as the son and heir of Kesiah Finch, late widow of Nathaniel Finch who served and was killed in the Revolutionary war, as set forth in the application hereto annexed Hereby authorizeing and empowering my said Attorney to obtain from the proper authorities at Washington City or elsewhere certified copies of any Army Records or documents, in which the said service of the said Nathaniel Finch may be recorded, and to procure any and all other evidence to establish this claim, as fully as he might or could do if personally present, and to receive from the department, at Washington any certifficates drafts, _______, warrants or other document in behalf of and relateing to said claim, hereby revokeing and countermanding all other authority that may have been given for the above specified purpose.

Solomon Finch

State of New York
Wyoming County ss. Before me the undersigned a Justice of the peace in and for said county and state this day personally appeared the above named Solomon Finch and acknowledged the foregoing power of Attorney to be his voluntary act and deed for the purposes therein specified. Taken and acknowledged before me this 28 day of June AD 1856

Anson Howard J.P.

State of New York
Wyoming County ss. I John H. Bailey Clerk of the County Court in and for said county and state do hereby certify that Anson Howard Esqr. before whom the foregoing acknowledgment was taken and who has thereto signed his name was at the time of so doing a Justice of the Peace in and for the county and state above named duly commissioned and sworn, that all his official acts as such Justice are entitled to full faith and credit, and that his Signature thereto is genuine In testimony whereof I have hereto signed my name and affixed my Official seal this 4th day of July AD 1856

John H. Bailey Clerk

Kesiah Raplee

State of New York

Yates County ss. On this 23rd day of July AD 1856 personally appeared before me a Justice of the Peace within and for the county and state aforesaid Kesiah Raplee aged Eighty three yeares a resident of the town of Milo in the county and state above named, who being duly sworn according to law declares that Kesiah Finch; late of the County of Orange in said state; deceased, was at the time of her death the widow of Nathaniel Finch, who held a commission, and served as adjutant in the Regiment of New York Militia commanded by Col. John Hathorn, as will appeare by his original commission hereto attached, and who also served as a Captain in said Regiment, as will appeare by reference to his original Muster Roll, hereto attached, and the muster, and pay Rolls at the office of the war department at Washington, in the war of the American Revolution and was killed at the battle of Minisink during said war. That she is unable to state the time when the said Nathaniel Finch enlisted or for what length of time he served in said war, that the said Kesiah Finch was maried to the said Nathaniel Finch in Orange County in said state, that the date of said mariage, by whom the ceremony was performed ~~and the maiden name of the said Kesiah Finch~~ are unknown to declarant That the maiden name of the said Kesiah Finch was Kesiah Allison and that no record either public or private, of the same can be found, that the said Kesiah Finch died at the said County of Orange in said State of New York in the year 1813 and that no public or private record of her death can be found, by declarant, and that the said Kesiah Finch was at the time of her death the widow of the said Nathaniel Finch, and unmaried and unmaried, as will appeare by proofs hereto attached. She further says that she; declarant; Solomon Finch a resident of the town of Castile in Wyoming County and State of New York , and Hannah Currin a resident of ~~the City and State of New York~~ Bradford County State of Pennsylvania, are the only surviveing children, and immediate heirs of the said Kesiah Finch, as will appeare by reference to proofs hereto attached.

She makes this declaration for the purpose of obtaining whatever pension or bounty land she may be entitled to as a child and heir of the said Kesiah Finch, under acts of Congress granting pensions and bounty land to the widows and children of deceased Revolutionary Soldiers.

And I do hereby constitute and appoint Solomon Finch of Fayett County in the State of Iowa my lawfull attorney to prosecute this claim, hereby authorizeing and empowering my said Attorney to obtain from the proper authorities at Washington or elsewhere certified copies of any army records or documents in which the said service of the said Nathaniel Finch may be recorded, and to procure any and all other evidence to establish my claim as fully as I might or could do if personally present, and to receive from the department at Washington, any certifficates, drafts warrants, or documents in behalf of and relateing to said claim, hereby revokeing and countermanding all other authority which may have been given for the above specified purpose.

Kesiah Raplee

State of New York

Yates County ss. Sworn to and subscribed, and the foregoing power of Attorney acknowledged before me the day and year first above named, and on the same day personally came before me Amos Ellis and William Van Orsdol residents of said County of Yates to me known as credible witnesses who being sworn according to law declare that they are personally acquainted with Kesiah Raplee who has made and subscribed the foregoing declaration, that they were present and saw her sign her name to said declaration, and that from the appearance of the applicant and their acquaintance with her, they believe that she is the identical person she represents herself to be, that they are disinterested in the claim and reside at the above named place.

Amos Ellis
Wm Van Orsdol

Sworn to and subscribed before me and I certify that I am not interested in the claim or concerned in its prosecution and I believe the claimant to be the person she represents herself to be.

George Van Orsdol a Justice of the peace in and for Yates County

---

State of New York

Yates County ss. I ~~Clerk of~~ I Lewis B Graham Clerk within and for said county and state do hereby certify that George Van Orsdol whose name is subscribed to the certifficates of the proofs of the annexed declaration, and before whom the foregoing affidavits and acknowledgments appeare to have been made and taken, was at the time of making the same, a Justice of the Peace in and for the county and state above named, duly commissioned and sworn, and that all his official acts as such Justice are entitled to full faith and credit In testimony whereof I have hereunto signed my name and affixed the seal of said court this twelfth day of September AD 1856

Lewis B Graham Clerk

Hannah Currin

State ~~and City of New York~~ of Pennsylvania

~~New York~~ Bradford County County ss. On this 28 day of July AD 1856 personally appeared before me the undersigned a Justice of the Peace in and for the County and State aforesaid Hannah Currin aged Seventy Nine years a resident of ~~said City of New York~~ the County of Bradford and State of Pennsylvania who being first duly sworn according to law declares that Kesiah Finch; late of the County of Orange in said state; decd, was at the time of her death the widow of Nathaniel Finch who held the commission, and served as Adjutant in the Regiment of New York Militia, commanded by Colonel John Hathorn, as will appeare by his original commission hereto attached, and muster and pay rolls at the Office of the War department at Washington, and who also served as a Captain in said Regiment as will appeare by reference to his Original Muster Roll hereto attached, and the records above refered to in the war the American Revolution, and was killed at the battle of Minisink during said war. That she is unable to state the time when the said Nathaniel Finch enlisted or for what length of time the said Nathaniel Finch served in said war; that the said Kesiah Finch was maried to the said Nathaniel Finch in Orange County in said State, that the date of said mariage, by whom the ceremonys was performed, ~~and the maiden name of the said Kesiah Finch~~ are unknown to declarant, that the maiden name of the said Kesiah Finch was Kesiah Allison and that no record either public or private of the same can be found; that the said Kesiah Finch died at the said County of Orange in said State of New York in the year 1813, and that no public or privat record of her death can be found by declarant. And that the said Kesiah Finch at the time of her death was the widow of the said Nathaniel Finch and unmaried as will appeare by proofs hereto attached, she further states that she; declarant; Solomon Finch a resident of the town of Castile, Wyoming County in said State of New York, and Kesiah Raplee of the town of Milo in Yates County in said state are the only surviveing children and immediate heirs of the said Kesiah Finch as will appeare by reference to proofs hereto attached.

She makes this declaration for purpose of Obtaining whatever pension or bounty land she may be entitled to as the child and heir of the said Kesiah Finch under acts of Congress granting pensions and bounty land to the widows and (________) children of deceased revolutionary soldiers

And I the said Hannah Currin do hereby constitute and appoint Solomon Finch of the County of Fayett in the State of Iowa my lawfull attorney to prosecute this claim hereby authorizeing and empowering my said Attorney to Obtain from the proper authorities at Washington or elsewhere certified copies of any Army Records or documents in which the said service of the said Nathaniel Finch may be recorded, and to procure any and all other evidences to establish my claim, as fully as I might or could do if I were personally present, and to receive from the pension office at Washington, any certifficates drafts, warrants or documents in behalf of and relateing to said claim, hereby revokeing & countermanding all or

any other authority which may have been given for the above specified purpose
Sworn & Subscribed Before me July 28, 1856
H. C. Baird J. Peace

Hir
Hannah X Corwin
Mark

Sworn to and subscribed, and the foregoing power of Attorney acknowledged before me the day and year first above named And on the same day personally came before me Cornelius Hunsiker and Miles G Rapolee residents of said ~~City~~ of ~~New York~~ County of Bradford and State of Pennsylvania, and to me known as credible witnesses who being first duly sworn according to law declare that they are personally acquainted with Hannah Currin who has made and subscribed the foregoing declaration that they were present and saw her sign her name to the same, and that from the appearance of the applicant, and their acquaintance with her they believe that she is the identical person she represents herself to be, that they are disinterested in the claim and reside at the above named place.

Cornelius Hunsiker
Miles G. Rapolee

Sworn to and subscribed before me and I certify that I am not interested in the claim or concerned in its prosecution, and that I believe the claimant to be the person she represents herself to to be.
Henry C. Baird J. Peace

State ~~and County~~ of of Pennsylvania
~~New York~~ County of Bradford ss. I Allen M Kean Clerk of the Court of __________ & Prothonotary of the Com Pleas within and for said county and state do hereby certify that H. C. Baird whose name is subscribed to the certif-ficate of the proofs of the foregoing declaration, and before whom the foregoing affidavits, and acknowledgments appeare to have been made and taken was at the time of Makeing the same a Justice of the Peace in and for said county duly commissioned and sworn, and that all his official acts as such are entitled to full faith and credit In testimony whereof I have hereunto signed my name and affixed the seal of said court this 31st day of July AD 1856

Allen M Kean Prothonotary

State of Ohio
Office of Probate Court
Seneca County ss. Be it Remembered that heretofore to wit on the 2d day of June AD 1856 an Entry was made in the Journal of said Court which reads as follows to wit

In the Matter of the Heirs of Kesiah Finch June 2 1856

On motion to the Probate Court within and for said County by the Heirs of Kesiah Finch deceased through Solomon Finch there attorney for the proof of heirship and upon producing in open Court Peter Eastman and Sarah Eastman two Credible witnesses who being duly Sworn questioned and examined the Court find from such testimony that Kesiah Finch deceased departed this life at the County of Orange in the State of New York some time during the year 1813 and at the time of her death was the widow of Nathaniel Finch who served and was killed in Battle during the Revolutionary War and that Solomon Finch a resident of Casteil in Wyoming County State of New York when last known to said witness Kesiah Raplee a resident of the Town of Milo in Yates County and State of New York when last known to witnesses and Hannah Corrin a Resident of the city and State of New York when last known to witnesses are the only surviving Children and immediate Heirs of the said Kesiah Finch and the Court further that the said Kesiah Finch remained unmarried from the time of her Husbands death up to the time of her death and that the above Named Heirs are the Children and Heirs of the said Nathaniel Finch by his said wife Kesiah Finch

---

State & County of aforesaid I J. M. Hord Deputy Clerk of the said Court do hereby Certify the foregoing to be a Correct Copy of the Journal Entry made in the above Case and entered upon the Journal of the said Court In Testimony whereof I have hereunto affixed the seal of said Court and subscribed my Name officially at Tiffin this 19 day of November AD 1856

J. M. Hord Dept Clk

Com. of Pensions
Dear Sir

Will you be so kind as to inform me whether the application of Solomon Finch, Hannah Currin and Kesiah Raplee for pension & c. have been rcd. and whether any action has yet been had on them. Any communication would be thankfully rcd.

Solomon Finch

Dec. 19th 1856

Tiffin Sen. Co. Ohio

The above ap was for pension &c for revolutionary services I have rcd no acknowledgement of the receipt of the papers & consequently cannot send the number. They were sent between the 10th and 15th of November, last. I am anxiously awaiting information. My home is in ________ if you could let me know nearly when the ap would be reached you would much oblige me.

Dec. 31st 1856

Respectfully,
Solomon Finch
Tiffin Ohio

Tiffin O. Jany 15th 57

Dear Sir

I received your communication concerning claim of Solomon Finch and others for Revolutionary pension I relied upon the law of 1798 giveing the widows of soldiers killed in battle 7 years half pay. I supposed this was due Kesiah Finch at the time of her death If you still consider that they are not under any circumstances entitled to relief, please return me the Original papers, or if this cannot be done certified copies of the same as I wish to retain them

Respectfully your obt servt
Solomon Finch

To the Com. of Pensions

Dear Sir

Will you please send me the papers relating to the application of Solomon Finch, Hannah Curin and Kesiah Raplee, for revolutionary pension, filed in your office in the month of Oct. last I think. I have not their number.

Respectfully
Solomon Finch
Tiffin Ohio
P.S. Send to me at Tiffin. The claim was rejected.
Sept. 2d, 1857 S. F.

R.3542

Nathaniel Finch New York

A list of men (in the militia ________ belonging to Capts. Blain, Bertolph, Holly and Miller) under command of Capt. Nathl. Finch, Ramapough, 6 May 177(illegible) returned from service May 20 has been removed from said claim to be sent to the War Dept.

Sent to War Dept. Jany. 16, 1913

A.C.M. Aug. 10, 1911

Rejected

Revolutionary War Pension and Bounty Land Warrant Application Files
Series M-804; Roll # 974

Henry Gunsaluss
New Jersey/New York (S.8,643)

Col. James L Edwards
Commissioner of Pensions
Washington D.C.

City of Cincinnati, Ohio
June 23rd 1834.

Sir/

I send you, inclosed, the declaration & certificates of Henry Gunsaluss, with a view to obtain a pension for him, under the Act of Congress of June 7th 1832.

The application was made to a former term of our Common Pleas court, near its close, to receive the proofs, as required by the War Department, and one witness, the Revd Wm Burke, was then sworn; but Wm Warrenton, and an old soldier of the Revolution, who had been spoken to, to attend, never presented himself until this day, when the proceedings were ________. This is the reason why the dates of the commencement and conclusion of the inclosed paper, are so far apart.

I am, sir, with great respect,
Your Obdt. Svt
H. P. Gaines

State of Ohio

County of Hamilton ss. On this 14th day of April 1834, personally appeared before the Court of Common Pleas of said County of Hamilton & State of Ohio, Henry Gunsaluss, a resident in the City of Cincinnati, in said County, aged seventy four years on the 8th day of December last, who being first duly sworn, according to law, doth on his oath, make the following declaration, in order to obtain the benefit of the provision made by Act of Congress, passed June 7th 1832.

That about the first of May 1777, he volunteered in Orange County in the State of New York, in the company commanded by Capt. ~~Dewitt~~ Newkirk in Col. McLaughlin's regiment of New York Militia, for a term of three months, and immediately marched to Newburgh, and in going to that place, passed up the Hudson river by land, that after remaining in Newburgh about one month, marched through Summerset County N. Jersey, and the town of Princeton in the same state to Trenton, at which two places last named, he performed the remainder of his services for said term of three months, that during this period of service, the applicant was engaged in a skirmish with the enemy at a place in New Jersey called Bondbrook, in which the latter were beaten, but the Americans however, lost several men, that several regular officers were with the applicant in said skirmish, but he cannot now recollect their names, that after the expiration of said term of three months, he was discharged by his captain before named, and immediately afterwards, at Trenton aforesaid, hired himself as a substitute for one John Shirts in the company commanded by Capt. (name not recollected,) in the Regiment commanded by Col. Forrest, of the New Jersey Militia, to serve three months, and served out said term, and was regularly discharged by said Captain last named, that during this period he was in the battle at Trenton on the morning of the 26th of December 1777, when Genl. Washington captured the Hessians, but was in no other battle or skirmish during said term last named, that after receiving his discharge as aforesaid, he returned to his father's residence in said County of Orange in the State of New York, and there remained until the month of Feby 1778, when he was drafted with others of Captain Dewitt's company of Militia, and marched under the command of Captain _______ in Col. Newkirk's regiment to guard the frontiers of said State of New York against the depredations of the Indians; that the period for which he was thus drafted, was six months, and he served out the same accordingly, and was regularly discharged by the Captain last named; that during this period of service he was engaged in several skirmishes with the Indians, in one of which Capt. Little of the said regiment was killed; that the scene of the said skirmishes was in Ulster County N. York; that he served with several regular officers during this tour, but cannot now recollect the names of any others than Capt. ______ and Majr. Blue of the New York Line; that the place at which he was discharged, was DeWitt's Mills, on Neversink river, & the time was in August 1778, that he was a private only during said three periods of service, and as to his discharges above mentioned, they have been lost long since; that he afterwards from August 1778 until the latter part of the year 1781, a ranger in guarding against the depredations of the Indians on the frontiers of said state, and during all of that long period, he was in said Captain Dewitt's Company of Militia, that he cannot specify his services as required at the War Department, as the business of rangers was to go out in scouting parties once or twice in every week according to emergencies, & they remained out not more than two or three days at each time making their observations on the movements of the enemy, that from the nature and character of this enemy, the company to which the applicant belonged was at all times obliged to be on the alert, that the captain who directly commanded said rangers, was named Alexander Woody, & the Col. was named Blue as before stated; that if this statement of his services while a ranger should not be deemed sufficiently certain to authorise it to be taken into account at the War Department, the applicant's claim for the same must be, and is hereby waived, & his claim is merely confined to the three periods of service first mentioned, amounting in the aggregate to one year.

The said applicant finally states, that he ~~knows of~~ has no documentary evidence, and that he knows of no person whose testimony he can procure, who can testify to his service. He hereby relinquishes every claim whatever to a pension or annuity, except the present, and declares that his name is not on the Pension roll of the Agency of any state.

Sworn to and subscribed the day and year aforesaid
Danl. Gano Clerk

his
Henry X Gunsaluss
mark

The following interrogatories were propounded by the Court to the above named applicant;

1st Where and in what year were you born?

Answer - I was born Ulster County in the state of New York on the 8th day of December 1759.

2nd Have you any record of your age, and if so, where is it?

Answer - I have a record of my age in my bible taken from the original record in my father's bible in his hand writing, more than thirty years ago.

3rd Where were you living when called into service, where have you lived since the Revolutionary War, and where do you now live?

Answer - I was living in Orange County, New York when ~~when~~ called into service, and remained there during, & until seven or eight years after the Revolutionary War, & then removed to N. Jersey & ultimately to Hamilton County about thirteen years past.

4th How were you called into service, were you drafted, did you volunteer, or were you a substitute? And if a substitute, for whom?

Answer - I first volunteered for three months, & next was a substitute for three months for John Shirts, and lastly, I was drafted for six months; as before stated, and I subsequently was a ranger until after the capture of Lord Cornwallis at Yorktown, Virginia, in which latter capacity I had to perform duty two or three days in each week, in scouting parties sent out by my Captain, to reconnoiter and give notice of the approach of the Indians.

5th State the names of some of the Regular officers, who were with the troops where you served, such continental and Militia regiments as you can recollect, and the general circumstances of your services.

Answer - I cannot recollect the names of any other regular officers than those already named; but I recollect serving with Capt. Moody of the New York Militia, in addition to the militia officers already named; & the general circumstances of my service have been fully stated already.

6th Did you ever receive a discharge from the service, and if so, by whom was it given and what has become of it?

Answer - I received a discharge at the expiration of each term of service, as above stated, signed by the respective Captains under whose command I served at the time, but all of said discharges have been lost long since.

7th State the names of persons to whom you are known in your present neighbourhood, and who can testify as to your ~~veracity~~ character for veracity, and their belief of your services as a soldier of the Revolution.

Answer - I am known to Ebenezer Hulse Esqr, sheriff of Hamilton County, Ohio, Danl. Gano Esq. Clerk of said County Common Pleas Court, Mr. Jno. Harris, Genl. James Findley; formerly a member of Congress, & numerous others, who would testify for me, as to my character for veracity, and also as to their belief of my services as a revolutionary soldier.

Sworn to & subscribed before me in open court June 23d 1834 Danl. Gano Clk

his
Henry X Gunsaluss
mark

We William Burke, a clergyman residing in the County of Hamilton and state of Ohio, and Wm. Warrenton, residing in the same County and state, do hereby certify that we are well acquainted with Henry Gunsaluss who has subscribed and sworn to the above declaration, that we believe him to be seventy four years of age, that he is reputed and believed, in the neighbourhood where he resides, to have been a soldier of the Revolution, & that we concur in that opinion.

Sworn & subscribed the day and year aforesaid

Danl. Gano Clk

William Burke

Wm. Warrington

And the said court do hereby declare their opinion, after the investigation of the matter, and after putting the interrogatories prescribed by the War Department, that the above named applicant was a Revolutionary soldier, and served as he states. And the court further certifies, that it appears to them that William Burke, who has signed the preceding certificate, is a clergyman, resident in the County of Hamilton Ohio, and that Wm. Warrenton, who has also signed the same is a resident in the same county and state, and is a credible person, and that their certificate is entitled to credit.

State of Ohio Hamilton County

★

I Danl. Gano, Clerk of the Court of Common Pleas in and for the ~~for the~~ County of Hamilton and state of Ohio, do hereby certify, that the foregoing, contains the original proceedings of the said court, in the matter of the application of Henry Gunsaluss for a Pension. In testimony whereof, I have hereunto set my hand and seal of office, this 23rd day of June 1834.

Danl. Gano Clk

★

Ohio

Henry Gunsaluss of Hamilton Co in the State of Ohio who was a Pr in the Company commanded by Captain Newkirk of the Regt commanded by Col Laughin in the New Yk Militia 1 yr $40.00 Inscribed on the Roll of Ohio at the rate of 40 Dollars Cents per annum, to commence on the 4th day of March, 1831. Certificate of Pension issued the 10 day of July 1834 and sent to H. P. Gaines, Cincinnati Ohio

Arrears to the 4th of Mch 34	120.00
Semi-anl. allowance ending 4 Sepr	20.00

140.00

{Revolutionary Claim, Act June 7, 1832.}
Recorded by J T Sprague Clerk,
Book E Vol. 8 Page 9

Invalid File No. 8,643
Henry Gunsaluss Priv. Rev War
Act: June 7th 1832
Index: - Vol. 1st, Page 154
[Arrangement of 1870.]

Revolutionary War Pension and Bounty Land Warrant Application Files
Series M-804; Roll # 1147

Benjamin Gunsolley
New York (R.4,389)

State of New York
County of Ontario ss. John Van Auken of Phelps Ontario County & State aforesaid being duly sworn deposes and says that he is Eighty four years of age, that in the time of the Revolution he was a Resident of Northampton County State of Pennsylvania & on the Banks of the Deleware River. That he was in the service of the United States during said Revolution as will appear by his papers on file in the War Department. That he is a Pensioner of the United States for said service.

That he is well acquainted with Benjamin Gunsolley an old man, now a Resident of the County of Shiwasee State of Michigan & has been acquainted with said Gunsolley ever since & During the Revolutionary War. That in the time of the said Revolutionary War, the said Benjamin Gunsolley lived on the New York side of the Deleware River in the County of Orange, it being then upon the then Indian Frontier, which was the most of the time of the Revolution infested by Indians. That at one time during said War, he saw the said Gunsolley in the Fort Called Kuykendalls Fort on the said Delware River doing duty as a Soldier. Saw him Several times doing duty as a soldier during the said War. That he understood that the prominent Officers who commanded the militia to which said Gunsolley belonged was Capt William Westfall & also one Capt Moses Cortright, there being at that time Several picket Forts or Block Houses erected for the protection of the Inhabitants on said Frontier & that all the able bodied militia were more or less in actual service during said War. That the said Gunsolley was a good & faithful soldier & served as a private soldier but in what Regiment he could not state nor can he state what the name of the Col was, he could not state the precise years or the months, or the number of months or years the said Gunsolley served as aforesaid. But as near as he can now Remember the said Gunsolley was like the deponent & most of the inhabitants in actual service in said Revolutionary War, the whole time of the Continuance thereof.

Subscribed and sworn this the 13th day of May AD 1851
Before me William Hildreth Justice of the peace of
Ontario County

John Van Auken

Ontario County ss. I William Hildreth a Justice of the peace in & for said County do hereby certify that I am acquainted with John Van Auken who subscribed his name to & who made the foregoing affidavit that he is a man of Good Moral Character & his Statements therein are Entitled to Credit.

Dated Phelps May 13th 1851.

William Hildreth J P

State of New York

Ontario County Clerk's Office ss. I, Reuben Murray, Jr., Clerk of said County, Do hereby Certify, That William Hildreth Esquire, before whom the annexed instrument was subscribed and sworn, was at the time the same purports to have been so subscribed and sworn, a Justice of the Peace in and for said County, duly elected, sworn and authorized to act as such. And further, that I am well acquainted with the hand-writing of such Justice and verily believe his signature subscribed to the jurat of said affidavit, to be genuine. In Testimony Whereof, I have hereunto subscribed my name, and affixed the Seal of the said County, this 14th day of May 1851.

Reuben Murray Jr. Clerk

State of Michigan

Shiawassee County ss. On this 4th day of Jun AD 1851 before the County Court for Shiawassee County being a Court of Record personally appeared in open court Benjamin Gunsolley of Caledonia in said county aged Ninety two years who being duly sworn and examined in and by said Court makes the following declaration in order to obtain the Benefits of the act of Congress passed in 1832 for granting Pensions to certain officers and soldiers who served in the war of the Revolution

That at the commencement of the Revolutionary war he resided near a place called "Mamacating" in the County of Orange in the State of New York

That previous to the declaration of Independence he was enrolled in the company of militia or state troops commanded by Captain William Westfall in the Regiment commanded by Col Levi Paulding

That during the whole of the war of the Revolution he continued a member of said company and was in actual service at the times and in the manner herein after stated

That during the summer season, viz. from the first of april in each and every year from & including the year 1776 until and during the year 1780 he was in actual service until late in the fall a term of eight months or about that as a member of said company

That his said services was as one of a scouting party organised in each year for the purpose of guarding the lines against the hostile Indians

That the "lines" so guarded extended from a place called "Napanee" to "Lurenkill" a distance of twelve miles in the County of Orange

That said scouting parties were not under the imediate command of said Captain Westfall but usually under the direction of inferior officers or other persons called "guides" who were well accquainted with the Country and the habits and manner of warfare of the Indians

That during the two first seasons said party was under the direction of James Chambers & John Stage as "guides" That after this other men of the party acted as guides that Anthony Van Auken Harman Rosecrants Garret Van Auken David Cuibec Elias Cuykendall William Cox William Ellsworth John Ellsworth Roswell Goff & John Goff were amongst the men that were engaged with him in said service

That during the said eight months or about that in each year he was he was in actual service all the time

That he engaged in no other business and did no other business during the time above stated

That from about the month of December to the month of April following in each year he remained at home there being then no danger from the Indians as was supposed He further says that during the said war he was in the Battle of "Mannesink at Beaver Brook" with the Indians under Brant

That he saw Col Tustan ~~who was killed~~ in said in said Battle who was killed in said Battle that he saw said Col Tustan lying dead on the ground that he also knew Captain Tyler who was also killed

That about the first of April 1781 he went again into service and was stationed in the fort at Paramus or Haverstraw and continued to Serve there in said Fort until the fourteenth day of November 1781 a term of seven and a half months. that he was discharged this year soon after the news of the taking of Corwallis was received That he received a written discharge from this term of service signed by Col Levi Paulding (which said discharge was written by Sergeant Thomas Murphy)

That during this term of Service he was under the command of Capt Moody of the regular army who had the command of said Fort, that he was drafted at this time to serve at said Fort

That his house was burned about eighteen years ago and his said discharge and all his other property therein was destroyed. That he did not Scout any during this ~~year~~ term of service, that he served all the time as a private soldier. That he has now no discharges or other evidence of his service except the affadavits of those who saw him and the affadavits of those who have long known him and understood him to have been in service

That a great many years ago he applied to a lawyer to get a pension for him but was told that he could not get one unless he would make oath that he was worth no property, that he was not willing to make said oath and gave up all idea of making application, that it is not many years since he was first informed that a law had been passed to give him a pension, that he was then old and had no one to take an interest in the matter untill lately when he came to live with his son James Gunsolley who has undertaken to get it for him

He relinquishes all claim to a pension except the present and hereby declares that his name is not on the Pension Roll of any State or Teritory

Sworn to in open Court the day & year first above written
Ebenezer F. Wade Clerk
witness
E Gould

his
Benjamin X Gunsolley
mark

★

State of Michigan

Shiawassee County I Hiram Smith of Corunna in said County a minister of the gospel do hereby certify that I am acquainted with Benjamin Gunsolley described in and who has made & sworn to the foregoing declaration, that I have made inquiry of the people in the neighborhood in which he lives and from information derived from such inquiry I am satisfied that he is aged Ninety two years and served as he has stated in said declaration

Sworn this 4th day of June 1851 in open Court
E. F. Wade Clerk

Hiram Smith

State of Michigan
Shiawassee County we the undersigned citizens of the County of Shiawassee do hereby certify on oath that we are acquainted with Benjamin Gunsolley who has sworn to the foregoing declaration that he is reputed in the neighborhood in which he lives to have been a soldier of the Revolution to be of the age and to have served as stated in said declaration and that we concur in that opinion
Sworn this 4th day of June 1851 in open Court
E. F. Wade Clerk

E N Young
Samuel Beatie

State of Michigan. In Shiawassee County of June Term one thousand eight hundred and fifty one
Present Hon A B Chipman Judge
Personally appeared in open Court on the fourth day of June AD 1851 before said Court being a court of record the above named Benjamin Gunsolley and was examined on oath and stated that the matters stated in the foregoing declaration by him signed are true and from such examination the Court are satisfied that the said Benjamin Gunsolley was a soldier of the Revolution and served as he states. It is made to appear to the satisfaction of said Court that Hiram Smith who has signed and sworn to the foregoing certificate is a minister of the gospel, and that he and the above named E N Young & Samuel Beatie who have signed and sworn to the above certificate are men of truth and veracity. In witness whereof the Court has caused its seal to be hereto affixed and the signature of the Clerk to be hereto signed

Ebenezer F. Wade Clerk

Owosso July 3 1851

Hon J E Heath
Sir
Enclosed is the application of Benjamin Gunsolley for a pension under the act

of 1832 Will you please give it an examination and communicate the result to me

Yours Very Respy
A Gould

State of Michigan
Shiawassee County ss. James Gunsolley of Caledonia in said County being duly sworn deposes and says that he is aged forty seven years and is the son of Benjamin Gunsolley whose application for a pension is hereto annexed. Deponent further says that when a boy and from that time to the present he always understood and it was generally understood and believed in the neighbourhoods where he lived that said Benjamin Gunsolley was a Revolutionary soldier. That he never heard it questioned or doubted. Deponent further says that about eighteen years ago the said Benjamin Gunsolley had his house burned and all or nearly all its contents. That he deponent has frequently from about the time of said fire spoken heard said Benjamin speak of having his discharge burned in his house, and deponent from what he knew of his father and what he then heard him and others say never doubted that he had such discharge and that the same was destroyed by said fire

Deponent further says that when about ten years old he lived with his father in the town of Phelps Ontario County New York that the deponent remembers being at the house of one Anthony Van Auken in said town of Phelps at which time the said Van Auken in a conversation with one Charles Roye stated that he said Van Auken was in service in the Revolutionary war with deponents father. That from that time to the present the remarks of people have been uniform that said Benjamin Gunsolley was a soldier in said war And further deponent saith not

Sworn this 3d day of July 1851 before me

Ira Merell Justice of the Peace

James Gunsolley

I certify that said James Gunsolley is a reputable man & a man of truth & veracity

Ira Merell Justice of the Peace

State of Michigan
Shiawassee County ss. I do hereby certify Ira Merell Esq whose name appears to the above proof & certificate was at the date of the same a Justice of

the Peace in & for said County, duly qualified, & that his signature thereunto is genuine In testimony whereof I do hereunto affix my name & the seal of the Circuit Court for said County at Corunna this 3d of July AD 1851.

Ebenezer F. Wade Clerk

Detroit April 6 1855

Hon L. P. Waldo

Sir

Please inform me at your earliest convenience whether Hannah Parson a widow of Joseph Parson a Revolutionary Soldier has ever filed her application for a Pension and if so was a Certificate issued and to whom sent by the Department.

Also whether the application of one Gunsolley a Revolutionary Soldier has ever been rec'd at the Department and whether Certificate is issued. If no Certificate has been issued in either case please inform me how each case stands. and what further proof is required

Yours Respectfully
T. H. Hartwell

James Gunsolly Power of Attorney

Thomas H. Hartwell of Detroit Michigan is hereby appointed to ~~be my~~ Atty. to procure for me the Pension money to which I may be entitled as one of the heirs of Benjamin Gunsolley a soldier in the Revolutionary War. In Testimony whereof I have hereunto set my hand and seal this 1st day of May A.D. 1855 at Corunna Shiawassee Co. Mich

Signed Sealed and delivered in presence of James Gunsolley

A. H. Beach

G W Wait

Detroit May 2 1855

Hon L. P. Waldo

Sir

Enclosed please find Powers of Attorney giving me authority to act in behalf of heirs of Benjamin Gunsolley and Joseph Parson soldiers in the Revolutionary War as instructed in your favor of the 16th ult.

I am Sir Respectfully

Your Obdt Serv't
T. H. Hartwell

Rejected

Revolutionary War Pension and Bounty Land Warrant Application Files
Series M-804; Roll # 1147

Benjamin and Sophia Hains or Haynes
New York (W.4,693 and B.L. Reg.312065 - Act 3-3-1855)

Pennsylvania
Wayne County ss. On this thirty first day of August AD eighteen hundred and twenty personally appeared in open Court of Common Pleas in and for the County of Wayne (being a Court of Record having been made such by the Laws of Pennsylvania See act of 22 May 1722 &c) Benjamin Hains aged about seventy nine years and resident in the County of Wayne who being first duly Sworn according to Law doth on his oath make the following declaration in order to obtain the provision of the acts of Congress of the 18th March 1818 and the first of May 1820, that he the said Benjamin Hains enlisted during the war on or about the month of May 1776 at a place called peen pack on the Delaware River in the State of New York in the company commanded by Captain Jacob Dewitt junior in the Regiment commanded by Col. Dubois who at that time resided near the Wall-Kill in the Town of Minisink and State of New York, in the Line of the State of New York in the Continental establishment and continued to serve in the said corps for upwards of nine months during which time his Captain was broken and Lieut. Bull took the command, that he was ordered from the said corps by his commanding officer, together with one Benjamin Brink on the Frontier as a pilot and Served in that capacity to the Troops stationed thereon, being well acquainted with the country between the Rivers Delaware and Susequehanna, along the River Lackawaxen &c. That he was ordered as a Spy by General Hand to go to Tioga Point on the Susquehanna River to ascertain the Situation of the enemy who had then encamped at that place under the command of one Butler as he beleives that on his return he was ordered as a pilot to General Sullivans army that then lay at Stroudsburgh near the Delaware River and conducted them through the wilderness to Wilkesbarre on the Susquehanna River, and from thence to Newtown in the State of Pennsylvania, and continued to Serve in the Service of

the United States untill the end of the war when he was discharged at McCarty's in the State of Pennsylvania. That he was in the Battle of Coschecton at the time a Major Wood was taken prisoner by the Indians, and also at the Battle of _____ and McCarty's with the Indians on the Delaware River and aslo at the taking of Fort Montgomery. That he at this time is not able to furnish any testimony other than than the affidavit of Walter Kimble which is hereto subjoined.

Sworn and subscribed in open court the 31st of Augt.
Sheldon Norton
Present N B Edlred

his
Benjamin X Hains
mark

Walter Kimble aged 64 years being duly sworn in open court as aforesaid doth declare and say that he is well acquainted with Benjamin Haynes, who makes the foregoing declaration and has been ever since before the year 1776 that he enlisted in the army of the United States under one Captain Jacob Dewitt Junior _____ to be acquainted that he knows the said Benjamin Haynes served as a Pilot to General Sullivans army to conduct the said corps from Stroudsburgh near the Delaware River through the woods to Wilkesbarre and from thence to Newtown on the Susquehanna River. That he served as a pilot to the Virginia Troops conducting them ~~to them~~ to the mouth of the Lackawaxen River and from thence to Coschecton. The commanding officer of this detachment this deponent does not recollect, but a person by the name of Alexander Patterson served as commissary, and further this deponent saith not

Sworn and subscribed in open court
Abisha Woodward

Walter Kimble

---

and in pursuance of the act of 1st May 1820 I do Solemnly swear that I was a resident citizen of the United on the the eighteenth day of March one thousand eight hundred and eighteen and that I have not since that time by gift Sale or otherwise disposed of my property or any part thereof with intent thereby so to diminish it as to bring myself within the provisions of an act of Congress entitled "an act to provide for certain persons engaged in the Land and Naval Service of the United States in the Revolutionary War" passed on the 18th day of March AD 1818, and that I have not nor has any person for me in trust any property or securities contracts or debts due to me, nor have I any Income other than what is contained in _____ schedule hereto annexed and by me subscribed.

Real property he has none - 1 Cow, I Straw Bed, 3 old benches used as chairs, 4 old _____ 1 Rifle Gun.

Present Nathan Eldred

his
Benjamin X Hains
mark

I am by occupation a common laborerer _____ afflicted with the Rheumatism in both my shoulders by which with other _____ for the most time rendered unable to discharge any business. I have a wife Rachel aged about 48 years, a healthy and industrious woman. I have six children none of whom reside with us and who are in such indigent circumstances as to be unable to contribute to any support

Sworn and subscribed in open court the 31 Augt. 1820 before

Sheldon Norton {Filed 31 Aug 1820}

Present N. B. Eldred

his
Benjamin X Hains
mark

Pennsylvania

Wayne County ss. I Sheldon Norton Prothonotary of the Court of Common Pleas in and for the County of of Wayne, do hereby certify that the foregoing oath and the Schedule thereto annexed are truly copied from the records of the said Court; and I do further certify that it is the opinion of the said court, that the total amount in value of the property exhibited in the aforesaid Schedule is twenty five dollars. In testimony whereof I have hereto set my hand and affixed the Seal of the said Court on this fourth day of September AD 1820

Sheldon Norton
Prothonotary of the Court of Com. Pleas of Wayne Co

Pike County ss. Aaron Westbrook aged about sixty five years being duly sworn saith that he has known Benjamin Hayns who made the foregoing declaration ever since before the year of 1776, that he enlisted in the service of the

United States in the month of May 1776 under one Captain Jacob Dewit junr of the Rangers during the war at a place called Pen Pack on the Delaware River in the State of New York in the Continental Service, that he served with him in the service of the United States for more than on year at one time, and was with him at the battle of McCarty's with the Indians on the Delaware River in State of Pennsylvania and was with him at the battle with the Indians at ~~Cochecton~~ Lackawaxen and at the close of the war the said Benjamin Hayns was honorably discharged at McCartys in the State of Pennsylvania, _____ further this deponent sayth not.

Sworn and subscribed before me this 19th day of November 1821

D Jayne J P

Aron Westbrok

Pike County ss. I do further hereby certify that the above named Aaron Westbrook is a credible witness.

D Jayne

Pike County ss. I certify that Daniel Jayne whose name appears above is a justice of the peace acting under the authority of the Commonweath of Pennsylvania as appears from the Records now in my keeping In Witness whereof I have hereto set my hand and seal of Office this 19th day of Nov. A.D. 1821

Jno Brink Recorder

Pike County ss. Philip McCarty being duly sworn saith that he is well acquainted with Benjamin Hayns who makes the foregoing declaration and has known him since before the year of 1776. That he knew him whilst he was in service of the United States in the Continental Service and have served with him as a soldier in the same I was with him at a Battle with the Indians at McCarty's which is my place of Residence and also went with him on a scouting party up the Laxawexen River just after the battle at the mouth of Lackawaxen in pursuit of the Indians and have known him serve in the Continental establishment more than one year at one time and further this deponent saith not.

Sworn and subscribed before me this 19th day of Nov. 1821

D. Jayne J P

his
Philip X McCarty
mark

Pike County ss. I do further hereby certify that the above named Philip McCarty is a credible witness and is considered a man of truth & veracity

D. Jayne

Pike County ss. I certify that Daniel Jayne whose name appears above is an acting justice of the peace under the authority of the Commonweath of Pennsylvania, as appears from the Records in my keeping and full faith and credit are to be given to his acts as such Witness my hand and seal of Office this 19th day of Nov. 1821

Jno Brink Recorder

Pennsylvania

Benjamin Hains of Wayne Co in the State of Pena who was a Private in the Regiment commanded by Colonel Dubois &c of the New York line, for the term of War Inscribed on the Roll of Pennsylvania at the rate of 8 Dollars per month, to commence on the 31 of August 1820 Certificate of Pension issued the 23d of Jany 1822 and Sent to Hon. T. J. Rogers House of Reps

Arrears to 4th of Sept 1821	97.31
Semi-anl. all'ce ending March 1822	48
	$145.31

{Revolutionary claim, Act 18th March, 1818.}

Pike County ss. Personally appeared before me the subscriber a Justice of the peace for said County Thomas Van Etten of the township of Montague County of Sussex and State of New Jersey, about two miles distant from the town of Milford in Pike County who being duly sworn according to law deposeth & saith that he is in the seventieth year of his age and that he was acquainted with Benjamin Hanes & Sophia Smith from his infancy and during the revolutionary war, that the said Sophia became the reputed wife of said Benjamin and that the said Benjamin & Sophia lived and cohabited together as man & wife before the battle was fought between the Americans and indians on the west side of the Delaware River in Pennsylvania about four miles distant from the place now called Milford in Pike County Penna. of which said battle this deponant has a distinct recollection and has no doubt but that the said Benjamin was in said engagement and this deponant further says that Sopherine Westbrook, by whom the said Benjamin & Sophia were said to be married was a Justice of the peace for the County of Sussex and State of New Jersey and that the said Sopherine Westbrook married the sister of this deponant that the said Sopherine Westbrook has been dead about eight or nine years, and that he did not before his death act as a Justice of the peace for about forty years or thereabouts and further this deponant says that the said Benjamin Hanes husband of said Sophia was reputed and believed by his neighbours to have been a soldier of the revolution and further this deponant saith not.

Sworn & subscribed before me this 27th Decr 1837
Francis A L Smith J. P.

Thomas Van Etten

Pike County ss. I Francis A L Smith a Justice of the peace for said County do hereby certify that I am well acquainted with Thomas Van Etten the deponant above named and that he is a credable person and, is entitled to be believed. In witness whereof I have hereunto set my hand & seal this 27th Dec 1837.

Francis A L Smith J. P.

Pennsylvania
Pike County ss. I Daniel W Dingman Jr Recorder of Deeds &c in and for said County and Prothonotary of the Court of Common Pleas of said County Do hereby certify that Francis A L Smith Esquire before whom the above and foregoing affidavit or deposition was made and taken was at the time thereof & still is an acting Justice of the peace duly commissioned by the Governor of the Commonwealth of Penna., and further that I am well acquainted with the hand writing of said Justice and know the signature to said affidavit annexed purporting to be his is genuine & further that full faith &credit is & ought to be given to his acts & doings as such. In witness whereof I have hereunto set my hand & affixed the seal of said Court at Milford this 27th December AD 1837.

D. W. Dingman Jr Proty

And the said court declair it their opinion that the said applicant is entitled to a pension That she is of good repute for honesty, and for truth and veracity, and that her statements are entitled to credit. Also that Roger Haynes who is this day examined in court as a witness is also of good repute for truth and veracity, and that his statements are also entitled to credit.

State of Pennsylvania

Luzerne County ss. I John Smith Prothonotary of the court of common pleas of Luzerne County do certify that the above and foregoing hereunto attached are the original proceedings had in open court in the matter of the application of Sophia Haynes for a pension In witness whereof I have hereunto set my hand and affixed the seal of the said court at Wilkes Barre this Third day of January in the year of our Lord one thousand eight hundred and Thirty eight

Jno Smith Prothonotary

Penna.

Wayne County ss. Personally appeared before me the subscriber a justice of the peace in and for said County James Manning who after being duly sworn deposeth and saith that he was acquainted with Benjamin Haynes in his life time that immediately or a short time preceding his death and during his last sickness the physician order'd wine to be administered to the sd Benjamin Haynes which was procured from deponent & which is charged to him on his books; the last item of which is under date of 12th July Anno Domini 1822 and to the best of his recollection the said Haynes died a short time afterwards not exceeding three or four weeks. Deponenent further saith that the said Benjamin Haynes at the time of his decease was a pensioner of the United States for services rendered during the revolutionary war and further saith not.

Sworn & subscribed before me the 30th day of Dec. 1837

Thos. Spangenberg Justice of the Peace

James Manning

Wayne County ss. I Thomas Spangenberg a justice of the peace in and for said county do hereby certify that I am well acquainted with James Manning and that he resides and for the last 20 years has resided in the Borough of Bethany County aforesaid, that he now holds a commission and is an acting justice of the court of Com pleas in & for said County and that full faith & credit may be given to the above deposition.

Thos. Spangenberg

State of Pennsylvania
Wayne County ss. I Paul S. Preston, Prothonotary of the County Courts of Common Pleas in and for said County do hereby certify that Thomas Spangenberg is an acting Justice of the peace in and for said County, duly commissioned and authorised by the laws of the state aforesaid to act as such and further do certify that the signature within purporting to be his is true and genuine. In testimony whereof I have hereunto set my hand and affixed the seal of said court at Bethany this thirtieth day of December A.D. 1837.

Paul S Preston Prothy.
per P. G. Goodrich

State of Pennsylvania
Pike County ss. Personally appeared before me the subscriber a Justice of the peace of the town of Milford in and for the County of Pike aforesaid Abraham Kittle a Revolutionary Soldier and Pensioner of the United States who resides in the township of Montague in the County of Sussex in the State of New Jersey about seven miles distant from said Town of Milford in Pike County and being duly sworn according to law, deposeth and saith that he was well acquainted with Benjamin Hanes a Revolutionary Soldier and Sophia his wife formerly Sophia Smith before their marriage and before the commencement of the Revolutionary war. That this deponant does not remember the month or year of the marriage of the said Benjamin Hanes and Sophia Smith but he is confident they were married during the Revolutionary war and sometime before the close and termination of said war - that this deponant understood at the time of their marriage and verrily blieves they were married by Sopherine Westbrook a Justice of the peace for the County of Sussex in the State of New Jersey. That the said Justice is now dead & has been dead for about ten years. That the said Benjamin Haines and Sophia lived and cohabited together as man and wife and were reputed as such by all who knew them in the neighbourhood where they lived during the Revolutionary war. And further this deponant recollects distinctly that the marriage of said Benjamin Hanes and Sophia took place before before the Engagement or Battle was fought on the Pennsylvania side of the Delaware River between the Americans under the command of Captains Van Etten, Ennis & Westbrook and the Indians about four miles distant from the town of Milford in said Pike County and that the said Benjamin Hanes was in said engagement or battle. And further said deponant recollects that the said Benjamin and Sophia were married before the battle was fought at ~~Half Way~~ Beaver Brook on the Delaware in the State of New York and this deponant has no doubt and has every reason to believe that said Benjamin

Hanes was in said Battle or Engagement at ~~Half Way~~ Beaver Brook, and further that this deponant's acquaintance with said Benjamin continued from the period of the Revolutionary war, until the time of his death and further this deponant saith not.

his
Abraham X Kittle
mark

Pike County ss. I do certify that the foregoing deposition of Abraham Kittle was duly taken sworn to and subscribed before me a Justice of the peace as above described and I do further certify that the said Abraham Kittle the above deponant is a person of good character & worthy of the utmost credit. In testimony whereof I have hereunto set my hand & seal this 23d day of December AD 1837.

Francis A L Smith J. P.

State of Pennsylvania

Pike County ss. I Daniel W Dingman Jr Recorder of Deeds &c in & for the County of Pike and Prothonotary of the Court of Common Pleas of said County Do hereby certify that Francis A Smith Esq. before whom the foregoing affidavit or deposition was taken, was at the time thereof and still is an acting Justice of the peace duly commissioned by the Governor of the Commonwealth of Pennsylvania, and further I do certify that full faith and credit is & ought to be given to his acts and proceedings as such ~~Justice~~ and that I am well acquainted with the signature of said Justice and that the said signature purporting to be his annexed to the foregoing deposition is genuine and in his own proper hand In testimony whereof I have hereunto set my hand & affixed the seal of said Court at Milford this 27th December 1837.

D. W. Dingman Jr Proty

Pike County ss. Personally appeared before me the subscriber a Justice of the peace in and for said County Samuel Helm a Soldier of the Revolution war and Pensioner of the United States, and being duly sworn according to law deposeth and saith that he was acquainted with Benjamin Hanes from his infancy, and during the revolutionary war, and that he was also acquainted with Sophia Smith who became the wife of said Benjamin Hanes and that the said Benjamin Hanes at the time of his marriage with said Sophia resided in the County of Sussex and State of New Jersey, and this deponant further says that, he was present and saw the said Benjamin Hanes and Sophia married and this deponant further says that said Sophia before her marriage to said Benjamin, resided & lived in the house of this deponant in the said County of Sussex, and the said Benjamin & Sophia were married by Sopherine Westbrook Esq. a Justice of the peace for the County of Sussex in the State of New Jersey. This deponant further says that he does not distinctly remember the month or year when the said Benjamin & Sophia were married, but he is quite sure that they were married before the battle was fought between the Americans under the command of Captains Westbrook & Van Etten, and the indians on the west side of the Delaware River now in the County of Pike and about four miles from the town of Milford in said Pike County in which said engagement this deponant was wounded by a shot from an indian which passed through both his thighs, and that the said Benjamin Hanes was with this deponant in said engagement, and that deponant thinks that this battle was fought in the month of March seventeen hundred & eighty as deponant blieves and that the marriage of said Benjamin & Sophia took place before said battle was fought, and this deponant further says that the said Benjamin Hanes was always reputed and blieved by his neighbours, to have been a soldier of the revolution and to have been a soldier in the battle fought at Bever Brook in York State on the Delaware River, and further this deponant says that he was out on several occasions as an indian spy with the said Benjamin Hanes and on one occasion started with said Hanes from ~~the fort~~ Westbrooks Fort in Sussex County New Jersey, and marched from thence through the woods to the mouth of the Lackawaxen River, and from thence to Coshecton, with a view to discover the position of the indian enemy who were about that time making frequent incursions upon our frontiers, and further saith not

Sworn & subscribed before me this 27th Dec 1837.
Francis A L Smith J. P.

Samuel Helm

Pike County ss. I Francis A L Smith a Justice of the peace for said county do hereby certify that I well acquainted with Samuel Helm the deponant herein before named, that I have known him for upwards of ~~forty y~~ thirty years past, that said Samuel Helm resides in the township of Milford in Pike County and that he is a person of good character and is entitled to credit & belief In witness whereof I have hereunto set my hand & seal this 27th Dec 1837

Francis A L Smith J. P.

Pennsylvania

Pike County ss. I Daniel W Dingman Jr Recorder & Prothonotary of Pike County Do hereby certify that Francis A L Smith Esq before whom the foregoing affidavit or deposition was made & taken was at the time thereof and still is an acting Justice of the peace ~~duly~~ for said county duly commissioned by the Governor of said Commonwealth, and further that I am well acquainted with the signature of said Justice and know the same to said affidavit annexed purporting to be his is genuine & in his own proper hand & further full faith & credit is & ought to given to his acts & doings as such Justice In Witness whereof I have hereunto set my hand & affixed the seal of the Court of Common Pleas of Pike County at Milford this 27th December AD 1837.

D W Dingman Jr Proty

J L Edwards Esq

Sir,

You will please examine the Enclosed, and give me such information as within your power in relation to the application

Very Respectfully Yours

David D Wagener

House of Reps

March 3 1838

House of Rep. March 21, 1838

J L Edwards Esq

Dr Sir,

I enclose you Mr Jones letter, which you will please notice. I leave the subject, with a single remark, that I have not a doubt, but there is a mistake in the Name I have no idea that any deception was thought of.

Yours Respectfully

David D Wagener

return the Enclosed Letter, with your Reply

James L. Edwards Esquire

Wilkes Barre
January 5, 1838

Dr Sir

I have this day deposited the proceedings had before the Court of Common Pleas of Luzerne County in the case of Sophia Haynes an applicant for a pension under the act of Congress of July 4, 1836, in the Post Office directed to the Secretary at war If the evidence is not complete, you will please address me at Carbondale my place of residence. If it is deemed sufficient you will please send me the certificate. Her husband was pensioned in 1821 or 2 his dockumentary evidence is on file in your office. Have you instructions as to the mode of obtaining pensions under the various acts of Congress, and could you furnish me? by so doing you would much oblige

Your Obedient Servant
Lewis Jones Jr

P.S. Mrs Haynes is in needy circumstances

⋆

State of Pennsylvania ss. On the third day of January AnDi 1838 personally appeared before the Judges of the Court of Common Pleas of the County of Luzerne Sophia Haynes, (widow of of Benjamin Haynes a revolutionary soldier and pensioner of the United States deceased) a resident of Carbondale Luzerne County Penna aged about Seventy eight years who being first duly sworn according to law doth on her oath make the following declaration in order to obtain the benefit of the provision made by the act of Congress passed July 4th 1836 That she is the widow of Benjamin Haynes who was a private soldier in the ~~militia~~ Revolutionary war. That at the time her said husband entered the army he was as near as she recolects a resident of Montague township, Sussex County New Jersey he first served as an enlisted soldier and continued in the army to the close of the war whether all the time under enlistments she does not know. That he served under Captains Van Etten Ennis and Westbrook and was as she recollects once under the immidiate command of General Washington. That he was in the battle of Trenton and in a battle fought between the Americans and Indians on the west side of the Delaware river in Pennsylvania about four miles from the place where Milford now is in Pike County he was at this time as she thinks under the command of Captain Van Etten. That he was also in the battle of Beever brook in the State of New York and as she thinks at that time under the command of Captain Westbrook she recollects also that he was in a skirmish with the indians at the mouth of the Lackawaxen river, that about this time her husband took a tory by the name of Lord to the Easton Jail whom he with some others took off his horse at or near Cochecton on the delaware. She further says that her husband was both before and after their marriage with the army as a soldier in diffirent parts of the country, but she does not recollect all the places he passed through, but that he was as near as she can recollect in Sulivan's army at the time he went up the Susquehanna river in pursuit of Indians She further declares that she was married to the said Benjamin Haynes in the year of our Lord one thousand Seven hundred and Seventy eight or nine she does not recollect certainty which that her husband continued in the army after they were married till its close and was in several engagements some of which she has mentioned after their marriage. That her husband the aforesaid Benjamin Haynes died on the 18th day of July 1822 and that she has remained a widow ever since that period as will more fully appear by ~~the~~ reference to the proof hereto annexed. She further says that she has no documentary evidence of her husbands services or of his discharge, but that he was himself a pensioner up to the time of his death and that the evidence upon which he was pensioned she presumes is on file in the pension office that he was pensioned as she thinks in the year 1821 or 1822.

Sworn and subscribed to in open court on the 3d day of January 1838
Jno Smith Prot

her
Sophia X Haynes
mark

State of Pennsylvania
Luzerne County ss. Personally appeared in open court this third day of January AoDi 1838 Roger Haynes who being duly sworn according to law deposeth & saith that he is the son of Benjamin Haynes died a pensioner of the United States and a soldier of the revolution. That his father died on or about the 18th of July AoDi 1822 That his mother Sophia Haynes ~~has~~ the present applicant for a pension has remained unmarried ever since the death of ~~the~~ her husband the aforesaid Benjamin Haynes, ~~that~~ to his certain knowledge

Sworn & subscribed in open court the 3d day of January 1838
Jno Smith Prot

Roger Haynes

Pennsylvania
Wayne County ss. Personally appeared before me the subscriber one of the Judges of the Court of Common Pleas in and for said County Thomas Spangenberg Esquire of the Borough of Bethany in said County aged sixty two years and

upwards who being duly affirmed according to Law doth depose and say, that he has resided in this County for more than fifty years last past. That he knew Benjamin Haynes who was a pensioner under the Laws of Congress and who died in said County about the year Eighteen hundred and twenty two. That he never knew or heard of any other man by the name of Benjamin Haynes That he knew the wife of the said Benjamin Haynes and has known her for more than forty years past and always knew her by the name of Sophia Haynes. That the said Benjamin Haynes had the reputation of being an indian spy and an indian killer in the Revolutionary war and that he had the particular reputation of killing an indian by the name of "Kanope." That he never knew of any other Haynes than the said Benjamin who had the reputation of being in the Revolutionary war or of having a pension

~~Sworn~~ affirmed and subscribed before me this 17th day of March AD 1838.

Thos Spangenberg

James Manning

Pennsylvania

Wayne County ss. Personally appeared before me the subscriber one of the Judges of the Court of Common Pleas in and for said County Eliphalet Kellogg of the Borough of Bethany and County aforesaid aged seventy two years and upwards who being duly sworn according to Law doth depose and say That he has resided in this County for more than forty years last past That he knew Benjamin Haynes who was a pensioner under the Laws of Congress in Eighteen hundred and twenty two, and that he never knew or heard of any other man by that name That he knew and has known the wife of the said Benjamin Haynes decd for more than Thirty years last past, and that the woman who is now making an application for a pension was the wife of the said Benjamin Haynes. That the said Benjamin Haynes had the reputation of taking a part in the war of the Revolution. That he never knew of any other man by the name of Haynes who had that reputation. That he believes the individual who drew up the application of the said Benjamin Haynes made a mistake in calling her "Rachael." That the deponent is informed that the papers and application of Benjamin Haynes in 1822 was drawn up by Nathaniel B Eldred who now resides in Warren County Penna. a distance of Three hundred miles from here.

Sworn and Subscribed before me this 17th day of March 1838

James Manning

E Kellogg

Pennsylvania

Wayne County ss. I Paul S Preston Prothonotary of the Court of Common Pleas in and for said County do hereby certify that James Manning before whom the foregoing depositions were taken is an acting associate Judge for said County duly commissioned and qualified and that the individuals who have subscribed to the foregoing depositions are men of integrity

In testimony whereof I have hereto set my hand and Seal of Office this seventeenth day of March Amo Domini 1838, and that the signatures of the said James Manning, and of Thomas Spangenberg and Eliphalet Kellogg attached to the foregoing depositions are true and genuine.

Paul S. Preston Prothontary
per P. G. Goodrich

House of Reps
Saturday May 27 1838

J. L. Edwards Esq.

My Dear Sir,

The Bearer is Mr Haynes Son of Sophia, widow of Benjamin Haynes who has come in with additional evidence in relation to the Identity of his mother. I hope they will be satisfactory. Judge Eldred (a Gentleman of the first respectability) has in his communication clearly made out that the Error was made by him, & that he has no doubt, that this is the same Lady, who was the wife of Haynes. About leaving the City for a few Days, prevents me from seeing you personally. I do trust that you will have no further difficulty in granting her the Pension.

Yours
David D Wagener

My Dear Sir

The brave Roger Hains will call on you he is an old acquaintance of mine and has some business in the Pension department which he will explain to you. I will return it a favour if you will render him such assistance as you can consistently.

In haste with Great Respect
Yours
N. B. Eldred

Honl. D. D. Wagener
A. __________

Genl. S. McKean

May 19th 1838

Smethport McKean County Penna
May 19th 1838

To J. L. Edwards
Dear Sir

I have been called on by the friends of the widow of Benjamin Haynes deceased formerly of Wayne County Penna to make some explanation in relation to the declaration of Benjamin Haynes for a pension whose papers were drawn up by me and in which he states the name of his wife to be "Rachel." I well recollect that I drew up the papers for Benjamin Haynes but I cannot now distinctly recollect the time but it was not far from the year Eighteen hundred and twenty. I then resided in Wayne County Penna. I became acquainted with Benjamin Haynes in A.D. 1813 or 1814 and frequently saw him untill his death I had seen his wife several times she was an aged woman at the time I drew her husbands declaration, but her name I do not recollect, but she was the same woman with whom he resided when I first knew him and resided with him at his death and was afterwards called his widow. I learn from Roger Hains the (son of Benjamin) and from the affidavits of E. Kellogg and T. Spangenburg Esqr that the widows name is "Sophia." I cannot account certainly for this variation in the name and whether the mistake is one of my own I cannot say, but I am confident that there is some mistake in drawing up the declaration as there was no other person of the name of Benjamin Haynes in the County of Wayne or elsewhere to my knowledge but the one whose declaration I prepaired. I know the whole family and saw the widow after the death of her husband, she was the same person who had passed for his wife and beleived to be such ever since I first knew the family. After the death of her husband Benjamin she called on me in relation to her husbands pension. She was then an old woman

Benjamin Haynes was ~~a~~ low dutchman and spoke very broken English and I recollect that I had great difficulty in understanding him sufficiently to draw up his declaration, besides he was a very ignorant man in feeble health and his mind at times was somewhat impaired by age and infirmity. I recollect the circumstance that he gave me his age at first quite too young and afterwards corrected himself by conversing with an old man who knew him from a boy. If my recollection serves me he could neither read nor write nor had he any record of his age - from the affidavits of Mr. Kellogg and Mr. Spangenburg (who I very well know) I cannot for a moment doubt but the widows name is inserted wrong in the declaration. I may have misunderstood it considering the broken manner in which he spoke or I may have stated her name wrong in the hurry of drawing up the papers something of this kind must have occurred if her name is "Sophia," for I cannot be but satisfied from my own knowledge that the woman now applying is the widow of Benjamin Hains whose pension papers I drew and who was placed on the pension list, having known her long before as well as since the death of Benjamin Haynes. I make this statement with pleasure and hope it may be satisfactory as I should regret extremely that the old lady should be deprived of her pension & particularly if it was occasioned by any mistake of my own

Very Respectfully Yours &
N. B. Eldred

Smethport McKean Co Penna
May 19th 1838

To J. L. Edwards Esq.
Dear Sir

I have been called on by the friends of the widow of Benjamin Hains deceased formerly of Wayne County Penna to make some explanation in relation to the declaration of Benjamin Hains for a pension whose papers were drawn up by me and in which he states the name of his wife to be "Rachel." I well recollect that I drew up the papers for Benjamin Hains but I cannot now distinctly recollect the time but was not far from the year of Eighteen hundred and twenty I then resided in Wayne County Penna. I became acquainted with Benjamin Hains (who made his application for a pension) in Eighteen hundred and thirteen or fourteen and frequently saw him untill his death I had seen his wife several times. She was an aged woman at the time, but her name I do not now recollect but the same woman with whom he resided when I first knew him, was living at his decease and called his widow. I understand from Roger Hains the son and from the affidavits of E. Kellogg and Thomas Spangenburg Esq that the widows name is "Sophia." I cannot account certainly for this variation in the name and whether the mistake is one of my own I cannot say but certain I am that there is some mistake in drawing up the original papers, as there was no other person of the name of Benjamin Haynes in the County of Wayne, or elsewhere to my knowledge but the one whose declaration I drew up. I know the whole family and saw the widow after the death of her husband Benjamin, she was the same person who had passed for his wife and believed to be such ever since I first knew the family, after the death of her husband she called on me in relation to her husbands pension, her age appeared to be grater than that mentioned in her husbands declaration. She was then and old woman. Benjamin Haynes was ~~a~~ low dutch, & spoke very bad English and I recollect that I have great difficulty in understanding him sufficiently to draw his declaration, he was a very ignorant man and in feeble health and could not give me truly his own age. I recollect the circumstance that he give me his age at first quite too young and corrected it afterwards by a conversation with an old man who knew him from a boy. If my recollection serves me he could neither read nor write nor had any record of his age - from the affidavits of Mr. Kellogg and Mr. Spangenburg I cannot for a moment doubt but the widows name is inserted improperly. I may have misunderstood Benjamin as to the name as his language was partly low dutch & partly English and a mixture of both, or I may have stated her name wrong in the hurry of business. Something of this kind must have occurred if her name is "Sophia" for I cannot be otherwise than satisfied from my own knowledge of her that she was his widow having known her so long before Benjamins death and also subsequent to that time. I make this statement with pleasure and hope it may be satisfactory to the department as I should regret extremely that the old lady should be deprived of her pension and particularly if it was occasioned by any mistake of my own

Very Respectfully Yours
N. B. Eldred

McKean County ss. N B Eldred being duly sworn saith that the facts set forth in the above and foregoing letter are true to the best of his knowledge and belief

Sworn and subscribed before me this 21st May 1838
John E _____ J P

N. B. Eldred

Pennsylvania

Sophia Haynes widow of Benjamin Haynes who died on the 22 July 1822 of Luzerne Co. in the State of Pennl. who was a Private in the commanded by Captain of the Regt. commanded by Col. Dubois in the N. York line for 7 months Inscribed on the Roll of Pennsylvania at the rate of 23 Dollars 33 Cents, per annum, to commence on the 4th day of March, 1831. Certificate of Pension issued the 31st day of May 1838 and sent to Roger Haines Esqr. Present

Arrears to the 4th of March '38	163.33 1/3
Semi-annual allowance ending 4 Sept '38	11.66 2/3
	$175.00

Recorded by D. Brown Clerk,
{Revolutionary Claim, Act July 4, 1836, Section the 3}

Book A Vol. 2 Page 140

J L Edwards Esq
My Dear Sir,
I enclose you additional Evidence in the case of the Widow Haynes, for an increase of Pension. Your prompt attention will oblige your friend _____
June 14 1838

David D Wagener
House of Rep

_____ Mr. Curtis _____ to which I also call your attention

D. D. W.

Penna Pike County ss. Samuel Helm of the township of Milford Pike County, Pensioner of the United States, in the seventy ninth year of his age being duly sworn according to law deposeth & saith that he was well acquainted with Benjamin Hanes a revolutionary soldier who served in Captain Peter Westbrook company of Col Jacob Rosencrance Regiment of New Jersey millitia that in the month of May ~~or June~~ in the year 1779 the said Benjamin Hanes together with this deponant enlisted in the company of said Peter Westbrook then stationed at a place called "Westbrooks fort," in the County of Sussex and State of New Jersey and the said Benjamin together with this deponant continued in the service of the United States as soldiers under the command of said officers until the month of November following that the said Benjamin & this deponant in the following year to wit 1780 in the month of March again enlisted in the service of the United States under the said Peter Westbrook and the said Benjamin and this deponant were in the battle or engagement fought by the Americans and the indians in the month of March in the same year on the west side of the Delaware River on the Coneshaugh Creek now in the County of Pike, in which engagement Captain Peter Westbrook & Lieutenant Ennis were both killed by the indians and at the same time this deponant was wounded in both thighs, and after the death of Captain Westbrook, deponant blieves Captain Peter Putman succeded to the command of the said "fort" and the said, deponant after his recovery from his wounds, rejoined his company to which the said Benjamin belonged and deponant further says the said Benjamin continued to serve as a soldier during the year last mentioned until about the month of November following, that in the year following to wit1781, said Benjamin together with this deponant entered in the month of April ~~entered~~ into the service of the United States under said Captain Peter Putman of Col.

Jacob Rosencrants Regiment then stationed at said Westbrooks fort, and the said Benjamin with deponant continued to do duty as soldiers that year until the month of October following, after which time deponant removed and went to a place called, Mamakating Hollow in New York State The services of said Benjamin & deponant during the first second & third years that they served as soldiers under the same officers, consisted in making frequent marches of exploration on the frontiers, in gaining & giving inteligence of the indian enemy and in doing the ordinary duties of a soldier and further saith not.

Sworn & subscribed before me this 9th June 1838.

D. W. Hagerty J. P.

Samuel helm

Pike County ss. I D W Hagerty Esq. a Justice of the peace for said county do certify that I am well acquainted with Samuel Helm the above named deponant, that he is a person of good character, & his statements entitled to credit & belief.

D. W. Hagerty J. P.

Pennsylvania

Pike County ss. I Daniel W Dingman Jr Recorder of Deeds &c, and Prothonotary of the Court of Common Pleas of said county Do hereby certify that David W Hagerty Esq. before whome the foregoing affidavit was taken and by whom the above certificate was made was at the time thereof and still is an acting Justice of the peace for said county duly commissioned &c and further I certify that I am well acquainted with the hand writing of said Justice & know the signature of said Justice purporting to be his to said affidavit & certificate annexed to be genuine In Witness whereof I have hereto set my hand & seal of said court this 9th June 1838.

D W Dingman Jr _____

Pennsylvania

Pike County ss. Personally appeared before me the subscriber a Justice of the peace for said County Jacob Lott of the township of Milford in said County and being duly sworn according to law deposeth and saith that he was well acquainted & knew Benjamin Haynes a soldier of the Revolutionary War, that said Haynes resided at that time in the County of Sussex in the State of New Jersey, that this deponent distinctly recollects that said Haynes served under Captain Abraham Shimer of Colonel Jacob Rosencrants Regiment of New Jersey Militia. This deponent cannot say that said Haynes was enlisted under said Shimer for any definite time nor can this deponant say in what year during the Revolutionary War said Haynes served under said Shimer, nor can this deponant say what length of time said Haynes served under said Shimer. Yet he this deponant is confident said Haynes done the duties of a soldier under said Shimer & other officers during the said war and said Haynes was stationed under said Shimer at a place in the township of Montague in said Sussex County New Jersey an this deponent further says that said Haynes served under said Shimer on the frontiers "one season" which generally consisted of five or six months and further this deponent saith not.

Sworn & subscribed before me July 23d 1838.
D. W. Hagerty J. P.

his
Jacob X Lott
mark

⋆

Pike County ss. I hereby certify that Jacob Lott who subscribed and sworn to, the foregoing affidavit is a person of good character, & worthy of belief and that I am well acquainted with said deponant. Witness my hand this 23d day of July 1838.

D. W. Hagerty J. P.

⋆

Pike County ss. I Daniel W Dingman Jr Recorder &c & Prothonotary of Pike County, do hereby certify that David W Hagerty Esquire before whom the foregoing proceedings were had was at the time thereof and still is an acting Justice of the peace duly commissioned &c and further the signatures thereto purporting to be said Justices to said affidavit & certificate affixed are genuine In testimony whereof I have hereunto set my hand & affixed the seal of the Court of Common Pleas of said County this 23d day of July 1838.

D. W. Dingman Jr.

⋆

State of New Jersey
Sussex County ss. On this tenth day of Jany in the year of our Lord Eigh-

teen hundred and thirty nine personally appeared before me the subscriber one of the Justices of the peace in and for said County of Sussex Abraham Kittle Eaged Eighty three yearse on the third day of March next, who being duly sworn according to Law, doth depose and say that he resides in the township of Montague in said County of Sussex that he was well acquainted with Benjamin Haynes, that he entered in the armey of the Revolutionary Ware under the command of Capt. Shime on the frunteere on the Delaware River in said County of Sussex in the month of March or Aprl. in the year 1776 that he entered for nine months Saw him in the servis severl severl timmes threw the coarse of the season but what time he quit the servis this deponent dose not say or know and further sayith not

Sworn and subscribed the day and year above said
John Hull & witness

his
Abraham X Kittle
mark

I certify that the above named deponent Abrm. Kittle is a credable person and full faith and credit may be Given to him

John Hull J. P.

State of New Jersey
Sussex County ss. I Joseph E Edsall Clerk of the said Court of Common Pleas in and for said County of Sussex do hereby certify that John Hull Esquire before whom the foregoing deposition appears to have been taken was at the time of taking the same and still is a Justice of the peace in and for said County duly commissioned and sworn and that his signature above written is genuine

In testimony whereof I have hereunto subscribed my name and affixed the seal of said Court at Newton in said County this seventeenth day of January A.D. Eighteen hundred and thirty nine

Jos. E. Edsall

Montage Susex Co. St of New Jersey
Mr Edwards Esq
Der Sir
I was called on By A friend of Sophia Haynes conserning howe long her hus-

ban Benjamin Haynes had served in the Revolution war. And I dewe know of his service _____ monts At one time under Capt Westbrooke _____ in the fort called Capt Courtright fort for one year But was called out _____ foure Day in the service for the _____ yeare this I Dewe Sertyfie to that I Am Not Abell to go Before A _____ And their is Not one within ___ or ___ mils Witness my hand _____

Witness MaryCrowel
_____ Decker

his
Abraham X Kitel
marke

☆

Treasury Department,
Second Comptroller's Office,
August 20th, 1839.

Sir:

Under the act of the 6th of April, 1838, entitled "An act directing the transfer of money remaining unclaimed by certain Pensioners, and authorizing the payment of the same at the Treasury of the United States," Sophia Haynes a Pensioner on the Roll of the Philadelphia Agency, at the rate of Twenty three Dollars and thirty three Cents per annum, under the law of the 4th July, 1836, has been paid at this Department, from the 4th of March, 1838, to the 4th Sept. 1838.

Respectfully, yours,
Albion K. Parris
Comptroller.

To the Commissioner of Pensions,
Present.

☆

Washington DC
Jany 23d/42

My Dear Sir
please examine & report to me on the within in order that I may reply to it

Most Respectfully
B. A. Bidlack

☆

Treasury Department,
Second Comptroller's Office,
Nov. 18th 1842.

Sir:

Under the act of the 6th of April, 1838, entitled "An Act directing the transfer of money remaining unclaimed by certain Pensioners, and authorizing the payment of the same at the Treasury of the United States," and the 3d section of the Act of August 23, 1842, extending the time within which pensioners may receive their pensions from the pension agents, Sophia Haynes, a Pensioner on the Roll of the Philadelphia, Pa. Agency, at the rate of Twenty three Dollars and 33 Cents per annum, under the law of the 4th July 1836, has been paid at this Department, from the 4th of March 1840, to the 4th Sept. 1841.

Respectfully, yours,
Albion K. Parris
Comptroller.

To the Commissioner of Pensions,
Present.

(stamped Feb. 23)

Hon W. H. Dimmick,

Dear Sir

You will find in this Evolope papers prepared to obtain the Bounty land due to the children of Benj. Haines decd. of this county a pensioner of the revolution. The papers show that Ben Haines was placed on Pension list many years ago and died soon after, this Pension was obtained through me perhaps you remember as I think you was in my office at the time

After I left here for the West in 1835 Ben Haines' widow obtained a Pension in right of her husband. They both recd their Pension untill their deaths, their children now apply to me for their Bounty land. They apply to me to assist them perhaps they know I do such business for nothing. I can find neither of the Pension certificates Roger Haines thinks they were burned up in a house at Carbondale. If I could find them they would aid the department in refering to their respective cases. Still they are all on file in Washington and establish the fact that Ben Haines was a Soldier of the Revolution & that his Widow drew a pension in his right, & the papers enclosed prove three (and only three) who are now applying for the Bounty land. You know them I presume Abraham & Roger Haines and a Daughter who married a Simpson. There is so little of this kind of business done here we have no case form. These persons are not only very poor but very ignorant as you know if you remember them, be kind enough to interest yourself for them

& if the Warrant is granted please send me the forms necessary to sell & transfer the same. I have not the law before me but I believe the Warrant may be issued to an Administrator - please give this matter your early attention as I want to get rid of this family Yet I want Justice done them

My health is poor and this letter is more than I have wrote in three months. I would like to say something about the crisis and the country but I cannot

All as usual
Yours very Respectfully
N B Eldred

P.S. I have seen that a ballance is due on both of Benj. Haines' & his widows pension if so will you inform me how it can be had

Yours
NBE

Geo C Whitney Es

Washington Feb 27

Dear Sir

In reference to the application of Hains children for bounty land I think you will find the word "minor" explained at the 1st Session of the 35 Congress.

Truly Yours
W H Dimmick

Pennsylvania

Wayne County ss. Personally appeared before me the subscriber Register for the Probate of Wills and granting of letters of administration in and for the County of Wayne Roger Hains who being duly sworn saith that he is the son of Benjamin Hains and Sophia Hains of Wayne County now deceased Benjamin Hains was a soldier in the Revolution, I believe in Capt. Westbrooks Company of New Jersey, of this the deponent is not certain but he was placed on the Pension Roll he died in the year 1822 in July or August. He left a widow Sophia Hains my mother my mother Sophia obtained through my agency a pension of half pay during her life and died in Carbondale in 1843 in July or August the papers in relation to these matters are in the Pension Office he believes and perhaps they may be found but at present he knows not where they are. Sophia Hains my mother the wife of Benjamin Hains died leaving three children who were her only heirs to wit ~~this~~ Roger Hains Deborah now intermarried with Peter Simpson and Abraham Hains He further swears that he he has for a valuable consederation transferred before this time all his Interest in the estate of his mother Sophia real or personal to Urial Simpson one of the children of my sister Deborah and therefore has no interest in the application now being made by the heirs of my mother Sophia.

Sworn and subscribed before me February 8, 1858.

Wm G. Arnold Register of Wills &c for Wayne County Penna.

Roger Haynes

The above affidavit was sworn and subscribed to before me on the 8th day of February A.D. 1858 and I hereby certify that I know the affiant to be a credible person

Wm G. Arnold Register of Wills &c

State of Penna.

Wayne County ss. Before me the subscriber Register for the Probate of Wills and granting Letters of Administration in and for the County of Wayne personally came Wm Hull aged sixty two years residing in Mount Pleasant Township Wayne County and after being duly sworn saith that he known Benjamin Hains then of Mount Pleasant Township Wayne County Penna. forty years ago and up to the time of his death He left a widow Sophia Hains who he understood and believes obtained a pension in right of her husband as the widow of a Revolutionary soldier. Benjamin Hains died about the year Eighteen hundred and twenty two, I assisted in ~~placing him in his coffin~~ laying him out after his death. His wife died at Carbondale in Luzerne County some six to ten years after the death of her husband Benjamin Hains leaving three children to wit Deborah now married to Peter Simpson Roger Hains and Abraham Hains

~~Sworn~~ Affirmed and subscribed before me this 8th day of February 1859

Wm G. Arnold Register of Wills &c, for Wayne County Penna.

William Hull

The above affidavit was sworn & subscribed to before me on the 8th day of February A.D. 1838 and I hereby certify that I know the affiant to be a credible person.

Wm G. Arnold Register of Wills &c.

Pennsylvania

Wayne County ss. Personally appeared before me the subscriber, Register for the Probate of Wills and granting Letters of Administration Charles Kennedy (aged sixty six years) a resident of Mount Pleasant Township Wayne County who being duly affirmed, doth depose and say that he knew Benjamin Haynes for a long time before his death (who lived in Mount Pleasant Township Wayne County) and died about the year 1822 or 1823, leaving a widow who lived some years after the death of her husband. At her death there were three children named Deborah (married with Peter Simpson), Roger, and Abraham Haynes.

Affirmed & subscribed before me this 8th day of Feb. 1859

Wm G. Arnold Register of Wills &c for Wayne County Penna.

Chas. Kennedy

The above affidavit was sworn & subscribed to before and I certify that I know the affiant to be a credible person

Wm G. Arnold Register of Wills &c.

State of Pennsylvania.

Wayne County, ss. I, Prothonotary and Clerk of the several Courts in and for Wayne County and State aforesaid, do hereby certify that Wm. G. Arnold Esq., is and was at the time of taking the foregoing acknowledgment an acting ~~Justice of the Peace~~ Register & Recorder in and for said County, duly qualified and commissioned as such, and by the laws of Pennsylvania is duly authorized to take such acknowledgment, and that the foregoing signature, purporting to be his, is genuine. In Testimony Whereof I have hereunto set my hand and affixed the Seal of said County, this 21st day of February one thousand eight hundred and ~~fifty~~ sixty one.

Jno K. Jenkins Prothonotary, Clerk, &c.

State of Pennsylvania

County of Wayne ss. On this 11th day of February A.D. 1861 before me a Justice of the peace in and for the said County, personally appeared Deborah Simpson aged sixty four years, who being duly sworn saith that she resides in Mount Pleasant Wayne County Penna and has resided there since her infancy that she is the oldest child of Benjamin & Sophia Haines deceased late of Mount Pleasant aforsaid, that her said father was a Soldier of the Revolution & was placed on the pension list about the year 1822 (but whether of the line of Pennsylvania or New Jersey this deponant cannot state) as by reference to records in pension office will appear.

And she further saith that her said father Benjamin Haines died about the year 1822 leaving a Widow Sophia Haines my mother who drew a pension in right of he said husband Benjamin Haines untill the time of her death which occured at Carbondale (where she resided for some years) about the year 1844. That the said Sophia Haines left three children (all heirs of said Benjamin Haines & of herself) to wit. Deborah ~~now m~~ (who was intermarried unto Peter Simpson who died February 6, 1861.) This deponent (who has on more than one occaision seen her mothers certificate of pension.) Also Roger Haines who, has conveyed all his interest in the Estate of my mother Sophia Haines unto my son Urial ~~Haines~~ Simpson as this deponent understands & believes & also, Abraham Haines. She further saith that she makes this declaration for the purpose of obtaining the Bounty lands to which she & the other heirs of Benjamin Haines are Entitled by acts of Congress

And she further declares that neither her father nor her mother have applied for or received Bounty land as she understands and beleives. And she further saith that neither herself or any of the heirs of Benjamin & Sophia Haines have applied for or received Bounty land under any act of Congress

Witness
C. F. Eldred
N. B. Eldred

her
Deborah X Simpson
mark

We Nathaniel B. Eldred & Thomas Spangenberg residents of Bethany Wayne County Pennsylvania, upon our oaths and affirmation declare that the foregoing declaration was signed & acknowledged by Deborah Simpson in our presence and that we beleive from the appearance and statements & from our own knowledge that she is the identical person she represents herself to be

N B Eldred
Thos Spangenberg

The foregoing declaration an affidavit were sworn to before me on the day & year above written & I certify that I know the affiants to be credible persons &

that the claimant is the person she represents herself to be & that I have no interest in this claim

B. T. West J. P.

⋆

State of Penna

Wayne County ss. Personally appeared before me a Justice of the peace in and for said County, Thomas Spangenburg of the County of Wayne and State of Penna and after being duly ~~sworn~~ affirmed saith, being Eighty five years old on June last That he has known Benjamin Haines deceased formerly of Wayne County and State of Pennsylvania since the year seventeen hundred and ninety five up to the time of his death. It was always understood that during the revolutionary war he acted with the United States Troops as an Indian Spy. That some forty years ago he was placed on the pension list as a Soldier of the Revolution

That after his death his widow Sophia Haines was placed on the pension list in the right of her said husband Benjamin Haines and drew his pension up untill the time of her death in Eighteen hundred & forty ~~forty~~ five when she died as this affirmant understands & beleives. This affirmant understands that the heirs of Benjam Haines and Sophia Haines are applying for the bounty land allowed them by act of Congress

That he has been familiar with the children of Benjamin Haines and Sophia Haines from their infancy. They left three children to wit. Roger Haines (whose deposition is taken in this case) also Abraham Haines and Deborah Haines now intermarried with Peter Simson. They all reside in this county and they are the only children of Benjamin and Sophia Haines, who were here placed on the pension list as above refered to. And this affirmant further saith that, he has no interest what ever in this aplication for bounty Land. He further beleives that neither of the children can either read or write except that it may be that Roger Haines may be able sign his name.

Thos Spangenberg

Thomas Spangenburg Affirmed & Subscribed to (before me this 18 day of February A.D. 1861.) this affidavit I certify that I know the affiant to be a credible person, & the claimant is the person she represents herself to be

B. T. West J. P.

⋆

State of Penna

Wayne County ss. Personally appeared before me a Justice of the peace in and for said County Nathaniel B Eldred of the County of Wayne & State of Pennsylvania and after being duly sworn saith being sixty six years of age in January last. That he has known Benjamin Haines deceased formerly of Wayne County Penna since the year of Eighteen hundred & Twelve up to the time of his death. It was always understood and beleived that during the revolution he acted with the United States Troops as an Indian Spy That in or about the year A.D. Eighteen hundred and twenty two he was placed on the pension list as a Soldier of the Revolution but whether he was in the line of Penna or New Jersey he cannot now state

That this deponent drew up his papers which doubtless are on file at the proper office at Washington & that his pension certificate he has seen. After the death of Benjamin Haines his widow Sophia Haines was placed on the pension list in right of her husband Benjamin Haines and drew her pension up untill the time of her death in Eighteen hundred and forty three, when she died as this deponent understands and beleives. This deponent understands that the heirs of Benjamin Haines and Sophia Haines are applying for the bounty land allowed them by act of Congress

That he has been familiar with the children of Benjamin and Sophia Haines from their infancy they left three children, to wit. Roger Haines, (whose deposition is taken in this case,) also Abraham Haines, and Deborah Haines, now intermarried with Peter Simson, they all reside in this county and are the only children of Benjamin and Sophia Haines, who heretofore were placed on the pension list as above refered to, and the deponent further saith that he has no interest in this aplication for Bounty Land, whatever. He further beleives that neither of the children can either read or write except it may be the Roger Haines may be able to sign his name

N B Eldred

N. B. Eldred sworn and subscribed to (before me this 18 day of February Eighteen hundred & sixty one) this affidavit and I certify that I know the affiant to be a credible person & the applicant is the person she represents herself to be

B. T. West J. P.

State of Pennsylvania,

Wayne County, ss. I, John K. Jenkins Prothonotary and Clerk of the several Courts in and for Wayne County and State aforesaid, do hereby certify that Benjamin T. West Esq., is and was at the time of taking the foregoing acknowledgment an acting Justice of the Peace in and for said County, duly qualified and commissioned as such, and by the laws of Pennsylvania is duly authorized to take such acknowledgment, and that the foregoing signature, purporting to be his, is genuine. In Testimony Whereof I have hereunto set my hand and affixed the Seal of said County, this 21st day of February one thousand eight hundred and ~~fifty~~ sixty one.

John K. Jenkins Prothonotary, Clerk, &c.

Widow, &c. File No. 4,693
Sophia Haynes Wid.
Benjamin Haynes Pvt. N.Y. line
Act: July 4th/36, Sec. 3rd
Index: - Vol. A, Page 303
[Arrangement of 1870.]
Revolutionary War Pension and Bounty Land Warrant Application Files
Series M-804; Roll # 1154

Abraham Harding
New York (R.4,601)

State of New York

Seneca County ss. I John Green of Romulus in the said county do solemnly swear that I do sincerely and truly believe that Abraham Harding was in the Revolutionary ~~ary~~ service and served in the Regiment belonging to Col. William Ellison was an officer and acted as such at the battle Called the Delaware Battle & also at the Battle of Fort Montgomery & have every reason to believe that the said Abraham Harding was in both the above mentioned Battles & acted as an officer & held a commission ~~this deponent says he~~ Don't recollect in particular of having seen Said Harding while he was in the service that I lived in the neighborhood at the time Harding was in the service & frequently heard from him as being in the service and serving in the above mentioned battles as an officer saw him frequently afterwards but not particularly while in the service to my recollection but have no doubt from what I knew of him at the time but that said Harding was an officer in the above mentioned Regiment as I lived at that time in the neighborhood.

Taken and subscribed before me this 5th day of March AD 1826
Stephen R Miller one of the Justices of the peace for the said County

John Green

State of New York
Seneca County ss. I Ernest A Dunlap Clerk of the Court of Common Pleas in & for the County of Seneca certify that Stephen R. Miller before whom the annexed affidavit appears to have been made was at the date thereof a magistrate as therein mentioned and further that I am acquainted with the hand writing of said Magistrate and verily believe the Signature of Stephen R Miller thereunto subscribed is genuine In testimony whereof I have hereunto subscribed my name and affixed the Seal of the said Court at Ovid this 22nd day of June 1826.
Ernest A. Dunlap Clerk of the Court
of Common Pleas for said County

State of New York
Yates County ss. I Stephen Amsbury of Benton in the said County do solemnly swear, that Abraham Harding was in the Revolutionary service and served as a captain of Militia, and truly & sincerely believe that he served under Colonel William Ellison in the battle called the Delaware Battle and also at the Battle of Fort Montgomery, and I was knowing to his being, and acting in the said Militia service as an Officer at a number of other times and places, and have always been acquainted with him since, and lived in the same neighborhood with said Harding at the time he performed the above mentioned services. I believe General Clinton was the commanding Officer - under whom Col. Ellison and said Harding served
Subscribed and sworn to before me this 8th day of March 1826 Saml. G. Gage One of the Justices of the Peace in and for the said county

Stephen Amsbury

State of New York
Yates County ss. I Abraham H. Bennett Clerk of the said county of Yates do hereby certify that Samuel G. Gage, before whom the affidavit of Stephen Amsbury appears to have been made and before by whom it was taken, was on the day of taking the same a Justice of the Peace in and for said county appointed, sworn and duly authorised by law to act as such. That I am acquainted with the hand writing of the said Justice and verily believe his signature to the certificate at the foot of the said annexed affidavit of Stephen Amsbury to be in the proper hand writing of the said Justice In witness whereof I have hereunto set my hand and affixed the seal of said county this 15th day of June 1826

A. H. Bennett Clerk

State of New York
Monroe County ss. I Samuel Finch of Parma in said county do solemnly swear that Abraham Harding was to my certain knowledge in actual service in the revolutionary war and he served as a captain in the regiment commanded by Col. William Ellison and afterwards commanded by Col James Thurston. That he was to my certain knowledge in the battle at fort Montgomery and as I verily believe in the Delaware battle And I was knowing to his continuing as an officer in actual service three years and I do verily believe he served in the revolutionary army at various other times and in the whole much more than three years I do further believe that General George Clinton was the commanding officer under whom Cols. Ellison & Thurston and the said Abraham Harding served

Subscribed and sworn before me this 13th day of March 1826
Peter F. Thatcher One of the Justices
of the Peace in and for said county

Samuel Finch

State of New York
Monroe County ss. I certify that Peter F. Thatcher whose signature appears to the above affidavit was on that day of the date thereof a Justice of the Peace in and for said county duly appointed and sworn. In testimony whereof I have hereunto set my hand and affixed the Seal of said county this 19th day of October 1826.

F. Cutler Dep clerk

In the Matter of A. Harding affdt of Smith Mapes

State of New York
Ontario County ss. Smith Mapes of the Town of Seneca in Ontario County being duly Sworn saith, That he has been acquainted with Abraham Harding before the Revolutionary war, that he served as Captain in the Militia in Colonel William Ellison's Regiment, he was at the battle called the Deleware Battle, and also in Fort Montgomery Battle, and that he was knowing to his being in actual service in the War for more than Three years, and that he served in the capacity of Captain the whole time, and that he believes General Clinton was the commanding officer.

Subscribed & sworn the fifteenth day of August 1826 before me Jarvis L Smith one of the Justices of the peace in & for Ontario County

Smith Mapes

State of New York

Ontario County ss. I certify that Jarvis L Smith whose signature appears to the above certificate was on the day of the date thereof a Justice of the Peace in and for said county duly appointed and sworn In testimony whereof I have hereunto set my hand and affixed the seal of said county this 23d day of August 1826

Ralph Lester Clerk

In The Matter of Abraham Harding Affdt. of A. Harding

State of New York

Monroe County ss. Abraham Harding being duly sworn saith that he was Eighty years of age the fifteenth day of May last past, that at the commencement of the Revolutionary war he was commissioned as a Captain of a company of volunteer Militia of the State of New York Commanded by Col. William Ellison, That he was in the immediate command of said company that he was in the Regular service nearly the whole time during the years 1777-1778 and 1779. And more than one half of the time previous and subsequent during the whole period of the Revolutionary war, was at Fort Montgomery with his company at the time it was taken in the year 1777, shortly after was at the battle on the Delaware River with the Indians and at several other battles. That he continued in said service until the termination of the war, and that a that time before the termination of said war he was stationed at New Windsor in Orange County of said State ~~for the purpose~~ under the command of Maj. Moses Hatfield for the purpose of restraining private excursions of a British party passing up the river Hudson, to obtain Forage and that this deponent was ordered from thence to Goshen and there disbanded his company upon the declaration of peace, and this deponent further saith that his commission was destroyed soon after the war by one of this deponents children who got possession of the same without his knowledge & used the same for Thumb papers at school, and further saith not

Sworn before me this 4th day of October 1826.

Timothy Childs Judge of Monroe County &

Counsellor at Law of the Supreme Court of said State

Abraham Harding

State of New York

Monroe County ss. I Abraham Harding do solemnly sware that I was a resident Citizen of the United States on the 18th day of March 1818, and that I have not since that time by gift sale or in any manner whatsoever disposed of my property, or any part thereof with intent thereby to diminish it or to bring myself within the provisions of an act of Congress Entitled "an act to provide for certain persons engaged in land and naval service of the United States in the Revolutionary war passed on the Eighteenth day of March one thousand eight hundred and eighteen & that I have not nor has any person in trust for me, any property or securities contracts or debts, due to me, nor have I any in come other than what is contained in the schedule hereto annexed and by me subscribed

signed Abraham Harding

I do hereby Certify that the above named Abraham Harding did on the fourth day of October 1826 personally appear in open Court & make Oath to the above affidavit by him subscribed.

F. Cutler Dep. clerk

"Schedule of all the property Estate and income possessed by Abraham Harding of Clarkson in the County of Monroe. Dated 4 Oct. 1826. viz:

one bed & Bedding, one chest one looking glass one tea kettle one _________ - two pails - a few articles of crockery four knives & forks one _______ and Hooks as Subscribed by me

A. Harding"

Monroe county ss. I Timothy Childs Judge of Monroe County Courts certify that Abraham Harding appeared in open Court at October term of said Court in the year 1826, and made oath to the truth of the above schedule and that in my opinion the value of the property therein set forth is ten dollars. Dated Oct 19. 1826.

Timothy Childs

I hereby certify that the above are the copies of the original oath, schedule & certificate as filed in this office this nineteenth day of October Eighteen hundred and twenty six. In testimony whereof I have hereunto set my hand and affixed the seal of said county this nineteenth day of October 1826.

F. Cutler Dep. clerk

Jany 15, 1827

Hon. P. Adams H of R

Sir: The documents forwarded by you to this Office in support of the claim of Abraham Harding to a pension on account of revolutionary service have been examined. He states that he entered the service at the commencement of the war as a Captain and was attached to the regiment commanded by Colonel Ellison and continued in service untill the year 1779 Colonel Ellison under whom he served did not belong to the Continental line of the Army and as the act of the 18th of March 1818 under which he claims provides for those only who served on the Continental establishment his case is not embraced by the provisions of that act. The papers in his case will remain on our files, agreeably to the regulations of this Department. W. G. Clark's letter to you is herewith returned

---

Revolutionary War Records Section.Department of the Interior,
Bureau of Pensions.
Washington, D.C., , 19

In reply to your request of Nov. 15, received Nov. 19, 1908 for a statement of the military history of Abraham Harding a soldier of the Revolutionary War, you will find below the desired information as contained in his (~~or his widow's~~) application for pension on file in this Bureau.

Dates of Enlistment or Appointment	Length of Service	Rank	Officers under whom Service was Rendered		State
			Captain	Colonel	
Served nearly all of 1777-1778 and 1779, and more than half the time prior and subsequent thereto, during the whole War.		Captain		William Allison	N.Y.

Battles engaged in, Fort Montgomery, Delaware River against Indians, and other battles not specified.

Residence of soldier at enlistment,

Date of application for pension, Oct. 4, 1826, in Monroe Co., N.Y.

Residence at date of application,

Age ~~at date of application~~, Eighty years May 15, 1826.

Remarks: Claim not allowed for the reason that the clmt. did not serve in the Contl. Line of the army as required by the provision of the act of Mch. 18, 1818 under which he applied.

Respectfully,
Commissioner.

1164 West Water St., Elmira, N.Y.
Friday, Oct 6, 1922

Department of Interior,
Bureau of Pensions
Washington, D.C.

Gentlemen:

Kindly send me military history of Abraham Harding, a soldier of the Revolution, who served with the rank of captain

Thanking you in advance for this information, I remain,

Respectfully yours,
Alice G. Fisher

---

22 Highland Ave.
Jersey City, N.J.
Feb. 8, 1924.

Commissioner of Pensions
Department of the Interior
Washington, D.C.,

Dear Sir,

Will you please send me the Revolutionary War record of Abraham Harding who was in an Orange Co. New York regiment. I think he applied for pension in 1832.

Thanking you, most sincerely
(Miss) Minniebel Knight

March 11, 1924.

Rev. and 1812 Wars Section.
Minniebel Knight,
22 Highland Ave.,
Jersey City, N.J.
Madam;

I have to advise you that from the papers in the Revolutionary War pension claim, R.4601, it appears that Abraham Harding was born May 15, 1746, place not stated.

He served nearly all of 1777-1778 and 1779, and more than half the time prior and subsequent thereto during the whole War, as Captain in Colonel William Allison's New York Regiment of Militia, and was in the battle of Fort Montgomery, Delaware River against the Indians and other battles not specified.

He applied for pension October 4, 1826, while a resident of Clarkson, Monroe County, New York, but the claim was not allowed as he did not serve in the Continental Line of the Army, as required by the provisions of the Act of March 18, 1818, under which he applied.

Children are referred to but no names are stated and there is no further family data on file.

Respectfully,
Commissioner.

Ray Roberts Knight, M.D.
Ralph Thos. Knight, M.D.
La Salle Building
Minneapolis
April 8th, 1927.

U.S. Pension Office,
Dept. of Revolution Pensions
Washington, D.C.
Gentlemen:

In making a study in genealogy I find an item in regard to the pension claim of Abraham Harding, R- 4601, Oct. 4, 1828, apparently not allowed because he was not in the Continental Line. I believe at that time pensions were not granted to former members of Colonial Militias. I would like to secure through this claim data as to place and date of birth of Abraham Harding and a brief as to his service. My information lists him in Orange County, New York. Will you please supply the information, and also, let me know what would be the expense of a photostat copy of the entire claim.

Yours truly,
Ray R. Knight, M.D.

RRK:E

April 13, 1927

Rev. and 1812 Wars Section.
Ray R. Knight, M.D.
LaSalle Bldg.
Minneapolis, Minn
Sir:

I have to advise you from the papers in the Revolutionary War pension claim R.4601, it appears that Abraham Harding was born May 15, 1746, place not stated.

He served nearly all of 1777-1778 and 1779, and more than half the time prior and subsequent thereto during the whole War, as Captain in Colonel William Allison's New York Regiment of Militia, and was in thr battle of Fort Montgomery, Delaware River against the Indians, and other battles not specified.

He applied for pension October 4, 1826, while a resident of Clarkson, Monroe County, New York, but the claim was not allowed, as he did not serve in the Continental Line of the Army, as required by the provisions of the Act of March 18, 1818, under which he applied.

Children are referred to, but no names are stated, and there is no further data on file as to family.

Photostat copies of the original papers in the above noted pension claim, will be furnished you under the Act of August 24, 1912, at the rate of fifteen cents a page, with fifteen cents extra for certification, if that be desired. The number of pages to be copied is eight.

The proper amount should be forwarded to this Bureau, by money order drawn to the order of the Secretary of the Interior.

Respectfully,
Winfield Scott
Commissioner.

Rejected

Revolutionary War Pension and Bounty Land Warrant Application Files
Series M-804; Roll # 1187

JASON AND MARY HORTON
NEW YORK (R.5,241)

BRIEF in the case of Jason Horton of Bridgewater Township Somerset County in the State of New Jersey
(Act 7th June, 1832.)

1. Was the declaration made before a Court or a Judge? Court

2. If before a Judge, does it appear that the applicant is disabled by bodily infirmity?

3. How old is he? Seventy four years

4. State his service, as directed in the form annexed.

PERIOD.	DURATION OF SERVICE	RANK.	NAMES OF GENERAL AND FIELD OFFICERS UNDER WHOM HE SERVED.
1776	Applicant says he cannot positively affirm that he served more than 2 years, nor would he dare certify that he served less but he is ready to declare that he served nearly 6 years	Enrolled in the NY Militia Private & Sergeant	Gen Clinton Col. Tustin Maj. Hatfield Capts. John Jackson, Sweezey, Jones, Sawyer, Little Lieuts. McDowell, Tryon

5. In what battles was he engaged? Battle with the Indians near the mouth of Lackawack

6. Where did he reside when he entered the service? Orange County N.Y.

7. Is his statement supported by living witnesses, by documentary proof, by traditionary evidence, by incidental evidence, or by the rolls? Traditionary

8. Are the papers defective as to form or authentication? and if so, in what respect? Reference to the accompanying Brief of objections

Certify that the foregoing statement and the answers agree with the evidence in the case above mentioned.

H. Dumas Examining Clerk.

Declaration in order to obtain the benefit of the Act of Congress, passed June 7th 1832.

State of New Jersey

Somerset County ss. On this thirteenth day of August in the year of our Lord personally appeared in open Court before the Judges of the Inferior Court of Common Pleas, now sitting at Bridgewater in the County of Somerset and State of New Jersey Jason Horton a resident of the said Township of Bridgewater in said County aged seventy four years in december next, who being duly sworn according to law, doth on his oath ~~say~~ make the following declaration in order to obtain the benefit of the Act of Congress passed June 7th. 1832.

That He entered the service of the United States under the following named officers and served as herein stated.

" I was born on Long Island, Suffolk County in the State of New York on the 18th December AD 1758 - my parents moved to Orange County near Goshen in the same state when I was a child, and soon after the commencment of the Revolutionary War, I arrived at Sufficient age to bear arms and was enrolled as a militia soldier in Capt John Jacksons Company at Chester a village near Goshen in said Orange County, the Regiment was commanded by Col B. Tustin where I continued to do duty. That most of the time during the war I acted as Sargent. I kept myself completely equipt for duty at my own expence and always turned out when called upon by my Officers & never deserted the post assigned me. In the year 1776 was on command at Fort Montgomery on the north River in the same county And laboured at erecting the Fort, was there many times doing soldiers duty, how long at each time I cannot precisely recollect but think it was generally two weeks. In the same & successive years until peace was declared I was many times on military service on the lines towards New York near forty miles from Goshen. I have been on service at New City, at Paramus at Tapan and several other places in the County now called Rockland in the State of New York, was at Hackansack and Bergen in New Jersey where the militia of Orange County kept Garison, was employed against the Savages at Minisink where I was many times on guard & once when we were alarmed by the Savages I was sent an Express to Col Phillips at the Walkill for assistance & performed the duty to the satisfaction of my officers. In the fall of 1777 the British army attacked Fort Montgomery, our regiment was immediately under marching orders, we had proceeded but a short way when news came that the Fort was taken, we were then ordered to turn our course toward New Windsor. The Enemys shipping soon appeared before that place, we were prepared for them had they landed. Some of our men took two strangers On horseback going with dispatches from Burgoyne to the Commanding Gen at New York, we were immediately under orders to march to

Esopus about 40 miles from New Windsor, under Gen Clinton, we had a considerable body of militia & one regiment of Continental troops commanded by Col Webb of New England. The shipping of the Enemy had proceeded up the North river and were Expected to land at Esopus, we endeavoured to get there as soon as they did, But when we were within a few miles we saw the smoke, which told us they had been there & it was in ruins. On the afternoon of the same day we heard the Official news of the surrender of Burgoyne to Genl Gates, we encamped at Esopus a few days until the British shipping returned down the river. The 2 men I have before mentioned we brought with us as prisoners under a strong guard, they were tried by Court Martial & one of them condemned to be hung as a spy. I saw him Executed near the Town, he was suspended on an apple tree. Soon after this we were under marching orders to return home. I think I served about one month at this time. In, I think July 1779 the Indians made an attack on the inhabitants of Minisink & made great destruction. A detachment the Goshen militia Commanded by Col B Tustin & the Warwick Regiment Commanded by Col J. Hartshorn were under marching orders in a few hours, arrived at the place of destruction in the after noon of the day in which the Indians had left. We pursued them, travelling all night some on foot & some on horseback. About break of day our Com Officer was informed that the Indians had Encamped about 4 miles in advance Soon after sun rise, we left our horses & baggage under a guard. We then hastened under the pilotage of Cap Duncan, who was well acquainted with the country. I think it was the intention of the Officers to take them by supprise. But the Enemy became apprised of our meditated attack and were prepared for us. Cap Duncan was well Acquainted with Indian warfare & much reliance was placed on him. He was the first man killed. We in a few minutes perceived we were surrounded. Their leader was Brandt. Our men fought until our amunishion failed. When the Officers gave orders for us to retreat in which many were killed. Col B Tustin was supposed to have fallen. & Cap Jones & Cap Little both of Col Tustins regiment. I think not more than one fourth of our men Ever returned home. This battle was near the mouth of the Lackawack. Capt John Wood was taken prisoner by them & conveyed to the British Garison at or near Detroit. On the retreat I fell in with two men, one a Capt Harker from Sussex County was badly wounded we assisted him to Minisink. We arrived at a fort two days after the battle much fatigued, having been nearly three days without provision. There was with us a small detachment from Sussex County Commanded by Maj Meeker, he was wounded in the sholder. There was also with us one Company of artillery Commanded by Cap Brown of the standing army. At another time when on duty at this place we were allarmed by the following Circumstance A family named Brooks, consisting of father mother, Daughter & daughters husband named Chapman was supprised by the indians about break of day. Chapman in attempting to escape was shot dead, they plundered & burnt the house, the family were taken prisoners but were soon permitted to return, they however came and informed us what had happened & I was one of those who buried Chapman. Soon after the war ended I moved into Somerset County New Jersey where I have mostly ever since which is about 40 years. I am in the 74 year of my age & unable to work for a support for myself & wife I cannot positively affirm that I served more than 2 years, nor would I dare certify that I served less but I'm ready to declare that I served a great number of times on military duty for nearly six years. I am at the distance of 70 or 80 miles from the place where I was enrolled. I cannot therefore have the benefit of the Testimony of those who were with me & who knew of my service to my Country. I was out under the following officers at different times, viz Cap Jackson, when Chapman was killed Lieut McDowell, Cap Sweezy Cap Jones, Cap Sawyer, Capt Little, Lieut Tryon, Maj Hatfield, has no documentary evidence of his said service. He hereby relinquishes every claim whatever to a pension or annuity except the present & declares that his name is not on the pension roll of the agency of any State

Sworn and subscribed the day and year aforesaid
Wm. B. Gaston

Jason Horton

Somerset County ss. Isaac Southard & Tho. Talmage Jr of said Township of Bridgwater being duly sworn according to Law on their oath say, that they are neighbors of the above named Jason Horton & have been for a number of years past. That he is a man of truth & veracity, & believe entire confidence would be placed in his declarations by all who know him

Sworn and subscribed the day & year aforesaid
Wm. B. Gaston

Isaac Southard
Thos Talmage Jun

We Isaac Southard and Thomas Talmage residing in said Township of Bridgwater hereby certify that we are well acquainted with Jason Horton who has subscribed and sworn to the above declaration. That we believe him to be seventy four years of age. That he is reputed and believed in the neighborhood where he resides, to have been a soldier of the Revolution, and that we concur in that opinion

Sworn and subscribed the day and year aforesaid
Wm. B. Gaston

Isaac Southard
Thos Talmage Jun

⁂

And the said court do hereby declare their opinion, after the investigations of the matter and after putting the interrogatories prescribed by the War department, that the above named applicant was a Revolutionary soldier, and served as he states And the court further Certifies that it appears to them that Isaac Southard & Thomas Talmage Jr. who signed the preceding certificate, reside in said Township of Bridgwater, & that Thomas Talmage Jr & Isaac Southard who signed the preceding affidavit, reside in the same place. That they are all credible persons, & that their Statement is entitled to credit, and the said court further certifies that, there is no clergiman at this time residing in the congregation where said Horton resides

Wm. B. Gaston
Danl. H Disborough
John Kirkpatrick

⁂

I Peter D. Vroom do hereby certify that I am well acquainted with Jason Horton who has made the above declaration. That he is a man of truth & veracity & I. believe his statement to be correct, not from my own personal knowledge, but because I am satisfied that the said applicant from his known and long established good character is incapable of testifying to or stating any matter in relation to his Revolutionary Services, that he does not believe to be strictly true.

P. D. Vroom

⁂

I William B Gaston the magistrate who administered the foregoing oaths do hereby certify that Isaac Southard, Thomas Talmage Jun, & Peter D Vroom, are persons of truth & credibility, and entitled to full credit and belief.

Wm. B. Gaston

⁂

State of New Jersey
Somerset County ss. Be it Known that on this 4th of February 1833 before me William B. Gaston presiding Judge of the Court of Common Pleas of

said county personally appeared the above named Jason Horton who being duly sworn deposeth and saith that by reason of old age and the consequent loss of memory he cannot swear positively as to the precise length of his different tours of duty annually, but according to the best of his recollection he served two full years in all as a militia soldier; and for such service I claim a pension.

Sworn and Subscribed before me the day and Year aforesaid

Wm B Gaston

Jason Horton

State of New Jersey

Somerset County ss. I John M. Mann Clerk of the Inferior Court of Common Pleas in and for said County hereby certify that William B. Gaston before whom the within Affidavit of Jason Horton was taken and who has made the within Certificate was at the time of the date thereof the Presiding Judge of said Court duly empowered to administer Oaths &c and that the proper signature of the said William B. Gaston is subscribed to said Affidavit & certificate. In witness whereof I have hereunto set my hand and Affixed the seal of said Court at Bridgewater this fourth day of February in the year of our Lord Eighteen hundred and thirty three

J. M. Mann

State of New Jersey

Somerset County ss. I, John M. Mann Clerk of the Court of Common Pleas in and for said county do hereby certify that the ~~original~~ foregoing contains the original proceedings of the Court in the matter of the application of Jason Horton for a pension In Witness whereof I have hereunto set my hand and seal of office this eighteenth day of August in the year of our Lord one thousand eight hundred and thirty two.

J. M. Mann

New Jersey

Jason Horton of in the State of N. Jersey who was a Private in the compy. commanded by Captain Jackson of the Reg commanded by Col Tustin in the N.Y. line for two years Private $80.00 Inscribed on the Roll of N. Jersey at the rate of 80 Dollars per annum, to commence on the 4th day of March, 1831. Certificate of Pension issued the 14th day of Feb. 1833 and sent to Hon J. Southard House Reps

Arrears to 4th of Sept. 1832 $120.00
Semi-anl. allowance ending 4 Mar 1833 110
$160.00

Recorded by ______ Luce Clerk, {Revolutionary Claim, Act June 7, 1832.}

Book D Vol. 4 Page 155

Application of Mary Horton widow of Jason Horton
of Somerset County New Jersey

Bradford Co. Pa.. ss. Personally appeared before the subscriber a Justice of the Peace in and for the county aforesaid, John Horton of the township of Asylum county of Bradford and state of Penna. who being duly sworn says that he was seventy five years old in July last, that he is personally acquainted with Mary Horton wife and widow of Jason Horton late of Somerville, Somerset county and state of New Jersey, who departed this life on the 28th of December 1837 aged seventy nine years. He further states that he is the brother of the said Jason Horton, and that he was present at the wedding of the said Mary and Jason Horton, that they were married by the rev. Mr. Seely, a presbyterian minister, at Sugar Loaf in orange county State of New York, in March A.D. 1783. He also states that the said Jason Horton was a soldier of the revolutionary war, that he was frequently out in the service of his country for several years during the course of the war, that he belonged to the Goshen Regiment, under the command of Col. Tustin, & in Capt. Jackson's company, that he recollects particularly of the said Jason Horton being called out to New Windsor and other places along the Hudson about the year 1777, also recollects that he was in the Indian battle at Minisink near the Delaware in Col. Tustin's regiment and thinks in a company under the command of Lieut. Mosher in the year 1779. He also recollects that the said Jason was out on general alarms on the frontiers at Minisink and other places in the year 1780 and was there in a company commanded by Capt. John McDowell in Col. Tustin's regiment. He also states that the said Jason Horton was always truly patriotic, and on all occasions took an active part against the enemy and was true and faithful in the service of his country.

The foregoing statement sworn to and subscribed
before me this 15th day of Oct. A.D. 1838
Wm Terry Justice of Peace

John Horton

State of Pennsylvania

Bradford County ss. I J C Powell Prothonotary of the Court of of the Court of common Pleas of Bradford County do certify that Wm. Terry is a Justice of the Peace in and for said County duly commissioned and Qualified and that his commission is dated January 24th AD 1825 and that the above Signature purporting to be his is genuine In testimony whereof I have hereunto set my hand and Official seal this 20th day of October AD 1838

J. C. Powell Prothonotary
A. S. Chamberlin Dep

Declaration of Mary Horton

Declaration in order to obtain the benefit of the third Section of the act of Congress
of the 4th July 1836.

State of New Jersey Somerset County ss. On this 18th day of February AD 1839 personally appeared before me the subscriber one of the Judges of the court of Common pleas, in & for said County, Mary Horton, resident of sd. County aged seventy Five years who being by me first duly sworn according to Law, doth on her oath make the following declaration in order to obtain the benefit of the provision made by the act of Congress passed July 4, 1836. That she is the widow of Jason Horton, late a pensioner of the United States, who was a private in the Revolutionary War. She refers to her husbands application on file in the pension department, & the proof hereunto annexed in relation to his services She further declares that she was married to the said Jason Horton on the fourth day of March Seventeen Hundred & Eighty Three by the Revd. Mr. Seely, a presbyterian minister at Sugar Loaf in Orange County in the State of New York. That her husband the aforesaid Jason Horton died on the 28th of December 1837, at Somerville in said County of Somerset, & that she has remained a widow ever since that period as will more fully appear by reference to the proof hereunto annexed. That after the death of her said husband she drew the balance of pension due to that time. She has no documentary evidence of her said marriage, except an entry thereof, in the hand writing of her Said husband made at the time, & herewith forwarded

Sworn to & subscribed the day & year above written before me
James Taylor

Mary Horton

Jason Horton his Book

I was born on Long Island Suffolk County the 18th of December 1758

Mary Terry was Born on Long Island Suffolk County South Hold Township on the 4 of July 1763

I was married March the 4 1783 to Mary Terry Daughter of Uriah Terry

our first child which was a Daughter Born January 20th 1784 which we Named Sarah

our 2d child which is a Son named Uriah Born August 29th 1786

our 3rd child Daughter named Elizabeth Born January 6th 1790

our 4th child Daughter named Mary Born December 31st 1791

our 5th child a Daughter born the 6th March 1794 Named Abigail

John ______ a frind of mine Departed this Life July 20th 1795 aged 39 years & 9 months and 4 days Hulda his widow born 16th June 1757

our 6th child a Daughter born 20 _____ 1796 Named ________

May 21 1796 I was Baptised by the Revd Robart Findley at Baskenridge the Next day being ____ ____ I and my wife were admited to the table of the Lord

June 28 1796 our children six in number were Baptised

our 7th child a Daughter named _______ born 15th Sept1799

our Eighth child a Son born the 9 of Jan 1805 named _______he Departed this world the 17 April aged 3 months and 7 Days

departed this life December the 28 1837 Jason Horton in the 80 year of his age

---

State of New Jersey Somerset County ss. On this 22d day of February 1839 before me James Taylor above named personally appeared Uriah T. Horton who being by me duly sworn on his oath saith that he is a son of the above named Jason Horton & Mary Horton. That he is now Fifty two years of a age that he has known the Book, or record from which the annexed leaves were taken as long as he can remember any thing. That it is the hand writing of his Said father, & the only record of marriages births & deaths of his fathers family. That his father died the 28th of December 1837 leaving deponents mother his residence & that she has remained a widow ever since Deponents father at the time of his death was a Revolutionary pensioner at the rate of Eighty Dollars per year Deponent, has subscribed his name upon the Said several leaves hereunto annexed

Sworn to & subscribed the day & year above before

James Taylor

Uriah T. Horton

I certify that the above named Uriah Horton is a person of credibility

James Taylor

State of New Jersey Somerset County ss. I John M. Mann Clerk of the court of Common pleas in & for said County do hereby certify that the foregoing is the original application & proceedings of Mary Horton for a pension. That the above named James Taylor Esq is & was at the time of taking said depositions one of the Judges of the court of Common pleas of sd. County In Testimony whereof I have hereunto set my hand & Seal of office at Bridgwater this 22d. Day of February 1839.

J. M. Mann

Additional evidence in the case of Mrs. Horton for a pension

State of New York

Orange County ss. Personally appeared before me one of the commissioners of deeds In and for for the County of Orange Jeremiah Horton, to me personally known and a compatant witness who being duly sworn by me according to law deposeth and saith, that at the commencement of the war of the Revolution Jason Horton belonged to Col Tustins Rigt, Capt Jacksons Comp of Militia, and that he went out on general alarms, that is when the Rigt was calld out In Mass and performed such duty as was ordered, the time of servis that was required at a general alarm was Uncertain but often from six to Thirty days and some times longer. Jason Horton performed Towers of duty which took place by classing the men and the men releiving each other at a particular time or on a set day, the time of servis of a tower was for the most part half of a month In actual servis but some times much longer. In the year 1778 or 1779, Colo Tustons Rigt was calld out In mass and marched to the Indian frontiers where with other Militia they followd a party of Indians commanded by the noted Brant, Into the wilderness not far from Coshacton near the delaware River, a battel took place the Militia gave way, it was reported at that time that the defeat of the militia was occationed for the want of ammunition Colo Tustin was kild, several Commissioned Officers, a number of respectable men and heads of families belonging to sd Rigt were left dead on the battel ground. Jason Horton was in that Ingagement and made a safe retreat by taking a different course from the way the militia entered the wilderness he did not git back to the encampment untill the second day after the action he was out two nights after the battel. Jason Horton served under a number of Officers which belonged to said Rigt at different times and places, deponant never knew or heard of Jason Horton staying at home or to evade going when calld out on general alarms, or neglecting to perform his Towers of duty as they had been previously arranged by his Officers. Deponant saith at this day he is not able to assertain with certainty the amount of time that Jason Horton was In actual military servis In that war, but as the Militia in Orange County for five or six years was often calld out In mass and otherwise the deponant hath reason to calculate that Jason Horton was out In the servis not less than fifteen Months nor to exceed Thirty months the deponant further saith that he can say with confidence and with the greatest asurence that Jason Horton performed as much Military duty as others belonging to the said Rigt further the deponant saith not Sworn and subscribed before me this 14 day of April 1840

Jeremiah Horton

Sworn before me this Eighteenth day of April one thousand Eight hundred and Forty

James Thorp Commissioner of of deeds
of Orange County St of New York

State of New York

Orange County ss. Personally appeared before me Fredk. H. Brewster, Justice of the piece in and for said county of Orange this 19th day of December 1840, David G. Drake personally known to me and a compatant witness, who being duly sworn by me according to law, Deposith and saith that he is in the Eightieth year of his age. In the war of the Revolution he was well acquainted with Jason Horton, and the Deponent and Jason Horton both of them belonged to Col. Tustins Rigt and Capt Jackson Company of the malitia, and Deponant saith that he was knowing to Jason Horton being out in the melitia in actual servis a number of different times. Deponant saith that he was out with Jason Horton in the militia performing millatiary duty under the following Officers on the Indian frontiers commanded by Col. Hathorn at Old Minisink flats, commanded by Capt Sweezy at peinpack and Other posts several times, commanded by Capt Jackson at a post on the Indian frontiers under the command of Lieut Tryon at a fort on the east side of Shoungunk Mountain commanded by Insine Moore. Deponant saith at this day his memory doth not serve him as to be able to assertain the length of time that they served under those Officers that hath been above named. Deponant further saith that he the Deponant was not personally in the battle which was fought by the militia with a party of Indians near Cosheston when Colon Tustin fell and Severel commissioned Officers and many of the militia were kild Deponant saith that Jason Horton he thinks was in that ____________ Deponant saith that he hath all the asurance that he could have unless he the Deponant was personally by his side the Deponant is not abel to state with certainty the number of months that Jason Horton was out in the militia in actual servis during that war, but what Occurs to his memory at this day that he was not out less than fifteen nor to exceed more than twenty-five months, the militia in Orange County was often calld out and part of them was, kept Out most of the time as the british lines lay on one side and the Indian frontiers on the other. Deponant saith that he never heard or knew of Jason Horton neglecting to go out with the militia and to perform such military duty as he was ordered to perform, further this Deponant saith not

Sworn & subscribed before me
Frederick H. Brewster Justice of the Peace of Orange
County State of New York this 19th day of Decr 1840

his
David G. X Drake
mark

(Date stamped 1897)

Atlanta, Georgia
December twelfth

Mr H. Clay Evans
Dear Sir,

Kindly let me know what your records have concerning Revolutionary services of following named persons - Uriah Terry, Southold L.I. wife's name - Abagail Cleaveland, Jason Horton N.Y. or N.J. wife's name - Mary Terry. The information is desired for historical and genealogical purpose and an early reply will greatly oblige.

Very truly yours,
Mary Bockover Foy
c/o 105 Forest Ave.
Atlanta, Georgia

Rejected

Revolutionary War Pension and Bounty Land Warrant Application Files
Series M-804; Roll # 1330

Peter and Elizabeth Kemble or Kimble
New York (W.20,311)

State of New York

Chemung County ss. Samuel Finch of the town of Elmira in the said county being duly sworn deposeth and saith that he was a soldier in the war of the revolution and that he belonged to a company of militia of Which James Sawyer was Captain Gilbert Vail Lieut and Samuel Webb was Ensign that in may of 1776 he was drafted in a company of which John Little was Captain and went to Haverstraw and along the north river for two months and that he knew Peter Kemble in the said service as a seargeant in a company of drafted militiamen under the command of Captain John Whittaker and that the said Peter Kemble served in the said company for the term of two months and was discharged in the month of July 1776 the service was under Col Wisner that ~~in the~~ in the month of July 1778 he was again in the service under Abraham Hardin Capt as a classman and that he knew Peter Kemble in the said company as a sergeant in the said company and that the said Kemble served in the said company as sergeant for one month at Paramus Hackensack and Ramapoo in the regiment commanded by Col Wisner

Subscribed & Sworn this 16th day of December 1837 before me John W Wisner Justice of the Peace Chemung County N.Y.

Samuel Finch

☆

I certify that I believe Samuel Finch who made and subscribed the foregoing affidavit to be a man of truth and veracity and his statements above made are entitled to full credit and beleif

John W Wisner JP

☆

State of New York

Chemung County Clerks office ss. I hereby certify that John W Wisner Esq whose name is subscribed to the within afft. was when the same bears date a Justice of the peace of said County only qualified, and that his signatures within subscribed are genuine. Witness my hand & Seal of Office the 16th December 1837

Th Maxwell Dep Clk

☆

State of New York
Chautaugue County ss. Personally appeared before me thee undersigned one of the Justices of thee peace in and for the said County this day Elias Clark and being duly sworn deposeth and saith that he was well acquainted with Peter Kembal in thee time of the Revolutionary war that said Kembal then lived in thee County of Orange & State of New York That said Kembal married one Elisabeth Clark he thinks in the year 1777. This deponent further saith that the said Kembal was in the Revolutionary service and that he thee said Clark served in the same company with thee said Kembal under Captain Rosecrantz Lieut Clark & Ensign Coley Saith that they were called out on service in the fall or winter of 1775 and went to Kaakiatt and from thence they went to Tappan & were gone from home about two months This deponent further saith that thee name of the Colonel was either Wesner or Thurston Further saith that the said Kembal was again called out he thinks in the spring of 1776 and went to Green Pond was gone as near as he can recollect about one month Does not know the name of thee officers under whom hee served as he the said Clark at this time was not in thee service

Subscribed and sworn before me this 18th day of Jan 1839
Horace Bronson Justice

his
Elias X Clark
mark

I Certify that I am acquainted with Elias Clark who made and subscribed the above affidavit and know him to be a man of truth and veracity and his statement therein made entitled to full credit

H Bronson Justice

Jesse Jay being duly sworn deposeth and saith that he has for many years been acquainted with Elias Clark who made and subscribed the above affidavit & knows him to a man of truth and veracity

Subscribed and sworn before me this 18th day of Jan 1839
H Bronson Justice

Jesse Jay

State of New York

Chautaugue County ss. George W. Tew. Clerk of said County do hereby certify that Horace Bronson Esq. whose name is subscribed to the above certificate and affidavit was on the 18 day of January 1839. and now is a Justice of the Peace in & for said County. duly elected and qualified according to law. and I further certify that the signature "H Bronson" above subscribed are genuine. In testimony whereof. I have hereunto set my hand and seal this 21. day of January. AD 1839.

G. W. Tew Clerk

State of New York

Orange County ss. On this twelfth day of April eighteen hundred and thirty nine personally appeared before the undersigned a Judge of the court of common please in and for Orange County Elisabeth Kemble a resident of the town of Minisink in the county of Orange and State of New York aged Seventy nine years who being duly sworn according to law doth on her oath make the following declaration in order to obtain the benefit of the provision made by the act of Congress passed July 4, 1836 That she is the widow of Peter Kemble who was a private and a Sergeant in the war of the Revolution and who according as she has been informed and believes served as follows. That in the year of 1775 her said husband Peter Kemble belonged to a company of Militia of which ~~James~~ James Rosecrantz was Captain

James Clark was Lieutenant and Samuel Cole was Ensign and in the Regiment commanded by Col William Allison in Orange County and State of New York That in the fall of the year 1775 and she believes in the month of October Captain Rosecrantz company was called out in service to a place called Kaakiatt somewhere near the Hudson River and served at that place and at Tappan and along the Hudson River and served two months and were discharged some time in December of the same year and that her said husband Peter Kemble went with the rest of the company and served in the said company the full term of two months but she is unable to state what Col he was under but believes it was either Col Wisner or Lieut Col Tusten That in the month of March 1776 her said husband was called out in the company of Captain Rosecrantz and in the Regiment under Col Allison and went to Green Pond somewhere along the Hudson River and served one month That in may of the same year her said husband was drafted out of Capt Rosecrantzs company and went to Haverstraw and along the north river for two months and served as a Sergeant under Capt John Whittaker and in the Regiment commanded by Col Wisner That in the month of November of the same year Capt Rosecrantzs company of Militia was called out with the rest of the Regiment commanded by Col Allison and Rendezvoused at Goshen on or about the 12th day of November and went from thence to a place called New City on the Hudson River and served at New City Tappan Herontown and other places for the term of six weeks and that her said husband Peter Kemble went with the said company and served in the said company as a sergeant for the full term of six weeks That in the month of July 1778 her said husband was called out in to the service as a classman and served as a sergeant at Paramus and other places along the frontiers for the term of one month in the company commanded by Capt Abraham Hardin and in the regiment commanded by Col Wisner That her said husband served as a classman &c and upon alarms a great many times along the Hudson River and on the western frontiers during the years 1779, 1780 1781 and 1782 but from old age and loss of memory she is unable to state the precise times or the length of service except that an alarm was made by the people on the western frontier along the Neversink river on the 19th day of July 1779 when the whole of the Goshen Regiment was called out to Minisink and her said husband Peter Kemble went with the rest in the company commanded by Capt Samuel Cole and under the command of Col John Hathorn and Lieut Col Tusten and Major Meeker that the said Regiment followed the Indians from Minisink to a place called beaver brook where a dreadfull engagement ensued that her said husband was in the said engagement and was wounded in the arm that in this service her said husband was absent ten days in this service her husband was also a sergeant

She further declares that she was married to the said Peter Kemble on the twenty fourth day of March Seventeen Hundred and Seventy Eight that her husband the aforesaid Peter Kemble died on the fourteenth day of March One Thousand Seven Hundred & Ninety Seven and that she has remained a widow ever since that period as will more fully appear by the proof hereunto annexed That she was married by the Rev Nathan Kerr pastor of the Presbeterian Church in Goshen That she has no documentary evidence of her said husbands services and that she is not a pensioner of any state or teritory in the United States

The figures "1778" were interlined before signing
Sworn to and subscribed on the day and year above
written before me - signed her name by making her Mark never
having been taught to write
Hulet Clark a Judge of Orange County Comm Pleas

Her
Elizabeth X Kemble
Mark

I Hulet Clark a Judge of the court of common please before whoum the foregoing declaration was made do certify that Elizabeth Kemble who made and subscribed the same is a woman of truth and her statements herein made ought to be believed and that from personal examination I believe she ought to be excused from attending the court of record in the said County

Hulet Clark a Judge of Orange County

State of New York

Orange County ss. George Kimber Being duly Sworn deposeth and saith that he was well acquainted with Peter Kemble in his lifetime and with Elizabeth his wife and that the said Peter Kemble died on the fourteenth day of March Seventeen Hundred Ninety Seven and that Elizabeth Kemble has ever since the death of her husband Peter Kemble remained a widow

I certify that the above affiant is a credible witness

The word "death" interlined before signing

Sworn and Subscribed the twelfth day of April 1839 before me

Hulet Clark a Judge of Orange County

George Kimber

State of New York

Orange County ss. Margaret Horton of the Town of Minisink in the said county being duly sworn deposeth & saith that she is eighty four years of age and that she was the Sister of Peter Kemble who was a soldier in the militia of Orange County and State of New York in the Revolutionary war and belonged to a company under the command of Captain Rosecrantz and in the Regiment commanded by Col Allison That her brother Peter Kemble served in the militia from time to time during the whole of the war but she is unable to state particular times except the following viz that in the winter of 1776 & 1777 the said Peter Kemble went from home under Capt Rosecrantz and in Col Allisons Regiment down towards New York and was gone about six weeks he afterwards at newburgh and other places along the north River and along the valley of the Neversink untill the end of the war That on the 19th of July 1779 the Indians came down upon Minisink and burnt and destroyed the same when a general alarm was made and the Goshen Regiment was called out in a body under Col Benjamin Tusten and went to Minisink and from thence to a place called beaver brook where a battle ensued and the whites were defeated in this alarm her brother Peter Kemble turned out with the rest of the company to which he belonged and she thinks the company was commanded by Capt Cole and went to Minisink and from thence to Beaver brook and was in the engagement that took place there between the whites and the Indians and was wounded in the abdomen & arm in the service she thinks he must have been at least ten days She further says that her brother Peter Kemble was married sometime in the war to Elizabeth Clark and that the said Peter Kemble is now dead and that the said Elizabeth still remains his widow

Sworn and subscribed by making her mark on
account of old age and infirmities and I certify the above is a
Credible Witness Hulet Clark a Judge of Orange County
Common Pleas

her
Margaret X Horton
mark

State of New York

Orange County ss. I Albert S Benton Clerk of the said county do hereby certify that Hulet Clark before whom the foregoing declaration of Elizabeth Kemble and the affidavits of George Kimber and Margaret Horton were subscribed and sworn was at the time of taking the same a Judge of the Court of Common Pleas in and for said county and further that I am well acquainted with the hand writing of the said Judge and verily believe that the signatures to the said declaration and affidavits are genuine. In witness whereof I have hereunto set my hand and affixed the seal of the said county the second day of November 1839

A. S. Benton Clerk

I do hereby certify, that I am in possession of the Marriage records kept by the late Rev. Nathan Kerr Pastor of the first Presbyterian Church in Goshen. Orange County & State of New York, And that I find the marriage of Peter Kemble and Elizabeth Clark, dated the fifth day of March One thousand seven hundred and seventy eight, recorded in said records in fair Legible writing and figures (viz) Married Peter Kemble and Elizabeth Clark, Mar. 5. 1778.

Sworn before me the 17 Decr 1839 before me & I certify that the declarant is a reputuble witness

H W Elliott Judge of Orange Com Pleas

H. M. Hopkins, Clerk of the
Session of said Church

State of New York
Orange County ss. I Albert S. Benton Clerk of the Court of Common Pleas in and for said county do certify that Horace W Elliott is a Judge of the said Court of Common Pleas as above, and that the foregoing signature purporting to be his is genuine In testimony whereof I have hereunto subscribed my name and affixed the seal of the said court this 8th day of January 1840

A. S. Benton Clerk

★

New York

Elizabeth Kemble widow of Peter Kemble N.Y. who died on the 14 March 1797 of Orange Co. in the State of N. York who was a Pr. & Sergt. in the company commanded by Captain Rosecrants of the regt commanded by Col. Allison in the N.Y. Militia line for

Pr. 3 months	10
Serg. 4 months & 22 days	23.66
	$33.66

Inscribed on the Roll of New York at the rate of 33 Dollars 66 Cents per annum, to commence on the 4th day of March, 1831. Certificate of Pension issued the 16 day of January 1840 and sent to Hon. N. Jones H.R.

Arrears to the 4th of Sep. '39	286.11
Semi-annual allowance ending 4 Mar '40	16.83
	$302.94

Recorded by D. Brown Clerk,
{Revolutionary Claim, Act July 4,1836, Section the 3}
Book A Vol. 2 Page 205

★

Widow, &c. File No. 20311
Elizabeth Kemble widow of
Peter Kemble Sergt Rev War
Act: July 4th 36
Index: - Vol. A, Page 393
[Arrangement of 1870.]
Revolutionary War Pension and Bounty Land Warrant Application Files
Series M-804; Roll # 1469

★

Jeremiah Kinner
New York (S.13,656)

Brief in the case of Jeremiah Kinner of Brookhaven, Suffolk Co in the State of New York
(Act 7th June, 1832.)

1. Was the declaration made before a Court or a Judge? Open Court.
2. If before a Judge, does it appear that the applicant is disabled by bodily infirmity?
3. How old is he? 74 years.
4. State his service, as directed in the form annexed.

Period when service rendered	Duration of claimant's service			Rank of the Claimant	Names and Ranks of the Field Officers under whom he served
	Years	Months	Days		Gen. Woodhul
Feby. 1776		6		private	Capt Selah Strong & Col Josiah Smith
		3		private	Capt McDowell Col Thurston
		3		private	Capt McDowell Col Thurston

5. In what battles was he engaged? A skirmish with Indians
6. Where did he reside when he entered the service? Brookhaven N York
7. Is his statement supported by living witnesses, by documentary proof, by traditionary evidence, by incidental evidence, or by the rolls? by a living witness & traditionary
8. Are the papers defective as to form or authentication? and if so, in what respect? correct

Certify that the foregoing statement and the answers agree with the evidence in the case above mentioned.

Richard Cutts Examining Clerk.

State of New York

County of Suffolk ss. On this second day of October in the year 1832 personally appeared in open Court before the judges of the Court of Common Pleas of the said County of Suffolk being a Court of record of the said County and State, now sitting, Jeremiah Kinner a resident of the Town of Brookhaven in the said County of Suffolk and State of New York aged seventy four years on the fourteenth day of July last, who being first duly sworn according to law doth on his oath make the following declaration in order to obtain the benefit of the act of Congress passed June 7 - 1832.

That he was born in the Town of Brooklyn in the County of Kings and said State of New York in the year 1758. That he has not any record of his age. That he was living in the said Town of Brookhaven when called into service as hereafter mentioned and has always resided there since that time except whilst serving in the Revolutionary war and still resides there. That he entered the service of the United States under the following named officers and served as herein stated. That some time in the month of February 1776 but the particular day he cannot recollect he enlisted as a volunteer to serve as a private soldier for six months and upwards but how much longer he does not distinctly recollect in a company of militia commanded by Captain Selah Strong and of which Lieutenant Caleb Brewster was the first lieutenant and Lieutenant Nathaniel Brewster was the second Lieutenant. That he enlisted in the said Town of Brookhaven. That he entered immediately into active service with the said company and was stationed at Mount Misery to guard the country from the apprehended invasions of the British and the tories. That he remained and served at Mount Misery which is in the said Town of Brookhaven until August following his enlistment when he went with his company to the west end of Long Island where they composed a part of the Regiment commanded by Colonel Josiah Smith in General Nathaniel Woodhull's Brigade. That he retreated with the American troops under General Washington into the City of New York where they remained some time under the command of Lieutenant Caleb Brewster, Captain Strong being sick at home with a fever. That this deponent remained in and about New York in the service until his said term of enlistment had expired when he was regularly discharged from service by Lieutenant Caleb Brewster. That he has not now any written discharge and can not state the exact time when he was discharged, but this deponent is positive that he was actually engaged in active service under the said enlistment for the term of six months and upwards. That he returned to Brookhaven aforesaid where he remained until the British took possession of that part of Long Island when he left there and went to Goshen in Orange County in the State of New York where he remained until Minisink was burnt by the Indians. That he then entered the service under Captain McDowell in Colonel Thurston's Regiment, as a volunteer. That this deponent in company with five others of whom Squire Bradley's and Priest Bradley's sons were two, left Goshen on the day when Minisink was burnt and rode that night as far as Howell's tavern where they staid all night that the next morning they rode to the fort at Minisink which was then invested by the Indians. That the Indians supposing that the Light horse were coming retreated into the woods when this deponent and his associates entered the Fort. That the militia commanded by Colonels Thurston, Haythorn and Wisner entered the Fort on the same day when this deponent joined Captain Mc-

Dowell's company. That this deponent in company with a detachment consisting of about one hundred and thirty three men on the same day followed in pursuit of the Indians who were commanded by Butler and Brandt, that they had several skirmishes with the Indians in one of which Colonel Thurston was killed close by this deponent's side, and this deponent was wounded by a ball in his knee. That the American forces were eventually defeated and dispersed and this deponent reached Minisink the next day. That this deponent then returned home to Goshen where he was confined by his wound and unable to do any business for about three months. That during this period this deponent was ordered by Captain McDowell to go to West Point but the sergeant who was directed to call on this deponent on seeing him reported him unfit for duty and he was thereupon suffered to remain at home. That soon after this deponents recovery he was directed by Captn McDowell to proceed to a log fort about ten miles North of Minisink near the residence of Captn DeWitt. That this deponent accordingly proceeded to that fort and remained there in the performance of his duty as a private soldier for a period of three months when his term of enlistment of service being up he went home. That this deponent has not any written discharge, and as the transactions are of a remote date he cannot state the exact times when the services performed by him subsequent to the retreat from Long Island commenced or terminated, but this deponent is positive that the times when he was engaged in fighting the Indians, was confined by his wounds and was stationed at the said last mentioned Fort as herein before mentioned taken together exceeded six calendar months and this deponent further says that he has no other documentary evidence, but that the Honorable Thomas S. Strong who is the son of Captain Selah Strong can prove his services in Captain Strongs company, and this deponent does not now recollect any person now living who served with him. He hereby relinquishes any claim whatever to a pension or annuity except the present, and declares that his name is not on the pension roll of the agency of any state

Sworn to and Subscribed in open Court the day and year aforesaid
Joseph L Huntting Clk

Jeremiah Kinner

I Thomas S. Strong residing in the Town of Brookhaven in the County of Suffolk and State of New York do hereby certify that I am the son of the late Selah Strong Esquire who was the Captain of the company mentioned in the foregoing deposition That this deponent was eleven years old when the said company was formed and resided with his father in the neighbourhood where it was formed That the company was a part of the Regiment commanded by Colonel Josiah Smith which was usually called Cowboys, that the said Selah Strong held a commission as Captain as aforesaid, that the said company was stationed at Mount Misery in Brookhaven aforesaid to guard the coast until near the time when the American forces retreated from Long Island when they went to the West end of Long Island and joined the other American forces there. That according to the best of this deponents recollection the said Jeremiah Kinner was a private in the said company, and this deponent further says that he has always understood and believes that the said Jeremiah Kinner was engaged in fighting the Indians ~~during~~ in the Revolutionary war

Sworn and Subscribed the day and year aforesaid
Joseph L Huntting Clk

Tho. S. Strong

We Samuel Davis Junior and Richard W. Smith both residing in the Town of Brookhaven in the County of Suffolk hereby certify that we are well acquainted with Jeremiah Kinner who has subscribed and sworn to the foregoing declaration, that we believe him to be seventy four years of age that he is reputed and believed in the neighbourhood where he resides to have been a soldier of the revolution and that we concur in that opinion

Sworn and Subscribed the day and year aforesaid
Joseph L Huntting Clk

Saml. Davis
Richard W. Smith

And the said Court do hereby declare their opinion after the investigation of the matter and after putting the interrogatories prescribed by the war department that the above named applicant was a revolutionary soldier and served as he states, and the said Court further certifies that Thomas S. Strong Esquire who subscribed the within affidavit was formerly First Judge of this Court and that it appears to them that Samuel Davis Esquire and Richard W. Smith Esquire reside in the Town of Brookhaven in the immediate neighbourhood and That they are credible persons and that their statement is entitled to credit.

I Joseph L Huntting Clerk of the Court of Common Pleas of the said County of Suffolk do hereby certify that the foregoing contains the original proceedings of the said Court in the matter of the application of Jeremiah Kinner for a pension.

In testimony whereof I have hereunto set my hand and seal of office this Second day of October in the year one thousand eight hundred and thirty two

Joseph L Huntting Clk

New York

Jeremiah Kinner of Suffolk Co in the State of New York who was a private in the Company commanded by Captain Strong of the regt commanded by Col Smith in the N.Y. mi line for 12 mo Inscribed on the Roll of New York at the rate of 40 Dollars 00 Cents per annum, to commence on the 4th day of March, 1831. Certificate of Pension issued the 2 day of May 1833 and Sent to J B Strong

Arrears to the 4th of March 1833	$80.00
Semi-anl. allowance ending 4 Sep	20.00
	$100.00

Recorded by ___ L. Williams Clerk,
{Revolutionary Claim, Act June 7, 1832.}
Book E Vol. 4 Page 23

Invalid. File No. 13,656
Jeremiah Kinner Pri Rev War
Act: June 7th 32
Index: Vol. 1, Page 208
[Arrangement of 1870.]
Revolutionary War Pension and Bounty Land Warrant Application Files
Series M-804; Roll # 1493

John Knapp
Continental/New York (S.23,292)

I Hezekiah West a clergyman residing in Orwell Township Bradford County Pennsylvania do certify that I have been acquainted with John Knapp of Franklin Township of the county and state aforesaid for the last fifteen years and that I believe him to be seventy five years of age and that he is reputed and believed ~~to~~ in the neighborhood where he resides to have been a soldier of the revolution, and that I concur in that opinion, and that the said Knapp has since I have been acquainted with him sustained a good character and is considered to be a person of truth and veracity

Sworn & subscribed before me the 10th day of Sept. A.D. 1832

Francis X. Homet Justice of the Peace

Hezekiah West

I, Eliphalet Mason a Justice of the peace in and for the County of Bradford Pennsylvania and a Commissioner of said county do certify that I am well acquainted with John Knapp of Franklin Township in the county and state aforesaid and have been so acquainted with him for twenty five years or more that I believe him to be about seventy five years of age and that he is reputed and believed to have been a soldier of the Revolutionary war in which opinion I fully concur and that during the whole time that I have been acquainted with him he has maintained an unblemished character for truth and veracity

Sworn & subscribed before me the 10th day of Sept 1832

John L Webb Justice of the peace

Elipt. Mason

State of Pennsylvania

Bradford County ss. I James P Bull Prothonotary and Clerk of the Court of Common Pleas of said county do certify and make known that Eliphalet Mason & Francis X Homet are and were at the time of their signing the foregoing acting justices of the peace duly commissioned as such in & for said county and that the official proceedings in this behalf are entitled to full faith & credit In testimony whereof I have hereto set my hand & seal of office the 13th day of Sept 1832

J. P. Bull Prothonotary

And the said court do hereby declare their opinion after the investigation of the matter and after putting the interrogatories prescribed by the War Department that the above named applicant was a revolutionary soldier & served as he states and the court further certifies that it appears to them that Hezekiah West who has signed this certificate (to the declaration of John Knapp annexed) is a clergyman resident in the township of Orwell in Bradford County Penna. and that Eliphalet Mason who has also signed the same is a resident in the Township of Monroe in said county and is a credible person and that their statements are entitled to credit

By the Court

J. P. Bull Prot.

★

I James P Bull ~~Clerk~~ Prothonotary of the Court of Common Pleas of Bradford County aforesaid do hereby certify that the foregoing contains the original proceedings of the said court in the matter of the application of John Knapp for a pension In testimony whereof I have hereunto set my hand and seal of office this 13th day of September 1832

J P Bull

★

State of Pennsylvania
Bradford County ss. I certify that Eliphalet Mason Esq whose deposition in the case of the application of John Knapp for a pension to which this certificate is appended was taken before me on the 10th day of September 1832 is reputed a man of truth & veracity & so considered in the neighborhood where he resides - certified the 12th day of Feby. 1833

John L Webb Justice of the peace

★

State of Pennsylvania
Bradford County ss. I James P. Bull Prothonotary of the court of common pleas of said county do certify that Eliphalet Mason who signed a certificate on a paper hereto attached on the 12th Feby instant, Abiram ______ who signed his name to the accompanying declaration on the 11th Feby instant and John L Webb who signed the foregoing were at the times of their respectively signing the same _______ Justices of the peace in & for the county aforesaid each duly commissioned as such & that their official acts & proceedings in this matter are entitled to full faith & credit. In testimony whereof I have hereto set my hand & seal of office the 12th day of Feby. 1833

for J P Bull Prot.
E L Goodrich

★

Bradford County ss. Selah Arnout being duly sworn says that John Knap now of Franklin township in said County of Bradford lived in Goshen township Orange County New York at the time deponent went to live with his (deponent's) brother which was he thinks in the year 1777. That in the years 1778 and 1779 ~~he thinks~~ (as near as he can recollect as to time) there were frequent alarms by the Indians approaching the frontiers and that said John Knapp was called & frequently went out with the militia to protect the frontiers. That said Knapp belonged to the same company of which deponents brother was the Orderly Sergeant. That said Knapp resided about a mile and an half from deponents brother with whom he (deponent) lived.

Sworn and subscribed before me Sept. 11th 1832

Elipt. Mason J. P.

Selah Arnout

State of Pennsylvania

Bradford Co I certify that the above named Selah Arnout is a man of good character for truth & veracity, so considered by in the vicinity where he resides

Feb. 12, 1833

Elipt. Mason Justice of the peace

State of Pennsylvania

County of Bradford ss. On this 13th day of September 1832 personally appeared in open court, before the Judges of the court of common pleas of said County of Bradford John Knapp a resident of Franklin township in the County of Bradford & State of Pennsylvania aged seventy five years who being first duly sworn according to law doth on his oath make the following declaration in order to obtain the benefit of the act of congress passed June 7th 1832

That he entered the service of the United States under the following named officers, and served as herein stated viz: ~~under Captain John Jackson's company~~ in the company commanded by Capt John Jackson the Battalion commanded by Major Moses Hatfield Regiment by Col Saml. Drake & Lieut. Col. Holliburt Brigade by Genl. Marine Scott in the New York line. That he enlisted in the month of June 1776 between the middle and he thinks the 20th of the month at Goshen township Orange County New York and served under the officers above named until the first of December following, at the time he entered the service he resided at ~~the~~ Goshen the place where he enlisted. That he was in the City of New York when one of Genl. Washingtons life guards was executed for conspiring to take his life also at the time the British fleet came into the Harbour. He was on parade in the City of New York when Independence was published to the army. At the time of the evacuation of New York he was at his fathers on a furlow sick with the camp distemper and joined the army again at Kings bridge. From Kings bridge the army marched to White Plains where he was when Genl. McDougals brigade had a battle with the British. From White plains the army went to Peekskill where he was taken out for the Genl. Paymasters Guard in which capacity he served until the last day of November. On the first of December he went to where the army lay at Haverstraw and went that night in company with Samuel Gardner who was also on the Paymasters Guard with himself to lodge with a brother of said Gardners about a mile from the camp and on returning next morning to the camp found the company to which he belonged had gone home when said Gardner & he ~~went~~ went on home after them without any certificate of discharge.

After he had been at home about nine days there was an alarm given and he volunteered to go with the militia ~~under~~ in the company commanded by first lieut. William Stewart in the regiment commanded by Col. William Allison (The Major Moses Hatfield being then ~~taken~~ a prisoner taken by the british at New York) and served a tour of six weeks and three days ~~during which~~ In this tour he was at Tappan near Haverstraw where he remained with the Regiment until he was dismissed. He received no certificate of discharge.

He further states that in the Spring of 1777 he was called out on a tour of duty & served as near as he can recollect half a month under Capt John Minthorn who belonged to the Regiment commanded by Col. John Haythorn of the New York Militia. His duties in this tour were keeping guard and scouting.

He further states that in June of the year 1777 he was called on a tour of duty to Fort Montgomery where he served about half a month ~~under~~ in the company commanded he thinks by Capt. John Little. His duties on this tour were keeping guard and serving "on fatigue."

He further states that at the time Fort Montgomery was taken he was called out on an alarm and served about ten days at Murdners Creek below New Winsor ~~under~~ in the Regt. commanded by Col. Benjamin Truxton.

He further states that in the Spring of 1778 he was called out with the militia to erect a garrison at West Point where he wrought at the Garrison one month under Capt John Little in the Regt. commanded by Lieut. Col. Benjamin Truxton (Col. Wm. Allison being then a prisoner at New York)

He further states that he afterwards served another tour at West Point working at the garrison about a week or ten days.

He further states that he volunteered to go to Newburg, in a company of Artificers, to repair waggons &c for the Continental service in which capacity he served as near as he can recollect for the term of three months under the ~~Capt~~ command of Captain James Young.

He further states that in the years 1778 & 1779 there were frequent & sudden alarms by the Indians approaching the frontiers & that he was called and frequently went out with the militia to protect them that he was at the ~~Delaware~~

battle ~~so called~~ with the Indians on the Delaware River about 30 miles in the wilderness called the "Delaware battle" - was out on an alarm when Peanpeck was burnt and on frequent alarms until the close of the war.

He has no documentary evidence of his service & there is no person living within his knowledge by whom he can prove his services except Selah Arnout of whose deposition is forwarded, his native place is in Goshen Orange County New York & he ~~lived~~ resided there during the war and for the term of five or six years thereafter, from thence removed to Luzerne Co Pa. & resided three years & from thence removed to Bradford Co & has lived there to this time, he has no record of his age He hereby relinquishes every claim whatever to a pension or annuity except the present and declares that his name is not on the pension roll of the agency of any state

Sworn and subscribed the day & year aforesaid
J. P. Bull Prot

John Knapp

⁂

Personally appeared before me the undersigned a Justice of the peace in and for the County of Bradford & State of Pennsylvania John Knapp who being duly sworn deposeth & saith that by reason of old age & ~~the consequent~~ the consequent loss of memory he cannot swear positively as to the precise length of his service but according to the best of his recollection he served not less than the periods mentioned below and in the following grades.

For five months and ten days I served as a private in Capt Jacksons company as set forth in the foregoing declaration For one month & fifteen days I served as a private in Col Allisons regiment as set forth in the foregoing For fifteen days I served as a private under Capt Minthorn as set forth in the foregoing For fifteen days I served as a private in a company commanded by Capt John Little as set forth in the foregoing For ten days I served as a private in Col Benjamin Truxtons Regiment For one month I served as a private under Capt John Little as set forth in the foregoing For ten days I served as a private at West Point For three months I served as a private in a company of artificers commanded by Capt James Young For two months in all I served at various times in alarms as set forth in the last paragraph of the foregoing, and for such service I claim a pension

Sworn and subscribed this 18th day of February 1833 before me
(two words erased before signing)
Abiram ________ Justice of the peace

John Knapp

Sir

USS chamber Feb 24 1834

I have received a letter from John Knapp of Penna. informing me that his pension documents have been sent to the War Dept. and desiring me to call the attention of the proper office to his case I have known Mr. Knapp for more than thirty years there is not a more honest man living he is quite aged and very poor I hope you will examine his case and Grant him that relief which his country has designed

Very respectfully yours
Saml. McKean

Mr. Knapps case is admitted Please send the certificate to me

Saml. McKean

★

Sir

Washington Feb 27 1834

within this you will receive the application of John Knapp of Bradford County Penna for a Pension and accompanying documents, and we desire thus unsolicited to add our testimony to the excellent character of the applicant Mr. Knapp and request that this may remain on the files of the department as such, we have both been acquainted with Mr. Knapp for nearly thirty years during which period he has been an exemplary member of the Baptist Church and universally considered to be a man of sincere and practical piety we have the most unlimited confidence in his probity and integrity and have often heard him relate his revolutionary services as ______ Mr Knapp is now aged desolate and destitute and has beyond doubt procured all the testimony within his reach in addition to which he has reiterated his services and attests it by his sollemn oath. The veracity of which will never be doubted by those who know him and it is all and the last that he can do to obtain in his last hours the __________ of his country

We are with Great respect your obt servts

Saml. McKean
John Laporte

Hon Lew Cass

★

Pennsylvania

John Knapp of Bradford Co. in the State of Penn who was a Pr in the Company commanded by Captain Jackson of the Regt. commanded by Col

Drake in the N.Y. Militia 1 Yr 2 Mo 15 Days $48.32 Inscribed on the Roll of Pennsylvania at the rate of 48 Dollars 32 Cents per annum, to commence on the 4th day of March, 1831. Certificate of Pension issued the 3d day of March 1834 and Sent to Hon Jno Laporte, H.R.

Arrears to the 4th of Sepr. 1833 120.80
Semi-anl. allowance ending 4 Mch '34 24.14½
144.94

Recorded by J. T. Sprague Clerk,
{Revolutionary Claim, Act June 7, 1832.}
Book C Vol. 5 Page 31

BRIEF in the case of John Knapp ~~of~~ County of Bradford in the State of Penna. (Act 7th June, 1832.)

1. Was the declaration made before a Court or a Judge? Court

2. If before a Judge, does it appear that the applicant is disabled by bodily infirmity?

3. How old is he? 75 years

4. State his service, as directed in the form annexed.

Period when service rendered	Duration of claimant's service			Rank of the claimant	Names and Ranks of the Field Officers under whom he served
	Years	Months	Days		
In 1776 June about 20th		4	10	private enlisted	Capt John Jackson Col Sam Drake - Gen M. Scott, N.Y. Line
last day of Nov		1 ½	3	Volunteer	Col William Allison
1777 Spring		½			Capt John Minthorn, Col Haythorn - N.Y. Line
1777 June		½			Capt John Little
			10		Col Truxton
1778 Spring		1			Capt John Little, Col Truxton
		1/4 or	10		
		3		Volunteer artificer	Capt James Young

5. In what battles was he engaged? Delaware Battle with the Indians

6. Where did he reside when he entered the service? Goshen

7. Is his statement supported by living witnesses, by documentary proof, by traditionary evidence, by incidental evidence, or by the rolls? Traditionary evi-

dence. One living witness not certified credible

8. Are the papers defective as to form or authentication? and if so, in what respect? Defective. Papers not attached Time of service not definitely stated

Certify that the foregoing statement and the answers agree with the evidence in the case above mentioned.

_______ Luce Examining Clerk.

BRIEF objections to the admission of Pension Claims, of which those answered in the negative, in red ink, will apply to the claim of John Knapp which require further proof or explanation.

I. Does the declaration show where the applicant resides?

II. Did the declarant make the declaration in the county where he resides?

III. If the applicant did not make the declaration in the county where he resides, is there any reason assigned for his not doing so?

IV. Has the declarant mentioned the period or periods of the war when he served?

V. Has he mentioned with precision the length of his service, and the different grades in which he served, in language so definite as to enable the department to determine to what amount of pension he is entitled?

VI. Has he given the names of the officers under whom he served, in conformity with the regulations?

VII. Has he made a relinquishment of every claim to any other pension than the one under the Act of June 7, 1832?

VIII. Has the Court given their opinion?

IX. Has the Clerk given his certificate?

X. Is the Clerk's seal affixed? And if so, has it a device or inscription by which it can be distinguished from any other seal?

XI. Has the applicant obtained the evidence of a clergyman and another respectable citizen as to their belief respecting his age and the general belief in his neighborhood relative to his revolutionary service?

XII. If the clergyman's affidavit has not been produced, is there any reason assigned for not obtaining it?

XIII. If the applicant has no documentary evidence, and has not obtained the testimony of at least one living witness, has he stated in his declaration that such proof cannot be had?

XIV. If the applicant's name is not found on the rolls of military service, has he produced a witness or witnesses as to his service?

If he has produced direct proof as to service, has the magistrate who administered the oath, certified to the credibility of the witness or witnesses?

Has the proper officer certified to the official character and signature of the magistrate as the regulations direct?

XV. Does it appear from any of the papers that the witness was in such a situation, or of such an age, as to have a personal knowledge of the applicant's service?

XVI. Are the papers authenticated as the regulations direct? no - see Quere 16

XVII. Are the seven interrogatories prescribed by the War Department answered satisfactorily? And, if not, which of them is not so answered?

EXPLANATORY NOTES.

QUERE V. In a case where the applicant cannot, by reason of the loss of memory, state precisely how long he served, he should amend his declaration by making an affidavit in the following words: "Personally appeared before me, the undersigned, a Justice of the Peace, &c. A. B. who, being duly sworn, deposeth and saith, that, by reason of old age, and the consequent loss of memory, he cannot swear positively as to the precise length of his service, but according to the best of his recollection he served not less than the periods mentioned below, and in the following grades: For year months, and days, I served as a For months and days, I served as a ; and for such service I claim a pension.

It is important in all cases to determine with precision the period for which each applicant served, and the particular rank he held, as the law directs the pension to be paid according to the grade of the pensioner and the length of his service. The use of the phrase about three or four months is too indefinite, and all such qualifying expressions are objectionable. Some persons who apply for pensions merely state they served two years in the militia, &c. without specifying the tours; the names of the officers, and other particulars respecting their service. This form of a declaration is highly objectionable. It must in every case be clearly shewn under what officers the applicant served; the duration of each term of engagement; the particular place or places where the service was performed; that the applicant served with an embodied corps called into service by competent authority; that he was either in the field or in garrison; and for the time during which the service was performed, he was not employed in any civil pursuit.

QUERE XIV. Proof of Service.

In a case where the name of the applicant is not found on the records of the Department, he must prove his service by two credible witnesses, who are required to set forth in their affidavits the time of the claimants entering the service, and the time and manner of his leaving the same, as well as the regiment, company, and line to which he belonged. The magistrate who may administer the oaths must certify to the credibility of the witnesses, and the official character and signature of the magistrate must be certified by the proper officer, under his seal of office.

QUERE XVI. Mode of authenticating papers.

In every instance where the certificate of the certifying officer who authenticates the papers is not written on the same sheet of paper which contains the affidavit, or other paper authenticated, the certificate must be attached thereto by a piece of tape or narrow ribbon, the ends of which must pass under the seal of office of the certifying officer, so as to prevent any paper from being improperly attached to the certificate.

* In every case where the clerk of the court, or other certifying officer, has no public seal of office, the certificate of a member of congress, proving the official character and signature of the certifying officer, should accompany the papers.

Henry E. Knapp.
Menomonie, Wisconsin

Oct 1-1913
Pension Bureau,
Washington, D.C.
Gentlemen:

There died in 1836 at Springfield Township, Pa. John Knapp who served in Rev War. probably from Orange Co. N.Y. He is said to have drawn a Pension under Act of 1832. Will you please give me copy of his record.

Yours truly
Henry E. Knapp

(His wife was Eunice Wilcox & children Samuel, Aaron, Mary, Eunice, Amos, John, Prudy, Betsy, William, Janet, Mahala. These may aid you in finding him)

Rev. War.

Commissioner of Pensions
Washington, D.C.

Will you please give me Pension Bureau report of John Knapp. He was born in Goshen, N.Y. in 1757. Died Franklin township, Bradford Co. Pa. (age 75 years.) in Feb. 4, 1836. He was married Nov. 6, 1780, to Eunice Wilcox in Goshen, N.Y. He applied for pension Sept. 1832.

Oct 17, 1913.

address, Mrs. W. E. Knapp,
925 College Ave.
Elmira, N.Y.

The Commissioner of Pensions
Washington D.C.
Dear Sir,

I have secured all the data necessary, in my Knapp line, to fill out application papers for membership to the Society of the Daughters of the American Revolution except positive proof that John Knapp (my grt-grt grandfather) was in the Revolutionary war, at time of Revolution John Knapp lived in Orange County N.Y. and removed to Bradford Co. Pa. in 1796. In "Bradford Pioneer & Patriot Families," page 274 I find he was granted a pension under act 1832. I wish to know if you can find records proving that he received a pension for service in Revolution. Thanking you in advance

Very Sincerely
Weltha P. Smith

Oct 24-1913.

Copy

Wilmot E. Knapp
Attorney and Counselor at Law
431 Realty Building
Elmira, N.Y.

Revolutionary War Records Section
M. B. H.
S. F. 23292
Rev. War.

Department of the Interior
Bureau of Pensions
Washington, D.C.
Oct. 24, 1913.

Mrs. W. E. Knapp
925 College Avenue, Elmira, N.Y.
Madam:

In reply to your request of October 17, received Oct. 20, 1913, for a statement of the military history of John Knapp, a soldier of the Revolutionary War, you will find below the desired information as contained in his application for pension on file in this Bureau:

Dates of Enlistment or Appointment	Length of Service	Rank.	Officers under whom Service was Rendered.		State
			Captain	Colonel	
June, 1776	Served at various times in all 11 ½ months.	Pvt.	John Jackson Lt. William Stewart John Minthorn John Little	Samuel Drake William Allison John Hathorn Benjamin Juston	N.Y.
	3 months	Artificer.	James Young		

Battles engaged in, With Indaina on Delaware river.
Residence of soldier at enlistment, Goshen Twp, Orange Co., N.Y.
Date of Application for pension, Sept. 13, 1832. His claim was allowed.
Residence at date of application, Franklin Twp., Bradford Co., Penn.
Age at date of application, 75 years, born in Goshen, Orange Co. N.Y.
Remarks, died Feb. 4, 1836.
There is no family data on file.

Respectfully
G. M. Saltzgaber,
Commissioner

Columbus Ohio
Jan 12,1914.

To The Pension Bureau;
Revolutionary war department
Washington D.C.
Gentlemen;

I am searching for Revolutionary war records of two ancestors, would you kindly tell me if they received a pension for service in the Revolution: and if so what their services were? One ancestor was John Knapp who enlisted from Goshen N.Y. and after the Revolutionary war removed to Bradford Co. Pa. where he died in 1836. The other ancestor was Dr. James Tunis of Hanover Pa. who was a surgeon in the Revolution. Thanking you in advance

Very Sincerely Yours
Weltha Penney Smith

Coudersport, Pa.
April 6, 1914.
Dear Sirs:

Will you please send me the military record of John Knapp, my great-great grandfather who married Eunice Wilcox, applied for pension in 1832 which he received. He was at that time living in Franklin township, Bradford Co. Pa. He died Feb. 4, 1836. At the time of enlistment he was living in Goshen, Orange Co. N. Y.

Respectfully,
Nellie A. Perkins.

Cloverdale Poultry Farm,
C. N. Reynolds, Proprietor

Canton, Pa., Sept 4 1917

Gentlemen,

Will you kindly give me the war record of John Knapp whose wife was Eunice Wilcox. I have no dates but he died at Burlington Pa.. Thank you very much in advance.

Yours truly
(Miss) Alice Reynolds

Wilmot E. Knapp
Attorney and Counselor at Law
431 Realty Building
Elmira, N.Y.

November 14, 1933

Department of the Interior
Bureau of Pensions
Washington, D.C.
Gentlemen:

The undersigned is the great grandson of John Knapp a Revolutionary soldier, who for many years before his death resided in Bradford County, Pennsylvania, and died on February 4, 1836 at the home of my grandfather.

I am about to apply for membership to the Newtown-Battle Chapter Sons of the American Revolution in this city, and consequently have to present data in regard to it. What I desire, is the usual Revolutionary war record of him from the Commissioner. I have in my possession at this time a certificate made by your Department, in answer to a letter written by my former wife, of which the enclosed is a copy for your information. I would like a certificate of the same character for the purposes mentioned.

WEK:GM

Encl:

Yours truly

W. E. Knapp

Revolutionary War Records Section.Department of the Interior,

Bureau of Pensions.

Washington, D.C., , 19

In reply to your request of , received for a statement of the military history of John Knapp a soldier of the Revolutionary War, you will find below the desired information as contained in his (or his widow's) application for pension on file in this Bureau.

Dates of Enlistment or Appointment	Length of Service	Rank.	Officers under whom Service was Rendered.		State
			Captain	Colonel	
June, 1776	Served at various times in all 11 ½ months.	Pvt.	John Jackson Lt. William Stewart John Minthorn John Little	Samuel Drake William Allison John Hathorn Benjamin Juston	N.Y.
	3 months	Artificer.	James Young		

Battles engaged in, With Indains on Delaware river.

Residence of soldier at enlistment, Goshen Twp, Orange Co., N.Y.

Date of Application for pension, Sept. 13, 1832. His claim was allowed.

Residence at date of application, Franklin Twp., Bradford Co., Penn.

Age at date of application, 75 years, born in Goshen, Orange Co. N.Y.

Remarks, died Feb. 4, 1836.

There is no family data on file.

Respectfully,

Commissioner.

Mr. Wilmot E. Knapp
431 Realty Bldg.
Elmira, New York

November 24, 1933
John Knapp - S.23292
BA-J/AWF

Dear Sir:

Reference is made to your letter in which you request the record of John Knapp, a Revolutionary War pensioner from Bradford County, Pennsylvania.

The data furnished herein were obtained from the papers on file in pension claim, S.23292, based upon the Revolutionary War service of John Knapp.

He was born in Goshen, Orange County, New York, date not stated, nor are the names of his parents given.

While residing in Goshen Township, Orange County, New York, John Knapp enlisted in June, 1776 and served at various times as a private in the New York militia, under Captains John Jackson, John Minthorn, John Little, Lieutenant William Stewart, Colonels Samuel Drake, William Allison, John Hathorn, Benjamin Tusten, and three months in Captain James Young's company of artificers. He was out in many scouting parties, was in a battle with the Indians on the Delaware River; length of all his service one year, two months and fifteen days.

He continued to live in Goshen until five or six years after the Revolution, then moved to Luzerne County, Pennsylvania, where he lived three years and moved to Bradford County, Pennsylvania.

John Knapp was allowed pension on his application executed, September 13, 1832, then a resident of Franklin Township, Bradford County, Pennsylvania. John Knapp died February 5, 1836.

There are no data in this claim pertaining to the family of John Knapp.

In order to obtain the name and address to whom the last payment of pension was made, you should apply to the Comptroller General, General Accounting Office, Records Division, Washington, D.C., citing all the following data -

John Knapp
Certificate No. 26044
Issued March 3, 1832
Rate $48.32 per annum
To commence March 4, 1831
Act of June 7, 1832
Pennsylvania Agency.

Very truly yours
A. D. Hiller
Assistant to Administrator

Invalid. File No. 23,292
John Knapp Pri Rev War
Act: June 7th 32
Index: - Vol. 2, Page 225
[Arrangement of 1870.]
Revolutionary War Pension and Bounty Land Warrant Application Files
Series M-804; Roll # 1500

Ezekiel Lane
New Jersey, New York, Penn. (S.23,760 and S.4,491)

State of New York

County of Erie ss. William Desha now of Buffalo in the State of New York in the County of Erie Being duly sworn deposes and says that he is of the age of sixty nine years and upwards and is well acquainted with Mr. Ezekiel Lane now of Cleveland of the State of Ohio who is nearly ten years older than this deponent and that he was acquainted with him and they were boys together in the same neighborhood Before the Revolutionary War and at the commencement of the War this deponent lived in ~~Yankeetown on the Pennsylvania~~ Sandytown on the New Jersey side of the state line opposite ~~Sandytown~~ Yankeetown in the State of ~~New Jersey~~ Pennsylvania and soon after the War commenced they the inhabitants on the Pennsylvania side were driven off to Jersey side that the second year of the war this deponant and the said Ezekiel Lane both entered the service of the United States by joining a company under Captain Peter Westbrook at Sandytown aforesaid about the middle of April of that year that the other officers of the company were Lieut. James Westfall & & Ensign Cornelius Westfall that they were engaged in scouting and guarding fording places and also in little garrisons which were kept up on and near the fording places on the Delaware River that soon after they entered the service they crossed over into Pennsylvania & while there and soon after they had a skirmish with Indians & Tories in which Capt Westbrook was killed that sometime after John Westbrook the uncle of the deceased Captain Peter Westbrook was elected Captain and took the command and Cornelius Westbrook was elected Lieutenant and under these officers the said ~~William DeShay~~ Ezekiel Lane continued in the service until winter doing active duty The next spring and he thinks in April and about the first of that month they again both entered the service under the same officers and both done the same duty as they had done the Season before or nearly the same and continued in active service until the approach of Winter. That again the next spring they both entered the service at the early part of the spring and he thinks about the first of April under the same officers & again done nearly the same duty until winter of that year, that during all that time they were both private soldiers and were either in the field or in some little garrisons and were employed in no civil pursuit whatever That their duty was generally about the same as above stated and they were called out and served only with one two or three companies together most usually but sometimes when their company went with other companies they were under the command of Col. Rosecrantz. Samuel Westbrook was also the Major of the Regiment. ~~After the above service was performed~~

Sworn & subscribed before me this 22d day of July 1833

Harry Slade J. P.

his
William X Desha
mark

I certify that the above named William DeShay is a credible person & that his statement is entitled to credit

Harry Slade J. P.

★

State of New York,

Erie County Clerk's Office, ss. I, Noah P Sprague Clerk of the County of Erie, do hereby certify that Harry Slade Esq. whose name is subscribed to the annexed certificate & proof of the annexed affidavit was at the time of the execution of the annexed instruments a Justice of the Peace in and for said county, duly elected and sworn and acting as such. And further, that I am acquainted with the hand writing of said Justice and verily believe the signature "Harry Slade" subscribed to said Instruments to be his genuine signature. In testimony whereof, I have hereunto set my name and affixed the seal of the Court of Common Pleas of said county, this First day of August A.D. 1833

Noah P Sprague Clerk

★

State of Ohio

Seneca County ss. On this fifth day of October A.D. 1833 personally appeared in Open Court before the court of Common Pleas of Seneca County in the State of Ohio at the October Term thereof Zekiel Lane of Pleasant Township in said county aged seventy eight years who being first duly sworn according to law on his oath makes the following declaration in order to obtain the benefit of the act of Congress passed June 7th A.D. 1832.

That he was born at Crum Pond on the east side of the North River and does not know the name of the County on the 13th day of December 1755 as he is informed and believes. He has no record of his birth.

About the middle of April 1777 then residing at Sandy Town in Sussex County in the State of New Jersey he entered into the service of the United States as a private soldier at said Sandy Town. He entered as a volunteer in a company of volunteers Commanded by Capt. Peter Westbrook, James Westfall was Lieutenant and Cornelius Westfall Ensign. He was engaged in scouting, guarding fording places along the Delaware river and in small garrisons near said River From one of those places he crossed over the River into Pennsylvania and joined another company under the command of Capt Vannatta and had a skirmish with the Indians and Tories in which Capt Westbrook, Lieutenant Benjamin Annis, and Alexander Rosegrants were killed. After whose decease John Westbrook was elected Captain and Cornelius Westbrook Lieutenant to fill the vacancies. Under these officers he continued to serve until in the winter when he was discharged, having served at least six months. He received a written discharge from his Captain, and has lost it

Again about the first of April 1778 he entered into the Service of the United States at Sandy Town aforesaid as a private soldier, a volunteer in the same Company and under the same Officers and was employed as in the year before in scouting guarding the fording places, and small garrisons along the Delaware river until in the winter following when he was regularly discharged having served at least six months He received a written discharge but has lost it.

Again in April 1779 he entered into the Service of the United States at Sandy Town aforesaid in the same Company and under the same Officers a volunteer private soldier. He was again employed in scouting, guarding, and in small garrisons as before, he served until winter and was again regularly discharged. Having served in this tour at least six months This time he did not receive a written discharge. He does not now recollect the names of any Regular Officers. He was generally in small parties under a Captains Command except in a few instances his company was joined by two or three others, then they were commanded by Col. Rosecrantz and Major Samuel Westbrook.

Since the Revolutionary War he has resided at _____ Creek and Caturagus in the State of New York. In May 1833 he removed into the State of Ohio and landed at Cleveland where he made a temporary Stay, and then settled in Pleasant Township Seneca County Ohio where he now resides. He has no documentary evidence of his services and knows of no person except William Desha whose deposition accompanies this declaration whose testimony he can procure who can testify to his services Ozias Hart and Martin Lane are persons in his neighborhood who can testify to his reputation as a soldier

He hereby relinquishes every claim whatever to a pension or annuity except the present and declares that his name is not on the Pension Roll of any Agency of any State

Sworn to and subscribed the day and year aforesaid in Open Court Joseph Howard Clk

his
Ezekiel X Lane
mark

We Martin Lane and Hosias Hart of Seneca county Ohio and residing in the neighborhood of the above named Ezekiel Lane, and formerly residents of the same neighborhood where he resided in the State of New York hereby certify that we are and in the State of New York were well acquainted with the said Ezekiel Lane who has subscribed and sworn to the above declaration that we believe him to be 78 years of age, that he was reputed and believed in the neighborhood where he resided in the State of New York last before he came to this state to have been a soldier of the revolution, and that we concur in that opinion.

Sworn to and subscribed the day and year aforesaid in open court
Joseph Howard Cllk

Ozias Hart
Martin Lane

And the said court do hereby certify their opinion after the investigation and after putting the interrogatories prescribed by the war department that the above named Ezekiel Lane was a soldier of the Revolutionary War and served as he states. And the Court further certifies that it appears to them that Martin Lane and Hosias Hart who have signed the preceeding certificate are residents in the neighborhood of said applicant that they are credible persons and that their statements are entitled to credit

D. Higgins Presd Judge

Benjn. Pittenger associates
Agreen Ingraham

I Joseph Howard Clerk of the Court of Common Pleas of Seneca County and State of Ohio do hereby certify that the foregoing contains the original proceedings in the matter of the application of Ezekiel Lane for a pension In testimony whereof I the clerk of said court have hereunto set my hand and affixed the seal of said court at Tiffin in said county this 9th day of October in the year of our Lord one thousand eight hundred & thirty three.

Joseph Howard Clerk

☆

Ohio Pittsburgh

Ezekiel Lane of Seneca Co in the State of Ohio who was a Private in the Co. commanded by Captain Westbrook of the Regt commanded by in the New Jersey line for 18 months. Inscribed on the Roll of Ohio Pittsbg. at the rate of 60 Dollars 00 Cents per annum, to commence on the 4th day of March, 1831. Certificate of Pension issued the 5th day of Nov. 1833 and sent to Joseph Howard, Tiffin, Seneca Co. Ohio

Arrears to the 4th of Sept. 1833	150
Semi-anl. allowance ending 4 Mar 34	30
	$180

Recorded by Wm. R. Palmer Clerk,
{Revolutionary Claim, Act June 7, 1832.}
Book E Vol. 8 Page 111

☆

Panama July 1842

Dear Sir

Enclosed are the papers of Mr Ezekiel Lane of this town, to procure a pension to which I am requested to ask your personal attention if need be. As Mr. Lane had his papers made out several years ago while he resided in Buffalo and for some reason they were neglected & he failed. If they should not be allowed will you have the goodness to state the reason or get the Commissioner to do so. In reference to the affidavits of H. P. Willcox & N. Sacket you are refered to Hon N. P. Talmadge U.S.S.

Yours &c
Abner Lewis

☆

House of Reps
Aug 8, 1842

Dear Sir

The enclosed papers relating to the services of Ezekiel Lane a Revolutionary soldier, you will please examine, and inform me whether they are sufficient to establish his claim to a Pension

Very respectfully Yr Ob Svt
S N Clarke

State of New York
Chautauque Co. ss. H. P. Willcox being Duly Sworn, before H. H. Hawkins Esq a Justice of the Peace of said county, Deposeth & says that he has for the last fourteen years been well acquainted with Ezekiel Lane, that he has been represented to me as one of the few remaining Revolutionary Soldiers, to which in common with other patriot Spirits we are indebted as a nation for our independence and Freedom.

That during my acquaintance with Mr. Lane, for the above period I have had considerable dealing, & have found him at all times Honest, & Honerable, & a man in whom I have allways regarded, possesing the most strict veracity.

Mr. Lane since my first acquaintance has been poore, & with great patience & fortitude has labored for that subsistance, which his merits as a soldier would fairly entitle him to, from this Government, so far as the pention law can afford that relief.

Sworn and subscribed before me this 15th day of July 1842
H. H. Hawkins J. P.

H. P. Willcox

State of New York
Chautauque County ss. Niram Sackett being duly sworn before H. H. Hawkins Esqr. one of the justices of the peace of said county deposeth and says that he has been acquainted with Ezekiel Lane for about ten years. That said Lane to appearance must be near eighty years of age, that he is poor and infirm And that he is reputed to be one of the few remaining veteran soldiers of the revolution His reputation for truth and veracity is never questioned by those who know him. In that respect he stands unimpeachable and further this deponent saith not

Sworn and subscribed before me at Hanover in said co. July 15th 1842
H. H. Hawkins J. P.

N. Sackett

Jamestown
July 23 1842

Sir

I am acquainted with Niram Sackett & H. P. Willcox within named. They are men of respectible character and talents & worthy of full credit

Commissioner of Pensions

R. P. Marvin Late _____

State of New York

Chautauque County ss. John K. Lothridge of the town of Hanover in said county being duly sworn before H. H. Hawkins Esqr One of the Justices of the peace of said county deposeth and says that he has been acquainted with Ezekiel lane for about twenty years that from his own statements and those of several revolutionary pensioners now Dead in relation to services of the revolution he has no doubt but he is one of the soldiers of the revolution. And he should fully concur in the statements _____ the above affidavit subscribed by N. Sackett And further this deponent saith not

Sworn and subscribed before me at Hanover July 15th 1842

H. H. Hawkins Justice Peace

J. K. Lothridge

State of New York

Chautauqua County ss. Amos Sottle being duly sworn before H. H. Hawkins a justice of the peace of said county deposeth & says that he has been acquainted with Ezekiel Lane for about forty six years That he first became acquainted with him in Buffalo. That he has resided since that time mostly in the western part of the State of New York and in Ohio. That he now resides as he is informed and believes in the Town of Harmony in said county of Chautaugue. That the said Lane is reputed to be a soldier of the Revolution ~~who~~ That he has freequently heard Old pensioners speak of said Lane as one of their associates in arms in the revolutionary war. That to all appearance said Lane must be near Eighty years of age. That he is poor and infirm, and that his reputation for honesty and integrity truth and veracity is good. That in the opinion of this deponent any statement that said Lane will make in relation to services rendered as a revolutionary soldier may be regarded as truth And further This deponent saith not

Sworn and subscribed before me at Hanover

in said county this 15th day of July 1842

H. H. Hawkins Justice of the Peace

Amos Sottle

State of New York

Chautauque County ss. I do hereby certify that I am well acquainted with the Individuals who made the foregoing affidavits viz Niram Sackett Amos Sottle John K Lothridge and Henry P Wilcox all residents of the County of Chautauque and that that are men of unquestionable integrity and candor

July 15th 1842

H. H. Hawkins Justice Peace

State of New York

Chautauque County ss. I, John G. Hinckley Clerk of said county and also Clerk of the Court of Common Pleas in and for said County, do hereby certify that H. H. Hawkins Esqr. Whose Signature is set to the Jurats of the foregoing depositions & also to the foregoing Certificate was on the day of the date of said Jurats & of said Certificate a Justice of the Peace in and for said county duly Elected & Sworn. And further that I am well acquainted with the hand writing of said Justice and verily believe the Several Signatures to said Jurats & Certificates "H. H. Hawkins" subscribed are genuine. In Testimony Whereof I have hereunto Set my hand and affixed the seal of the said court & county July 28, 1842.

John G. Hinckley Clerk

Cattaraugus County ss. William DeShay being duly sworn deposeth and saith that he is Eighty one years of age that he was born in Sussex County in the State of New Jersey where he continued to reside until about two years after the close of the Revolutionary War. That this deponent was well acquainted with Ezekiel Lane, was brought up with him in the same neighborhood in the State of New Jersey, that the said Ezekiel Lane entered the the continental service with this deponent as a volunteer near the commencement of the Revolutionary War the year not reccollected that the said Lane and this deponent were minute men and served together in the same company four or five summers about seven months each summer the precise time this deponent cannot reccollect. Our Captains name was Peter Westbrook our Major was Samuel Westbrook he and the Captain were brothers Capt Westbrook was killed the second season we served, I think, in a skirmish with the Indians in the State of Pennsylvania on the Delaware River John Westbrook then came in Captain of our company and commanded the company as long as we served. John Westbrook was related to Peter and Samuel but was not their brother. Said Lane and this deponent served during the whole time above mentioned in the New Jersey Militia as volunteers, further this deponent saith not

Subscribed and sworn this 21st
day of July 1842 before me
Thos. J. Wheeler a Judge of Catt County Courts

his
William X DeShay
mark

I hereby certify that William DeShay who made the within affidavit is reputed and believed to have been a soldier of the Revolution that his character for truth and veracity is good. That in my opinion he is not able, in consequence of bodily infirmity, to attend and give his evidence in open court. In witness whereof I have hereunto subscribed my name this 21st day of July 1842

Thos. J Wheeler a Judge of Catts Co Courts

I Thomas J Wheeler the above named Judge do also further certify that the within named William Deshay declared that he was unable to write his name in consequence of bodily infirmity and I wrote his name & saw him affix his signature by making his mark. In witness whereof I have hereunto subscribed my name the day and year above written

Thos. J. Wheeler

State of New York

Cattaraugus County Clerk's Office ss. I George W. Gillet clerk of the county of Cattaraugus and also clerk of the court of Common Pleas in and for said county Do hereby Certify that Thos. J. Wheeler Esqr whose signature is set to the Jurats of the foregoing depositions and also to the foregoing Certificate was on the day of the date of said Jurats & of said certificate a Judge of Cattaraugus County courts duly commissioned and sworn And further, that I am acquainted with his handwriting and verily believe that his signatures to said Jurats & certificates are genuine. In Testimony Whereof I have hereunto set my hand and affixed the seal of the said court & county this 2d day of August A.D. 1842.

G. W. Gillet Clerk

State of New York

Chautauque County ss. Ezekiel Lane of the town of Harmony in the County of Chautauque and State of New York, aforesaid being duly sworn doth depose and say that, as near as he can recollect, he was Eighty one years of age on the thirteenth day of December last past and that according to his best recollection he entered the Revolutionary army Sometime in the forepart of July in the year Seventeen hundred and Seventy six or Seventeen hundred and Seventy seven, cannot now say definitely which; Previously to this period, this deponent resided with his father in the town of Upper Smithfield in the State of Pennsylvania. He, (this deponent) entered the service, in the company commanded by Peter Westbrook or Westebrook. The Lieutenant's name was James Westfall and the Ensign's name was Cornelius Westfall, who resided in the county of Sussex in the State of New Jersey, the regiment was at that time commanded by Colonel John Rosecrantz of Sussex County aforesaid. The first skirmish we had was on the Neversink River about a mile above its junction with the Delaware River. At this place a small body of Indians were collected and burning the houses in which the inhabitants lived. The contest at this place last between four and five hours when we succeeded in dispersing the Indians, and they left the place. As soon as a sufficient number could be collected, this deponent joined the company of Captain Abraham Westfall whose Lieutenant's name was Elijah Van Auker, we then went up the Delaware River in pursuit of the Indians who left Never Sink and overtook them about the fifty miles up the River where they had been joined a number of Torys. This detachment of Indians and Torys was commanded by Captain Brant The number under his command as Captain Brant has since told me was about two hundred & fifty. Here we got most dreadfully flogged Heard Captains Tyler & Little Colonel Thurston and Lieutenant Wisner were killed with a large number under their command. I was engaged in the army this year about seven months. The next spring I again entered the company of the said Peter Westbrook and remained with him until Sometime in the ensuing Summer when in an engagement with the Indians Captain Peter Westebrook was killed, and John Westebrook was appointed Captain, _____ Westebrook was appointed Lieutenant and Cornelius Westebrook Ensign under these officers in Colonel Rosecrants Regiment, this deponent remained until the close ~~of the close~~ of his term of service in the army. I remained in the army five years, but not in actual service during the winter seasons. I served in the same company with William Deshay, whose affidavit I have procured, and which is hereunto annexed I served in each year about seven months and at all other times held myself in readiness to go into the service at a moment's warning. The above named William Deshay is the only man that I know of living, that was with me in the army of the Revolution, him, I supposed to be dead, until about a week since my friends succeeded in finding him in the County of Cattaraugus County in this state My business in the army was in guarding the American settlements against the ravages of the merciless Indian Savages, and was therefore prevented from engaging in any business profitable to myself during the whole term of five years after I entered the service

Sworn and subscribed this 26th day of July 1842 before me
Asahel Clark J P

his
Ezekiel X Lane
mark

State of New York

Chautauque County ss. I Asahel Clark the above named justice of the peace do certify that the said Ezekiel Lane who made the above affidavit is reputed to be a man of undoubted veracity, whose statement is entitled to full faith and credit. I do further certify that the said Ezekiel Lane from his age and from being unaccustomed to writing his name, as he informed me is unable to write his name, and that by his direction I wrote his name and he made his mark in my presence.

July 26, 1842

Asahel Clark J. P.

State of New York

Chautauque County ss. I, John G. Hinckley Clerk of said county and also Clerk of the Court of Common Pleas in and for said county do hereby certify that Asahel Clark Esqr Whose signature is set to the Jurats of the foregoing Deposition & also to the foregoing Certificate was on the day of the date of said Jurat & Certificate a Justice of the Peace in and for said county duly Elected and Sworn. And further that I am well acquainted with the hand writing of said Justice and verily believe the Signatures "Asahel Clarke" Subscribed to said Jurat & Certificate are genuine. In Testimony Whereof I have hereunto Set my hand and affixed the Seal of the Said Court & County this 28 day of July 1842

John G. Hinckley Clerk

Pen Off
Aug 11 42

Sir

I have the honor to inform you that the claim of Ezekiel Lane has been examined & his papers filed under the act of 7 June '32. He is required to make a formal declaration under the act of 7 June 32 before a court of record & in all respects comply with the regulations.

If he has heretofore taken any measures to assert a claim he should afford full information on that point and transmit any papers heretofore executed with that view.

Each term of service however short should be distinctly set forth in the following particulars - the period, length & grade, where marched or stationed, distance from home, probable number of men embodied. How paid, subsisted & quartered. The best attainable proof from surviving comrades & from the archives of New Jersey at Trenton should be adduced

Hon. S. N. Clarke H. R.

Panama Nov. 2d 1842

Dear Sir

Enclosed is the declaration of Ezekiel Lane the other papers pertaining to his pension are with the department

Yours &c
A Lewis

State of New York

Chautauque County ss. On this 13 day of October in the year of our Lord, one thousand eight hundred & forty two, personally appeared in open court being a court of Record for the County of Chautauque, Ezekiel Lane, resident in said county, aged Eighty one years on the thirteenth day of December last past, who first being duly sworn according to law, doth on his oath make the following declaration in order to obtain the benefit of the act of Congress passed June 7th 1832.

That he entered the service of the United States under the following named officers and served as herein stated. This declarant entered as a private soldier in the company commanded by Captain John Van Sickle, the name of the Lieutenant in said company was Lewis Mead, and the name of the Ensign was James Wells. The name of the Colonel who commanded the Regiment, to which said Van Sickle's company belonged, was John Stroud. This declarant entered into said Van Sickle's company as a volunteer about the middle of the month of May, in the year ~~eighteen hundred~~ Seventeen hundred and seventy six as this declarant verily believes, as according to best of his recollection and belief it was the year after he was fifteen years old, and the reason he believes it was about the middle of May of that is, that it was at the time of planting corn, some had planted their corn, and some that planted, had not yet commenced, and that he remained on the said company in the service of the United States until some time on the month of ~~October~~ November of that year the precise day of that month on which he left he cannot now state it being so long ago, and has no memorandum by which to refresh his recollection. When I resided in the town of Upper Smithfield the county not _____ This was in the State of Pennsylvania.

This declarant further states, that in the Spring of the next year, which according to his best recollection and belief, was Seventeen hundred and seventy seven, he entered in the same capacity into the service of the United States at an earlier period and according to the best of his recollection & belief was quite the last of April or the first of May, into the company commanded by Captain Moses Cordtright whose Lieutenant's name was Elijah Van Auker. The name of the Colonel commanding the regiment to which Captain Cordtright's company belonged he does not now recollect, that he remained in the company of Captain Cordtright, until sometime in the month of November of that year the day of the month he does not recollect for the reasons above stated, at this time this declarant had no settled place of residence being driven from Smithfield in Pennsylvania, where I resided the year before by the Indians, and my father had broken up house keeping. It was near Goshen in the County of Orange in the State of New York where I entered the company of Captain Cordtright this year (1777)

This declarant further states, that the next year, being the year Seventeen hundred and seventy eight he entered into the company commanded by Captain Peter Westbrook, whose Lieutenant's name was James Westfall, and whose Ensign's name was Cornelius Westfall. The name of the Colonel who commanded the regiment, at this place was John Rosecrants at this time I resided with my Brother Solomon Lane in the town of Sandyston, in the County of Sussex, in the State of New Jersey This year, this declarant entered in the service of the United States in the company of the said Peter Westbrook as early as the first of April, as he recollects, that it was the first rise of water that spring which was high enough to enable the owners of boards and other lumber on the Delaware River, to run them to market, we were called out at this time, to repel the Indians who waylaid the teamsters who were hauling their lumber from the mills to the Delaware River. This year also (1778) this declarant remained in the army until late in the month of November when the snow came on.

This declarant further states that he remained in the same company until the fall of the year and sometime in the month of November in the year Seventeen hundred and eighty, though he was not actually engaged in the service of ~~the service~~ of the United States during the winter seasons, and while the snow was on the ground for the reason that during the times of year when the snow was on the ground, the Indians were not out to disturb the settlements of the inhabitants. In the spring of Seventeen hundred and seventy nine he again went into actual service in said company, in the forepart of the month of April, and continued said service till into the month of November when he again retired from the service until the Spring of Seventeen hundred and eighty when as early as the first April he again entered the service in said company, and remained in the same until after the first of November of this year (1780) when he left and did not return any more during the war of the Revolution This declarant further states, that that in the fall of the second year, after he entered the company of Peter Westbrook, Peter Westbrook was killed by a party of Indians and Tories where the Lieutenant of another company that we fell in with after we crossed the Delaware River into Pennsylvania was also killed. This Lieutenants name was William Annis. There was a number more belonging to our party killed. Then John Westbrook took the command of the company to which belonged & retained the command until I left the company and the army. The first engagement we had with enemy after I entered the service was at the Fort on the Neversink River in the State of New York, at this engagement none of our party was killed, except a man by the name of James Van Auker. This was the second year after I entered the service, at just at the close of wheat harvest. About three days after this engagement, I went with the rest of the company to which I belonged about a mile from the Fort and there we joined several other companies making a detachment of about two hundred and fifty men who all together went in pursuit of Indians and Tories above mentioned The names of the officers that I now recollect other than those of our own company above mentioned were Colonel Thuston, Major Meeker, Captain Little, Captain Tyler, Lieutenant Wisner. On the second day after we started we over took, the Indians and Tories at a place called Beaver Brook, where a contest took place in the forepart of the day, and continued for some two or three hours, according to the best of my recollection in this engagement Col. Thuston, Major Meeker, Captain Little, Captain Tyler and Lieutenant Wisner were killed. Captain Brant was the one who command this detachment of Indians and Tories. After this skirmish we returned again to the fort. These were the only engagements we had that year worth mentioning

The next engement we had in which I was engaed was the one in which

Captain Peter Westbrook was killed which was the second year after I went into New Jersey and joined Captain Westbrook's Company, we had several other small skirmishes during the time I was in the army but none in which any of our party was killed.

This declarant some nine or ten years ago had his declaration and papers made out for the purpose of procuring a pension, and delivered them to a man by the name of _____ a lawyer who resided in Tiffin on the Sandusky River in the State of Ohio who said he was going to the City of Washington since which this declarant has heard nothing on this from the papers so made out, or the person to whom they were delivered.

Sworn & subscribed before me this 13th day of October 1842
John G. Hinckley Clerk of Chautauque County
Witness, J. G. Hinckley

Ezekiel X Lane

State of New York
Chautauque County ss. I, John G. Hinckley Clerk of Said county do hereby certify that the Signature "Ezekiel Lane" subscribed to the foregoing Deposition was set thereto by me at the request of the Said Ezekiel Lane, he the Said Ezekiel Lane being unable to write on account of nervous trembling of his hand, that the mark thereto set was made by him in my presence & is his mark In testimony Whereof I have hereunto Set my hand and Affixed the Seal of the Court of Common Pleas in and for Said County at Mayville in said County this 13th day of October 1842.

John G. Hinckley Clerk

And the Court do hereby declare their opinion after an investigation of the matter, and after putting the interrogatories prescribed by the War Department, that the above named applicant was a revolutionary Soldier and served as he states. I, John G. Hinckley Clerk of the Court of Common Pleas of the County of Chautauque do hereby certify that the foregoing contains the original proceedings of the court in the matter of the application of Ezekiel Lane for a pension. In testimony Whereof I have hereunto Set my hand and Seal of Office this 13th day of October 1842

John G. Hinckley Clerk

Pen Off
Nov. 18 42

Sir

The addl. statement of Ezekiel Lane is unsatisfactory in its specifications and unsupported by the proofs indicated in my letter of 11 Aug last as attainable from surviving comrades and the public archives of New Jersey & Pa.

A. Lewis Esq.
Panama Chautauqua Co NY

J. L. Edwards Esq
Comr. of Pensions

House of Reps.
Jany. 17, 1843

Dear Sir,

The enclosed letter from Abner Lewis Esq. asking information in regard to the application of Ezekiel Lane for a pension, I beg leave to refer to you, and have to request you, if in your power, to give the information he desires.

Very respectfully Yr. Obt. Servt
S N Clarke

House of Representatives
Jany. 2nd 1844

J. L. Edwards Esqr
Dear Sir

Enclosed I send you the letter of Abner Lewis Esqr of Chautauque Co (NY) in relation to the Pension of Ezekiel Lane I believe the letter is sufficiently explicit, without any comment from me Please give me an answer as soon as the business will allow

Very Respectfully Yours
Asher Tyler

Pension Office,
Jan 4 1844

Sir,

I herewith return Mr Lewis' letter in the case of Ezekiel Lane, and have the honor to state that the letter from this Office to the Hon. S. N. Clarke, dated the 25th of January 1843, a copy of which is herewith enclosed, will shew what is required of the claimant

Hon Asher Tyler H R

Pen Off
Jany 25th 43

Sir

I have the honor to inform you in reply to Mr. Lewis letter respecting the case of Ezekiel Lane that Mr. Lewis has been informed that the terms of service were not specified in detail as the rules required and were not sustained by the proofs which the archives of New York and the knowledge of surviving comrades can furnish.

Hon. S. N. Clarke H. R.

Office of Secy of State
Trenton N.J.
April 5, 1844.

Sir

Your letter of the 4th Inst. is Just recd. I have carefully examined the Margins for Sussex County without finding the name of Ezekiel Lane. I have also Examined the Rolls of the 9 mos men without success The muster rolls of those who served on the Frontiers (The militia and State Troops) are not in this office nor are they known to be in existence. It is presumed they were destroyed in the Treasury dept. by fire Maj Westbrook commanded at Minisink in 1778. Jany, June, July, Augt. Sept. & Decr. in 1779, October. Maj Rosekrants commanded at Minisink in Apl. & May 1780 & 81. In April 1780 Capt. J. Shiner had a company of 30 men Capt. Hover had 31 men and in May 1781, Lt Marsh had ~~had~~ 16 men all stationed at Minisink on the Delaware, in 1777 Col Rosinkranz in Octr. under Brigr Forman in Pa. recd. £480.7.0 for service in 1778 he was almost constantly out. One dollar recd. covers the expenses.

Your Obt Sevt
Charles G. McChesney

Seth L Cole Esqr

Comptrollers Office
Albany Apl. 6th 1844

Sir

After a careful examination of the books documents &c remaining in this office, pertaining to the Revolutionary period, and a like examination in the office of the Secretary of State, the name of Ezekiel Lane is not found.

I am very respectfully Yr. Obdt. Servt.
Philip Phelps
Dep Comptroller

Seth L. Cole Esq.
Washington City
D.C.

Application for a Pension under the Law of the 18th March 1818.

State of New York

Erie County ss. Personally Appeared before me James G Dickie a Justice of the Peace in and for the said County, Ezekiel Lane who being duly sworn according to law, depose and says, that he is Eighty five years of age, that he Volunteer'd as a Private, in the service of the United States during the Revolutionary War, soon after the Battle of Bunker Hill in the year 1775, in the State of Pennsylvania in Captain Vansickles Company, Colonel John Strouds Regiment as a minute-man, that he join'd his Company and was stationed on the Delaware River, at a Place then call'd Upper-Smithfield-Town and while there stationed, a little over two years, was engaged in skirmishes several times, with the Indians, and Tories, at last we were driven by them, into Jersey State, Sussex County, we made our retreat during the night their numbers being far superior to ours. I was then order'd to join Captain Kortrights Company, Major Decker, don't remember our Colonels name, we were stationed on the Navesink River in York State, about two years, and there we encountered our old enemy again Indians and Tories, in two verry severe skirmishes in which we defeated them and kept our Fort, was also out in many scouting parties day and night, during our stay there about two years I then got a permit to go and see my friends in Jersey State. I had not been at home only four or five days, before I was call'd upon to join Captain Peter Westbrooks Company, Colonel Rosenkrants Regiment. Join'd our Company at the word, our old enemy again show'd themselves for fight, we pursued them from Jersey to Pennsylvania State, came up with them and went into action, and drove them off, and in that engagement our Captain Peter Westbrook was killed. John Westbrook after that took command of our Company, which then rais'd from fifty, to seventy men, don't remember the number of men embodied in the Regiment, that he continued in Westbrooks Company, until about one year after the Surrender of Cornwallis at York-Town, that they were disbanded then, and return'd to their homes, and friends, in 1782 while in the service, we sometimes drew our Rations, and often went without them, our scout was out in the Wood, in pursuit of the Indians, three days at one time, without anything to eat. We drew our Clothes sometimes & Shoes, but were often out of shoes, and bare footed Our pay was Eight dollars p month, we rec'd Continental-Money when we were paid, furthermore the facts stated above are substantially true, as near, as, he now remembers

Witness David Sleeper

his
Ezekiel X Lane
mark

Personally appeared before me this 13th day of April 1844 the above named Ezekiel Lane to me known & ~~subscribed~~ swore to the foregoing affidavit and made his mark being unable to write

James G. Dickie Justice of the Peace

State of New-York,

County of Erie, ss. I, Manly Colton, Clerk of the County of Erie, and of the Court of Common Pleas thereof, do hereby certify that James G Dickie Esq. whose name is subscribed to the annexed _____ affidavit was at the time of subscribing the same a Justice of the Peace in and for said County, duly Elected and sworn, and acting as such. And further, that I am acquainted with the hand writing of said Justice and verily believe the signature to said certificate to be his genuine signature. In Testimony Whereof, I have hereunto set my name and affixed the Seal of the Court of Common Pleas of said County, this 13th day of Apl A.D. 1844

Manly Colton Clerk

Albany New York

Ezekiel Lane of Chautauqa Co. in the State of New York who was a private in the company commanded by Captain Westbrook of the regiment commanded by Colonel in the Rev War Penn line for Nine Months. Inscribed on the Roll of Albany N.Y. at the rate of 30 Dollars Cents per annum, to commence on the 4th day of March, 1831. Certificate of Pension issued the 26th day of April 1844 and Sent to S L Cole Present.

Arrears to the 4th of Mar 1844	390.00
Semi-anl. allowance ending 4 Sep 44	15.00
	$405.00

Recorded by Fleming __________ Clerk,
{Revolutionary Claim, Act June 7, 1832.}
Book E2 Vol. 4 Page ___

Buffalo July 28th 1845

Jas. L. Edwards
Dear Sir

I want you to inform me what you have done, in the case of the application of one Ezekiel Lane a Revolutionary soldier for a pension. The papers were made out and forwarded to, the pension office by one Sleeper. The friends of Lane have requested me to ask the department, for this information, and I should be glad to hear from you by return mail as no information can be had from Sleeper. If the certificate has been granted please let me know when & for what amount.

Yours &c
J. D. Gros

N. B. If Mr Edwards is superseded inform me who the present Commissioner of Pensions is

Gros

Pension Office,
Aug 15, 1845

Sir,

In answer to your letter of the 28th ulto I have to inform you that a pension certificate was issued for Ezekiel Lane on the 26th of April 1844, at the rate of $30 per annum, which appears to have been handed to Mr Seth L. Cole, then in this city.

J D Gros. Esq
Buffalo N York

Effingham, Ill.
February 8th 1904

E. S. Ware

Commissioner of Pensions.

Dear Sir,

Will you please have the records searched for the Revolutionary service of Ezekiel Lane, Buffalo, New York. He died at Buffalo, in 1847 or 1848; was drawing two pensions at the time of his death, which occured in his 102d year. It was claimed at the time of his death, that he was the first white man on the site where Buffalo now stands and built the first house there. I am very anxious to get his Revolutionary service to use for a young boy who wishes to become a member of the Children of the American Revolution and who is a gr. gr. gr. grandson of Ezekiel Lane. I hope that it may be possible for me to get this service at an early date.

Very Truly,

Mary Crooker Lloyd Reegent

"Ann Crooker St. Clair" Chapter, D.A.R.

Revolutionary War Records Section.

Department of the Interior,

Bureau of Pensions.

Washington, D.C., , 19

In reply to your request of , for the history of Ezekiel Lane, a soldier of the Revolutionary War, the following is taken from the soldier's claim for pension.

Dates of Enlistment or Appointment	Length of Service	Rank.	Officers under whom Service was Rendered.		State
			Captain	Colonel	
May 1775	to Nov. 1775	Pvt	John Van Sickle	J. Stroud	Penn.
Apr. or May 1777	to Nov. 1777		Moses Kortwrigh	Major Decker	N.Y.
Apr. 1778	to Nov. 1778		Peter & John Westbrook	John Rosekranz	N.J.
Apr. 1779	to Nov. 1779		Peter & John Westbrook	John Rosekranz	N.J.
Apr. 1780	to Nov. 1780		Peter & John Westbrook	John Rosekranz	

Battles engaged in, Several skirmishes with Indians

Residence of soldier at enlistments, Upper Smithfield Pa. & Sandistown N.J.

Date of application for pension, Oct. 5, 1833. His cl. was al.

Residence at date of application, Pleasant Township, Seneca Co. Ohio

Age at date of application, b. Dec 13, 1755 or 1760.

Remarks: It is not stated whether he was married The date of his death is not on record he was a res of Buffalo NY in 1848

Respectfully,
Commissioner.

State of New Jersey,
Office of Adjutant General.
G W
Trenton, May 2, 1905

Honorable Charles M. Bryant,
Old War and Navy Division,
Pension Bureau, Washington, D.C.
Dear Sir:

Mr. Godfrey has reported a couple of Revolutionary cases to this office of Ezekiel Lane, one bearing "Invalid #23,760;" the other "Invalid #4,491."

The services of these two men appear to be about the same, and the identity of them is so nearly alike that I desire to have an expression of information from your Bureau as to whether these are two distinct men.

I would thank you very much for the courtesy of an early reply.

Very respectfully,
R. Heber Breintnall
Adjutant-General

O. W. and N. Division.
Inv. File No. 23760
Ezekiel Lane Rev. War

___. S. C. Ex'r.
Department of the Interior,
Bureau of Pensions,
Washington, D.C., May 5, 1905.

Sir:

For use in the above-entitled claim for pension, you are requested to furnish this Bureau with a report in the case of Ezekiel Lane pensioned under the Act of June 7, 1832 by certificate No. 32131 issued Apl. 26, 1844, Albany N.Y. Agency, showing date of last payment of pension, date of death and to whom arrears were paid, for the purpose of identifying the above soldier with one Ezekiel Lane certificate 22,456 herewith.

Very respectfully,
V. Warner

Commissioner.

The Auditor For The Interior Department,
Treasury Department.

Treasury Department,
The Commissioner of Pensions.
Washington, May 8, 1905

Sir: In reply to your letter of May 5th in case of Ezekiel Lane Certificate 32131 Albany Agency, Revolutionary Roll, you are informed that, according to the latest account rendered by the U.S. Pension Agent, the records of this office show last payment to have been made at 30 per ~~month~~ annum in March 1848 to March 4, 1848. P.O. address in 1848: Buffalo, N.Y. Date of death not known. Letter herewith returned.

Respectfully,
Robert S. Person Auditor.
By _____ Dunham Deputy Auditor.

O. W. and N. Division.
S. Inv. File No. 4491
Ezekiel Lane Rev. War

___. C. Ex'r.
Department of the Interior,
Bureau of Pensions,

Sir:
Washington, D.C., May 5, 1905.

For use in the above-entitled claim for pension, you are requested to furnish this Bureau with a report in the case of Ezekiel Lane pensioned under the Act of June 7, 1832 by certificate No. 22,456 issued Nov. 5, 1833, Pittsbg Ohio Agency, showing date of last payment of pension, date of death and to whom arrears were paid, for the purpose of identifying the above soldier with one Ezekiel Lane certificate 32,131 herewith.

Very respectfully,
V. Warner
Commissioner.

The Auditor For The Interior Department,
Treasury Department.

Treasury Department,
Washington, May 8, 1905

The Commissioner of Pensions.

Sir: In reply to your letter of May 5th in case of Ezekiel Lane Certificate 22456 Pittsburg Agency, Revolutionary Roll, you are informed that, according to the latest account rendered by the U.S. Pension Agent, the records of this office show no payment to have been made at per month Letter herewith returned.

Respectfully,
Robert S. Person Auditor.
By _____ Dunham Deputy Auditor.

817 West Clinton Ave.,
Effingham, Ills
Dec. 4th, 1911

U.S. Commissioner of Pensions,
War Department, Washington, D.C.
My Dear Sir

I very much desire the Rev. War service of Ezekiel Lane, born 1746, died April 6th 1848, at Buffalo, N.Y. and I was advised by the Record and Pension Office to apply to your department for information. He was in both wars with Great Briton, and was drawing double pension at the time of his death so he was in the War of 1812, as as well as in the Rev. War. I have a picture of him taken after death that was sent me recently from the Historical Society in Buffalo. Ezekiel Lane was the first white man on the site where Buffalo now stands, and he and his fatherinlaw, Martin Middaugh, or Middagh, built the first house there. Lane fought in the battle of the Minnisink, in 1779. Another Lane, inferentially, the son of the "Settler," born in Buffalo in 1786, was interpreter in a council with the Seneca and Tuscarora Indians at Buffalo and Cattaraugus in 1838-39. The sons name may have been Ezekiel Lane, too, but I want the service of the Lane born 1746, for the Rev. War and for 1812. The "Buffalo" chapter of Daughters of the American Revolution, wish to mark his grave. As he was a pensioner his service must be of record in your department. Please do not confound the record of father and son. I would also like the service of Martin Middaugh who died in 1822 ~~and~~ at an advanced age. I am asking a great deal, and hope you can give it for which I will be truly Grateful.

[Mrs] Mary Crooker Lloyd

Record Division.
Mrs. Mary Crooker Lloyd,
817 West Clinton Avenue,
Effingham, Illinois.

December 9, 1911.

Madam:

In reply to your letter dated the 4th and received the 7th instant, you are advised that from the papers in the claim of Exekiel Lane, (Sur. File No. 23,760 Rev. War) whose record was furnished you under date of March 2, 1904, it appears that he applied for pension October 5, 1833, in Pleasant Township, Seneca County, Ohio, and his certificate was issued November 5, 1833 on the Pittsburg, Ohio, Agency, but on October 13, 1843 in Chautauqua County, New York, he applied again, stating that he had not received said certificate, and as no many had been paid thereon, a certificate was issued in liew thereof, April 26, 1844, on the Albany, New York Agency. From this it will be seen that he did not draw a double pension.

The soldier stated that he was born December 13, 1755, at Crompond (New York), the date of his death is not shown, nor is there any reference to service in the War of 1812.

You are further advised that the above is the only Ezekiel Lane found on the Revolutionary War pension records, which fail to afford any information in regard to Martin Middaugh.

Very respectfully,
Commissioner.

---

514 E. A. Ave.
Hutchinson Kansas
May 17, 1923

Dept. of the Interior
Bureau of Pensions:
Honorable Sirs:

In any historical research work I find it will be extremely helpful to have coppies of the Pension Papers of Ezekiel Lane who died in Buffalo N.Y. 1847, having passed his 100 birthday. He was a revolutionary soldier & was drawing two pension (for his services to his country) when he died.

Very Respectfully
Mrs. J. S. Simmons

N. B. This is patriot work & is not paid for.

Commissioner of Pensions,
Washington D.C.
Gentlemen,

For the purpose of D.A.R. records, I need information concerning Ezekiel Lane, born 1746 died 1847, who was an early resident of Buffalo, New York and who received a pension for Revolutionary services. Any information you can give me will be most gratefully received.

Most sincerely yours,
Alma W. Cobb
Newark, Delaware

Rev. & 1812 Wars Section.
Alma W. Cobb,
Newark, Delaware,

September 28, 1925.

Madam:

I have to advise you that from the papers in the Revolutionary War pension claim S-23760, it appears that Ezekiel Lane was born December 13, 1755 or 1760 (place not stated).

While residing in Upper Smithfield, Pennsylvania he enlisted in May 1775 and served to November 1775 as a private in Captain John Van Sickle's Company, Colonel J. Stroud's Pennsylvania Regiment and was in several skirmishes with the Indians.

He enlisted in April or May 1777 and served to November 1777 in Captain Moses Kortwright's Company under Major Decker in the New York Troops.

While residing in Sandistown, New Jersey he enlisted and served in Captains Peter Westbrook and John Westbrook's Companies, Colonel John Rosekranz' New Jersey Regiment as follows:

From April 1778 to November 1778; from April 1779 to November 1779 and from April 1780 to November 1780.

He was allowed pension on his application executed October 5, 1833 while a resident of Pleasant Township, Seneca County, Ohio.

It is not stated whether he was married. The date of his death is not on record. He was a resident of Buffalo, New York in 1848.

Respectfully,

E. W. Morgan,
Acting Commissioner.

Invalid.
File No. 23,760
Ezekiel Lane Pri Rev War
Act: June 7th 32
Index:- Vol.2, Page 234
[Arrangement of 1870.]

Invalid.
File No. 4,491
Ezekiel Lane Privt Rev War
Act: June 7th 1832
Index:- Vol.A, Page 214
[Arrangement of 1870.]
(Consolidated with Inv. File #23,760.)
Revolutionary War Pension and Bounty Land Warrant Application Files
Series M-804; Roll # 1519

Jacob List
New York (S.31,214)

Jacob List Dec for Pension 7th June 1832

State of Kentucky

Henry County ss. On this 30 day of december 1833 personally appeared before me David Adams a Justice of the peace in and for Henry County and one of the Judges of the County court for said County of Henry Jacob List aged 73 years who being first duly sworn according to law doth on his oath make the following declaration in order to obtain the benefit of an act of Congress passed the 7th of June 1832

This declarant states that he was residing in Orange County State of New York when he entered the service of the United and served under the following named officers he states that sometime in the summer of the year 1776 he enlisted for a period of twelve months he enlisted under one James Boyle a recruiting officer and was commanded by Capt William Blain he states that the company rendezvoused at Goshen in Orange County State of New York and under the command of Capt Blain he states that he marched to Albany where he states he remained but for a short period he was then marched to the frontiers under his said Capt Blain for the purpose of protecting the exposed part of the state of repelling the Indians and British from our borders he states that after spending some time upon the frontiers the company to which he was attached was ordered to march to Saratoga to unite with the American army that had come from Quebeck he states that there they were put under the Command of General Stark and remained with him until after the surrender of Burgoyne he states that he did assist in guarding said prisoners taken from Burgoyne as far as Goshen when his twelve months for which he had enlisted expired and was discharged he further states that some time in the month of June or July which he does not recollect he volunteered for a tour of three months under Capt Samuel Jones and Lieutenant Peter Gayle he states they marched from the drownded lands in Orange County to Lackawanna Creek in the frontiers here we were met by a tribe of Indians Commanded by a Captain Brant in the forks of the creek Lackawanna a dreadful engagement ensued in which every officer was killed except a Capt John Wood who was taken prisoner and kept by the Indians for five years and after this dreadful defeat I returned to Orange County in New York after which I was employed a great part of the war in defending the frontiers in scouting and spieing parties until the year of 1781 but how long I was actually employed in the service of the United States he does not distinctly recollect he states that he has no documentary evidence of his service and knows of no living witness by whom he can prove said service he hereby relinquishes every claim whatever to a pension or annuity except the present and declares that his name is not on the pension roll of the agency of any state

Sworn to and subscribed this date and year aforesaid

David Adams Justice of the peace for Henry County

Jacob List

Interrogatories put by the magistrate

Ques Where and in what year your Born

ans I was born in Pensylvania near the Delaware river in the year of 1760

Ques Have you any record of your age

ans My mother had a record of my age in her Bible I some time ago saw and read it

Ques where you living when called in to the service

ans I was living in Orange County State of New York and in the year of 1782 I moved to New Jersey where I lived till the year of 1803 I then moved to Henry County Kentucky where I have lived ever since

Ques How were you called into the service were you drafted did you volunteer or were you a substitute and if a substitute for whom

ans I first enlisted for twelve months I then volunteered for three months I then was employed in scouting and spieing on the frontiers for a considerable time

State the names of some of the regular officers who were with the troops where you served and such continental and militia regiments as you can recollect

ans General Stark General Schuyler Coln Hanes Coln ______ and Coln Turston which last was killed at Lackawanna creek

Ques Did you ever receive a discharge and by whom was it signed

ans I did receive a written discharge handed to me by Coln Hamilton which discharge I got burnt up

State the names of some of your neighbours who can testify as to your character for veracity and their belief of your service as a soldier of the revolution

ans I will name Mr. Josep Reeves and Mr Stephen Shaw

We Joseph Reeves and Stephen Shaw residing in the neighbourhood of Jacob List the above named applicant say that we are well acquainted with him believe him to be 73 years old he is reputed and believed in his neighbourhood to be a man of strict veracity and that he was a soldier of the revolution and we concur in the belief

Sworn and subscribed this date and year aforesaid

David Adams Justice of the peace for Henry County

Joseph Reeves
Stephen Shaw

I David Adams a Justice of the peace in and for Henry County do certify that Joseph Reves and Stephen Shaw who subscribed and sworn to the above certificate are credible men and that their statement may be relied upon. I further certify that there is no Clergyman residing in the neighbourhood of the aboved named Jacob List and that from bodily infirmity he is unable to appear in court to make his declaration and after putting the above interrogatories prescribed by the war department and investigating the matter I am satisfied that he was a soldier of the revolution and served as he has stated in his declaration

David Adams Justice of the peace for Henry County

I Edmund P Thomas Clerk of the county court of Henry County do certify that David Adams before whom this declaration was made is a Justice of the peace in and for Henry County and one of the Judges of the Court for which I am Clerk and that full faith and credit is and ought to be given to his official acts In tesamony whereof I have hereunto subscribed my name and affixed my seal of office this date and year aforesaid

Edmd P Thomas

Kentucky

Jacob List of Henry Co in the State of Kentucky who was a private in the com commanded by Captain Blain of the regt commanded by in the New York line for 9 months Inscribed on the Roll of Kentucky at the rate of 30 Dollars Cents per annum, to commence on the 4th day of March, 1831. Certificate of Pension issued the 20 day of Jany 1834 and Hon R M Johnson

Arrears to the 4th of Sept	75
Semi-anl. allowance ending 4 March	15
	$90

Recorded by Danl Boyd Clerk,
{Revolutionary Claim, Act June 7, 1832.}
Book E Vol. 7 Page 17

Treasury Department,
Second Comptroller's Office,
Nov. 30, 1838

Sir:

Under the act of the 6th of April, 1838, entitled "An act directing the transfer of money remaining unclaimed by certain Pensioners, and authorizing the payment of the same at the Treasury of the United States," Jacob List a Pensioner on the Roll of the Kentucky Agency, at the rate of thirty Dollars and Cents per annum, under the law of the 7th June, 1832, has been paid at this Department, from the 4th of Sept. 1837 to the 4th of March, 1838

Respectfully, yours,
Albion K. Parris
Comptroller.

To the Commissioner of Pensions,
Present.

Treasury Department,
Second Comptroller's Office,
September 19th, 1839.

Sir:

Under the act of the 6th of April, 1838, entitled "An act directing the transfer of money remaining unclaimed by certain Pensioners, and authorizing the payment of the same at the Treasury of the United States," Jacob List, a Pensioner on the Roll of the Kentucky Agency, at the rate of Thirty Dollars per annum, under the law of the 7th June 1832, has been paid at this Department, from the 4th of March 1838, to the 4th Sept. 1838.

Respectfully, yours,
Albion K. Parris
Comptroller.

To the Commissioner of Pensions
Present.

Treasury Department,
Second Comptroller's Office,
March 14th, 1842.

Sir:

Under the date of the 6th of April, 1838, entitled "An act directing the transfer of money remaining unclaimed by certain Pensioners, and authorizing the payment of the same at the Treasury of the United States," Jacob List, a Pensioner on the Roll of the Kentucky Agency, at the rate of Thirty Dollars per annum, under the law of the 7th June 1832, has been paid at this Department, from the 4th of Sept. 1840, to the 4th March 1841.

Respectfully, yours,
Albion K. Parris
Comptroller.

To the Commissioner of Pensions,
Present.

I*nvalid. File No. 31, 214*
Jacob List Pri Rev War
Act: June 7th 32
Index: Vol. 2, Page 243

[Arrangement of 1870.]
Revolutionary War Pension and Bounty Land Warrant Application Files
Series M-804; Roll # 1570

Daniel Miers
New York (S.13,933)

BRIEF in the case of Daniel Miers of Orange County in the State of New York (Act 7th June, 1832.)

1. Was the declaration made before a Court or Judge? Court

2. If before a Judge, does it appear that the applicant is disabled by bodily infirmity?

3. How old is he?

4. State his service, as directed in the form annexed.

Period when service rendered	Duration of claimant's service			Rank of the Claimant	Names and Ranks of the Field Officers under whom he served
	Years	Months	Days		
May 1776		4		Sargt	Capt Jackson, Genl
Apl 1777		1		Sargt	
June 1777		1			
Apl 1778		6		Private	Capt Westfall
May 1781		6		Sargt	Capt Westfall

Names of General and Field Officers under whom he served.

5. In what battles was he engaged? in New York when Washington was driven out. Minisink

6. Where did he reside when he entered the service? Orange County New York

7. Is his statement supported by living witnesses, by documentary proof, by traditionary evidence, by incidental evidence, or by the rolls? Traditionary

8. Are the papers defective as to form or authentication? and if so, in what respect? Defective Names of Officers under whom he served reqd

Certify that the foregoing statement and the answers agree with the evidence in the case above mentioned.

J Wilson Examining Clerk.
Jy 18 1833

State of New York

Orange County ss. On this fourth day of ~~September~~ December 1832 in the year of our Lord one thousand eight hundred and thirty two personally appeared in open Court before the Judges of the Court of Common Pleas (being a Court of Record) now sitting, in and for the said County of Orange, Daniel Miers a resident of the town of Minisink Orange County NY County and State aforesaid who being first duly sworn according to Law doth on his oath make the following declaration in order to obtain the benefit of the act of Congress passed June 7th 1832

That he entered into the service of the United States under the following named officers and served as herein stated

Who States That in May 1776 he entered the service of the United States in the five ~~mon~~ months service as a volunteer under Captain John Jackson of Goshen in the aforesaid County of Orange, Vance and Miller Lieutenants and Deponent was appointed 2d Sergeant. Went to New York, staid in New York until the American army was driven form that place, received a furlough of two months by reason of sickness staid home one month, being better, returned to his Company at Peekskill, where he remained until the five months had expired, when he was discharged and returned home. Gen Washington and Gen Putnam were both at New York. Gen Putnam was also at Peekskill

That in April 1777 he was drafted into the service for one month and was ordered to Ramapo on the route to New York. Deponent acted as sergeant during this term & served out the whole term. The Officers at Ramapo were strangers to Deponent, he does not recollect their names, at the end of the month deponent was discharged and returned home

That in June 1777 and shortly after his return from Ramapo he was drafted to go to fort Montgomery. Deponent thinks every fourth man was drafted, he went to the fort with several others where he remained one month as Sergeant

That in April 1778, or in May 1778, Deponent volunteered for six months under Captain Abraham Westfall, Lieutenants English and Rose was stationed at Peenpack on the frontiers, west of the Shawngunk Mountains at Dewits Fort and at Chambers fort. Deponent was sent out frequently sent out to reconoitre the woods to ascertain if the Indians were out. Deponent remained in the service at that time six months, whilst we were at Dewits fort there were several houses burnt by the Indians about six miles off

That Deponents next service was in the year 1781 under the same officers at the same place, that he was in the service at this time six months, was drafted and acted as Sergeant, went out about the first of May

Deponent further saith that he was in the battle of Minisink on the 22d of

July 1779, but was out at that time only three days, a great many officers and men were killed, and also many Indians, our troops were commanded by Colonel Hathorn. John Davis was Deponents Captain, Samuel Cole Lieutenant, Colonel Benjamin Tustin was there and Major John Woodhull - and many others

1st That he was born in Morris County New Jersey in 1757

2d Has a record of his age at home in the Town of Minisink

3d That he lived in Minisink when called into service, that he has ever since resided and still resides there

4th That he was called into the service as above stated

5th That does not now recollect any other officers than those already stated

6th Never received a written discharge and has no documentary evidence

7 That is known to Caleb Howell Ebenezer Mead and Peter Holbert who can testify as to his truth and veracity and their belief of his services as a soldier of the revolution

He hereby relinquishes every claim whatever to a pension or annuity except the Present and declares that his name is not on the Pension roll of the Agency of any State

Subscribed & Sworn the 4th day of ~~September~~ December 1832
Asa Dunning Clerk

Daniel Miers

The above declarant being now sworn doth further depose and say that except Thomas Waters whose affidavit is annexed he does know of any person who can prove his revolutionary services, all his old companions in arms, being either dead or having removed to parts unknown to this Declarant, and he further says that he cannot procure the Certificate of a Clergyman as to his services as a soldier of the Revolution derived from reputation. That the clergyman who now officiates in the neighbourhood of Declarant, is a young man & has officiated as such clergyman only three or four months & that he can not speak of the reputation of this Declarant as such soldier as aforesaid

Subscribed & sworn in open Court the
4th day of December 1832 before me
Asa Dunning Clerk for Com Pleas

Daniel Miers

BRIEF objections to the admission of Pension Claims, of which those answered in the negative, in red ink, will apply to the claim of Daniel Miers which require further proof or explanation.

I. Does the declaration show where the applicant resides? Yes

II. Did the declarant make the declaration in the county where he resides? Yes

III. If the applicant did not make the declaration in the county where he resides, is there any reason assigned for his not doing so?

IV. Has the declarant mentioned the period or periods of the war when he served? Yes

V. Has he mentioned with precision the length of his service, and the different grades in which he served, in language so definite as to enable the department to determine to what amount of pension he is entitled? Yes

VI. Has he given the names of the officers under whom he served, in conformity with the regulations? No

VII. Has he made a relinquishment of every claim to any other pension than the one under the Act of June 7, 1832? Yes

VIII. Has the Court given their opinion? Yes

IX. Has the Clerk given his certificate? Yes

X. Is the Clerk's seal affixed? And if so, has it a device or inscription by which it can be distinguished from any other seal? Yes

XI. Has the applicant obtained the evidence of a clergyman and another respectable citizen as to their belief respecting his age and the general belief in his neighborhood relative to his revolutionary service? No

XII. If the clergyman's affidavit has not been produced, is there any reason assigned for not obtaining it? Yes

XIII. If the applicant has no documentary evidence, and has not obtained the testimony of at least one living witness, has he stated in his declaration that such proof cannot be had? Has

XIV. If the applicant has produced a witness or witnesses as to his service, has the magistrate who administered the oath certified that he, she, or they are persons of credibility? Has the official character and signature of the magistrate been certified by the proper officer under his seal of office?

XV. Does it appear from any of the papers that the witness was in such a situation, or of such an age, as to have a personal knowledge of the applicant's service?

XVI. Are the papers authenticated as the regulations direct? Are

XVII. Are the seven interrogatories prescribed by the War Department answered satisfactorily? And, if not, which of them is not so answered? 5th Not

EXPLANATORY NOTES.

QUERE V. In a case where the applicant cannot, by reason of the loss of memory, state precisely how long he served, he should amend his declaration by making an affidavit in the following words: "Personally appeared before me, the undersigned, a Justice of the Peace, &c. A. B. who, being duly sworn, deposeth and saith, that, by reason of old age, and the consequent loss of memory, he cannot swear positively as to the precise length of his service, but according to the best of his recollection he served not less than the periods mentioned below, and in the following grades: For year months and days, I served as a . For months and days, I served as a ; and for such service I claim a pension."

It is important in all cases to determine with precision the period for which the applicant served, and the particular rank he held, as the law directs the pension to be paid according to the grade of the pensioner and the length of his service. The use of the phrase about three or four months is too indefinite, and all such qualifying expressions are objectionable.

QUERE XIV. Proof of Service.

In a case where the name of the applicant is not found on the records of the Department, he must prove his service by two credible witnesses, who are required to set forth in their affidavits the time of the claimants entering the service, and the time and manner of his leaving the same, as well as the regiment, company, and line to which he belonged. The magistrate who may administer the oaths must certify to the credibility of the witnesses, and the official character and signature of the magistrate must be certified by the proper officer, under his seal of office.

QUERE XVI. Mode of authenticating papers.

In every instance where the certificate of the certifying officer who authenticated the papers is not written on the same sheet of paper which contains the affidavit, or other paper authenticated, the certificate must be attached thereto by a piece of tape or narrow ribbon, the end of which must pass under the seal of office of the certifying officer, so as to prevent any paper from being improperly attached to the certificate.

In the matter of the application of Daniel Miers for a Pension under the act of Congress passed June 7 1832

State of New York

Orange County ss. Isaac B. Overton and David Christie both of the town of Minisink county and State aforesaid being duly sworn doth depose and say, the said Isaac that he has resided in the neighbourhood of the said Daniel upwards of thirty years and the said David that he has resided in the said nieghbourhood upwards of forty years. That they have been well acquainted with the said Daniel for the said periods. That the said Daniel has always sustained a good character for truth & veracity. That it is generally understood and believed in the said neighbourhood that the said Daniel was a soldier in the service of the United States in the revolutionary war in which they concur. It is generally understood & believed in the said neighbourhood that the said Daniel was a brave man & had rendered very important services as a soldier in the revolutionary war. That they would suppose him to be seventy four years of age & upwards That they never have heard the revolutionary services of the said Daniel doubted in the said neighbourhood or in any other place

Subscribed & Sworn in open Court the 4th day of September 1832
Asa Dunning Clerk

Isaac B Overton
David Christie

In the matter of the application of Daniel Miers for a Pension under the act of Congress passed June 7 1832

State of New York

Orange County ss. Orange County ss. Thomas Waters of the town of Goshen county & State aforesaid being duly sworn doth depose and say that he was acquainted with the said Daniel Miers in the revolutionary war and knows that he was in the battle at Minisink with the Indians on the 22d July 1779, deponent has no recollection of being out with said Miers on any other occasion, but said Miers was always reputed to have done much service in the revolutionary war and bore the character of a brave & good soldier, deponent also says that said Miers is reputed to be a man of integrity & truth

Subscribed & Sworn in open Court the 4th day of September 1832 before me
Asa Dunning Clerk

Thomas Waters

And the said Court do hereby declare their opinion after the investigation of the matter and after putting the Interrogatories prescribed by the War Department, that the above named applicant was a revolutionary soldier and served as he states.

And the court further certify that it appears to them that Isaac B Overton, David Christie and Thomas Waters who have signed and sworn to the preceding affidavits are residents of the places mentioned in their respective affidavits, that the are persons of credibility and that their statements are entitled to credit

I Asa Dunning Clerk of the Court of Common Pleas in and for the County of Orange do certify that the foregoing contains the original proceedings in the matter of the application of Daniel Miers for a pension. In testimony whereof I have hereunto set my name and Seal of Office the thirteenth day of December 1832

Asa Dunning Clerk

In the Matter of The Application Of Daniel Miers for a Pension under the Act of Congress Passed June 7th 1832

State of New York

County of Orange ss. Personally appeared before me the undersigned a Justice of the Peace in and for the County of Orange aforesaid, who being duly sworn, deposeth, and saith, that by reason of old age and the consequent loss of memory, he cannot swear positively as to all the particulars of his service, but according to the best of his recollection he served the periods mentioned below in all cases as a Sergeant ~~That in the year 1776~~ and under the Officers mentioned below. That in the year 1776 he served for the Period of five months as a volunteer, under Lieutenants Vance and Miller, and Captain Jackson, while in New York under Gen Washington, while at Peekskill under Gen Putnam (as he thinks). That in the year 1777 he and in the spring, he served a period of one month, (having been Drafted), at Ramapo, under Captain Suffern, the Officers at that place were strangers and he cannot now recollect the names of any others. That in the year 1777 and in the summer he served a Period of one month at Fort Montgomery on the north river, cannot recollect the name of his captain. Colonel McGlanery commanded the regiment, and Gen James Clinton had command of the Fort. That in the year 1778 he served at Peenpack and other places on the Indian Frontiers the Period of six months under Lieutenants English and Rose and Captain Westfall. Colonel Hetfield commanded the regiment under Gen Hathorn. That in the year 1781 he served a Period of six months on the same Frontiers at Peenpack and under the same officers. And for such service he claims a Pension

Sworn Before Me this 6th day of April 1833

H W Elliott J Peace

Daniel Miers

State of New York

Orange County ss. I Asa Dunning Clerk of the said County Do Certify that Horace W Elliott Esq before whom the foregoing affidavit was sworn was at the time a Justice of the Peace in and for said County and that I am acquainted with his hand writing and that the signature subscribed to the said affidavit is his genuine signature Witness my hand and Seal of office this 6th day of April 1833

Asa Dunning Clerk

New York

Daniel Miers of Orange in the State of N. York who was a Serg in the Comp commanded by Captain Jackson of the Reg. commanded by Col in the N. York line for 19 months Inscribed on the Roll of N. York at the rate of 95 Dollars per annum, to commence on the 4th day of March, 1831. Certificate of Pension issued the 27 day of July 1833 and sent to Hon. S. J. Wilkin Goshen

Arrears to the 4th of Ma '33	$190.00
Semi-anl. allowance ending 4 Sepr '33	42.50
	232.50

Recorded by Geo C. Stiles Clerk,
{Revolutionary Claim, Act June 7, 1832.}
Book E Vol. 3 Page 68

Treasury Department,
Second Comptroller's Office,
July 23, 1838.

Sir:

Under the act of the 6th of April, 1838, entitled "An act directing the transfer of money remaining unclaimed by certain Pensioners, and authorizing the payment of the same at the Treasury of the United States," The children of Daniel Miers deceasd a Pensioner on the Roll of the New York City Agency, at the rate of ninety five Dollars per annum, under the law of the 7th June 1832 has been paid at this Department, from the 4th of September 1834 to the 4th of March 1835.

Respectfully, yours,
Albion K. Parris
Comptroller

To the Commissioner of Pensions,
Present.

W. J. Coulter
Genealogy for Enjoyment
Phone xxx 4929

86 Watkins Avenue
Middletown, N.Y.
July 1st, 1935.

Pension Bureau,
Revolutionary War Section
Washington, D.C.

Gentlemen:

Some time at your convenience, would you be good enough to give me the information that you have on the pension record of Thomas Northup of Mt. Salem, Sussex Co., N.J., whose wife Lydia Doyle Northrup was receiving a pension on his services in the War of 1812, according to newspaper records here of 1883.

Also, at your convenience would you send me the pension record of Daniel Myers who died about 1835 and is buried on the Smith Farm in Wantage, Sussex Co., N.J. but who lived just over the line in the Town of Minisink, Orange Co., N.Y. Daniel Myers was at the Battle of Minisink in 1779

Thomas Northrup happens to be my great grandfather and Daniel Myers is the great, great, great grandfather of my wife.

Very truly yours,
W. J. Coulter

No hurry about this, any time at your convenience

Mr. W. J. Coulter
86 Watkins Avenue
Middletown, New York

BA-J/EEL
July 16, 1935
Daniel Miers - S.13933
Thomas Northrup - W.C.13422

Dear Sir:

Reference is made to your letter in which you request information in regard to Thomas Northrup, of Mount Salem, Sussex County, New Jersey, whose wife, Lydia Doyle Northrup, received a pension on account of his services in the War of 1812, and Daniel Myers, of Minisink, Orange County, New York, who served in the Revolutionary War, was in the battle of Minisink, and died about 1835.

The records of those soldiers are furnished you herein.

Daniel Miers

S.13933

The data which follow are obtained from the papers on file in Revolutionary War pension claim, S.13933, based upon the military service of Daniel Miers in that war.

Daniel Miers was born in 1757, exact date not shown, in Morris County, New Jersey. The names of his parents are not shown.

While residing in Minisink, Orange County, New York, Daniel Miers volunteered and served with the New York Troops, as follows: from May, 1776, five months in Captain John Jackson's Company, marched to New York City, where he remained until after the evacuation of the city by the American army: in the spring of 1777, one month at Ramapo, in Captain Suffern's Company: in June, 1777, one month at Fort Montgomery, under Colonel McGlanery (possibly meant for McClaughry): from April, 1778, six months in Captain Abraham Westfall's Company, Colonel Hetfield's Regiment; this service was rendered on the frontier against the Indians: in 1779, three days in Captain John Davis' Company, Colonel Hathorn's Regiment and was in the battle of Minisink on July 22, 1779: from May 1, 1781, six months in Captain Abraham Westfall's Company, Colonel Hetfield's Regiment: all of his service was rendered as a sergeant.

Daniel Miers was allowed pension on his application executed December 4, 1832, while residing in Minisink, New York. He signed Miers.

Children are referred to: their names and the name of soldier's wife are not shown.

In order to obtain the date of last payment of pension, the name and address of person paid and possibly the date of death of the Revolutionary War pensioner, Daniel Miers (S.13933), you should address The Comptroller General, General Accounting Office, Records Division, this city, and cite the following data:

Daniel Miers, Certificate No. 20008, issued July 27, 1833, rate $95 per annum, commenced March 4, 1831, Act of June 7, 1832, New York Agency.

Thomas Northrip

W.C.13422

Thomas Northrip was born near Blooming Grove? (not clearly written), New York. The date of his birth and the names of his parents are not shown.

Thomas Northrip enlisted at Wantage, Sussex County, New Jersey, and served from September 8, 1814 to December 5, 1814, as a private in Captain Thomas Teasdale's Company of New Jersey Militia.

In 1850, he was residing in Wantage Township, Sussex County, New Jersey and then gave his age as sixty years.

He was allowed pension on his application executed April 29, 1872, at which time he was a resident of Mount Salem, Sussex County, New Jersey, and gave his age as eighty years.

He died February 15, 1875, at Mount Salem, New Jersey.

The soldier married August 17, 1816 or 1817 (both dates are shown) at Wantage, New Jersey, Lydia Doyle. Neither had been previously married. The date and place of her birth and the names of her parents are not shown.

Soldier's widow, Lydia, was allowed pension on her application of May 23, 1878, at which time she was aged eighty-four years and resided at Sparrow Bush, Orange County, New York, to which place she had moved after the death of her husband.

She died September 14, 1882, at the home of her son-in-law, William C. Van Sickle, in Sparrow Bush, New York.

In 1878, the widow's daughter, Emily Van Sickle, was aged fifty-four years, and her granddaughter, Carrie Van Sickle, was aged nineteen years. Both resided in Sparrow Bush, New York. No other names of children are shown.

Very truly yours
A. D. Hiller
Executive Assistant to the Administrator.

Sherburne, New York
January, 17, 1940

Veterans Administration
Washington, D.C.
Gentlemen;

I believe you have custody of the U.S. pension records. Could you furnish me with the war record of Daniel Miers of Minisink, Orange County, N.Y., who served in the Revolutionary War in the Battle of Minisink, and died March 6, 1835?

Very truly yours,
Maynard Mires

January 31, 1940.
BA-J/___
Daniel Miers
S.13933

Mr. Maynard Mires
Sherburne
New York
Dear Sir:

Reference is made to your letter relative to Daniel Miers, a soldier of the Revolutionary War.

The data which follow were obtained from the papers on file in the Revolutionary War pension claim, S. 13933, based upon the military service ~~in that war~~ of Daniel Miers in that war.

Daniel Miers was born in 1757, exact date not shown, in Morris County, New Jersey. The names of his parents are not shown.

While residing in Minisink, Orange County, New York, Daniel Miers volunteered and served with the New York troops as follows: from May, 1776, five months in Captain John Jackson's Company, marched to New York City, where he remained until after the evacuation of the city by the American army: in the spring of 1777, one month at Ramapo, in Captain Suffern's Company: in June, 1777, one month at Fort Montgomery, under Colonel McGlanery (possibly meant for McClaughry): from April, 1778, six months in Captain Abraham Westfall's Company, Colonel Hetfield's Regiment: this service was rendered on the frontier against the Indians: in 1779, three days in Captain John Davis' Company, Colonel Hathorn's Regiment and was in the battle of Minisink on July 22, 1779: from May 1, 1781, six months in Captain Abraham Westfall's Company, Colonel Hetfield's Regiment: all of his service was rendered as a sergeant.

Daniel Miers was allowed pension on his application executed December 4, 1832, while residing in Minisink, New York. He signed Miers.

Children ~~are~~ were referred to: their names and the name of soldier's wife are not shown.

In order to obtain the date of last payment of pension, the name of person paid and possibly the death date of soldier, you should address the Comptroller General, General Accounting Office, Records Division, Washington, D.C., and cite the following data: "Daniel Miers, Certificate No. 20008, issued July 27, 1833, rate $95.00 per annum, commenced March 4, 1831, Act of June 7, 1832, New York Agency."

Very truly yours
A. D. Hiller
Executive Assistant to the Administrator

Invalid. File No. 13,933
Daniel Miers Sergt Rev War
Act: June 7th 1832
Index: Vol. 1, Page 256
[Arrangement of 1870.]
Revolutionary War Pension and Bounty Land Warrant Application Files
Series M-804; Roll # 1720

Johannes and Eve Miller
New York (W.16,650)

Miller, John

His name appears on a list of applicants for invalid pension returned by the District Court for the District of New York, submitted to the House of Representatives by the Secretary of War on April 25, 1794, and printed in the American State Papers, class 9, page 94.

Rank: Private

Regt: Colonel Hawthorn's

Disability: Received several wounds in an action with the Indians at Minisink.

When disabled: July, 1779

Residence: Warwick

Remarks: There are no militia rolls in this office.

Evidence transmitted by the District Court incomplete, viz: No examination of surgeons. No evidence of his leaving the service.

State of New York

Orange County ss. Thomas Willson of the Town of Newburgh in said county being duly sworn says that he is aged seventy eight years. That he was a soldier in the army of the Revolution that in the latter part of the month of June in the 1776 this deponent enlisted into a company of militia at Newburgh aforesaid of which company James Milliken was captain James Mourrell was first & Alexander Beaty was second Lieutenant said company belonged to Colonel Pauldings regiment that said company was on duty at Kingsbridge and at White Plains & Long Island & along the Hudson River, that said company was enlisted for a term of five months. That this deponent was before & until his death acquainted with Johannes Miller that said Miller enlisted in said company about the same time that this deponent did, and remained therein until the latter part of the month of December in the year aforesaid that during all the term of service this deponent & said Miller were messmates, and this deponent well knows that he faithfully served in said company from the latter part of the month of June until the latter part of the month of December in the year aforesaid having served nearly six months under an enlistment for five months that this deponent & said Miller were both discharged on the same day at Fort Constitution on the Hudson river that they were discharged within a few days of Christmas, that this deponent knew said Johannes Miller until the time of his death and that said Millers widow is now living, that this deponent knows that said Miller served in the aforesaid company more than five months & fifteen days but can not be positive as the number of days more than the aforesaid term of five months

Subscribed & sworn before me this 1st day of December 1836

H Armstrong Justice of the peace

Thos Willson

I the undersigned Justice of the Peace before whom the foregoing affidavit was made & subscribed, Do certify that I am well acquainted with Thomas Willson & believe him to be a man of truth & veracity & his statements entitled to full credit

H Armstrong Justice

⋆

State of New York

Orange County ss. Samuel S Crawford of the Town of Montgomery in said county being duly sworn says that he was aged seventy four years in the month of July last past. That he was a soldier in the army of the revolution. That ~~in~~ about the month of September in the year 1779, he was called out in the service of the United States That he belonged to Captain William Telfords company in which Henry Neely & John Barber were Lieutenants & William Crist ensign, that said company belonged to Colonel Hathorn's regiment & General Wm. Malcomb brigage, that said company was called out for the term of three months, that said company first mustered at Newburgh in Orange County & crossed the river & marched to Fishkill Village in Dutchess county & remained at said village & were quartered in the old stone church in said village & remained about one month that then they drew tents & were encamped within one half a mile from said village where the company remained ~~in said~~ until they were discharged in the early part of the month of November in said year that during all the aforesaid service this deponent well knew Johannes Miller who was a sergeant in said company & done duty in said company as a Sergeant, and that said Miller remained there until said company was discharged that this deponent cannot now recollect from ~~the~~ loss of memory the precise length of service in said Tour, but he know that said Miller was a Sergeant in said company & served until the same was discharged, but that said company did not remain to the recollection of this deponent the whole of the term of three months for which they were called out ~~but~~ and furfurther says not

Sworn before me this 12th day of December 1836

H Armstrong Justice of the Peace

Saml. S. Crawford

I the undersigned Justice of the peace before whom the foregoing affidavit of Samuel S Crawford was made & subscribed Do certify that I am acquainted with him & that he is a man of truth & veracity & his statements entitled to full credit

H Armstrong

State of New York
Orange County ss. John Newkirk of Newburgh in said county being duly sworn deposes & says that that he is aged eighty three years on the seventeenth day of January last past. That in the year 1778 according to the best of the recollection of this deponent, he knew Johannes Miller late of the Town of ~~of~~ Montgomery deceased that said Miller then belonged to a company of Militia of the State of New York of which Henry Van Keuren was captain, said company was during a part of the months of July & August in the year 1778, stationed at a place called Hunk on the west side of Shawangunk mountain in Ulster county for the term of one month, in defending the inhabitants against the Indians, that this deponent was then a captain of a company of Militia of the Town of Mamakating in Ulster county but now Sullivan that on the twenty second day of August in said year this deponent left his home with his company, composed ~~with~~ of men from several classes in Col John Cantine's regiment to relieve Captain Van Keuren's company that when this deponent arrived at Hunk he there found Johannes Miller in the service as a Sergeant under Captain Van Keuren in said company at Hunk and that said company was called out for one month & at the end of said month, they were relieved by this deponent that this deponent well reccollects that said Miller was in said months service as a sergeant, & that he well reccollects the day of the month which this deponent left his home for the purpose of relieving said Van Keuren, but he may be mistaken as to year but he knows that it must have been as late as the year 1778, that he was acquainted with said Miller both before & after the revolutionary war until the time of his death & that he was in the service as before stated.

Sworn before me this 13th day of December 1836
H Armstrong Justice of the Peace

John Newkirk

I the undersigned Justice before whom the foregoing affidavit of John Newkirk was subscribed & made Do certify that I am well acquainted with him & believe him to be a man of truth & veracity & his statements entitled to full credit

H Armstrong

State of New York
Orange County ss. Benjamin Van Keuren of Montgomery in said county being duly ~~duly~~ sworn says, that he was aged seventy one years on the fifteenth

day of May last, that he is the son of Henry Van Keuren who was a captain of a company of Militia in the army of the Revolution. That he was well acquainted with Johannes Miller & Eve his wife that the said Johannes Miller lived with the father of this deponent until about one year before he was married, that this deponent well recollects that in the year 1777, the father of this deponent was called to Fort Montgomery a few days before the fort was captured by the british to take the command of a company drafted from Col McClaughry's regiment for one month & that his said father was at said battle. That a few days after the taking of said fort this deponent went to New Windsor for the purpose of taking clothes to his said father & said Miller that he found his said father & said Miller both in the service at New Windsor, and that this deponent cannot recollect the precise term of service which his said father & said Miller done at that time but believes that he served the term of one month at New Windsor in guarding the Hudson River & preventing the british troops from Landing at the time they passed up to Kingston & burnt that village & then returned to New York, and this deponent well recollects that said Miller served with his said father in alarms during the year 1777 & 1778 that he knows that they were frequently out on duty but does not know the precise length of the tours of service. That this deponent recollects that he was present & saw the said Johannes Miller & Eve his wife married by the Rev Andrew King at the house of Johannes Moule the father of said Eve that they were married in the spring of the year, but this deponent cannot now recollect the year, but knows that they were married long before the close of the revolutionary war, & that he has heard his said father relating circumstances in relation to the service of said Johannes after he was married to his said wife & that the said Johannes used to keep the rolls of his said father while in service but does not know of his own knowledge that he was a sergeant that he know said Miller died in the month of December 1834, & that the said Eve Miller has remained his widow ever since & that the said Johannes Miller was usually called Hons by his neighbours & acquaintances & this deponent further says that it has always been reputed in the neighbourhood where he resided that said Miller ~~was~~ had been a soldier in the army of the revolution & of his own knowledge that he often went with his said father as a soldier on alarms, & further says not

Sworn before me this 2nd day of February 1837
H Armstrong Justice of the Peace

Benjamin Van Keuren

I the undersigned justice of the peace before whom the foregoing affidavit of Benjamin Van Keuren was made & subscribed Do certify that I believe him to be a man of truth & veracity & entitled to full credit

H Armstrong

State of New York

Orange County ss. John Miller of Montgomery in said county being duly sworn says that he was aged fifty six years on the eleventh day of December last past, that he is the son of Johannes & Eve Miller, that he has often heard his said father speak of his being in the five months service in the year 1776 and also that he had served under Captain Van Keuren at Hunk and on different tours along the western frontiers of Ulster County. That the family record here with transmitted is in the hand writing of this deponent, that he copied the same from a record in the hand writing of his father many years since that said old record was on the leaf of an old family bible belonging to his said father that the leaf became separated from said bible & that he copied the record transmitted herewith at the direction of his said father & that the same was in the hand writing of his said father & was then an old paper, & this deponent verily believes the entries were made of the marriage & birth of his said father & mother were made soon after their said marriage that the entry of the births & marriages on the paper herewith transmitted are all in the hand writing of this deponent except the one under the date of 1817, which is in the handwriting of his said father. That he knows that his mother has remained a widow ever since the death of his said father & that his said father was usually called Hons Miller instead of Johannes, & further says not

Subscribed & sworn before me this 2nd day of February 1837

H Armstrong Justice of the Peace

John Miller

I the undersigned justice of the peace before whom the foregoing affidavit of John Miller was made & subscribed do certify that I am acquainted with John Miller & believe him to be a man of truth & veracity & his statement entitled to full credit

H Armstrong

State of New York

Orange County ss. I, Lebbeus L. Vail Clerk of the Court of Common Pleas for said County do certify, that Horace Armstrong before whom the foregoing affidavit was subscribed and sworn was at the time of taking the same a Justice of the Peace in and for said County, and further that I am well acquainted with the hand writing of said Justice of the Peace, and verily believe his signature subscribed to the said affidavit and the foregoing Certificate to be genuine. In testimony whereof I have hereunto set my hand and affixed the Seal of the said County of Orange this 15th day of February 1837

Lebbeus L. Vail, Clerk

State of New York

Orange County ss. On this fourteenth day of February in the year one thousand eight hundred & thirty seven personally appeared before the undersigned a Judge of the county of Orange Eve Miller a resident of the Town of Montgomery in the county of Orange aforesaid & state of New York aged ~~sixty~~ seventy six years on the eighth day of October last who being first duly sworn according to law doth on her oath make the following declaration in order to obtain the benefit of the provision made by the act of Congress passed July 4, 1836. That she is the widow of Johannes Miller late of said Town deceased who was a private & sergeant in the service of the United States

That as this deponent has been informed & believes her said husband in the latter part of the month of June in the year 1776, enlisted into a company of Militia of the State of New York of which company James Milliken was captain James Morrell was first & Alexander Beaty was second Lieutenant said company belonged to Colonel Pauldings regiment said company was mustered at Newburgh in said county in the month of June aforesaid for the term of five months that said company was on duty at White plains Long Island & along the Hudson river and were discharged at New Windsor in said county about the first of January in the year 1777, having served near six months under an enlistment for five months.

That as she is also informed & believes, her said husband was again called into the service of the United States in the month of October in the year 1777, at the time of the taking of Fort Montgomery by the British on an alarm there in a company under the command of Captain Henry Van Keuren Robert Morrell & John Barber were Lieutenants said company was stationed at New Windsor in said county to guard against the British army which was then passing up the Hudson river & that she believes her said Husband remained at New Windsor until after the British army returned down the river to New York which service continued about one month

That as she is also informed & believes that her said husband was called into the service in the month of July in the year 1778 for the Term of one month, in a company of militia of which Henry Van Keuren was captain said company was composed of several classes of men from Col Cantine's regiment that said service of one month was done at Hunk on the west side of Shawangunk mountain in the then county of Ulster that her said husband served the term of one month as Orderly Sergeant and was discharged in the latter part of the month of August in the year last aforesaid

That as she is also informed & believes that her said husband on the first of the month of September in the year 1779, her said husband was again called into the service in a company of militia of said State under the command of Captain William Telford of which company Henry Neely & John Barber were Lieutenant & William Crist was Ensign said company belonged to Colonel Hawthorn's regiment & General Malcolm's brigade, for the term of three months said company was mustered at Newburgh & were marched to Fishkill in Dutchess county & remained at Fishkill until the first of November in said year when said company was discharged at said place having served but two months & during said service her said husband was a sergeant & done duty as such during the term last aforesaid

And that as she has also been informed & believes he said husband served several tours of service under Captain William Cross who was a captain of a company of militia under the direction of the committee of safety & belonged to Colonel Johannes Jansons regiment during the year 1780 & 1781, which said tours were from five to fifteen days at each time, but which said service will more fully appear from the rolls of Captain Cross ~~here with transmitted~~ & now on file in the pension office That her said husband was born on the twenty first day of January in the year 1760, as appears from the family record here with transmitted, that she was married to her said husband by the Rev Andrew King on the ~~tenth~~ sixteenth day of March in the year 1779, at Montgomery aforesaid as will appear from the record aforesaid. That her oldest child by her said husband whose name is John Miller was born on the eleventh day of December in the year 1779, that her said husband died on the seventeenth day of December in the year 1834, at Montgomery aforesaid and that her said husband was usually called amongst his neighbours & acquaintances "Hons Miller" instead of Johannes Miller & that she has remained a widow ever since as will more fully appear by the proof hereto annexed but that she has no other documentary evidence of her said husbands service.

Sworn to & subscribed on the day
& year above written before me
H W Elliott Judge of Orange Com Pleas

Eve Miller

I the undersigned judge before whom the foregoing declaration of Eve Miller was made & subscribed do certify that I believe her to be a woman of truth & veracity & her statements entitled to full credit & also that I am satisfied from personal inspection that she is unable from old age & feeble health to attend the sittings of the court of common Pleas of said county.

H W Elliott Judge of Orange Com Pleas

Johannes Miller born January 21st 1760 & was joined in Marriage to Eve Mould March 16th 1779, who was born October 8th 1760.

Johannes Millers family

John Miller son of Johannes was born 11th of December 1779

Easter Miller was born the 22d of Feb 1785 & died 17th of Sept. following

Hiram Miller was born the 11th Day of December 1788 & died in November following

---

Maria Miller was born 23 day of November 1793 died 21 of March 1817. Leaving two children viz ____________ & Johannes Miller Hunter

Easter a twin sister to Maria died the 2d day of December 1793

Johannes Miller departed this life December 17th 1834

☆

State of New York

Orange County ss. I, Lebbeus L. Vail Clerk of the Court of Common Pleas to said County Do certify that Horace W. Elliott before whom the foregoing affidavit was subscribed and sworn was at the time of taking the same First Judge of the Court of Common pleas for said County and further that I am well acquainted with the hand writing of said Judge and verily believe his signature subscribed the said affidavit and foregoing Certificate to be genuine. In testimony whereof I have hereunto set my hand and affixed the seal of the said County the fifteenth day of February 1837

Lebbeus L. Vail, Clerk

☆

Newburgh

March 17th 1837

Sir

Hon J W Brown has just handed me a letter from you under date of 10th of March instant, in relation to the claim of Eve Miller widow of Johannes Miller, in which I am informed that her case has been examined & the papers placed on file. That the decision is suspended on the ground that there is no evidence of her husband having been in the service after Dec 1779 the date of the marriage. When I transmitted this case to the department through Col J W Brown I anticipated a speedy decision on the ground that when I was in Washington last winter I placed in the hands of Dr ________ in your department _____ roll of Captain William Cross by which it appeared that Johannes Miller had served ten days under Captain Cross in short services ~~of~~ If I now recollect correcttly one tour of three days & one of seven days the last of which was as late as October 1780 or 1781, by those rolls it appears he was called Hons Miller, and to that fact I called the attention of Doct ______ who assured me that it was well known & admitted in your department that that Hons was but a contraction of the name Johannes, & all the affidavits show conclusively that he was generally called by the name of Hons. The rolls of Captain Cross refered to were brought by me to the department to support the claim of ________ widow of John Gillespie who had done service under Captain Cross. I trust that those rolls have not been lost or mislaid by any person in your department as I adverted to the fact that they would be required to support the claim of Mrs Miller when I left them with Doct ______, & he assured me that they would be placed on file & could at any time be refered to when those rolls are discovered I trust it will satisfactorily appear that Johannes Miller was in the service & done duty as a soldier ~~until~~ as late as October 1780, please give the claim a review and send any communictions to me on the subject

Yours Truly
H Armstrong
Newburgh O C New York

J L Edwards Esqr.
Washington

War Department.
Revolutionary Claim.

I certify that, in conformity with the 3d section of the law of the United States, of the 4th July, 1836, Eve Miller of the State of New York widow of Johannes Miller who was a Private in the War of the Revolution is entitled to receive Thirty five dollars and thirty cents per annum, during the time she may remain unmarried, commencing on the 4th of March, 1831, and payable semi-annually, on the 4th of March and 4th of September. Given at the War Office of the United States, this twenty second day of March one thousand eight hundred and thirty seven.

Examined and Countersigned.

Mechanics Bank
New York 10th April 1837

Sir

The name of Eve Miller, widow of Johannes Miller, a Private at $35 pr annum under the 3d Section of the act of July 4th 1836, was reported to this Agency, by your Letter of the 22d Ultimo.

The certificate of Mrs Miller is issued for $35.30 per annum.

I am very Respectfully

your Obt Servt
Jno. Fleming Presdt & pens agt

James L Edwards Esq
Commissioner of Pensions

Mechanics Bank
New York 14th April 1837
Sir
As requested by your Letter of the 12th Inst. I enclose for correction the Notification in the case of Eve Miller, widow of Johannes Miller deceased.

I am very Respectfully
your Obt Servt
Jno. Fleming Presdt & pens agt

James L Edwards Esq
Commissioner of Pensions
Washington

Albany

Eve Miller widow of Johannes Miller who died on the 17th December 1834 of Orange Co. in the State of N. York who was a Private in the company commanded by Captain Milliken of the Regt. commanded by Col. Paulding in the N.Y. mil. line for 10 months & 17 days Inscribed on the Roll of Albany at the rate of 35 Dollars 30 Cents per annum, to commence on the 4th day of March, 1831. Certificate of Pension issued the 22d day of March 1837 and sent to H. Armstrong Newburgh N.Y.

Arrears to the 4th of March '37	211.80
Semi-anl. allowance ending 4 Sep '37	17.65
	$229.45

Recorded by D. Brown Clerk,
{Revolutionary Claim, Act July 4, 1836, Section the 3.}
Book Vol. 1 Page 134

New York

Eve Miller widow of Johannes Miller who died on the 17 Decr. 1834 of Orange Co. in the State of N. York who was a Private in the company commanded by Captain Milikin of the Regt. commanded by Col. Paulding in the N. York mil. line for 14 months & 17 days N.Y. Inscribed on the Roll of New York at the rate of 48 Dollars 55 Cents per annum, to commence on the 4th day of March, 1831. Certificate of Pension issued the 17thday of Augt. 1837 and sent to H. Armstrong Newburgh, N.Y.

Arrears to the 4th of March '37	291.30
Semi-anl. allowance ending 4 Sep. '37	24.27
	$315.57

Recorded by D. Brown Clerk,
{Revolutionary Claim, Act July 4, 1836, Section the 3}
Book A Vol. 1 Page 134

Payments to be made at New York by John Fleming Esqr. Agent for paying Pensioners, in the Agency of N. York Recorded in the Pension Office, in Book Vol. 1 Page 134 by D. Brown Clerk.

Eve Miller widow of Johannes Miller Pri Rev War
Widow, &c. File No. 16,650
Act: July 4th 36
Index: - Vol. M, Page 51
[Arrangement of 1870.]
Revolutionary War Pension and Bounty Land Warrant Application Files
Series M-804; Roll # 1726

JOHN MILLER
NEW YORK (S.28,128)

Jno. Miller's appln. for a new certificate

J. C. Calhoun Esqr.

Florida May 5th 1820

Sir

Having lost or mislaid my original certificate as an invalid pensioner of the United States, my pension is witheld agreeable to a new regulation in the loan Office of the State of New York until I can obtain a coppy from the war department. Mr Calhoun will very much oblige me by causing a coppy to be made from the original record and sent on as soon as convenient to me at Florida Orange County and state of New York. I was placed on the pension list agreeable to an Act of Congress of the 23rd March 1792.

Yours &c
John Miller

State of New York
Orange County ss. Be it known, that before me James W. Wilkin Clerk ~~in the~~ of the County aforesaid personally appeared John Miller in the within letter mentioned and made oath that he is the identical John Miller who was placed on the pension list agreeable to an Act of Congress passed the 23rd March 1792. That he has had an original certificate in his possession from the war department. That in consiquence of the same he has been considered intitled to and has received a pension for a number of years past for wounds and disabilities incurred in the service of the United States during the revolutionary war. That that certifficate is now lost or mislaid and that in consiquence of his being able to produce the original certificate or a coppy thereof his pension has been suspended or witheld untill he can obtain such coppy. In Testimony whereof I have hereto signed my name and affixed the Seal of said county this Eighth day of May in the year 1820.
Jas. W. Wilkin Clerk

New York
John Miller, private, In the Army U. States N.Y. troops Inscribed on the roll of New York at $2.50 increased to $4 per mo. by Law of the 24th April, 1816. Certificate issued 17th May 1820 & sent to the Pensioner at Florida, Orange Co. New York

---

Invalid. File No. 28,128
John Miller Pri. Rev War Act:
Index: - Vol. 2, Page 277
[Arrangement of 1870.]
Revolutionary War Pension and Bounty Land Warrant Application Files
Series M-804; Roll # 1727

John and Anna Poppino
New York (R.3,680)

State of New York

Orange County ss. On the 5th day of March 1839 Personally appeared Daniel Poppino who being duly sworn doth depose & say that he is eighty years old That his father John Poppino had a majors commission in Col Hathorn's Regiment during the Revolutionary War That he had three sons John, William & witness that for three seasons of the war he verily believes one of the three sons was in the service nearly the whole time & that in his opinion his brother John now deceased performed his share of active service in the war That from the length of time which has elapsed and the want of documentary evidince he can only speak with certainty of the following service In 1776 John Poppino was a private in Capt John Sayre's company in Col John Hathorn's Regiment of Militia That he was at Fort Montgomery upon an alarm (which however proved a false alarm) for one week In the winter of the same year in the same company he was at Closter in New Jersey near Tappan Bay for two months In the summer of 1777 he was at Paramus in the same company one week. The Spring of the year after Fort Montgomery was taken He was at West Point assisting in its erection & doing guard duty in the same company commanded by Capt Richard Bailey in Col Henry Wisners Regiment for two weeks The same season that Burgoyne was taken in 1777 He was at Ramapo & was marched from there to General Washingtons head quarters & was in that vicinity when a spy on his way from New York to Burgoyne was taken & hung & went from there as a guard having charge of the arms & ammunition taken with Burgoyne to Easton in Pennsylvania this service occupied three weeks. The day before the Battle of Minisink he was called out & with Isaac Jennings were placed as a guard to take care of the horses & effects of those who went into the Battle This service was under Captain Baily & took four days

his
Daniel X Poppino
mark

Sworn & subscribed the day & year first above written & I certify that the witness is a reputable witness & that he makes his mark from inability to write in consequence of Palsy

H W Elliott Judge of Orange Comm Pleas

* * *

State of New York

Orange County ss. On this third day of October Eighteen Hundred and

Forty personally appeared before the undersigned a Judge of the Court of Common Please being a court of reccord in and for the said County Anna Foster a resident of the town of Goshen in the County of Orange and State of New York aged Eighty eight years who being duly sworn according to law doth on her oath make the following declaration in order to obtain the benefits of the provisions of the act of Congress passed July 4th 1836 and the act explanatory of said act passed March the 3d 1837. That she was married to John Poppino who after her marriage was a Soldier in the Militia of the County of Orange and State of New York and in the beginning of the war she thinks he belonged to the Regiment commanded by Col William Allison Lieut Col John Hathorn and after the division of the Regiment he belonged under Col John Hathorn and Lieut Col Wisner and Major John Popino and in the Company commanded by Capt John Sayres and afterwards by Capt Richard Bailey That in the Spring of the year of 1776 her said husband John Poppino was called out to help build Fort Montgomery along the Hudson River under Capt Sayres and was out one month under Col Allison or Clinton That her said husband was out several times to Fort Montgomery as a drafted Militia man during this season and once was out one week upon alarm at the Fort under Capt Sayres and Col Allison That in January 1777 a call was made by the committee of Safety for the State of New York for several companies of men to guard the frontiers of the State along the Hudson River and along the New Jersey line and her said husband was drafted to serve in the said service she thinks under Capt Minturn or Sayres in the Regiment under either Col Allison or Hathorn from about the Middle of January of the said year untill the first of April being a term of at least two months and a half which service he performed at Tappan and a place called Closter in the State of New Jersey and was discharged about the first of April 1777 That in the month of August or September the Regim under Col Allison was called out to Fort Montgomery and was out one month and a part of them was discharged at the expiration of one month with orders to be ready whenever called upon that she believes that her husband John Poppino was out under Capt Sayres one month and was discharged Just previous to the Capture of Fort Montgomery that immediately after the Fort was taken the Militia was again called out to Murderers Creek under Col Hathorn and that her husband was again out along the Hudson River under Capt Sayres at least two weeks That after Burgoyne was taken her husband was called out to Rammapoo and was marched to the head quarters of General Washington and was in that vicinity when a spy was taken when on his way from New York to inform Gen Burgoyne of the movements of the British Army about New York from there he went as a guard with the arms and ammunition taken with Burgoyne to Easton in Pensylvania she thinks that he was at this time under Capt Sayres and was out at least three weeks That after the the Militia was classed her husband always discharged his duty in his turn during the war untill the close of the war as a class man and upon alarms under the diffrent officers belonging to Col Hathorns Regiment but she is unable to all the particular times that her said husband was out but he was out in the years 1778 1779 1780 1781 and 1782 at Ramapoo West Point and other places along the Northe River and at Dewitts fort and other places along the frontiers against the Indians nor is she able to state the diffrent Officers of the Regiment that he served under except the following in the Spring of the year 1778 her said husband was out under Capt Richard Baily and Col Wisner at least two weeks at West Point helping to build the Fort at that place that in the month of July 1779 an alarm was made by the appearance of the Indians on the western frontiers of Orange County and Col Hathorns Regiment was called out Capt Bailey company went in the said Regiment and her husband John Poppino went in the said company to Minisink in pursuit of the Indians and was out one week that during the year 1780 her said husband went to Minisink several times a class man and in 1781 he went as a class man across the Mountain to Peenpack and along the valley of the Neversink River and was out again on the same frontier in 1782 but from old age and loss of her memory she is unable to make her case any plainer but relies on the witness that may be produced to substantiate her claim That she believes that her said husband John Poppino done more than Eighteen months service in the militia during the Revolutionary war that she has no documentary evidence to sustain her claim She further declares that she was married to the said John Poppino at her fathers house in the town of Goshen by the Rev Nathen Kerr Pastor of the Presbiterian Church in the said town on the thirty first day of December Seventeen Hundred and Seventy two and that the said John Poppino died on the 15th day of August 1790 that her name before Marriage was Anna Jackson That she was afterwards married by the Rev Mr Sergeant at Ridgebury in the said County to Aaron Clark on the 13th day of March 1802 and that her husband Aaron Clark died on the 6 day of April 1807 that she was married to Albert Foster on the 2d day of October 1820 and that the said Albert Foster died on the 19th day of June 1827 and that she was a widow on the 4th day of July 1836 and that she still remains a widow that the accompyning reccord of her marriage with John Poppino and his death and her marriage to Aaron Clark is the family reccord as kept in her family and has been kept in the family for a great many years that she has no reccord of her last marriage but relies on the affidavit of those who were present at the wedding And further she says not

Sworn & subscribed the day & year above written before me
H W Elliott Judge of Orange Com Pleas

her
Anna X Foster
mark

I certify that Anna Foster who made and subscribed the forgoing declaration is a woman of truth and veracity and her statements is entitled to full credit and belief and further that from personal examination I think she should be excused from attending the court of the said County & that from partial blindness she cannot write her name

H W Elliott Judge of Orange Com Pleas

State of New York

Orange County ss. William Poppino of the town of Goshen in the said County being duly sworn deposeth and saith that he is Sixty one years of age and has known the accompyning reccord for more than thirty years as the family reccord of the widow of John Poppino and Aaron Clark and has no doubt of the correctness of the same and that he was present at the marriage of the said widow of Albert Foster and that she was married to Albert Foster by the Rev Charles Cummins near Florida in the County of Orange on the seccond day of October 1820 and that the said Albert Foster died on the 19 day of June 1827 and that she still remains the widow of Albert Foster

Sworn and subscribed this 3d day of October 1840
and I certify that the above is a creditable witness
H W Elliott Judge of Orange Com Pleas

Wm. Poppino

John Poppino was born May 15th 1751
Anna Poppino, wife of John Poppino was born May 18th 1752
John Poppino and Anna Jackson were married 31st Dec 1772
John Poppino Son died August 15th 1773
James Poppino, Son of John & Anna was born Sept. 26 1773
John Poppino, Jun was born June 6th 1775, Died Jan 8th 1786
Daniel Poppino was born Jan 1st, 1777.
William Poppino was born May 30th 1779.
Julia Poppino was born June 16th 1781, Died March 3d 1784
Anna Poppino was born March 3d 1784 Deceased 8th November 1829
Elizabeth Poppino, was born Jan. 27th 1786. Died May 28th 1811
Jesse Poppino was born Oct. 2d 1787. Deceased November 8th 1819
Maria Poppino was born May 22d 1789.

Aaron Clarck was married to Mrs Anna Poppino March 13th 1802. Aaron Clarck Died April 6th, 1807

State of New York,

Orange County, ss. I, Lebbeus L. Vail, Clerk of the Court of Common Pleas, in and for the County of Orange, Do Hereby Certify, That Horace W. Elliott before whom the foregoing Affidavits was subscribed and sworn, was at the time of taking the same, a Judge of Orange Com Pleas in, and for said county: and further, that I am well acquainted with the hand writing of the said Judge and verily believe that the signature to the said Affidavits & Certificates is genuine. In Witness Whereof, I have hereunto set my Hand and affixed the Seal of the said County, the 4th day of May 1842

Lebbeus L. Vail Clerk
By L. L. Lockwood Dep Clerk of Orange County

In pursuance of a law passed the 27th of April, 1784, entitled "An act for the settlement of the pay of the Levies and Militia, for their services in the late War, and for other purposes therein mentioned," the following certificates have been issued, bearing interest at five percent per annum, viz:

1795 June 15	John Poppino	9	7	

State of New York,
Comptroller's Office, Albany, April 8th 1842.

I hereby certify the above to be true extracts from Book No. 1 now remaining in this office, and that the service for which the certificates above named were issued, were performed under the following officers, viz:

First - under Capt. John Sayres - in ~~Col~~ Major Jno. Poppino's Regt.

Philip Phelps
Dep. Comptr.

Dr Sir I have obtained the above in time for the mail
April 8th 1842

Yours &c
Robert Denniston

Pen Off
May 17 '42
Sir
I have the honor to inform you that the claim of Mrs Anna Foster who claims

for the services of her first husband John Popino has been examined & her papers filed. The evidence of her alleged marriage in 1772 is satisfactory. With respect to the service very imperfectly set forth by claimant the only witness is the brother of her late husband and deposes to but 1 mo 19ds only from spring of 76 to 79. After that date the Comptrs books ought to shew any service rendered in ~~Ha-thorns~~ the Orange County Militia. His certificate would appear to be for service ~~in~~ ending 15 June 1795 under Capt. Sayres of Major Poppino's regt. This requires some explanation. The amount if it be allowed would be but 3 mo 15 ds which with the forgoing makes an agregate of 5 mo 4ds only.

Hon. J. G. Clinton H. R.

Pension Office, April 30, 1856.

Sir,

In reply to your letter of the 28th instant asking to be informed of the condition of the claim of Anna Foster, I have the honor to enclose the copy of a letter addressed to Hon. J. G. Clinton dated May 17, 1842, which furnishes the ground of the rejection of a claim, presumed to be the one referred to by you. Six months' service in the revolutionary war is the minimum duration for which pensions are allowed under any act of Congress.

Very respectfully,
Your Ob't Servant,
J. Minot Commissioner.

Hon. J. A. Hughston,
House of representatives

Mrs. John Jacob Fields
110 High Street
Milford, Pa.

Veterans Administration
Washington D.C.
Dear Sirs,
Nov 22,1933

Please send in any information on the Revolutionary War service of the following men who were all in that war.

Wilhelmus Westbrook - Sussex Co. N.J.
Daniel Ennes - Sussex Co. N.J.
John Poppino - Orange Co. N.Y. militia

Thomas Sayre - Orange Co. N.Y.
Samuel Seely - Orange Co. N.Y.

Very truly yours
Mrs J. J. Fields

January 17, 1934
John Poppino - R.3680
Samuel Seely - W. 19335
BA-J/AWF

Mrs. John Jacob Fields
110 High Street
Milford, Pennsylvania
Dear Madam:

Reference is made to your letter in which you request the Revolutionary War records of Wilhelmus Westbrook of Sussex County, New Jersey, Daniel Ennes of same county, and John Poppino, Thomas Sayre and Samuel Seely all of Orange County, New York, Militia.

There are no claims for pension or bounty land on file, based upon service in the War of the Revolution of Wilhelmus or William Westbrook, Daniel Ennes or Innis, or Thomas Sayre, Sayres. Such claims are the source of Revolutionary War data furnished by this office.

The record of only one Samuel Seely has been found, who served in the New York troops, and of only one John Poppino. Their records are given herein as found in the respective claims for pension based upon their service in the War of the Revolution.

John Poppino - R.3680

John Poppino was born May 15, 1751; he was the son of John Poppino, the name of his mother is not shown, nor is it stated where he was born.

John Poppino married, December 31, 1772, Anna Jackson. She was born May 18, 1752 and they were married at her father's house in Goshen, Orange County, New York; the name of her parents are not given. This John Poppino died August 15, 1790. His widow, Anna Poppino, married, March 13, 1802, Aaron Clark, who died, April 6, 1807. She married, October 2, 1820, Albert Foster, who died June 19, 1827 or 1829.

The widow, Anna Foster, on October 3, 1840, then living in Goshen, New York, applied for pension, on account of the services of her husband, John Poppino, in the Revolutionary War.

It is alleged in the claim that John Poppino served from the spring of 1776 in the Orange County, New York, militia, on several alarms against the Indians, that he assisted in building Fort Montgomery and the fort at West Point, and guarded the frontier; he served under Captains John Sayres, Richard Bailey, Minturn, Major John Poppino (his father), Colonels William Allison, Clinton, John Hathorn, and Wisner.

The claim was not allowed as proof of service as required by the pension law was not furnished.

The children of John Poppino and his wife, Anna -

James	born September 26, 1773
John	born June 6, 1775, died January 8, 1786
Daniel	born January 1, 1777
William	born May 30, 1779
Julia	born June 16, 1781, died March 3, 1784
Anna	born March 3, 1784, died November 8, 1829
Elizabeth	born January 27, 1786, died May 28, 1811
Jesse	born October 2, 1787, died November 8, 1819
Maria	born May 22, 1789

Daniel Poppino, brother of the soldier, John Poppino, was eighty years of age in 1839; he then stated that his father, John Poppino was major in Colonel Hathorn's regiment, during the Revolution, and that he (Daniel) and his brothers John and William all served in the Revolutionary War.

Samuel Seely - W.19335

The date and birth of Samuel Seely are not shown in this claim, nor are the names of his parents given.

Samuel Seely, while residing in Orange County, New York enlisted and served with the New York troops, as follows -

From October or November, 1775, three weeks as private in Captain Thomas Moffat's company, Colonel McLaughlin's regiment; in 1776, three and one half months as orderly sergeant in Captain Ebenezer Woodhull's company of dragoons; from October, 1777, six weeks in Captain Woodhull's company of dragoons; from October, 1778, one month as cornet in same company, under Colonel Cooper; from December, 1778, three weeks as teamster, no officers given; from July, 1779, one month as cornet in Captain Woodhull's company of dragoons.

The soldier, Samuel Seely, married January 4, 1779 in Orange County, New York, Mercy Bartlett. They moved about 1790 to Tioga (later Chemung) County, New York. He died January 10, 1827.

The widow, Mercy Seely, was allowed pension on her application executed, September 11, 1838, then a resident of Elmira, Chemung County, New York, and aged seventy-seven years.

It is not stated that they had children.

Haynes Bartlett, brother of the widow, Mercy Seely, was eighty-one years of age in 1838 and a resident of Romulus, New York, and her brother, Ebenezer Bartlett, was seventy-nine years old in 1838 and living in Elmira, New York. They were both soldiers in the Orange County, New York Militia, War of the Revolution. Mercy Davis and Ebenezer Sayre, both stated that the said Ebenezer Bartlett married their sister, her name not shown.

Very truly yours
A. D. Hiller
Assistant to Administrator

Rejected

Revolutionary War Pension and Bounty Land Warrant Application Files
Series M-804; Roll # 1952

James Reeve or Reeves
New York (S.49,273)

James Reeve came before me   in the State of New York and maid oath that he is the same James Reeve to whom the original certificate in his possession was given of which the following is a copy - the certificate given by the State

By Richard Varick and Richard Platt of New York Esquies This is to certify that in pursuant of the Law of the State of New York Entitled an act making provision for offeciers Soldiers and Seamen who have been disabled in the Service of the United States Passed the 22nd of Aprel 1786, we have Examined James Reeve Junior of Goshen precinct in the county of orange Farmer and find that during the Late war he was a private in Captain John Littles company in the Regiment of orange county militia commanded by Colonel William Allison That on the twentieth day of July in the year one thousand Seven hundred Seventy Nine he was ordered out into actual Serves to repel an Invasion of the Savages at minisink in the Same county that on the twenty Second day of the Same month in the Same year being in the Services of the United States in the company and Regiment aforesaid in an Engagement with the Enemy he received a wound by a musket or rifel Shot from the Enimy through the upper bone of the Left arme above the Elbow by means whereof the Bone was fractured and his arm became Stiff by which he is rendered Incapable in Some measure of obtaining his Livelihood by labor and that on the twenty fourth day of aprel Last he was twenty Six years of age wherefore in pursuance of the Law afore Said we do further certify that upon the principles of the act of Congress of the Seventh day of June 1785 the Said James Reeve Junior is Entitled to receive from the Treasure of this State a yearely pension of thirty Six Dollars from the firs day of January one thousand Seven hundred and Eighty if not before paid given under our hand this Seventh day of aprel 1788

Richard Varick
Richard Platt

and that he Served in the Regiment of orang county militia commanded by Coll Willam Allison at the time he was disabled and that he now resides in the county of orange and has for the Last twenty one years previees to which he resided in the county of Ulster

---

Orrang County James Reeve came before me one of the _____ of the county afore Said in the State of New York and maid oath that he is the Same James Reeve to whome the orriginal certiffieat now in his possession was given of which the following is a coppy to witt that he was a prvete in Captain John Littles company in the Regiment of orrang county Melitia commanded by Colonel William Allison that on the twentieth day of July in the year one Thousand Seven hundred and Seventy nine he was ordered out into actual Servies to repel an Invation of the Savages at Minisink in the Same county that on the Twenty Second day of the Same month in the Same year being in the Service of the United States in the company and Regiment afore Said in an Engagement with the Enemy he received a wound by a musket or Rifele Shot from the Enemy through the upper Bone of the Left arm above the elbow by meanes which whereof the bone was fracturd and his arme became Stiff by which he is rendered uncapable in Some measure of obtaining his Livelihood and that he now resids in the Town of Minisink in the county of orrange and has resided there for the Last fifteen years previous to which he resided in the county of ulster in the State afore Said Sworn this 5th Day March 1790 Before Me

Jer Clark Justus peace

New York March 29 1824

Sir

I take the liberty of forwarding my affidavit of Identity with a view of obtaining a Certificate in the usual form which please to send to the care of Wm Ovington Branch Bank New York

I am very respectfully

Sir your obedt servt
James Reeve

J. L. Edwards Esq

By Richard Varick and Richard Platt of
the City of New York Esquires

This is to certify that in Pursuance of the Law of the State of New York entitled "an act making Provision for Officers Soldiers and Seamen who have been disabled in the Service of the United States" Passed the 22nd of April 1786 We have examined James Reeve Junior of Goshen Precinct in the County of Orange Farmer and find that during the late war he was a Private in Captain John Littles company in the Regiment of Orange County militia commanded by Colonel William Allison; That on the Twentieth Day of July in the year one thousand seven hundred and seventy nine he was Ordered out into actual Service to repel an Invasion of the Savages at Minisink in the same County; That on the twenty second Day of the same month in the same year, being in the Service of the United States in the Company and Regiment aforesaid, in an Engagement with the Enemy he received a wound by a musket or rifle Shot from the Enemy through the upper Bone of the Left Arm above the Elbow, by means whereof the Bone was fractured and his Arm became Stiff by which he is rendered incapable in some measure of obtaining his Livelihood by Labour and that on the twenty fourth Day of April last he was twenty six years of Age: Wherefore in Pursuance of the Law aforesaid we do further certify that upon the Principles of the Act of Congress of the seventh Day of June 1785 the said James Reeve Junior is entitled to receive from the Treasurer of this State a yearly Pension of Thirty Six Dollars from the first Day of January one thousand seven hundred and Eighty if not before Paid Given under our Hands this seventh Day of April 1788.

Richd. Varick
Richd. Platt

James Reeve Came before me one of the Justes of orange County in the State of New York and maid oath that he is the Same James Reeve to whom the original Certificate in his possession was given of which the following is a Copy - the Certificate given by the State

By Richard Varick and Richard Platt of New York Esquirs

This is to Certify that in pursuant of the Law of the State of New York Entitled an act making provision for offeciers Soldiers and Seamen who have been disabled in the Service of the united States Passed the 22nd of Aprel 1786 we have Examined James Junior of Goshen precinct in the County of orange Farmer and find that during the Late war he was a private in Captain John Littels Company in the Regiment of orange County militia Commanded by Colonel Willam Allison That on the twentieth day of July in the year one thousand Seven hundred Seventy Nine he was ordered out into actual Serves to repel an Invasion of the Savages at minisink in the Same County that on the twenty-Second day of the Same month in the Same year being in the Services of the united States in the Company and Regiment aforesaid in an Engagment with the Enemy he received a wound by a musket or riphel Shot from the Enimy through the uper bone of the Left arme above the Elbow by means wherof he is rendered Incapablle in Some measure of obtaining his Livelihood by Labar and that on the twenty fourth day of aprel Last he was twenty Six years of age werefore in pursuance of the Law afore Said we do further Certify that upon the principles of the act of Congress of the Seventh day of June 1785 the Said James Reeve Junior is Entitled to receive from the Treasure of this State a yearely pension of thirty Six Dollars from the first day of January one thousand Seven hundred and Eighty if not befor paid given under our hand this Seventh day of aprel 1788

Richard Varick
Richard Platt

and that he Served in the Regiment of orang County militia Commanded by Col Willam Allison at the time he was disabled and that he now resides in the County of orange and has for the Last forty five years previees to which he resided in the County of ulster

Sworn this twenty fifth day of March 1824 before me
Peter Holbert Justice peace

James Reeve

City of New York ss. James Reeve came before me and made oath that he is the pensioner to whom the above Pension was Granted that he never received a Certificate under the seal of the war department but has heretofore received payment on presentation of a certified coppy of the above Sworn before me this 29th March 1824

Wm Ovington N. P.

Orange County ss. Be it known that on the eleventh day of June 1824 before me Joseph Morrell a Commissioner &c in the County aforesaid, personally appeared James Reeve, and made oath that he is the Identical James Reeve, named in an original Certificate, now in his possession, of which the following is the original. In testimony whereof I have hereunto set my hand and Seal of Office the day ~~day~~ and year above mentioned.

Joseph Morrell

James Reeve

New York July 2d 1824

Sir

I take the liberty of again forwarding the application of James Reeve for a formal Certificate having obtained the old one as required by you on the 19th of April last.

I am very respectfully
Sir your obedt Servt
Wm Ovington

J. L. Edwards Esq

War Department.
Invalid Pension.

I Certify That, in conformity with the Laws of the United States, James Reeve, who was a Private in Captain John Little's Co. of ~~late a~~ Col. Wm. Allison's regiment of the N.Y. Militia, during the revon. was, on the seventh day of April 1788 inscribed on the Pension List, Roll of the New York Agency, at the rate of three dollars per month; and that his name is now on the roll of the same Agency, at the rate of $4.80 per month; commencing on the 24th day of April one thousand eight hundred and sixteen. This certificate is issued in lieu of one dated on the seventh day of April 1788. (cancelled.) Given at the War Office of the United States, this 6th day of July one thousand eight hundred and twenty four

J C Calhoun
Secretary of War.

New York

James Reeves, Private U.S. Army N.Y. Mil - Rev. Inscribed on roll of New York, at $3, and increased to $4.80, by law 24 Apl. 1816. Formal certificate issued 6 July 1824, and sent to Wm. Ovington, New York.

Phila January 14th 1845

Dr Sir

I herewith enclose a declaration for a pension under the act of the 7th June 1832, and also the amount of arears of a pension due to the two Surviving children of James Reeves, you will please issue the warrant for the amount ~~and send it~~ in the names of the parties ~~of~~ themselves, and forward it to me or to the County Town of Orange NY. & answer me as to the claim for a pension under the act above, and oblige your humble servant.

W H Todd

State of New York

Orange County On this fifth day of December 1844 came in open court before the court of ~~Oyer & Terminer~~ common Pleas in for said County, James H Reeves aged fifty four resident of said County who on his solem oath makes the following declaration, in order to obtain the benefits of the provisions of the act of congress of the 7th of June 1832.

That affiant is one of the surviving children of James Reeve decd who was a Revolutionary Soldier in the New York militia that he received a pension of $4.80 ___ per month as an invalid Soldir for wounds received in the service in July 1779, that his certificate is dated 6th July 1824. That said James Reeve departed this life on the 15th of January 1837, intestate & ~~this affiant~~ noone took _____ estate That your affiant is informed & believes that his deceased father never applied for the benefits of the above act of congress of the 7th June 1832, your affiant would further state that he sd James Reeve ~~never~~ died without leaving any widow, your affiant would therefore most respectfully make application for the benefit of the above act. He states that he is informed by his Said father that his father was drafted in the spring of 1776 as a militia Soldier for 2 months and Served as such the term for which he was called out that in 1777 he was called out in the summer of that year for two months and Served at fort montgomery, he thinks under capt Little & col Allison, that in the Summer of 1778 he was again drafted & and marched accordingly and Served out Two months Tour against the Indians across the ~~Indians at~~ Pean Pack or Shawangunk mountain that he was besides these enlist drafts often out at othertimes for Shorter tours on emergines of Sudden alarms, that in July 1779, he was again drafted for two months & while in the Service was wounded in the arm on the Delaware river at which time Capt Little, Major Tustan & others fell in Battle, that said Capt Little was sd Reves Capt & Col Allison the Col For the above wound recd at sd Engagement he was pensioned your affiant only Speaks as to the amount of service done by his father from information & of course cannot certainly know of his own knowledge anything he states that he has detailed the Service from the information of his father as honestly as if it now is in his power to do, to give it as favorable a construction as the case will admit of that his father was raised in this neighbourhood & county allways untill his death. Sworn to & subscribed this day & year aforesaid affiant would state that he has no documentary evidence or witness to sustain the above claim he would refer to George Woods declaration on the files as a pensioner who served with his father

Subscribed & sworn this 5th day of December 1844 before me
J. C. Wallace Clerk of the Court of
Common Pleas of Orange County

James H. Reeve

★

The court having read the foregoing and having heard other testimony of the reputed Service of said James Reeve state that said James Reeve was reputed to have Served his country in the war of the Revolution, and it is the Opinion of the court from the Evidence heard that he was a Soldier & must have Served Six months in the different tours which is ordered to be certified

★

We the undersigned having been for many years acquainted with James Reeve decd, upon our Oaths State that said James Reeve who departed this life in 1837, in Orange County NY was universally reputed to have Served his country in the war of the Revolution, we have Severaly signed our names & our ages affixed to each name

Subscribed & sworn this 5th day of Dec 1844 before me
John C. Wallace Clerk of the Court of
Common Pleas of Orange County

Squire Lee age 63
his
John X Little aged 63
mark

★

I hereby certify on oath that I have been acquainted with James Reeve for five years and said Reeve allways bore the reputation of having been a Soldier. I further state that I am now & have been an office minister of the gospell for thirteen years & said James Reeve was known to me to be a man of fair reputation & his Son is equally entitled to my favorable opinion & regard as a worthy citizen. Sworn to & subscribed by the undersigned this 5th day of December 1844

Subscribed & sworn this 5th day of Decr 1844, before me
J. C. Wallace Clerk of the Court of
Common Pleas of Orange County

Gabriel Conklin

On this 5th day of Decr 1844 It was proven in open court to the satisfaction of the court, that James Reeve a Revolutionary pensioner of the United States for wounds recd in service, departed this life the 15th day of January 1837, without leaving any widow but the following children towit James H Reeve of Orange Cty NY and Abigail widow of Jonathan Carpenter decd of Pa. The two above are all the children of sd James Reeve, the pensioner the original pension certificate of whom is here to the court Shewn all of which is ordered to be certified to the war department In Testimony whereof I John C Wallace Clerk of the court aforesaid, hereby certify that the foregoing declaration of James Reeve was sworn to in open court and the several witnesses also made oath to their respective depositions attached, and the foregoing order of court certify the children of said James Reeve was also made by the court, as witness my hand and the Seal of my said office at Goshen in & for said county this day above

John C. Wallace Clerk of the Court of
Common Pleas of Orange County

State of New York
County of Orange ss. James M Reeve of the Town of Minisink County of Orange aforesaid being duly sworn says that James Reeve deceased departed this life the fifteenth day of January in the year 1837. That he is the same referred to in the annexed depositions

Subscribed & sworn this 5th day of December 1844, before me
John C. Wallace Clerk of the Court
of Common Pleas of Orange County

James M Reeve

Pension Office
March 26 1845

Sir

The papers in the case of James Reeve decd. under the act of 7 June 1832 have been examined and filed.

This claim is asserted for 4 tours of 2 mo each in the militia of New York, without having procured any proof whatever to sustain the same James Reeves was an Invalid Pensioner up to the time of his death which took place in 1837, no attempt was made during that time to obtain the benefit of the act of 1832. In his declaration he mentioned no other service than the one term in 1779, in which he was wounded. Before this claim can be admitted it must be shewn by record or documentary evidence or by the testimony of those who were themselves engaged in the same service with him, that he performed in a regularly organized corps in field or garrison as much as 6 months service.

W. H. Todd Esq
Philadelphia Pa

Minisink Feby. 6th 1846

Sir

I wish to enquire if Wm. H. Todd of Philadelphia or his agent has presented at the war office, papers from the Orange County court of common pleas of the State of New York, bearing date December 1844, claiming the pension arrearges of James Reeve deceased of said county And, if so, what disposition has been made of them? Has he received the claim - when - and how much? Please inform me at your earliest convenience and Oblige yours Etc

Hon Wm L. Marcy
Sect. of War

James H Reeve

P.S. Address. Wells Corner Post Office Orange County N York

Pension Office
Sir,
May 13 1846

In answer to your enquiry respecting the case of James Reeve decd. I have to inform you that the letter from this Office to Mr. W. H. Todd of the 26th March 1845, a copy of which is herewith enclosed, will shew what proof has been re-

quired
Mr James H Reeve,
Wells Corners, N. York

To the Pension Bureau,
Dear Sirs,

Allow me to introduce myself as an amateur genealogist, who having heard of your courtesy in answering letters, am going to throw myself on your clemency. First let me state that I have no interest in pensions, whatsoever, but I do want to find out about the military service of a certain James Reeve to enable his descendants to join Revolutionary Societies.

James Reeve is given in N.Y. in the Revolution p.p. 171, 172, & the Jr. James p.273. I would be everlastingly obliged if you could send me the name of his wife, & the ages - of the James Reeves in connection with the 3 references given, & their residences if you have the data, anything that would enable me to decide what James Reeve it was - for there were several living at _______ at the time, & whether the references are to 1, 2 or 3 James Reeves, or if 2, which one got the pension, the Ensign of p.172, or the enlisted man of p.171.

Then also, I want to know if you can tell me anything to determine whether Garret Dyckman of pp.204; 211, & Garrett Dyckman were one & the same man, & which one served, or if both did? I have two Garret Dyckmans - one b. 3 Mch. 1756, d. 7 May, 1816, the other baptized at Sleepy Hollow Dutch Church, in 1741, June 6. The former married Joanna Odell; of the latter I know no more save that it is stated he served in the Revolution. That the former lived in Westchester Co. at some time is sure, for one of his sons was born there 1787, & I suppose him to be the Garret D. of Cortlandtown 1790. (Census.)

His son Jacob Garret Dyckman had a land right 83,719, 3 Sept. 1858. I suppose his own. I have sent to Albany for the information, but their records (what the fire left them), wont be available for 2 yrs. at least, they say, so I have ventured to ask you.

Knowing how kind you always are, I thank you in advance for your kindness. I am sure if you help me, you will get it back a thousand fold

Very Gratefully,

(Miss) Lucy D. Akerly
550 Park Ave.
New York City

24 Jan. 1912.

I do not send you a stamp, as this official business.

February 6, 1912.

War.
Miss Lucy D. Akerly,
550 Park Avenue,
New York City.
Madam:

In response to your letter dated the 24th and received the 25th ultimo, you are herewith furnished the history of James Reeve, Sur. File No. 49,273, Rev. War, and advised that the pension records of this Bureau fail to afford any information in regard to Garret Dyckman of the Revolutionary War, or Jacob G. Dyckman of the War of 1812.

Very respectfully,
Commissioner.

Revolutionary War Records Section.Department of the Interior,
Bureau of Pensions.

Washington, D.C., , 19

In reply to your request of , received for a statement of the military history of James Reeve Jr.___________ a soldier of the Revolutionary War, you will find below the desired information as contained in his (or his widow's) application for pension on file in this Bureau.

Dates of Enlistment or Appointment	Length of Service	Rank.	Officers under whom Service was Rendered.		State
			Captain	Colonel	
July 20, 1779		Private	John Little	William Allison	

ordered to Minisink and in engagement July 22, 1779
he was wounded in left arm above the elbow.

Battles engaged in, Minisink
Residence of soldier at enlistment, Orange Co., N.Y.
Date of application for pension, Mch 25, 1824
Residence at date of application, Orange Co., N.Y.
Age at date of application, born Apl. 24, 1762
Remarks: He was pensioned for said disability he died in Orange Co. Jany 15,

1837 leaving no widow & only two ch., James H. at 54 yrs. of Orange Co., N.Y. and Abigail wid. of Jonathan Carpenter of Pa.

Office note: service alleged by son James H. not substantiated & therefore not used. See Colonial N.Y. p.453 James Reeves Jr.

Very respectfully,

Daughters of the American Revolution
Los Serranos Chapter

Mrs. May H. Smith
Registrar
222 East E Street
Ontario, California
Feb.16/26

To the Pension Bureau
Dept. of Interior
Washington, D.C.

Will you kindly send me Pension records of James Reeve Jr. Orange Co. N.Y. also - Did his farther James Reeve Sr apply for pension? He may have _____ Orange Co N.Y. or Suffolk Co. War of Rev.

Yours truly
May H. Smith

also of David Burns of Orange Co. N.Y. War of Rev.

Daughters of the American Revolution
Los Serranos Chapter

Mrs. May H. Smith
Registrar
222 East E Street
Ontario, California

To the Pension Bureau
Washington, D.C.

Mar 20/26

Will you kindly give me the pension record of James Reeve of Southold Suffolk Co. He served in the Revolution

Yours truly
May H. Smith

April 15, 1926

May H. Smith
222 East E Street
Ontario, Calif.
Madam:

I have to advise you that from the papers in the Revolutionary War pension claim S.49273 it appears that James Reeve was born April 24, 1761.

While a resident of Orange County, New York (formerly of Ulster County, New York) he enlisted July 20 1779 as a private in Captain John Little's Company, Colonel William Allison's New York Regiment, was ordered to Minisink and in engagement on July 22, 1779, was wounded, being shot through the bone in his left arm above the elbow.

He was pensioned from January 1, 1780 on account of disability resulting from said wound. He died January 15, 1837 in Orange County, New York, leaving no widow (name of wife not given), and only two children, James H. Reeve, born about 1790, of Orange County, New York and Abagail, widow of Jonathan Carpenter, of Pennsylvania.

The above noted is the only soldier named James Reeve found on the Revolutionary War records of this Bureau. & said records fail to afford any inf in re to a John Reeves of Pa. However if he served from any other state & you will furn. as much as poss of the foll data concerning him - another search will be made in your behalf.

Respectfully,
Winfield Scott
Commissioner.

Nashville Tenn.
Oct 8, 1928

Pension Bureau
Dept of the Interior
Washington D.C.
My Dear Sirs:

Will you please advise me whether a pension was ever granted to my great - grand father James Reeves, who inlisted from N. Carolina, Rutherford Co. later living at various times in Virginia, Ky, Pa.. & Ohio.

I wish to become a member of a patriotic society.

Yours sincerely,
M. E. Dorner

May 4, 1929

Rev and 1812 Wars Section
Marian E. Dorner
Care Ward Belmont
Nashville, Tennessee
Madam:

I advise you from the papers in the Revolutionary War pension claim, S.49273, it appears that James Reeve or Reeves was born, April 24, 1761.

While a resident of Orange County, New York (formerly of Ulster County, New York), he enlisted, July 20, 1779, as a private in Captain John Little's Company, Colonel William Allison's New York Regiment, was ordered to Minisink and in an engagement on July 22, 1779, was wounded, being shot through the bone in his left arm above the elbow.

He was pensioned from January 1, 1780, on account of disability resulting from said wound.

He died, January 15, 1837, in Orange County, New York, leaving no widow (name of wife not given), and only two children, James H. Reeve, born about 1790, of Orange County, New York, and Abigail, widow of Jonathan Carpenter, of Pennsylvania.

The above noted is the only soldier named James Reeve found on the Revolutionary War records of this bureau, and said records fail to afford any information in regard to a John Reeves of Pennsylvania. However, if he served from any other state and you will furnish as much as possible of the following data pertaining to him - approximate dates of birth and death, places of residence at enlistment and after service and given name of widow - another search will be made in your behalf.

Respectfully,
E. W. Morgan
Acting Commissioner.

Honesdale Pa
Febry 12th 1935

U.S. Commissioner of Pensions
Washington, D.C.
Dear Sir,

Kindly, inform me of the war and family record of my Revolutionary ancestor, James Reeve or Reeves, Jr. He served as a private in Captain John Little's company & was wounded in the battle of the Minisink. Thanking you.

Yours truly,
Mrs. J. B. Canfield
Registrar Wayne Chapter D.A.R.
Honesdale Pa.

Mrs. J. B. Canfield
Honesdale
Pennsylvania

April 4, 1935
JAMES REEVE
S.49273
BA-J/MLB

Dear Madam:

Reference is made to your letter in which you request the Revolutionary War record of James Reeve or Reeves, Junior, who served as a private in Captain John Little's company and was wounded in the battle of the Minisink.

The record of James Reeve or Reeves follows as obtained from the papers on file in pension claim, S.49273, based upon disability incurred during his service in the Revolutionary War.

James Reeve or Reeves was born April 24, 1761; the place of his birth and names of parents were not stated, but he was reared in Orange County, New York.

While residing in Orange County, New York, he enlisted in the spring of 1776, and served two months on an alarm, names of officers not given; he also served on two other alarms of two months each - one in the summer of 1777 at Fort Montgomery, and the other in the summer of 1778 on an expedition against the Indians "across the Pean Pack" on Shawangunk Mountain, in both tours he served in Captain John Little's company, Colonel William Allison's New York regiment. He enlisted July 20, 1779, served as private in Captain John Little's company, Colonel William Allison's New York regiment, on an expedition against the Indians at Minisink, was in the battle at that place on July 22, 1779, in which he was wounded by a rifle shot from the enemy through the bone of his left arm above the elbow.

On account of disability resulting from the above-noted injury, he was pensioned from January 1, 1780, while a resident of Goshen Precinct, Orange County, New York. In 1790, he was a resident of the Town of Minisink, Orange County, New York. At the time he applied for pension, he was referred to as James Reeve, Junior. In 1824, the soldier stated that he had resided in Orange County, New York, for forty-five years and that previous to his residence there he lived in

Ulster County, in the same state.

James Reeve or Reeves died January 15, 1837, in Orange County, New York.

Soldier's wife, whose name was not given, did not survive him.

The only names of children of the soldier shown were:

James H. Reeve, who in 1844 was aged fifty-four years and a resident of Minisink, Orange County, New York, with post office at Wells Corner, in the same county.

Abigail, who in 1844 was the widow of one Jonathan Carpenter, of Pennsylvania, her age not stated.

In 1844, one James M. Reeve was residing in Minisink, Orange County, New York, his relationship to soldier's family not shown.

Very truly yours
A. D. Hiller
Executive Assistant to the Administrator

Rej. book - He was an invalid pens. & d. in 1837, & made no effort to obtain recognition under Act June 7, 1832. He claimed for service in 1779, when he was wounded.

James Reeve Payable on the 4th March and 4th September, at New York by the Pres. of U.S. Branch Bank. Registered in Book Vol.5 page 45 by Wm. Gordon Clerk. Exd. J. L. Edwards

Rejected

Revolutionary War Pension and Bounty Land Warrant Application Files
Series M-804; Roll # 2020

John Ried or Reed
New York (S.31,927)

Brief in the case of John Reed of Canandaigua
in the State of N. York

(Act 7th June, 1832.)

1. Was the declaration made before a Court or a Judge? in open Court

2. If before a Judge, does it appear that the applicant is disabled by bodily infirmity?

3. How old is he? Seventy Seven years

4. State his service, as directed in the form annexed.

Period.	Duration of Service			Rank.	Names of General and Field Officers under whom he served.
	Years	Months	Days		
July 1776			15	Private	Colo. Thurston
October 1776		2		Private	Colo. Thurston
Decemr 1776		3		Private	Lieut. Gale, Colo. Thurston
April 1777		1		Private	Capt Cortwright, Colo. Thurston
June 1777		1		Private	Capt Jones
August 1777			15	Private	Capt Little
October 1777		1		Private	Lieut. Gale
April 1778		1			officers not remembered
July 1778			15	Private	officers not remembered
Septemr 1778			15	Private	Capt John Little
July 1779			15	Private	Capt Jones

~~5. In what battles was he engaged?~~

6. Where did he reside when he entered the service? Goshen, N. York

7. Is his statement supported by living witnesses, by documentary proof, by traditionary evidence, by incidental evidence, or by the rolls? Traditionary and living evidence

8. Are the papers defective as to form or authentication? and if so, in what respect? the papers are correct in form and authentication

Certify that the foregoing statement and the answers agree with the evidence in the case above mentioned.

Wm. Gordon Examining Clerk.

State of New York

Ontario County ss. On this twenty seventh day of August 1832 personally appeared in open Court before the Judges of the County Court now sitting, John Reid a resident of the town of Canandaigua County and State aforesaid, aged seventy seven years, who being first duly sworn according to Law, doth on his

Oath make the following declaration, in order to obtain the benefit of the act of Congress passed June 7 1832

That he entered the service of the United States under the following named Officers, and served as herein stated.

1st At Goshen County of Orange and State of New York in the fore part of July 1776 in Captain Thompsons Company in Colonel Thurstons Regiment was marched to New Windsor and went on board a Sloop to Fort Montgomery, General James Clinton was in Command, and was on fatigue for half a month and was then discharged and went home.

2nd On the first October then next was drafted for two months and mustered at Goshen as aforesaid in Captain Sawyers Company in Colonel Thurstons Regiment and was marched to Fort Montgomery and Fort Clinton and done fatigue duty, helped to build the Fort, and mounted guard. General Clinton was in Command. Saw Governor George Clinton there whom he well knew, was discharged first December and went home, he helped to put the chain across the River at West Point

3rd On the ninth day of December at Goshen as aforesaid was again called out and mustered under Lieutenant Gale, and marched to Haverstraw and then to Tappan, remained there until March, and completed three months service

4th In the fore part of April 1777 was mustered in Captain Cortrights Company, same Regiment and was marched to Ramappo and there done duty for one month.

5th In the month of June following was called out to go to Minisink, Captain's name not remembered. Believes it was Captain Jones, served one month. The Indians had been there

6th In August following called out under Captain John Little and marched to Fort Clinton, and there done duty for half a month, and then dismissed and went home to Goshen

7th In October same year under Lieutenant Peter Gale was called out and marched to New Windsor was ordered to Wallkill by Governor Clinton to guard the Artillery and then returned to New Windsor, and having served one month returned home. This was the time Fort Montgomery was taken

8th In April 1778 was called out again and was ordered to West Point. Went in a Sloop from New Windsor and served one month.

9th In July 1778 in Captain Jones Company was ordered to Minisink and served half a month

10th In September same year in Captain John Littles Company in Goshen as aforesaid was ordered to Peen Pack and done guard duty half a month

11th In July 1779 in Captain Jones Company was ordered to Minisink, and served half a month Captain Jones was killed at this time, and also Colonel Thurston Captain Little and Captain James Knapp. This was in an excursion against the Indians. Has no record of his age. Was born in Scotland 29th January 1755 and removed to New York and arrived there 22d July 1774, and then removed to Schenectady and lived there fifteen months and then to Goshen in Orange County until February 1798 when he removed to Canandaigua where he now resides. Jacob Hulse of Canandaigua was in the service with him at Fort Clinton. He knows Phineas P. Bates of Canandaigua

He hereby relinquishes every claim whatever to a pension or annuity except the present and declares that his name is not on the Pension Roll of the agency of any State. Sworn and subscribed the day and year aforesaid in open Court.

John Ried

We Phineas P. Bates and Jacob Hulse residing in the town of Canandaigua in the Country and State aforesaid, hereby certify, that we are well acquainted with John Reid who has subscribed and sworn to the above declaration, that we believe him to be Seventy seven years of age, that he is reputed and believed in the neighborhood where he resides to have been a Soldier of the Revolution and that we concur in that opinion. And the said Jacob Hulse further certifies that he was in the service with the said John Reed at Fort Clinton. Sworn & subscribed the day & year aforesaid in open court

Jacob Hulse
P. P. Bates

And the said court do hereby declare their opinion, after the investigation of the matter, and after putting the interrogatories prescribed by the war department, that the above named applicant was a Revolutionary soldier and served as he states. And the Court further certifies that it appears to them that Phineas P. Bates and Jacob Hulse are both residents of the town of Canandaigua, are credible persons, and that their statement is entitled to credit.

I, Charles Crane, Clerk of the said court, do hereby certify that the foregoing contains the original proceedings of said court in the matter of the application of John Reid for a pension. In testimony whereof, I have hereunto set my hand and seal of office, this first day of September A.D. 1832.

Charles Crane

State of New York

Ontario County ss. In the matter of John Reids application for a Pension. Personally appeared before me, the undersigned a justice of the Peace in and for said county, John Reid, who, being duly sworn according to Law doth on his oath depose and say that the reason why he did not obtain the attendance of a clergyman to support his application for a Pension before the court on the 27th day of August 1832, was, that being deaf, and living remote from any place of worship, he has not attended for several years past, and has but little acquaintance with the clergyman of his town, and therefore did not request his attendance at the court as aforesaid, and has never conversed with him on the subject of his Revolutionary service. He further saith he has no documentary evidence, nor never had any. He is required to answer the interrogatories prescribed by the War Department. He has answered in his original declaration the three first and he now says in answer to the 4th Interrogatory he now says that he was always drafted, and always answered the call, and served as required.

To the 5th Interrogatory he now says, that by reason of Old age and consequent loss of memory, and being at this time sick and in a low state, he cannot give any other answer than what he gave in his original declaration before referred to. The 6th and 7th Interrogatories he has answered - He served in all, as a private soldier, not less than Fourteen months for which he claims a Pension

Sworn to and subscribed this 9th day of January 1833 before me

Henry Chapin J.P.

John Ried

State of New York

Ontario County ss. In the matter of John Reids application for a Pension. Personally appeared before me, the undersigned a Justice of the peace in and for said county Jacob Hulse, who, being duly sworn doth depone and say that he was born in the town of Goshen county of Orange and State of New York, and knew John Reid before the commencement of the Revolutionary War at Goshen as aforesaid, and he further saith that in the year 1776 he, this deponent was embodied in a company under Lieutenant Stewart in Colonel Ellisons Regiment, and served at Fort Clinton at the same time that John Reid was there, and as this deponent believes, in Captain Sawyers company in Colonel Thurstons Regiment. He is confident the said John Reid was there and in the service, but in consequence of old age and loss of memory, he cannot state the period of his service as aforesaid. But he knew the said John Reid, was a faithful soldier and from his acquaintance with him, he believes that he was in the service nearly all the war on short tours

Sworn to and subscribed this 21st day of January 1833 before me
Henry Chapin J.P.

Jacob Hulse

I certify the above named Jacob Hulse is a credible Witness.

Henry Chapin J.P.

State of New York

Ontario County ss. I, Charles Crane, Clerk of said county, do hereby certify that Henry Chapin, whose name is subscribed to the several certificates of the affidavit of John Reid and the affidavit and character of Jacob Hulse, all upon this sheet written, was, at the respective dates of the said certificates, a Justice of the peace in and for said county; and that I am well acquainted with his hand writing, and verily believe the said signatures respectively to be his proper act and deed. In testimony whereof, I have hereunto set my hand and seal of office, this twenty-first day of January A.D. 1833.

Charles Crane

New York

John Reed* of Canandaigua in the State of New York who was a Private in the company commanded by Captain Thompson of the Regt. commanded by Col Thurston in the New York line for 11 ½ months Inscribed on the Roll of New York at the rate of 45 Dollars per annum, to commence on the 4th day of March, 1831. Certificate of Pension issued the 1st day of March 1833 and sent to Hon J Dickson Ho Reprs

Arrears to the 4th of Sept 1832	$ 67.50
Semi-anl. allowance ending 4 Mar '33	22.50
	$ 90.00

Recorded by Jno Cromwell Clerk,
{Revolutionary Claim, Act June 7, 1832.}
Book D Vol. 5 Page 132

*Ried

---

14 _____ Ave.

Yonkers Park
Westchester Co. N.Y.

The Bureau of Pensions
Washington D.C.
Gentlemen,

Will you please send me a copy of the pension papers of John Reid who was a Revolutionary Soldier and who received his pension when a resident of Seneca, Ontario Co N.Y. I should also like to know if Barnet Lewis who was a soldier of the Rev. War from N. Jersey was a pensioner Thanking you in advance

I am yours very truly,
Caroline L. Wrenn
(Mrs. H. B. P. Wrenn)

August third.
(stamped U.S. Pension Office August 5, 1908)

Revolutionary War Records Section.
Department of the Interior,
Bureau of Pensions.
Washington, D.C., , 19

In reply to your request of , received for a statement of the military history of John Ried a soldier of the Revolutionary War, you will find below the desired information as contained in his (~~or his widow's~~) application for pension on file in this Bureau.

DATES OF ENLISTMENT OR APPOINTMENT.	LENGTH OF SERVICE	RANK.	OFFICERS UNDER WHOM SERVICE WAS RENDERED.	OFFICERS UNDER WHOM SERVICE WAS RENDERED.	STATE
			CAPTAIN.	COLONEL.	
July 1776	½ mo	Private	Thompson	Thurston	N.Y.
Oct 1 1776	2 mos	Private	Sawyer	Thurston	N.Y.
Dec 9	3 mos	Private	Lieut. Gale		N.Y.
April 1777	1 mo	Private	Capt. Cortwright		N.Y.
June 1777	1 mo	Private	Jones		N.Y.
Aug 1777	½ mo	Private	John Little		N.Y.
Oct 1777	1 mo	Private	Lieut. Peter Gale		
April 1778	1 mo				
July 1778	½ mo		Jones		

DATES OF ENLISTMENT OR APPOINTMENT.	LENGTH OF SERVICE	RANK.	OFFICERS UNDER WHOM SERVICE WAS RENDERED.	OFFICERS UNDER WHOM SERVICE WAS RENDERED.	State
Sept 1778	½ mo		John Little		
July 1779	½ mo		Jones		

Battles engaged in,

Residence of soldier at enlistment, not stated, entered ser. at Goshen, Orange Co. NY

Date of application for pension, Aug. 27, 1832.

His clm. was allowed

Residence at date of application, Canandaigua, Ontario Co., N.Y.

Age at date of application, b. Jan 29, 1755 Scotland.

Remarks: He died Jan. 27, 1833. No data as to family

Respectfully,
Commissioner.

Invalid. File No. 31,927
John ~~Reed~~ Ried Pri Rev War
Act: June 7th 1832
Index: Vol. 2, Page 337
[Arrangement of 1870.]
Revolutionary War Pension and Bounty Land Warrant Application Files
Series M-804; Roll # 2046

Nathan Stewart
New York (S.11,493)

City, County &

State of New York ss. On the nineteenth day of March in the year of our Lord one Thousand eight hundred & thirty three before me Richard Riker Recorder of the City of New York, appeared Nathan Stewart now residing in said City of New York, aged Seventy four years; who by reason of bodily infirmities is unable to attend in open Court, being duly sworn according to the form of the act in such case made & provided - Doth on oath make the following Declaration to obtain the Benefit of the Act of Congress passed June 7th 1832.

That he was born in the Town of Goshen in the County of Orange in the State of New York, situated twenty miles from the Town of New Burgh (which said Town of New Burgh is located on the Bank of the Hudson River) on the sixth day of November A.D. One Thousand Seven Hundred & fifty eight, where he resided until the year one Thousand Seven Hundred & Seventy five, when the War of the Revolution broke out, at which time he was enrolled as a Private in Captain Sweseys Company of Militia, and belonging to a Regiment known and designated in said county as the "First Regiment" of Militia, under the command of Colonel Thurston, and Brigadier General Elliston. And in the month of September One Thousand Seven Hundred & Seventy Six, he was ordered out by said Captain Swesey as a Private in his said Company of Militia, and employed in building Fort Montgomery (distant eighteen miles from said Goshen) and in defending the same against the British Troops from the City of New York, as they were continually subject to their attacks, where after remaining one month he returned to Goshen aforesaid. And in the month of January one Thousand Seven Hundred & Seventy Seven, in like manner he was ordered out and engaged as a Private in Captain Jones Company of Militia of the said "First Regiment," in constructing Fort Montgomery & Fort Clinton, where he continued one month & then returned to Goshen. And in the month of April One Thousand Seven Hundred & Seventy Seven was again ordered out as Private to join Captain Sawyers Company of the said "First Regiment" to garrison Fort Montgomery where he remained one month & returned to Goshen aforesaid. On the   day of August One Thousand Seven Hundred & Seventy Seven, an Express arrived at Goshen that Col Brandt with his Indians were approaching the Town of Pin Pack, he was then ordered out as a Private together with the company under the command of Captain Swesey aforesaid to repair to the Fort of Pin Pack. upon arriving there, they discovered the Town had been fired & nearly consumed, the Fort was strongly attacked by Tories & Indians, but was Defended with the loss of but one man, and after garrisoning the Fort at said place, & repelling the incursions of the enemy for a period of one month he returned to Goshen. And on the twenty third day of September One Thousand Seven Hundred & Seventy Seven, he was ordered out as a Private in Captain Littles Company of said "First Regiment," to garrison said Fort Montgomery, and remained there about two weeks, until it was captured on the sixth day of October One Thousand Seven Hundred & Seventy Seven, the same being under the immediate command General James Clinton & his brother Gov. George Clinton commanding Fort Clinton. He the said Nathan in company with one William Satterly escaped from the Fort in the night & went over the Sterling Mountain to New Windsor up the North River pursuant to Governor Clintons orders, who previous to the attack directed his men (in case the Fort was taken) to Rendezvous if possible at New Windsor. While the British were removing the chain, and Chevaux De Frise in the River Governor Clinton had collected Troops (Militia) at New Windsor from Ulster & Orange Counties, as the British proceeded in Sloops accompanied by the Roebuck Man of War & Spit Frigate (towards Kingston or Esopus) The Troops under Gov. Clinton marched forward to oppose them, but on arriving in sight of Kingston, it was discovered to be in flames & was soon reduced to ashes after remaining in the vicinity about three weeks, and protecting the inhabitants from the inroads of the enemy he returned to Goshen. And in March One Thousand Seven Hundred & Seventy eight, he was Drafted as a Private under Captain Horton of the said "First Regiment," and sent to West Point to garrison the Fort at that place, & assist in constructing Water Batteries, where he remained one month & returned to Goshen. And in June One Thousand Seven Hundred & Seventy eight he was ordered out as a Private in said Captain Sweseys Company & sent to Ramapoe New Jersey, where he remained one week, & was then ordered with the rest of the Company to march to Haverstraw on the Hudson River, where he was engaged for one month in protecting the neighbouring Towns from the attack of the enemy. And on the twenty second day of July One Thousand Seven Hundred & Seventy eight, he was ordered out as a Private, together with said Captain Sweseys Company, then under the command of Lieutenant Gale (the said Captain Swesey being at the time sick) to march to the Town of Minisink (distant eighteen miles from Goshen aforesaid) in consequence of an attack of Col. Brandt with 300 Indians & 75 Tories, upon their arrival the Town was nearly consumed, they had murdered many of male Inhabitants, and carried off the women & Children into the Wilderness. The said Nathan together with said Sweseys Company pursued them nearly Twenty miles & fell into an ambuscade, when an action commenced which continued about four hours, until the ammunition was expended. Col. Thurston our Commandant received Three wounds, and seeing our fire cease for want of ammunition exclaimed, "My brave fellows, every man for himself, and the Lord for us all," - I never will leave this ground and was afterwards found Dead. He the said Nathan with those who escaped returned to Minisink, and receiving recruits returned to the scene of action to Bury the Dead, but were obliged to abandon it, and after staying in the vicinity about Three weeks in watching the movements of the Indians he returned to Goshen. After the death of Col. Thurston; Col. Hatfield succeded in command, and on the twenty first day of July One Thousand Seven Hundred & Seventy nine; He the said Nathan (with one half of Captain Sweseys Company) by orders of said Col. Hatfield was Drafted as a Private and sent to garrison Stoney Point, after the same was captured by Genl. Wayne, and remained in said Fort one month & returned to Goshen And in the month of April One Thousand Seven Hundred & eighty, He the said Nathan with one Third of the said company under the command of said Captain Swesey was Drafted & Sent to the Town of Peekskill, to defend that as well as the adjacent Towns from the threatened attacks of the enemy, and remaining there two months, were releived by Regular Troops. And the said Nathan would here on Oath State, that in all the before described Services by him performed, he was duly Drafted by orders issuing from his Colonel to the Captain of the Militia Company of which he was a private, and that he was in actual & Special Service as before described, at the following named places, & for the following named periods viz.

Fort Montgomery	3 1/2 months	Pin Pack	1 month
Kingston	3/4 month	West Point	1 month
Ramapoe N.J.	1/4 month	Haverstraw	1 month
Minisink	3/4 month	Stoney Point	1 month
Peekskill	2 months		

Total 11 months & 3/4making in the whole Eleven months and Three weeks in which he was engaged in actual service, the particulars of the above statements are taken from memorandums in his possession, but the said Nathan would further state, that he was engaged at various other times and places for a much longer period as forming one of the Gaurd in taking Prisoners from Goshen Goal to Esopus to put on board the Prison Ships, and in keeping Sentry at various times at Goshen Goal aforesaid; but from his advanced age, impaired health, and consequent loss of memory is unable to particularize the times and durations of such service; and for which undefined service he disclaims all right to any allowance. After rendering the above described services He the said Nathan remained in Goshen aforesaid, until the evacuation of New York by the British Troops, the Twenty fifth day of November One Thousand Seven hundred & eighty three at which time he accompanied General Washingtons Army to New York (but not in a Military capacity) from which place he went to the Town of Blandford in Massachusetts, and remained for Twenty Years when he removed to the Town of Worthington in Ohio and continued about Four Years, when he moved to the Town of Manlius in New York State, where he resided, until he came to this city which was about Six Years since. And he the said Nathan further declares that he possesses no documentary evidence, and he knows of no person whose testimony he can procure as to his service. He Hereby relinquishes every claim whatever to a Pension or annuity except the present, and Declares that his name is not on the Pension Roll of any Agency or State.

Sworn to & Subscribed in my presence the day & Year
first above written. To wit, 19th day of March 1833.
R Riker Recorder of the City of New York

Nathan Stewart

We William A. Clark a Clergyman residing in New York, & Charles Edwards

Counsellor at Law of the same City hereby certify that we are well acquainted with Nathan Stewart who has subscribed & sworn to the above Declaration, we believe him to be Seventy four years of age; and is reputed & beleived in the neighbourhood where he resides, to have been a Soldier of the Revolution, and we concur in that opinion.

Sworn before me this 19th day of March 1833 - by
on the 21st March 1833.
William A. Clark & R. Riker

William A. Clark &
Charles Edwards

And the said Richard Riker Recorder of the City of New York doth hereby declare his opinion, after investigating the matter and putting the Interrogatories prescribed by the War Department, that the above named applicant was a Revolutionary Soldier and served as he states. And the said Recorder further certifies that it appears to him that William A. Clark, who has signed the above certificate is a Clergyman resident in the said City of New York; and the certificate of Charles Edwards Esq which is above inserted, who is a resident of said City of New York, is a credible person and that their Statements are entitled to credit.

New York 21st March

R Riker Recorder of the City of New York
H. Meigs, President of the Board of Aldermen

State of New-York,

City and County of New-York, ss. I, Abraham Asten, Clerk of the City and County of New-York, and Clerk of the Court of Common Pleas for the said City and County, Do hereby certify, That Richard Riker, Esq. whose name is subscribed to the annexed certificate, and Jurat, is Recorder of the City of New York duly commissioned & sworn and that Henry Meigs, Esq. whose name is also subscribed to the said certificate is an Alderman of the said City duly elected and sworn. In testimony whereof, I have hereunto set my Hand, and affixed the Seal of the said Court and County, the 21st day of March 183

Abrm Asten, Clk

New York 21 March 1833

J. L. Edwards Esqr

Sir

For the third time I trouble you with the aplicants claim Nathan Stuart my Father I hope This will be received as a satisfactory document. I am now here on business. My residence is in Baltimore. when you will have the kindness to communicate with me on This subject as early as you can after an examination. & if admited direct the steps to take to obtain The amount dew The aplicant

Yours Truly

F. A. Stuart

★

New York

Nathan Stewart of NY City in the State of New York who was a Private in the Company commanded by Captain Swesey of the Regt. commanded by Col Thurston in the New York line for 11 months and 21 days. Inscribed on the Roll of New York at the rate of 38 Dollars 97 Cents per annum, to commence on the 4th day of March, 1831. Certificate of Pension issued the 29 day of Mar 1833 and sent to F A Stewart Baltimore Md

Arrears to the 4th of March '33	$77.94
Semi-anl. allowance ending 4 Sep '33	19.48
	$97.42

Recorded by Jno Cromwell Clerk,

{Revolutionary Claim, Act June 7, 1832.}

Book E Vol. 4 Page 31

★

Invalid. File No. 11,493

Nathan Stewart Pri Rev War

Act: June 7th 32

Index: Vol. A, Page 360

[Arrangement of 1870.]

Revolutionary War Pension and Bounty Land Warrant Application Files

Series M-804; Roll # 2291

★

Anthony and Elizabeth Swarthout
New York (R.10,344)

War Department,
Pension Office,
May 23 1832

Sir:

The evidence in support of your claim, under the act of June 7, 1832, has been examined, and the papers are herewith returned. The following is a statement of your case in a tabular form. On comparing these papers with the following rules, and the subjoined notes, you will readily perceive that objections exist, which must be removed, before a pension can be allowed. The notes and the regulations will shew what is necessary to be done. Those points to which your attention is more particularly directed, you will find marked in the margin with a brace, (thus: }). You will, when you return your papers to this Department, send this printed letter with them; and you will, by complying with this request, greatly facilitate the investigation of your claim.

A Statement, shewing the Service of Anthony Swarthout, Steuben county N. York

Period when service rendered	Duration of claimant's service			Rank of the Claimant	Names and Ranks of the Company Officers under whom he served	Names and Ranks of the Company Officers under whom he served
	Years	Months	Days			
Enlist 1776		5		Private	Capt. John Wood	Maj. Moses Hatfield, Genl. James Clinton
Pressed 1776		about 2		Teamster		
1777		not less 5		Private	Capt. John Jackson	Col. Nichols

He states that from his last discharge he was in active service exclusive of___________ not less than 2 years

Battles in which the applicant engaged	Country through which he marched	Place of abode when he entered service, and age at the present period	Evidence by which the declaration is supported
Flatbush, Harlem Heights, York Island, White Plains, Beaverbrook	Chester	73 years Orange County N. York	Living witnesses & Traditionary

I am, respectfully,
Your obedient servant,

J. L. Edwards,
Commissioner of Pensions.

After the statement of service a copy of the regulations under the Act of June 7, 1832 is included but not reproduced here. An identical copy is included in the chapter entitled ***A Declaration of Service****. At the bottom of the regulations is a note that says that* "Notary Public not proper officer to authenticate papers to be used in support of claims for pensions under act of June 7, 1832"

State of New York
Ontario County ss. Be it known that before me Richard M. Bayly a Notary Public in and for said County and State, Personally appeared Aaron Swarthout of the Town of Tyrone in Steuben County and made Oath in due form of Law that, he is a Cousin to Anthony Swarthout of the Town of Wayne in said county of Steuben and knew him in Orange County in New York from boyhood, knew that said Anthony Swarthout joined the army as he believes in 1776, and entered for five months, saw him at home after his term of service had expired. Knows that he was in the service of the United States, during the Revolutionary War for several years, more or less, was at the Fort at Minisink and saw said Anthony there, saw him march with the Troops to the Delaware, and saw him also in said service on their return from the Battle of Beaver Brook, in which said Anthony was engaged as was understood at the Fort at that time, and as this deponent has no doubt whatever, that said Battle, as this deponent believes was in the year 1779.
Sworn to and subscribed this 7th day of July 1832 before me
R M Bayly Notary Public

Aron Swarthout

I Richard M Bayly the Notary Public before whom the above affidavit was taken, have to my satisfaction ascertained that Aaron Swarthout is a person of Credibility

R M Bayly

State of New York
Ontario County Clerk's Office I, Flavins J. B. Crane Deputy clerk of said county do hereby certify that Richard M Bayly Esq whos name is subscribed to

the several foregoing certificates, was at the date thereof a Notary Public in and for said county; and that I am well acquainted with the hand writing of the said Richard M Bayly and verily believe the said several signatures to be his proper act and deed. In testimony whereof I have hereunto set my hand and Seal of Office this 24 December A.D. 1833.

F. J. B. Crane

Geneva Ontario County NY
J any 2d 1834

J L Edwards Esquire Sir

I herewith reenclose Anthony Swathouts declaration properly corrected as I suppose

Yours Respectfully
R M Bayly

State of New York

Steuben County ss. On this Sixteenth day of October 1832 personally appeared in Open Court, before the Judges of the Court of Common Pleas now sitting, Anthony Swarthout a resident of the Town of Wayne in the County and State aforesaid, aged Seventy three years, who being first duly sworn according to Law, doth on his Oath make the following declaration, in order to obtain the benefit of the act of Congress, passed June 7, 1832.

That he entered the service of the United States under the following named officers, and served as herein stated: towit, That in July 1776 at Chester in Orange County New York where he then resided, he enlisted in Captain John Woods Company, in the Regiment of which Moses Hatfield was Major, but the name of the Colonel he has forgotten, in Genl. James Clintons Brigade, for the term of nine months. That he served out the Five months and was discharged at said Town of Chester by Captain Wood, and had a written discharge which has been lost; That during said service he was in the Battles of Flat Bush on Long Island, under Lord Sterling, at Harlaem Heights, at York Island and at White Plains; That after said discharge he was pressed with his team and went with a part of the army to to the Pensylvania line, and was then in service about Two months; That in the year 1777 the year that Burgoine surrendered & when the British shipping went up the North River to Esopus and burnt that place, he was ordered out in Captain John Jacksons Company in Colonel Nichols Regiment of New York Militia, and was in service the greater part of the summer and fall.

That he joined his Company and was discharged at said Town of Chester and was then in service not less than five months; That he continued in the service in said Captain Jacksons Company and Colonel Nichols Regiment and was on duty more or less every year untill the close of the war or untill the British evacuated New York. That he cannot particularise the days and months of his said services as he was more or less on duty every year; That one year he thinks it was in 1780 he was sent to West Point and worked there several weeks; That in 1779 he went out to the Frontiers with said Captain Wood and was engaged with the Indians in the Battle of Beaver Brook, when said Captain Wood was made prisoner; That he was repeatedly out on alarms from the Indians, Tories, British and Cowboys; That he frequently was sent in pursuit of Claudius Smith a notorious Tory who had joined the British, and was in the habit of committing depredations on the neighbourhood stealing cattle, horses &c; and That in all he was in actual service not less than Two years exclusive of the service rendered in 1776 and 1777; That he was born in Chester in Orange County, New York and lived there untill immediately after the war when he removed to Pensylvania, from thence to Seneca County New York and then to his present residence where he has resided for Twenty two years; That he has no documentary evidence of his services; that he hereby relinquishes every claim whatever to a pension or annuity except the present, and declares that his name is not on the pension Roll of the Agency of any State or Territory

Sworn to and subscribed in open Court the day and year aforesaid
M. S. Ramsey Dep Clk Steuben County

Anthony Swarthout

We David B Coryell a clergyman residing in Urbana near Wayne and ~~Aaron Swarthout~~ Asa Swarthout a resident of the town of Wayne hereby certify, that we are well acquainted with Anthony Swarthout, who has subscribed and sworn to the above declaration; that we believe him to be seventy three years of age; that he is reputed and believed, in the neighbourhood where he resides, to have been a Soldier of the Revolution, and that we concur in that opinion.

Sworn and subscribed the day and year aforesaid
M. S. Ramsey Dep. Clerk Steuben County

David B. Coriell
Asa Swarthout

And the said Court do hereby declare their opinion, after the investigation of the matter, and after puting the interrogatories prescribed by the War Department, that the above named applicant was a Revolutionary Soldier and served as he states. And the court further certifies that it appears to them that David B. Corrill and Asa Swarthout who have signed the preceding certificate are residents in Urbana & Wayne in said County that they are credible persons and that their statement is entitled to credit

State of New York
Steuben County ss. I, Martin S Ramsey Deputy clerk of the Court of Common pleas of said County do hereby certify that the foregoing contains the original proceedings of the said Court, in the matter of the application of Anthony Swarthout for a pension. In testimony whereof, I have hereunto set my hand and Seal of Office, this Sixteenth day of October 1832

M S Ramsey Dep. Clerk Steuben County

State of New York
Ontario County Anthony Swarthout of the Town of Wayne Steuben County the within named declarant hereby on his Oath further declares that he was born on the 4th day of July 1759 of which he has a record in his Bible. That in 1776 he served under Genl Clinton Col Nichols Major Hatfield Captain Wood Ensign Webb with Capt Wisner, Lieut Blain, Col Thurston in 1777 under Col Nichols Major Meeker who was afterwards killed Captain Jackson Lieut McDowell with Capt Wisner who was broke for Cowardice. That in 1777 when he was pressed into the service it was by a French Officer whose name he has forgotten.
Sworn and subscribed this 4th day of June 1833 before me
Eurotas Hastings J P

Anthony Swarthout

State of New York
Ontario County ss. I certify that Eurotas Hastings whose name is subscribed to the foregoing certificate was at the date thereof a Justice of the Peace in and for said county, and that I am well acquainted with ~~the~~ his hand writing and verily believe the said signature to be his proper act and deed. In testimony whereof I have hereunto set my hand and affixed my Seal of Office this 24 day of December A.D. 1833.

F. J. B. Crane Dep. Clerk of said county

New York

Anthony Swarthout of Steuben Co. in the State of New York who was a Private in the Compy. commanded by Captain Wood of the Regt commanded by Col Nichols in the N.Y. ma. line for10 months Inscribed on the Roll of New York at the rate of 33 Dollars 331/3 Cents per annum, to commence on the 4th day of March, 1831. Certificate of Pension issued the 7th day of March 1834, and sent to R. M. Bayley Geneva, Ontario Co. N.Y.

Arrears to the 4th of March 34	$100.00
Semi-anl. allowance ending 4 Sep. 34	16.66
	$116.66

Recorded by E W Brown Clerk, {
Revolutionary Claim, Act June 7, 1832.}
Book E Vol. 4 Page 122

Ovid July 31, 1843

Hon J. L. Edwards

Dr Sir

I enclose an application with all the proof in the case of Elizabeth Swarthout, widow of Anthony Swarthout of Steuben Co. New York.

Respectfully Your very Obt. Sevt
Arad Joy

Declaration

In order to obtain the benefit of the act of Congress of the 7th July 1838 entitled "an act granting half pay and pensions to certain widows"

State of New York

County of Steuben ss. On the 26 day of July 1843 personally appeared before the Subscriber a Judge of the court in and for said county Elizabeth Swarthout of the town of Wayne in said county and state, aged Eighty four years, who being first duly sworn according to law doth on her oath make the following declaration, in order to obtain the benefit of the provision made by the "act granting half pay and pensions to certain widows" (and also a further

Declaration

In order to obtain the benefit of the act of Congress of the 3d March 1843, granting pensions for one year to certain widows who have received pensions under the act of July 7, 1838)

That she is the widow of Anthony Swarthout who was a pensioner of the United States, and a Soldier in the War of the Revolution. That her said husband was in the battle of White plaines & did much service in the war of the Revolution & recd. a pension of thirty three dollars and thirty three cents per annum. She further declares that she was married to the said Anthony Swarthout in the year (1778) Seventeen hundred and seventy eight, that her said husband, the aforesaid Anthony Swarthout died on the 11th day of October 1842, that she was not married to him prior to his leaving the service, but that he was in the service before her said marriage to him, & that she cannot from want of memory recollect, whether he was, or was not in the service after her said marriage but the marriage took place previous to the first of January 1794 seventeen hundred & ninety four, viz, at the time above stated.

Subscribed & sworn before me this 26 day of July 1843

J Larrow First Judge of Steuben County

her
Elizabeth X Swarthout
mark

I certify that the above named Elizabeth Swarthout is a respectable person, that I consider her of the age represented by her & that from her great age, infirmity & being blind, she ought not to be required to appear in open court to make her declaration & application.

J Larrow First Judge of Steuben County

State of New York

County of Steuben ss. Personally appeared before the subscriber a Judge of the court in & for said county Asa Swarthout aged forty one years & upwards, having been born in Oct 1802, & after being first sworn according to law deposeth & saith that widow Elizabeth Swarthout above named & who is applying for a pension as the widow of Anthony Swarthout late a pensioner of the United States, is this deponents mother, & that the aforesaid Anthony Swarthout the said pensioner of the United States was this deponents father, & that this deponent is the twelfth child of his above named father and mother, that his said father Anthony Swarthout a pensioner of the United States & his said mother Elizabeth

Swarthout above named & who is applying for a pension, as the lawful fruits of their marriage had eleven children older than this deponent, all of single births & all of the same father & mother, the eldest of which was born 1779 agreeable to the old family record of the ages of all the family, that the above named Elizabeth Swarthout is the identical widow & who was the wife of the aforesaid Anthony Swarthout as early as 1778, as the facts well show from the family record of the ages of their children, that his said father, the aforesaid pensioner Anthony Swarthout died on the eleventh day of October 1842 eighteen hundred & forty two and that his said mother has not since married, but remains his widow

Subscribed & sworn before me this 26 day of July 1843

J Larrow First Judge of said county

Asa Swarthout

I certify that the above named Asa Swarthout is a respectable person, & entitled to credit in his deposition above

J Larrow First Judge of Steuben County

State of New York

County of Steuben ss. Personally appeared before the subscriber a Judge of the Court in & for said county Andrew Swarthout of Wayne in said county aged forty three years, & on being duly sworn according to law, deposeth & saith that Widow Elizabeth Swarthout within named & who is applying for a pension as the widow of Anthony Swarthout late a pensioner of the United States, is this deponents mother, & that the aforesaid Anthony Swarthout the said pensioner was this deponents father; that his aforesaid father Anthony Swarthout a pensioner & his said mother the aforesaid Elizabeth Swarthout, had as the legal fruit of their marriage ten children older than this deponent & that the eldest of their said children was born in 1779 agreeable to the old family record of all the dates of the birth of all their twelve children, that this deponents mother the aforesaid Elizabeth Swarthout is the identical mother of the twelve children above mentioned & that she is the identical widow of the aforesaid Anthony Swarthout late a pensioner of the United States, & that she must have been his wife in 1778, agreeable to the family record of the fruits of their marriage, that the said twelve children, own brothers & sisters of this deponent, were all of single births, that his said father the aforesaid Anthony Swarthout died on the 11th day of October 1842 & that his said mother has not since married, but remains his widow

Subscribed & sworn before me this 26th day of July 1843 Andrew D. Swarthout

J Larrow First Judge of said county

I certify that the above named Andrew Swarthout is a respectable person & fully entitled to credit in his deposition above.

J Larrow First Judge of said county

State of New York

County of Steuben ss. Personally appeared before the subscriber a Judge of the court in & for said county Ezekiel Swarthout of Wayne in said county aged sixty one years nearly having been born Augt. 1782 after being duly sworn according to law deposeth & saith that Widow Elizabeth Swarthout who is by the annexed, applying for a pension as the widow of Anthony Swarthout, late a pensioner of the United States, is this deponents mother, & that the aforesaid Anthony Swarthout late a pensioner of the United States was this deponents father, that his said father Anthony Swarthout, a pensioner, & his said mother Elizabeth Swarthout lived together as husband & wife from the first of this deponents recollection & for more than fifty five years to this deponents recollection, that they had as the lawful fruit of their marriage two children older than this deponent, that the eldest a daughter & own sister of this deponent was born in the year 1779, that his aforesaid father & mother, the said Anthony Swarthout & Elizabeth Swarthout had twelve children all of single births & all of the same father & mother, that this deponent is the third child, that the aforesaid widow Elizabeth Swarthout is the identical person & widow of Anthony Swarthout late & a pensioner of the United States, & who was his said wife as early as 1778, agreeable to the family record of their said children that this deponents father & the late husband of this deponents mother viz, the said Anthony Swarthout late a pensioner of the United States, died on the 11th day of October 1842, & that his said mother the aforesaid Elizabeth Swarthout has not since married, but is & has remained his widow.

Subscribed & sworn before me this 26 day of July 1843

J Larrow First Judge of said county

Ezekeil Swarthout

I certify that the above named Ezekiel Swarthout is a respectable person & entitled to full credit in his deposition above subscribed

J Larrow First Judge of said county

I have been acquainted with Ezekiel Swartout, one of the within deponents since the year 1820, and believe him to be over sixty years of age.

J S Ellsworth

Anthony Swarthout	born July 4th 1759
Elizabeth Swarthout	Born Oct 17th 1758
Catharine Swarthout	B March 9th 1779
Anna Swarthout	Born Jan 4th 1781
Ezekiel Swarthout	Born Augst 29th 1782
John L Swarthout	Born Oct 30th 1784
Caleb Swarthout	Born Dec 24th 1786
Elizabeth Swarthout	Born Oct 13th 1788
Henry Swarthout	Born April 13th 1790
Lovina Swarthout	Born April 9th 1793
Mary Swarthout	Born April 10th 1795
George F Swarthout	Born Oct 28th 1796
Andrew D Swarthout	Born Jan 16th 1800
Asa Swarthout	Born Oct 3d 1802

State of New York

Steuben County ss. I, William Hamilton, Clerk of said County, do Certify that Jacob Larrowe Esquire, before whom the preceeding affidavits purports to have been Sworn, was at the date of the same, First Judge of the Court of Common Pleas in and for said County, and that I am well acquainted with his hand writing, and verily believe his name Subscribed to the four preceeding affidavits, and 4 Certificates, to be his proper hand writing and Signatures. Witness my hand and Seal of Office this Twenty ninth day of July A.D. 1843.

Wm. Hamilton Clerk

Will Mr Hanks oblige me so much as to impress his seal & write his certificate in its proper place, turn the sheet inside out, direct & mail it to ~~me~~ Hon J. L. Edwards Comr. of Pensions on Monday morning I hope it will not fail of being done. Oblige yours truly

Arad Joy

Hon J. L. Edwards, Dr Sir I send this in support of the marriage of Elizabeth Swarthout, whose application under the law of July 1838, you have, from Steuben Co. New York.

Respectfully your Obt Svt
Arad Joy

State of New York
County of Seneca ss. Personally appeared before the subscriber a Justice of the peace in & for said county Ralph Swarthout of Lodi in said county aged seventy nine years, & on being duly sworn according to law, deposeth & saith that Anthony Swarthout late a pensioner of the United States from Steuben County New York was this deponents brother, and that Elizabeth Swarthout of Steuben Co New York who is applying for a pension as the widow of said Anthony Swarthout is the identical person & widow of the afforesaid Anthony Swarthout the late pensioner of the United States, and that she is the identical person who was as early as 1780 Seventeen hundred and eighty the wife of the aforesaid Anthony Swarthout, and that they were as early as 1780 living together as husband & wife, & that as the legal fruit of their marriage, they had six children before the year 1790 Seventeen hundred & ninety & that the aforesaid Anthony Swarthout died in October 1842 & that his widow the aforesaid Elizabeth Swarthout is still his widow
Subscribed & sworn before me this 21st day of ~~September~~ October 1843
John B. Young Justice

Ralph Swarthout

I certify that the above named Ralph Swarthout is a highly respectable person
John B Young Justice

State of New York
Seneca County ss. I do hereby certify that John B. Young Esq whose signatures appear to the foregoing Jurat & certificate was on the day of the date of said Jurat a Justice of the Peace in & for said County duly commissioned and sworn and authorised to take and certify the foregoing Deposition in manner within

contained, and further that I am well acquainted with the handwriting of the said Justice and verily believe that his signatures to the said Jurat & certificate are severally Genuine & in his own proper handwriting In Testimony whereof I have hereunto set my hand & affixed the seal of the Court of common pleas of said County this 30th day of October AD 1843.

C D Hanks Clerk

Ovid Dec 12th 1843

Hon A Dana

My Dear Sir

No man has a stronger desire to comply with all the rules of Hon Col Edwards in the pension office than I have, and when I have once fully learned the rule, I never break over it, but in making application for pensions under the law of 1838, I embraced in it, the law of March 1843, as in the inclosed paper, but after about the 20th Oct, I learned that the addition of the application of March 3rd was not acceptable, not in one instance since that time, have I put the objectionable clause in, the inclosed paper is made out in the same form of that of Mary Covert widow of Luke P Covert & Hannah Doughty widow of Skelleman Doughty, upon each of which a pension was granted without objection, I now, desire to with draw all claim, to that part of the application under the act of March 3rd 1843, that part being inclosed in brackets & hoops, I return the paper, as under the law of 1838 as it is complete under that law, & as I believe agreeable to the form of Hon Col Edwards as laid down to be complied with under that law

On the 30th Oct, 43 I mailed at this office the deposition of Ralph Swarthout in support of the marriage, which together with the proof attached to the application is fully equal in support of it, to either of the cases alluded to in this communication, the poor old woman is blind I was at her house at the day of taking the appl. and it having been already delayed since the 31st July last, when I mailed the inclosed sheet, I would be much pleased with as early a decision on it, as is consistent with Hon Col Edwards, I have intended to state the facts just as they are, & I think I have not deviated one word from the truth; if there is not a pension due to the widow, I dont want it, I ask only for the widows right

I am Sir truly Your Obt Sevt

Arad Joy

State of New York

County of Steuben ss. Personally appeared before the subscriber a Justice of the peace in and for said county Asa Swarthout of said county aged forty one years, after being duly sworn according to law, deposeth and saith that widow Elizabeth Swarthout who made application for a pension, as the widow of Anthony Swarthout late a pensioner of the United States, is this deponents mother, & that the aforesaid Anthony Swarthout was this deponents father, that the attached old family record, is the original family record, kept by his aforesaid father & mother, & that it is the old record, of the ages of the children of this deponents aforesaid father & mother, and contains the ages of this deponents brothers & sisters, that it is the old record that was in use in this deponents aforesaid father and mothers family twenty or twenty five years since.

Subscribed & sworn before me this 9 day of January 1844

Wm Birdsall J. P.

Asa Swarthout

I certify that the above named Asa Swarthout is a respectable person.

Wm. Birdsall

State of New York

Steuben County ss. I William Hamilton Clerk of the said County, do Certify, that William Birdsall Esquire, before whom the above Affidavit purports to have been Sworn was at the date of the same One of the Justices of the Peace in & for the said County, and that I am acquainted with his hand writing and verily believe his name Subscribed to the said affidavit, and also his name Subscribed to the above Certificate, to be his proper hand writing and Signatures. Witness my hand and Seal of Office this Eleventh day of January 1844.

Wm Hamilton Clerk

Washington City D.C.
House of Reps N.Y.
14 March 1844

My dear Sir

In this I had your evidence of marriage of Elizabeth Swarthout, which was enclosed to me to present, let it take its time with transient business as its not from my constituancy

Respectfully Yr. Mo. Obt. Svt.
Wm. S. Hubbell

Col. J. L. Edwards

Commissioner of Pensions

State of New York
County of Seneca ss. Personally appeared before the Subscriber a Justice of the peace in and for said county, Abraham P. Covert of Ovid in said county aged sixty nine years the 7th March 1844, after being duly sworn according to law, deposeth and saith that he has been for forty eight years personally acquainted with Anthony Swarthout late a pensioner of the United States from Steuben County New York and with his widow Elizabeth Swarthout who made application for a pension from Steuben County, as the widow of the said Anthony Swarthout, this deponent deposeth and saith, that at his first acquaintance with the said parties forty eight years since, that they were living together as husband and wife and that they had at that time as the legal fruit of their marriage ten children and that their two eldest were daughters and then were young women and that the eldest son was a young man, that this deponent well recollects that at the time of his first acquaintance he was but twenty years of age or not twenty one, and that he is at this time sixty nine years all but six days.
Subscribed and sworn before me this first day of March 1844
James Foster Justice of the Peace of said County

Abraham P. Covert

I certify that the above named Abraham P. Covert is a respectable person.
Dated March 1st 1844

James Foster Justice

State of New York
County of Seneca ss. Personally appeared before the Subscriber a Justice of the peace in and for said county, Thomas Covert of Ovid in said county aged eighty nine years, after being duly sworn according to law, deposeth and saith, that he has for fifty one years and over been personally acquainted with Anthony Swarthout, late a pensioner of the United States from Steuben County and with his widow Elizabeth Swarthout, who applied for a pension as his widow from Steuben County, this deponent deposeth and saith, that at his first acquaintance, more than fifty one years since, with the above named parties that they were living together as husband and wife, and that they had at that time eight or nine

children, that this deponent has a data, by which he arrives at the exact time or date of his first acquaintance with the said parties and that is, that this deponent, moved into the State of New York, and near his present residence in the fall of 1792 and settled down a neighbor to the above named parties, and was intimately acquainted with them and their family, and that their eldest children (daughters) were at that time nearly young women grown.

Subscribed and sworn before me this second day of March 1844
James Foster Justice of the Peace of Seneca County

Thomas Covert

I certify that the above named Thomas Covert is a respectable person.
Dated March 2nd 1844

James Foster Justice of the Peace of Seneca County

State of New York
County of Seneca ss. Personally appeared before the Subscriber a Justice of the peace in and for said county ~~Rachel~~ Jane Covert aged seventy seven years, after being duly sworn according to law, deposeth and saith that she has for fifty one years and over been personally acquainted with Anthony Swarthout late a pensioner of the United States from Steuben County and with his widow Elizabeth Swarthout who applied for a pension as his widow from Steuben County, this deponent deposeth and saith that at her first acquaintance, more than fifty one years since, with the above named parties, that they were living together as husband and wife, and that they had at that time eight or nine children, that this deponent has a data, by which she arrives at the exact date of her first acquaintance with the said parties, and that is, that this deponent moved into the State of New York and near her present residence in the fall of 1792 and settled down a neighbour to the above named parties, and was intimately acquainted with them and their family and that their eldest children (daughters) were at that time nearly young women grown.

Subscribed and sworn before me this second day of March 1844 James Foster Justice of the Peace of Seneca County

her
Jane X Covert
mark

I certify that the above named ~~Rachel~~ Jane Covert is a respectable person.
Dated March 2nd 1844

James Foster Justice of the Peace of Seneca County

State of New York
Seneca County ss. I Cyrus D. Hanks Clerk of the County of Seneca do hereby certify that James Foster Esq was on the several days of the dates of the within proceedings, Jurats and Certificates had before and certified by him a Justice of the peace in and for the said County duly commissioned and sworn, & further that I am well acquainted with the hand writing of said Justice of the peace & verily believe that his several signatures to said proceedings, purporting to be his, are severally genuine, and in his own proper hand writing. In Testimony whereof I have hereunto affixed my seal of office, and subscribed my name this fourth day of March 1844.

C. D. Hanks Clerk

Ovid May 3d, 44
Hon Wm S Hubbell
My Dear Sir

I am very much obliged for yours of the 27th ult. inclosing a letter from Hon Col Edwards, I beg leave to say, that I have not presented any such case as that letter refers to. I add a few lines of that letter - "The additional evidence sent through you by Mr Joy in the case of Sarah Swartwout affords no evidence that she was married as she alleges in 1774." The 3 depositions that I troubled you with, by mailing them to you on the 6th March, related to the marriage of Elizabeth Swarthout who claim to have been married in 1778 under the law of 1838, widow of Anthony Swarthout late a pensioner, & whose service I have not had disputed, be pleased to give notice of this case again, & ask that the proof that may now be filed with Sarah Swartwouts case be placed & filed & apply in the case of Elizabeth Swarthout. The 3 depositions were viz - Abm. P. - Thomas & Mrs Covert. This case having been delayed, in consequence of what I suppose to be an error, I presume it will at once be called up for decision

I am truly Your Obliged friend & Sevt
Arad Joy

Ovid April 15th 1845

Dear Sir

I may be allowed very respectfully to say that no person submits with more deference to any decision of yours, than myself, yet there has been not a few cases, where a decision was made against me, when after my explanation a different decision was made in my favor, thus encouraged by the past, I beg leave respectfully to call your attention to the case of Elizabeth Swarthout widow of Anthony Swarthout, late a pensioner under the law of 38 - her application was accompanied with the affidavit of 3 Sons aged 62 43 & 41 additional affidavit of Ralph Swarthout, a brother to the late pensioner, & subsequently by the family record and another affidavit, all of the witnesses certified to be respectable persons all this proof was on file & under went an examination before the 30th April 1844. The proof was far more, if I may be allowed to give my judgement to it, than in the case of Mary Howell, or Mary Covert both pensioned, without a word of objection under the same law, may I ask you to please re-view this case & decide whether a pension was not due and upon the proof on file, on your examination the 2d April 1844, before the case was cut off by the law of the 30th April; I ask nothing but justice to be done in this case, but I feel a confident assurance that if you will review the papers, without reference to any former opinions, as of 2d April 1844, that you will see it to be just to grant a pension as of that date, without reference to the law of the 30th April. I say again, I want nothing but the just deserts of the proof the 2d April 1844, which I believe would entitle to a pension at that date.

Pardon me for being this plain, as I shall cheerfully submit to your better judgement in a decision after you have fairly examined the case

Respectfully Your Most Obt Servt
Arad Joy

Hon J L Edwards
Comr of Pensions
Washington DC

Ovid N.Y. Oct 16, 46

Sir

The Proof of the marriage, only, having been wanting, & that having been established by a certificate of Hon J. S. Ellsworth I suppose there will be no further objection to granting a pension to Elizabeth Swarthout, widow of Anthony Swarthout, late a pensioner of the U.S., her papers are under the law of 1838 &

of New York.

From the decision of Hon J. Y. Mason on the 80 & 81 Page of the "Pension laws now in force" there can be no mistake as to the right of a pension, I judge

Respectfully Your Obt Servt
Arad Joy

Hon J L Edwards Comr of Pensions
Washington DC

Ovid N.Y. Nov 5th '46

Sir

Yours of the 31 Oct in the Elizabeth Swarthout case, does not explain fully, that case to me, now I suppose the completion of the proof of her marriage, having as I judge been made in Jany 1844 before the law of the 30th April '44 she would not be affected by that law, this is the only difference. I regret to trouble you, but I hope you will find me correct

Respectfully Your Obt Servt
Arad Joy

To the Hon Department of the Interior

You will Pleas to Allow my Agent Thomas Lumpkin to Examine the Papers on file in your Office of the Application Of the Widow Elizabeth Swartwout Suspended.

Yours Most Respectfully
Andrew Follet

Penn Yan 3d July 1847

Hon J L Edwards
Comr Pensions
Sir,

Mr Asa Swartout desired me to enquire of you if a pension had been allowed to Mrs. Elizabeth Swartout. I recollect being called on in the winter of 1846 while at Washington by Arad Joy Esq who desired me to go with him to the Pension office and state the aged in my opinion Ezekiel Swartout one of her elder children, which I did, my statement I understood to remove all difficulties. If the pension was granted who drew it.

Mrs Elizabeth Swartout was the widow Anthony Swartout a revolutionary soldier

Your early reply will greatly oblige both Mrs Swartout and yours truly

J S Ellsworth

Asa Swartout is also a son of Mrs E Swartout

Ho Reps March 1, 1848

Sir

I am requested in behalf of me of my constituents to ask you for a copy of that part of the papers of Anthony Swarthout in which the services of Swarthout are set forth and described. The papers I think were presented on Swarthouts application for a pension.

If the rules of your office admit I will thank you to send me the copy of the part of the papers reference to.

I am very respy yr obt Servt
Hon J L Edwards

D ______
Com Pensions

Pages Corners P.O. N.Y.
April 20 1857

Hon Geo C. Whiting
Commissioner of Pensions
Dr Sir

I am requested by the heirs of Elizabeth Swartwout widow of Anthony who applied for Pension under the act of June 7th 1838, (& whose marriage & the service of said Anthony being admitted as stated the Commissioners letter) to enquire why the arrears of Pension due her in her lifetime may not now be claimed by them. Mrs Swarwout died January 10th 1845. May I ask an Early reply to my Enquiry

Your Obt Servt
R W Crain

O. W. and N. Division.

Department of the Interior,
Bureau of Pensions,

Washington, D.C.,

In reply to your request for a statement of the military history of Anthony Swarthout, a soldier of the Revolutionary War, you will find below the desired information as contained in his (or his widow's) application for pension on file in this Bureau.

DATES OF ENLISTMENT OR APPOINTMENT.	LENGTH OF SERVICE	RANK.	OFFICERS UNDER WHOM SERVICE WAS RENDERED.	OFFICERS UNDER WHOM SERVICE WAS RENDERED.	STATE
			CAPTAIN.	COLONEL.	
July 76	5 mos		John Wood	Nichols	N.Y.
	2 mos	Teamster	officers not stated	officers not stated	
77	5 mos.		John Jackson	Nichols	

continued in service at various times every year succeeding during the war amounting to 2 yrs. Was under Capt. Wood

Battles engaged in, Flat Bush Harlem Hghts. York Is. & Wh. Pl. & Beaver Brook
Residence of soldier at enlistment, Chester, Orange Co., N.Y.
Date of application for pension, Oct. 16, 1832
Residence at date of application, Wayne, Steuben Co., N.Y.
~~Age at date of application~~, b. July 4, 1759 Chester, N.Y.
Remarks: Sol. mar. 1778 Elizabeth He d. Oct. 11, 1842 Wid. apply July 26, 1848 at 84 yrs Wayne N.Y. She died Jany 10, 1845 (Sol. was cousin to Aaron Swarthout S. File #11497)

Very respectfully,
Commissioner.

Jan. 30, 1907.

Miss Georgia Swarthout,
Canisteo, N.Y.
Madam:

In addition to the military history of Anthony Swarthout (R. File #10,344), you are furnished the following copy of a record of births on file with the other papers:

Anthony Swarthout	born July 4, 1759.
Elizabeth Swarthout	born Oct. 17, 1758.
Catharine Swarthout	born Mch. 9, 1779.
Anna Swarthout	born Jan. 4, 1781.
Ezekiel Swarthout	born Aug. 29, 1782.
John L. Swarthout	born Oct. 30, 1784.
Caleb Swarthout	born Dec. 24, 1786.
Elizabeth Swarthout	born Oct. 13, 1788.
Henry Swarthout	born Apl. 13, 1790.
Lovina Swarthout	born Apl. 9, 1793.
Mary Swarthout	born Apl. 10, 1795.
George F. Swarthout	born Oct. 28, 1796.
Andrew D. Swarthout	born Jan. 16, 1800.
Asa Swarthout	born Oct. 3, 1802.

Very respectfully,
Acting Commissioner.

Waterloo, N.Y.
Oct 23, 1916

Commissioner of Pensions
Washington D.C.
Dear Sir,

Have you on your pension list of the Revolutionary war soldiers the name of Anthony Swarthout Sr.? He entry lived or enlisted at Goshen, Orange Co. N.Y. If you have his name would you kindly send me what information you have concerning heirs. Thanking you I am

Respt Yours,
Cora Belle Clark
423 Mary St.
Waterloo, N.Y.

Rushville, New York
October 9, 1925

Commissioner of Pensions
Washington, D.C.
Dear Sir:

Will you kindly tell me whether the name Anthony Swartout, or his son Anthony Jr., appears upon your list of Revolutionary soldiers? It has been handed down through our family that one of these ancestors came through this section with Sullivan's army, ~~and~~ we are very desirous of finding their war records in order to obtain membership in the Daughters of American Revolution. Any assistance you can give in the matter will be much appreciated.

Respectfully yours
(Miss) Mabel Blodgett

Rev. & 1812 Wars Section
Miss Mabel Blodgett
Rushville N.Y.

November 10, 1925

Madam:

I have to advise you that from the papers in the Revolutionary War pension claim, R.10344, it appears that Anthony Swarthout was born July 4, 1759, in Chester, Orange County, New York.

While a resident of said place he enlisted in July, 1776, and served five months as private in Captain John Wood's Company, Colonel Nichols' New York Regiment; he was in the battles of Flat Bush, Harlem Heights, York Island and White Plains. He then served two months as teamster, officers not stated. He enlisted in 1777 and served five months as private in Captain John Jackson's Company, Colonel Nichols' New York Regiment. He continued to serve at various times until the British evacuated New York, amounting to two years in all, under Captains John Jackson and Wood and Colonel Nichols; he was in the battle of Beaver Brook.

He was allowed pension on his application executed October 16, 1832, at which time he was a resident of Wayne, Steuben County, New York. He died October 11, 1842.

He married in 1778 Elizabeth (her maiden name not stated) who was born October 17, 1758. She died January 10, 1845.

Children of Anthony and Elizabeth Swarthout:

Catharine	born March 9, 1779.
Anna	born January 4, 1781.
Ezekiel	born August 29, 1782.
John L.	born October 30, 1784.
Caleb	born December 24, 1786.
Elizabeth	born October 13, 1788.

Henry	born April 13, 1790.
Levina	born April 9, 1793.
Mary	born April 10, 1795.
George F.	born October 28, 1796.
Andrew D.	born January 16, 1800.
Asa	born October 3, 1802.

The above noted is the only Anthony Swarthout found on the Revolutionary War records of this Bureau.

Respectfully,
Winfield Scott
Commissioner

⭑

Nepera Park N.Y.
March 11 1932

War Department
Washington D.C.
To the Director of Pensions

Dear Sir; I am a man in my seventy-sixth year. My daughter is anxious to join the D.A.R. I have a letter from the adjutant's office advising me to write to you.

My greatgrandfather was Anthony Swarthout. There is no doubt he was a soldier in the American Revolution under Steuben & was a member of the body guard of the General at least for a time. My father remembered him well & I can remember talking with old men when I was a boy who knew him well. After the war he settled about a mile south of what is now Keuka Landing on the east side of Crooked Lake (now Keuka) & he & his wife are buried on the farm "He was a great big dutchman." The name may have been spelled differently at that time but on his tombstone it is spelled as above. Some spell it "Swartwout" now. His first name may be "Anton" on the records. My grandfather was left rather wealthy for that time & spent his time trying to reduce the _____ of whiskey. He died comparitively young. I despised the drunken old cuss & I never heard him say that he had a father even.

For my daughter's sake I would like to verify the fact. Personally I have no doubt. Can you help us?

Respt.
R. B. Swarthout
Nepera Park
Westchester Co. N.Y.

⭑

Mr. R. B. Swarthout
Nepara Park
New York

May 2, 1932
BA-J/MCS
Westchester County
Anthony Swarthout

R.10344

Dear Sir:

Reference is made to your request for the record of Anthony Swarthout, a soldier of the Revolution.

The data furnished herein were obtained from papers on file in the pension claim, R.10344, based upon the Revolutionary War service of Anthony Swarthout.

He was born July 4, 1759, in Chester, Orange County, New York.

While living there he enlisted in July, 1776, and served five months as private in Captain John Wood's Company in Colonel Nichols' New York Regiment and was in the battles of Flat Bush, Harlem Heights, York Island and White Plains. He then served two months as teamster, officers not stated. He enlisted in 1777 and served five months as private in Captain John Jackson's Company in Colonel Nichols' New York Regiment. He continued to serve at various times until the British evacuated New York, amounting to two years in all, under Captains John Jackson and Wood and Colonel Nichols and was in the battle of Beaver Brook.

After the Revolution he moved to Pennsylvania, from there to Seneca County, New York, and from there to Wayne, Steuben County, New York.

He was allowed pension on his application executed October 16, 1832, at which time he was living in Wayne, New York.

He died October 11, 1842.

He married in 1778 Elizabeth (her maiden name not stated) who was born October 17, 1758. She died January 10, 1845.

Children of Anthony and Elizabeth Swarthout:

Catharine	born March 9, 1779
Anna	born January 4, 1781
Ezekiel	born August 29, 1782
John L.	born October 30, 1784
Caleb	born December 24, 1786
Elizabeth	born October 13, 1788
Henry	born April 13, 1790
Levina	born April 9, 1793
Mary	born April 10, 1795
George F.	born October 28, 1796

Andrew D. born January 16, 1800
Asa born October 3, 1802

This is the history of the only soldier named Anthony Swarthout, searched under all spellings of the name, that is found in the Revolutionary War records of this office.

Very truly yours,
A. D. Hiller
Assistant to Administrator

Rejected

Revolutionary War Pension and Bounty Land Warrant Application Files
Series M-804; Roll # 2328

Levi Totten
New York (S.21,538)

Geo. Thurston & Son vs. Daniel Willcox

Washington ___ To the Honble Court of CommonPleas February Term AD 1811

George Thurston & Jeremiah Thurston of Hopkinton in the County of Washington, Merchants trading in Company under the Firm of George Thurston and Son, complain of Daniel Willcox of North Stonington in the County of New London in the State of Connecticut Cooper, in the Custody of the Sheriff in an action of the Case, for the Recovery of money due by Note which the Deft neglects to pay and is for that the Deft at Hopkinton aforesaid on the twenty third Day of October AD 1807 by his Note in Writing made of that Date and by the Deft signed for Value received on settlement of Book accounts, promised and on himself assumed to pay the ~~Plffs~~ or Order twenty seven Dollars & ninety eight Cents on Demand with Interest.

Yet the Deft tho often requested the aforesaid Sum of money and Interest to the Plffs to pay have ever refused & still refuse to pay the same or any Parts thereof to the Damage of the Plffs, sixty Dollars as laid in their Writ dated the third Day of January AD 1811, and hereon they sue by their attry

Levi Totten

Honr James L Edwards

Washington March 18th 1824

Sir

I return to you the papers of Mrs Totten, with such further evidence as I hope will be satisfactory Please to forward me the Certificate to send to her.

I have the honor to be Sir
with high respect your Obt Sert
Tristam Burges

⋆

John Hall's Affidavit

State of New York

Orange County ss. Be it known that on this 20th day of February 1833, Before me came John Hall of Warwick in said county, who being duly sworn according to Law, deposeth and saith that he was Born in Westchester in said State, but has resided in Orange County since he was twelve Years of age; that he, deponent is now seventy four years of age and a pensioner under the act of June 7. 1832. Saith that he has known Levi Totten before, during and a short time since the revolutionary War, and saith that the said Totten resided in said county of Orange. Deponent verily believes the said Levi Totten is now seventy four Years of age, that he deponent hath often seen the said Totten in the service as Militia private in Captain John Minthorns Company & in John Hathorn's Regiment at Ramapo Bridge, and at Ramapo he being in the same company. Deponent believes this service of said Totten was, during the Year 1777. and 1778. Deponent saith that he verrily believes that said Totten was a good and faithful soldier in the War of the revolution, and further saith not

Subscribed and sworn to the day and year first aforesaid before me
Nathaniel Jones Commissioner of Deeds &c.

John Hall

⋆

State of New York

Orange County ss. On the fourth day of March 1833 before me came Michael ~~Denton~~ Jackson of ~~Minisink~~ Walkill in said county a revolutionary pensioner under the act of 18th March 1818 & May 1st 1820 as appears by his certificate endorsed Recorded by "Benjn Lloyd ____ Clerk" in Book D Vol.5 Page 188 who upon his oath administered in due form of law declares that he was well acquainted with Levi Totten & belonged to the company of infantry commanded by Capt Daniel Denton in Col. Richmore's or Richmans Regt of troops of the New York line in which company the said Totten was a private & done duty at least nine months in the year 1776

Sworn & subscribed the 4 March 1833

H W Elliott Just Peace

Michael Jackson

I certify that I am acquainted with Michael Jackson & believe him worthy of credit

H. W. Elliott J Peace

State of New York

Orange County ss. I Asa Dunning Clerk of the County of Orange Do Certify that Horace W. Elliott Esq before whom the preceding affidavit purports to have been Sworn and whose name is Subscribed to the foregoing Certificate was at the date of the same a Justice of the peace in and for the said County and that I am acquainted with his hand writing & verily believe his name Subscribed to the said affidavit & Certificate to be his proper handwriting and Signature. Witness my hand and Seal of Office this 4th day of March 1833

Asa Dunning Clerk

I hereby Certify that I was Called on to Assist in Laying out the Dead boddy of Levi Totten Esqr of Scituate on monday February 25th AD 1833 and was with him at intervals for some days ~~Afterwar~~ previous his Death and on wensday morning following Assisted in putting the Corpse in the Coffin and formed the Procession for the funeral

Warwick October 7th AD 1833

Ambrose Vaughn

Kent ___ at Warwick this 7th day of October 1833 Personaly came Ambrose Vaughn Subscriber to the above and made Oath to the truth of the same

Before Me William B Spencer Just Peace

Kent ss. State of Rhode Island & Providence Plantations

I Silas Weaver clerk of the Supreme Judicial Court within & for the county of Kent and state aforesaid hereby certify that William B Spencer, whose name is within written is a Justice of the peace, within & for the town of Warwick in said county. that the Signature within written, purporting to be his is in the proper hand writing of said William B Spencer and that full faith & credit are due to the official attestations of said Spencer In witness whereof I hereunto set my hand and the seal of said court this twenty Second day of October AD 1833 at East Greenwich aforesaid

Silas Weaver

Declaration Susanna C Totten

State of Rhode Island Provedence Plantations

County of Kent ss. On this 8th day of October AD 1833 personally appeared in open court before the Justices of the Supreme Judicial Court now sitting at East Greenwich within & for said county of Kent, Susanna C. Totten widow of Levi Totten Esqr, late of Scituate in the County of Provedence in the state of R Island & Provedence Plantations now residing in Warwick in said State & on Solemn oath ~~declares~~ saith that from declarations often made by the said Levi while in full life & from memorandums & other writings made by him & left among his papers at the time of his death & hereunto annexed she doth verily believe that he the said Levi was ~~a~~ an inlisted soldier in the revolutionary ~~army~~ War & that he afterwards served various tours in the Militia during said war that is to say that he the said Levi ~~enlisted did~~ at Goshen in the County of Orange in the State of New York, did, on or about the 20th of february 1776 inlist as a private in the company of Capt Daniel Denton for one year that said company belonged to the 3d Regiment commanded by Col Rhudolph Ritzma or Ritchmore & was stationd near New York City that he was in the battle of White Plains & continued in service ~~un~~ in said company untill about the 7th of Decr 1776 when much out of health he went home on furlough & not regaining his health untill his year expired he did not join the Regiment again; but so soon as health did permit joined the service in the Militia. That in the fore part of the summer of 1777 he served two months under Capt John Minthorn afterwards the same season under the same Capt two months a part of the time near fort Montgomery In the fall of the same year he served two months under ~~Col John Hathorn~~ same Captain stationed near Fishkills. The same year one month under Col John Hathorn.

In 1778 he served two tours at West Point under Major Poppen equal to one month two other tours under Capt Millar at Makackamack or Neversink amounting to ~~one~~ two months & one month afterwards under Capt Sayer In

1779 he served one tour beginning in January under Lieut Armstrong of two months. After one month under Capt Miller & was afterwards ~~un~~ two months under Col Hathorn stationd at Poughkepsie

She the said Susanna on oath as aforesaid doth State that she often ~~often~~ heard her late husband tell the story of his revolutionary service, that he always was reputed to have been in the army; & that the papers hereunto annexed being various Memorandums of his service marked (~~here marked A~~ - A, B, C, D, E are in the hand writing of the said Levi & that she doth verily believe they contain the truth & that he really served as therein stated

And she further states that the said Levi was at the time of his death seventy three years old & that he died at said Scituate on the 25th day of February 1833.

Wherefore she prays that the amount due to said Levi ~~under~~ at the time of his death under the law of June 7 1832 may be allowed & paid to her

Susanna C. Totten

The above affidavit was subscribed and sworn to by the said Susanna C Totten on the said Eighth day of October 1833.

Witness Silas Weaver Clerk

I Silas Weaver clerk of said Supreme Judicial Court within & for said county of Kent do hereby certify that the foregoing contains the original proceedings of the said court in the matter of the application of Susanna C Totten for a pension for the service of Levi Totten deceased In testimony whereof I have hereunto Set my hand and Seal of office this twenty second day of October in the year 1833 at East Greenwich aforesaid

Silas Weaver

Kent ss. Supreme Judicial Court October Term AD 1833. on the third day of said term of said Court. Wilkins Updike of South Kingstown in the county of Washington & State of Rhode Island &c Esquire Says on oath that he was well acquainted with Levi Totten formerly of said South Kingstown and late of Scituate in said State deceased, that this affiant has frequently heard the said Levi Say that he Served in the revolutionary army both in the line and militia, and from the frequent conversation which this affiant had with said Levi about said Service, and from conversation which he has heard the said Levi have with other persons relative to said Service, this affiant believes that said Levi served in the revolutionary war as is Set forth in the declaration of Susanna C. Totten aforesaid, and I also believe that the ~~said~~ declarations of said Levi are entitled to credit

Sworn to & subscribed by the said Wilkins
Updike Esq this 9th day of October 1833 Witness.
Silas Weaver Clerk

Wilkins Updike

And the said Supreme Judicial Court do hereby declare their opinion after the investigation of the matter that the above named Levi Totten was a revolutionary soldier and served as the said Susanna C. Totten has stated as aforesaid.

And said court are also of the opinion that the character of Wilkins Updike Esquire for truth is good.

A

1776 in Service & on furlough
1777 3 Tours under Minthorn & one under Hathorn
1778 2 Tours at West Point under Major Poppino
2 Tours under Capt. Miller at Mahackamack otherwise called Minisink one in Decem under Capt. Sayer
1779 one in Jan. und. Lieut Armstrong
1779 one under Miller in July, fight
one under Col Hathorn at fishkill

B

Levi Totten to Azariah A Flagg
Scituate Rhode Island &c.

To the Honble Azariah A. Flagg Secretary of State for the State of New York, or to the acting Secretary of said State. Sir extreme Indigenie causes me to trouble you with my present Petition. I was a Soldier in the Revolutionary War & request of you to make or cause to be made Search of the Records in your Office for Evidence of my having performed such Service in the Revolutionary Army as I shall herein State. I was born in the Precinct of Goshen in the County of Orange, State of New York. my Father was a native of Hempstead on Lond Island, emigrated from thence & settled in Goshen where I was born on the 20th or 22nd Day of February in the year 1776 I inlisted a private Soldier in a Company commanded by Capt. Daniel Denton, Nathan Strong & Miller were his Subaltern Officers

Capt. Denton marched us to the City of New York where we were assosiated with 7 other Companies, 8 in the whole in a Regiment; numbered 3d Regiment commanded by Rhadolph Reitzma Col. often pronounced Col. Richmore, we were at first quartered in Houses, afterwards, encamped in Tents on Delansea's Land up the Bowry Lane

C

Memorandum of Petition to Congress - Written by Levi Totten

To the Honorable, the Senate, and House of Representatives, of the United States, in Congress assembled, on the first Monday in December, in the Year of our Lord 1830.

Levi Totten of South Kingstown, in the County of Washington, in the State of Rhode Island and Providence Plantations, humbly petitioning Shews, That in the Year, 1776 he was a Soldier in the Revolutionary Army and served more than nine Months of that Year in and about the City of New York; at various Stations on York Island; and at White Plains, and was in the Battle at White Plains, in October of that Year, & afterwards at Peekskill. At different Times he was under different Generals but General Washington Commander in Chief. Your Petitioner was born in the Precinct of Goshen, in the County of Orange, in the Province, now State of New York on the second Day of May AD 1759, Was the Son of a Farmer of that Place. His Father was born at Hempstead on Long Island about twenty two miles from the City of New York, and married and settled in Goshen before mentioned. Your Petitioner further states that when he was yet a Youth, the Acts of the British Parliament for blocking up, or shutting the Harbor of Boston, and for seizing and carrying Persons accused or suspected of Crimes, to Places beyond Sea for Trial, not by a Jury of their Peers, or the Law of the Land, together with other arbitrary Decrees of the British Government, raised in the mind of your Petitioner, a Spirit of Opposition which never forsook him. That on Reading an Account of the Commencement of Hostilities at Lexington, by a Detachment from the British Army at Boston. Your Petitioner, with Intent to assist, if necessary, in defending the Rights and Liberties of his Injured Country, diligently learned the Use of the Firelock and Bayonet. And on or about the twenty second Day of February AD 1776, He left a Peaceful and comfortable Home, against the Consent of his Parents, and inlisted for one Year, a private Soldier in a Company, of which Daniel Denton of Goshen was Captain; Nathan Strong first, James Miller second Lieutenants, and Cadmiel Moore Ensign. Which Company with others, as your Petitioner believes, were raised by the Advice & Direction of the Provincial Congress holden at the City of New York

Memorandum of Petition to Congress - Written by Levi Totten

D

Minutes of Letter to Azariah A Flagg

Scituate Rhode Island &c.

To the Honble Azariah A. Flag Secretary of State for the State of New York. Sir extreme Indigence causes me to request you to make Search of the Records in your Office for Evidence of my having served in the Army of

Military Services - 1776 1 Year Capt Denton

1777 2 Months forepart Summer under Capt John Minthorn Militia Stationed at Ramanpough Latter Part Summer 2 Months under do, same Station Lieut Col Wisner commanded part of the Time ~~while~~ we were near Fort mongomorry when it was taken In the Fall of the year 1777 2 Months Tour at Fishkills under Col. John Hathorn Militia Col.

E

Petition to War Depart Memd of Service

Susanna Totten of Levi Totten in revly War

That in the Year 1776 he was a Soldier in the Revolutionary commanded by General Washington and served most of that year in his Army at and about the City of New York and the various stations on York Island, opposite Hurl Gate & near Fort washington Also at the white Plains & was in the Action with the Enemy at that Place October 28th 1776, and continued in Service till on or about the 7th Day of December following, when being out of Health he obtained a Furlough to retire into the County till his Health should be recovered. That he did not recover his Health till after his Year had elapsed & did not return to Join the Army again under that enlistment. That after his Engagement had expd. he performed divers Tours of Duty in the Militia of the State of NY, & in New Jersey & was more than Half of the Time in actual service in the revolutionary Army during the years 1777 & 1778 ~~and performed one Tour of two Months in 1778~~ was a Volunteer in a Battle between the N.Y. & N Jersey ~~Volunteer~~ Militia Volunteers ~~with~~ a Party of Indians commanded by Captain Brant as he was called on the 25th Day of July AD 1779 about 35 Miles above the Fork of Delaware River & in which about 130 of the Volunteers were engaged & beaten of which number forty eight were missing believed to be ~~slain~~ killed as many of their Bodies were afterwards found & buried by another Party of Volunteers from the same Militia & one of the Party of the Volunteers was known to be taken Prisoner & carried to Canada & given up to the British.

I Daniel Watson of South Kingstown in the county of Washington and State of Rhode Island &c Physician say on oath that I was acquainted with Levi Totten late of South Kingstown aforesaid and have frequently heard the said Levi Say that he Served in the army ~~during~~ in the revolutionary war and have heard him talk with other persons about said Service, and I believe it is ~~generally~~ believed and understood by people in the neighborhood in said South Kingstown, where the said Levi Totten resided, that he the said Levi was a soldier in the revolutionary army.

Daniel Watson

Kent ss. At East Greenwich October 16th 1833 Personally appeared Daniel Watson Subscriber to the foregoing affidavit and made oath to the truth of the same and I certify that the character of said Watson is good for truth

Before Me Christopher Johnson
Justice of the Peace

Kent ss. State of Rhode Island & Provedence Plantations

I Silas Weaver clerk of the Supreme Judicial Court, within and for the county of Kent and State aforesaid hereby certify that Christopher Johnson whose name is above written is a Justice of the Peace, within & for the town of East Greenwich in said county, that the signature above written purporting to be his is in the proper hand writing of said Christopher Johnson and that full faith & credit are due to the official attestations of said Johnson In witness whereof I hereunto set my hand & the seal of said court this twenty Second day of October AD 1833 at East Greenwich aforesaid

Silas Weaver

I Charles Brayton of Warwick in the county of Kent and state of Rhode Island and Providence Plantations Esquire Say on oath, That I was acquainted with Levi Totten late of Scituate, and formerly of South Kingstown deceased for more than forty years previous to his decease, That I have frequently heard the said Levi Say that he served in the army of the United States during the revolutionary war, and have heard him converse with others on said Subject, and I have no doubt, but that said Levi did service in the army of the United States during said war, from the fact of my having heard him assert it frequently, and from general reputation And I further say that the character of said Levi for truth was good.

Charles Brayton

Kent ss. State of Rhode Island &c.

At East Greenwich October 19 1833

Personally appeared Charles Brayton subscriber to the above affidavit and made solemn oath to the truth of the same and I ~~also~~ certify that the character of said Charles Brayton for truth is good.

Job Durfee one of the Justices of
the Supreme Judicial Court

Kent ss. State of Rhode Island & Providence Plantations

I Silas Weaver Clerk of the Supreme Judicial Court, within and for the county of Kent, and State aforesaid hereby certify that Job Durfee Esquire, whose name is within written is one of the Justices of said Supreme Judicial Court, that the signature within written, purporting to be his is in the proper hand writing of said Job Durfee, and that full faith and credit are due to the official attestations of said Durfee. In witness whereof I hereunto set my hand & the seal of said court this twenty second day of October AD 1833 at East Greenwich aforesaid

Silas Weaver

War Department
Pension Office
Decr. 26. 1833

Sir

The documents in support of the claim of Susanna Totten, widow of Levi Totten decd. have been examined and are herewith returned Before a final decision can be had in her case the Clerk of the Court must certify that she is the widow of the deceased and that Totten died on the day set forth. The widow must amend her declaration by relinquishing all other pension except the one now claimed

I have the honor to be
Very Respectfully Yr. Mo. Obt. St.

J. L. Edwards

Hon T. Burges
House of Representatives

⁂

State of Rhode Island &c

County of Kent ss. On this tenth day of March AD 1834 personally appeared in open Court before the Court of Probate of the Town of Warwick Susanna C Totten widow of Levi Totten late of Scituate in the County of Providence in said State now residing in said Warwick and on solemn oath says that she is the widow of the said Levi Totten deceased. That the said Levi died in said Scituate on the 25th day of February AD 1833 and she hereby relinquishes all and every claim to any annuity or pension or arrear thereof except the present and declares that the name of the said Levi is not nor is her own on the pension roll of the agency of any State

Susanna C Totten.

Subscribed and sworn to by the said Susanna C Totten in open Court this 10th day of March AD 1834

Witness Geo A Brayton Clk

⁂

I Emma Ann Staunton of Coventry in the County of Kent & State of Rhode Island &c depose and say that I have been personally acquainted with Susanna C. Totten and Levi Totten now deceased. That I was present at their marriage and saw them married and that the said Susanna was and continued to be the wife of the said Levi untill his death and that I was at the dwelling of the said Levi at the time of his death and during his last sickness and that the said Levi died on the 25th day of February AD 1833

Emma Ann Stanton

Subscribed and sworn to in open Court by the said Emma Ann Staunton this 10th day of March AD 1834

Witness Geo A Brayton Clk

⁂

I Martha Totten now residing in Warwick in the County of Kent depose and say that I am the daughter of the said Levi Totten deceased and of Susanna C Totten and resided with my parents at the time of the death of my father that he died on the 25th day of February AD 1833

Martha Totten

Subscribed and sworn to by the said Martha Totten in open Court this 10th day of March AD 1834

Witness Geo A Brayton Clk

★

And the said Court of Probate do hereby declare their opinion from the evidence adduced and which is here to annexed that the Susanna C Totten the within applicant is the widow of the said Levi Totten late of Scituate deceased and that the said Levi Totten died at said Scituate on the 25th day of February AD 1833 and do hereby certify the same accordingly and the said Court do further certify that it appears to them that Ambrose Vaughn Emma Ann Staunton and Martha Totten who have subscribed the preceeding affidavits are credible person and that the said Susanna C. Totten is also a credible person In Testimony Whereof I George A Brayton the present Clerk of the said Court have hereunto by order of said Court set my hand and affixed the seal of the said Court of Probate this Tenth day of March AD 1834

Geo. A. Brayton

★

Rhode Island

Levi Totten decd. of Scituate in the State of R.I. who was a Private in the Company commanded by Captain Denton of the Regt commanded by in the R.I. line for 2 years Inscribed on the Roll of R. Island at the rate of 80 Dollars Cents per annum, to commence on the 4th day of March, 1831. Certificate of Pension issued the 31st day of March 1834 and sent to Hon T. Burges House of Reps

Arrears to the 4th of Sept 1832	$120.00
amt. due up to the day of his death	38.00
Feb 25, 1833	$158.00

Recorded by D D Addison Clerk,
{Revolutionary Claim, Act June 7, 1832.}
Book E Vol. 2 Page 132

★

Invalid. File No. 21,538
Levi Totten Pri Rev. War
Act: June 7th 32

Index: - Vol. 2, Page 420
[Arrangement of 1870.]
Revolutionary War Pension and Bounty Land Warrant Application Files
Series M-804; Roll # 2402

James Van Fleet
New York (R.10,864)

State of New York

County of Cayuga ss. Personally appeared before ~~me the undersigned a Justice of the peace of said~~ County James Van Fleet of the town of Haight in the County of Alleghany in the State aforesaid after being duly sworn makes oath and says that he is aged Seventy nine years on the fifth day of August last. That he resided in the town of Goshen in the County of Orange in said State of New York during the Revolutionary War. That in the fore part of the Summer of the year 1775 he entered the Service of the United States in a Company of Militia commanded by Capt. Jones, & Lieut. Martiness Decker, and continued in such service for the term of about one month in the building of fortifications at Fort Montgomery on the bank of the Hudson River in the County of Orange. That he again entered the said service early in the winter of the year 1776 or 1777 (which of the two years he can not now recollect) under Capt. Moses Cortright Lieut. Martiness Decker, and Ensign Ephraim Middaugh in the Regt. commanded by Col. Ellison, they marched to Tappan in the State of New Jersey, they were at Tappan when Gen. Washington surprised and captured the Hessians at Trenton. That he continued in said service for the term of about five weeks. That he again entered said service in the Summer of the year 1777 and marched to Ramapo in New Jersey under Capt. Sweesey & Lieut. Tryon, in Col. Gilbert Cooper's Regt. and continued in said service three weeks. That he again entered the service in October 1777 under the command of Capt. Moses Cortright & Lieut. John Van Tuyl, they marched for the purpose of Reinforcing the garrison at Fort Montgomery and arrived too late, the British having captured said Fort before they could arrive there, they then marched to New Windsor on the Hudson River & from thence to Bank's bridge and continued in said service for the term of about three weeks. That in a few days thereafter he again entered the said service and marched to a place called Murderers creek on the Hudson River, and remained in such service about two weeks in building fortifications at Murderers creek, under Lieut. Stuart, the other officers he does not now recollect.

That he again entered the service under the command of Major John Decker early in the Spring of the year 1778 and marched to Sussex in the State of New

Jersey to relieve the inhabitants (there) against the Refugees, and continued in said service about one week. And this deponent further saith That from the opening of the spring of the year 1778 to the end of the War the militia in the west part of Orange County were ordered to hold themselves in readiness to march at a moments warning and to guard the frontiers against the Indians and that this deponent as also the whole of the Militia aforesaid did so hold themselves in readiness. That he frequently marched with scouting parties (or Indian spies) to the distance of from one to thirty miles. That during the whole of said time from the spring of the year 1778 to the end of the War in each year from the opening of the spring to the beginning of winter this deponent as well as all the other militia in that vicinity carried his gun daily and almost every day under the commands of Lieut. John Van Tuyl, Capt. Wilhelmus Westfall ~~Lieut. Levi Van Etten, Ensighn Elijah Van Auken~~, and other officers. That this deponents service was of such a nature that he cannot now tell the precise length of time, but is confident that he did serve at the times and at the several places aforementioned at least three years and for the greater part of that time he was a Sergeant

Subscribed and sworn to this twenty fifth day of Sept. 1832 before me
Cornelius Cuykendall Justice of the peace

James Vanfleet

State of New York

County of Cayuga ss. Henry Van Etten of the town of Sempronius in said County and Daniel Van Fleet and Martin Cuykendall of Owasco in said County each being duly sworn before me the undernamed Justice of the peace of said County severally depose and say, That they were well acquainted with James Van Fleet now residing in the town of Haight in the County of Alleghany, during the Revolutionary war, that they are now well acquainted with him That they resided during the Revolutionary war in the same town and vicinity that said James Van Fleet resided in, that during the revolutionary war they have frequently seen the said James Van Fleet in the service of the United States, that they know that he served as a Sergeant in said service, and these deponents each further state that they have examined an affidavit hereunto annexed and this day made and sworn to before Cornelius Cuykendall a Justice of the peace of Cayuga County and that from the personal knowledge each of these deponents have of most of the facts therein stated the deponents severally believe the facts therein set forth and sworn to by the said James Van Fleet are substantially true

Subscribed and sworn to this twenty
fifth day of Sept. 1832 before me
Cornelius Cuykendall Justice of the peace

Daniel Vanfleet
Henry Van Etten
Martin Cuykendall

State of New York

County of Cayuga ss. I Cornelius Cuykendall, Justice of the Peace as aforesaid in and for the County aforesaid do further certify that Henry Van Etten, Daniel Van Fleet and Martin Cuykendall, who have subscribed and sworn to the preceding affidavit are personally known to me, and that their deposition is entitled to full credit. Dated this seventeenth day of June 1833.

Cornelius Cuykendall Justice

State of New York

Cayuga County ss. I Michael S. Myers Clerk of said county do certify that Cornelius Cuykendall Esquire before whom the preceding affidavits purport to have been sworn, was at the date thereof a Justice of the Peace in & for said county, and that I am acquainted with his hand writing and verily believe his name subscribed to said affidavits to be his proper hand writing and signature. Witness my hand & seal of Office this twenty sixth day of September A.D. 1832

M. S. Myers Clerk
pr _____ Myers Dy Clerk

State of New York

Allegany County ss. On this thirty first day of October in the year of our Lord one thousand eight hundred & thirty two, personally appeared before the Judges of the Court of Common Pleas in & for said County in open Court, James Van Fleet aged seventy nine years, two months and twenty six days, who now resides in the town of Haight in said County of Allegany & State aforesaid, who being first duly sworn according to law, maketh oath in order to obtain the benefit of the provisions of the act of Congress, passed June seventh 1832.

That he resided in the town of Goshen in the County of Orange & state aforesaid at the commencement of the Revolutionary war. That in the forepart of the summer of the year 1775, thinks about the commencement of July, this declarant entered the service of the United States in a Company of Militia ~~of~~

~~Militia~~ commanded by Capt. Jones, who was afterwards killed at Beaver Creek on the Delaware River about the 20th of July in the year 1779. Martiness Decker was Lieutenant of the Company, declarant was a volunteer, he was in the service at this time one month, & was employed under Govenor George Clinton in the building of fortifications at Fort Montgomery on the bank of the Hudson River in Orange County. That he again entered the service early in the winter of the year 1776 or 1777 (which of the two years he cannot now recollect, but thinks late in the 1776) under Captain Moses Cartwright & Lieutenant Martiness Decker & Ensign Ephraim Middaugh. Decker was Second Lieutenant, & John Van Tuyl was first Lieutenant of the Company. Colonel Ellison commanded the Regiment. Brigade Major John Hearn formerly commanded declarant's Regiment. Major Hearn lived in Tappan, at this time declarant was marched to Tappan in New Jersey & was at this place when General Washington surprised & captured the Hessians at ~~Washington~~ Trenton. Declarant continued in service at this time about five weeks. Colonel Haghthorn was present with his Regiment. Colonel McClarkson was present with his & there was also present a Regiment composed principally of Dutch commanded by Major Cantine. During this term of service declarant with the troops present had a skirmish with the enemy consisting of tories, & c, who after a few shots retreated, and during this term of service declarant and the troops with him surprised a party of the enemy consisting of Hessians and Regular troops, at a small village about three miles from Hackinsack, who fled and left behind them a quantity of stores consisting of spirits - Brandy, Jamaica Rum, Tea, Biscuits, &c, declarant was this time drafted.

That he again entered said service in the summer of 1777, about the last of June and marched to Ramapo in New Jersey under Captain Sweesey and Lieutenant Tryon in Colonel Gilbert Cooper's Regiment and continued in said service three weeks, declarant was this time drafted, there was a body of regular troops at Ramapo under the command of Captain Black, as declarant thinks, & thinks part of Col. Haghthorn's Regiment was also present. Declarant at this time went with the baggage of some general officers, among whom was General Pearson to Kings ferry. General Sullivan was then at King's ferry with his division of the army. General Stone & some other general officers were present, & while declarant was there General Washington with his army marched in.

That he again entered the service in October 1777, was drafted under the command of Captain Moses & Lieutenant John Van Tuyl, they marched for the purpose of reinforcing the garrison at Fort Montgommery but arrived too late, the enemy having already captured the fort, they then marched to New Windsor on the Hudson River & from thence to Banks bridge on Walkill Creek, declarant was in service at this time about three weeks. Other regiments of militia were present. Govenor George Clinton commanded in person & drew declarant and the other soldiers up in the form of battle, but the enemy did not land, the

American regiment of red coats were at this time present, & declarant & his Major, John Decker was present on the spot when the spy was taken & swallowed the box containing the letter written by General Henry Clinton to General Burgoyne, declarant remembers reading this letter which he thinks was to this effect "I have now the key of the North River - can now go where I please & can meet you any day you please to appoint"

That a few days afterwards he again entered the service, under Lieutenant Stuart, the other officers he does not now recollect, & marched to a place called Murderers Creek, where he was employed in building fortifications, remained in service this time about two weeks, there were several other regiments present, & an Engineer whose name he does not recollect, these works were afterwards abandoned & the troops went to work at West Point.

That he again entered the service under the command of Major John Decker, the other officers he cannot recollect, in the spring of 1778, & marched to Sussex in the State of New Jersey to releive the Inhabitants against or from the Refugees & continued in said service about one week. There were some Regiments of the New Jersey militia present, during the said several terms of service & until the year 1780 or 1781 this declarant was sergeant.

And this declarant further states that in June or July of 1778 this declarant went on a scouting party up the Delaware River under the command of Captain Moses Cartwright & Lieutenant Martiness Decker & was in service one week.

And this declarant further states that from and after the commencement of the fall of 1778 he acted until the close of the war as a minute man in the western part of Orange County to guard that frontier against the Indians, that he very frequently went on scouting parties & acted as an Indian spy, that he was called out daily with his rifle & other implements of war & was during this time under the command of Captain Wilhelmus Westfall & Lieutenant John Van Tuyl, & several other Officers.

And this declarant further states that his service was of such a nature that he cannot specify the precise length of time he was in the field, but he is confident he did serve at the times & at the several places aforementioned at least three years and for the greater part of the time he was sergeant. And this declarant further states that he was born in Manasink (now Goshen in Orange County New York) then in New Jersey, in the year one thousand seven hundred & fifty three on the fifth day of August, that this declarant obtained this information from a record in the family bible & from the parish register kept by the Rev. Mr. _______ in the aforesaid town, that he was living in the town of Goshen, since called Deer Park at the time he entered service, that he continued to live there until the fall of 1822 when he removed to Marcellus, now Skeneateles, in Onondaga County in this state, where he continued until the month of May of the present year, since which time he has lived in the town of Haight as aforesaid.

That he was well acquainted with Henry Van Etten, Daniel Van Fleet, & Martin Cuykendall at Goshen aforesaid, & at Marcellus aforesaid, who now live at Owasco in Cayuga Co. Van Etten & Cuykendall were captains, and all of them have testified as to their personal knowledge of this declarants services during the revolutionary war, & whose affidavit accompanies these papers, that he is acquainted with ~~Westfall~~ George Westfall who now resides in Haight, & who was his near neighbor in Goshen from the close of the Revolutionary War untill declarant left there. Westfall's affidavit accompanies these papers.

Here hereby relinquishes every claim whatever to a pension, or an annuity except the present, & declares that his name is not on the pension roll of any agency in any state. Sworn to & subscribed the day & year aforesaid.

James Vanfleet

I George Westfall residing in the town of Haight do hereby certify, that I have been acquainted with James Van Fleet the person Spoken of in the above writings and he who hath subscribed and sworn to the above declaration

Subscribed and sworn this 30th day of October 1832 before me

H Miller J. P.

George Westfall

State of New York

Allegany County Clerk's Office I, Judson W. Sherman Clerk of said County do certify that Henry Miller Esquire before whom the within affidavit was sworn, was at the time the same purports to have been sworn a Justice of the peace in & for said County duly authorized by law to take the same, that I am well acquainted with his hand writing, and verily believe his signature to the same to be his proper hand writing. In Witness whereof I have hereunto set my hand & affixed the seal of said County at Angelica this 31st day of October 1832.

J W Sherman

And the said Court do hereby declare their opinion after putting the interrogatories prescribed by the War Department that the above named James Van Fleet was a revolutionary soldier and served as he states & that the several papers accompanying this, as this declarants affidavit, Henry Van Etten's, Daniel Van Fleet & Martin Van Fleet's affidavit and George Westfalls affidavit were produced in court.

John Griffin Judges of
Asa Allen the Court
Josiah Utter of Com pleas

I, Judson W. Sherman Clerk of said County do certify that the foregoing ~~is a true copy~~ contains the original proceedings before said Court in the matter of the application of James Van Fleet for a pension In Witness whereof I have hereunto set my hand & affixed the seal of said Court at Angelica this thirty first day of October, 1832.

J W Sherman

State of New York
County of Allegany ss. Personally appeared before me, the undersigned, a Justice of the Peace in and for said County James Van Fleet, who being duly sworn, deposeth and saith that by reason of old age, and the consequent loss of memory, he cannot swear positively as to the precise length of his service; but according to the best of his recollection, he served not less than the periods mentioned below, and in the following grades: For one month commencing in July 1775 I served as sergeant. For one month and seven days commencing in the winter of 1776 or 77 I served as a sergeant. For twenty one days commencing in the summer of 1777 I served as a sergeant. For twenty one days commencing in October 1777 I served as a sergeant. For fourteen days commencing late in 1777 I served as a sergeant. For seven days commencing in the spring of 1778 I served as a sergeant. For seven days commencing in July 1778 I served as a sergeant, during all the preceding terms of service I was second sergeant. For three years commencing in 1778 I served as a minute man and was an orderly sergeant & for such service I claim a pension.

And I further declare that there is no clergyman residing in the Town of Haight with whom I am acquainted whose testimony could be procured as to my character and services. And furthermore this deponent saith not.

Sworn & subscribed this 29th day of July 1833 before me
B. F. Smead J.P.

James Vanfleet

State of New York

Allegany County Clerks Office ss. I, Judson W. Sherman clerk of said County do certify that Benjamin F. Smead before whom the preceding affidavit was sworn, was at the time the same purports to have been done a Justice of the Peace in & for said County ~~and further that~~ duly authorized by law to take the same, & further that I am well acquainted with the hand writing of the said Benjamin & verily believe his signature to the same to be genuine. Witness my hand & the seal of said County this twenty ninth day of July, A.D. 1833.

J W Sherman Clk

War Department
Pension Office
Oct. 23 1833

Sir

The papers of James Van Fleet an applicant for a pension have again been examined In his Supplemental Declaration the claimant has Specified, only four months & seventeen days service in the manner required under note of the printed sheet accompanying his papers in being sent back for amendment. The requisition not having been complied with the case is suspended and the papers placed in file

Anthony S. Chew Esq
Angelica
New York

War Dept
Pension off
Feby 26 34

Sir

In reply to Mr. Chews letter respecting the case of James Van Fleet I have the honor to state that a letter was addressed to Mr. Chew on 23 Oct. suspending the claim which appears to have been recd by him & with which it will be necessary to comply before that decision can be reversed. The allegation of a continuous service of 3 years in the militia can not be ~~allowed~~ admitted and the regulations require each term of an occasional service to be circumstantially recited as which has been communicated to Mr. Chew.

With respect to the complaint of Mr. Raymond in relation to the decision of this office requiring claimants to obtain certificates of service from the archives

_______, it is proper to remark that the regulations call for the best evidence in the power of a claimant to produce. The difficulties in obtaining those certificates are much regretted but there are no means within the reach of the Dept. to remove them. Their letters are herein retd

Hon P. C. Fuller

State of New York

County of Allegany ss. It is hereby certified that satisfactory evidence has been exhibited before me the County Judge & Surrogate of said County being a Court of Record, that James Van Fleet a reputed ~~was a~~ Revolutionary ~~Pensioner~~ Soldier of the United States ~~at the rate of $ per annum and that~~ died ~~on~~ in the month of July ~~day of~~ in the year eighteen hundred and forty one leaving surviving him the following named children viz: the deceased left no widow but left Solomon Van Fleet & Daniel Van Fleet who are now living and are children of the deceased and that the said Solomon Van Fleet & Daniel Van Fleet are the only surviving children of the aforesaid James Van Fleet deceased In Testimony Whereof, I have hereunto set my hand and affixed my seal ~~of office~~ of the Surrogates Court this thirtieth day of April one thousand eight hundred and fifty-three

Lucian P Wetherly Co. Judge & Surrogate of said County

Washington May 9th 1853

Hon L P Waldo

I herewith enclose order of _____ & power of atty, in case of James Van Fleet of New York act 1832 suspd. You will perceive that is declaration is sustained by the affidavits of Henry Van Etten, Danl Van Fleet and Martin Cuykendall all of whom are Revolutionary pensioners and testify most positively to his service

Your obt svt

A H Markland

State of New York

County of Allegany ss. Be it known, that on this twenty third day of November A.D. 1852 before me, the undersigned, a Justice of the Peace in and for the County and State aforesaid, personally appeared Solomon Van Fleet resident of New hudson in said County, and made oath according to law that he is the legal heir of James Van Fleet who was a Soldier in the war of the Revolution, and who has heretofore made application for the benefit of the Pension Act of that he is directly interested as a claimant in said pension and makes this affidavit to be filed with such additional evidence or arguments as my Agent may use in prosecuting said claim. Countermanding and revoking all Powers of Attorney, and authority heretofore given, I hereby constitute and appoint C. W. Bennett, of Washington City, my true and lawful Attorney, to prosecute the claim of James Van Fleet for any amount of Revolutionary Pension or increase of Pension that may be due, and I hereby authorize my said Attorney to examine all papers or documents in relation to said claim on file in the Department at Washington City, or elsewhere, to file additional evidence or arguments, and to receive for my exclusive benefit, the certificate which may be issued for said claim, which certificate I wish made payable at the Washington City, D.C. Agency; to appoint one or more substitutes under him for the purposes herein expressed, and to do all things that I might or could do were I personally present. Hereby ratifying and confirming all that my said Attorney and Agent shall lawfully do in the premises.

Solomon Van Fleet

Sworn to, Subscribed and Acknowledged before me the day and year above written, and I hereby certify that I believe the said Solomon Van Fleet to be directly interested in said claim as set forth, and to be the identical party therein alleged.

Orlin Marsh Justice

State of New York

County of Allegany ss. It is hereby Certified, that satisfactory evidence has been exhibited before me, Clerk of the Court of Allegany County that Solomon Van Fleet the party who has sworn to, and acknowledged the foregoing Declaration and Power of Attorney is the ~~Identical~~ claimant for the reputed soldier aforesaid. I further certify, that Orlin Marsh Esq. before whom the preceding affidavit and Power of Attorney were made and acknowledged, was, at the time of so doing, a Justice of the Peace, in and for said County, and that the signature purporting to be his, I believe to be genuine. In Testimony Whereof, I have hereunto set my hand and affixed my seal of Office, this 24th day of Novr 1852

John J. Rockafellow of said ___________

Pension Office
29th July 1853.

Sir,

The case of James Van Fleet was suspended in 1834 for want of proof of his services, no material evidence has been adduced, since that time, and no reason can be discovered for reversing the decision heretofore made

A. H. Markland Esqr.
Washington D.C.

Power of Attorney to examine Claim

Know all men by these presents that I Solomon Van Fleet of the County of Allegany and State of New York do hereby constitute and appoint Charles C. Tucker of the City of Washington D.C. my true and lawful agent and Attorney to prosecute the claim of James Van Fleet of said County of Allegany for any amount of Revolutionary pension or increase of Pension that may be due under any act or Resolution of Congress. And I hereby authorize my said Attorney to examine all papers and documents in relation to said Claim on file in the departments at Washington City or elsewhere to file additional evidence or arguments and receive the certificate which may be issued for said Claim and to appoint one or more substitutes under him for the purposes herein expressed and to do all things that I might or could do were I personally present. Hereby ratifying all that my said Attorney may lawfully do in the premises. Witness my hand and seal the 24 day of February 1855

In presence of
C. P. Halsey

Solomon Van Fleet

State of New York
Allegany County On this 24 day of February 1855, before me a justice of the peace in and for said County personally appeared Solomon Van Fleet and acknowledged the foregoing power of Attorney to be his act and deed for the purposes therein expressed.

Solomon Van Fleet

State of New York
Allegany County ss. On this 24 day of February 1855 Before me the ~~sub-scriber personally appeared Solomon Van Fleet~~ undersigned a justice of the peace

in and for the county and state aforesaid personally appeared Solomon Van Fleet a resident of said County and made oath according to Law, that he is the identical Solomon Van Fleet who executed the foregoing power of Attorney and that he is interested in said Claim and makes this affidavit to be filed with such additional evidence and arguments as my said Attorney may act in prosecuting said Claim

Sworn to and subscribed before me this 24 day of February 1855

Solomon McKeen J.P.

State of New York

Allegany County ss. I James M Mott Clerk of the County courts in said County of Allegany do hereby certify that Solomon McKeen before whom the foregoing acknowledgment and affidavit were made and who has thereunto signed his name was at the time of so doing a justice of the peace in and for the county and state aforesaid duly commissioned and sworn, and that all his official acts as such are entitled to full faith & credit In witness whereof I have signed my name and affixed my official seal this 7 ~~day of February~~ day of March AD 1855

James M Mott Clerk

---

44 Morgan Rd., Buffalo, N.Y.
Sept 22, 1938

Veteran's Administration
Wash. D.C.
Dear Sir

I would greatly appreciate any information concerning the Revolutionary War or pension record of a James or Jacobus Van Fleet. He was born about 1753 and died July 7, 1841. His service was rendered in and about Deerpark & Minnisink Orange Co. The name may have been spelled Van Fleet, Van Vliet or Van Vleat. Could I also obtain your records concerning Major John Decker? He was 2nd Major of the Goshen Regiment in Orange Co., of which Wm. Allison, Esq., was Colonel.

Thanking you for any information I may receive, I am greatfully yours

Mrs. N. E. Metz
44 Morgan Rd
Buffalo, N.Y.

November 2, 1938.
BA-J/ady
James Van Fleet
R.10864

Mrs. N. E. Metz
44 Morgan Road
Buffalo, New York
Dear Madam:

Reference is made to your letter relative to James Van Fleet and Major John Decker who served in the Revolutionary War.

Revolutionary War data furnished by this office are obtained from claims made to the United States for pension and bounty land based upon military service of soldiers in that war.

A careful search of the records fails to show such a claim on file on account of the services of John Decker as described by you, all spellings of the names searched.

The data contained herein were obtained from the papers on file in the Revolutionary War claim for pension, R. 10864, based upon the military service of James Van Fleet in that war.

James Van Fleet stated in 1832 that he was born August 5, 1753, in "Manasink (now Goshen in Orange County New York) then in New Jersey." The names of his parents are not shown.

In the fall of 1822 he moved to Marcellus; that part which was later called Ska neateles, Onondaga County, New York, and in May 1832 he moved to Haight, New York.

James Van Fleet applied for pension September 25, 1832, at which time he was a resident of Haight, Allegany County, New York. He stated then that while a resident of Goshen, that part which was later called Deer Park, New York, he enlisted in July 1775, served at various times, about four months and three weeks in all, as sergeant in the New York troops under the following officers, Lieutenant Stuart, Captains Jones, Moses Kortwright, Swazy, and Major John Decker, and Colonels Allison and Gilbert Cooper. That from the fall of 1778 to the close of the war he served as orderly sergeant and Indian spy in Captain Wilhelmus Westfall's New York Company in the western part of Orange County and guarded that frontier against the Indians.

His claim for pension was not allowed as he failed to furnish proof of his service as required by the pension law.

He died in July 1841, place not stated, and left no widow, but the following children survived him, Daniel Van Fleet and Solomon Van Fleet, who were living in 1853. In 1852 said Solomon, was residing in New Hudson, Allegany County, New York, and was living in said county in 1855, his age not given.

In 1832 one Daniel Van Fleet was living in Owasco, Cayuga County, New York, his age or relationship to soldier not stated. There are no further family data.

Very truly yours
A. D. Hiller
Executive Assistant to the Administrator

Rejected

Revolutionary War Pension and Bounty Land Warrant Application Files
Series M-804; Roll # 2445

Daniel Vanfleet
New York (S.14,763)

BRIEF in the case of Daniel Vanfleet of Cayuga Co
in the State of N York
(Act 7th June, 1832.)

1. Was the declaration made before a Court or a Judge? open Court
2. If before a Judge, does it appear that the applicant is disabled by bodily infirmity?
3. How old is he? 77 years
4. State his service, as directed in the form annexed.

Period.	Duration of Service			Rank.	Names of General and Field Officers under whom he served.
	Years	Months	Days		
April 1776		3		Private	officers not remembered
July 1776		3		Private	Capt Tilford
Spring 1777		9		Sargeant	Capt Faulkenter Genl Clinton
Spring 1778		1		Sargeant	Capt Cudderback
		about 4 in 1779			Capt Westfall Col Tuston
		about 3 in 1776			
Period.	Duration of Service			Rank	Names of General and Field Officers under whom he served.

					many other times and officers not stated

5. In what battles was he engaged? none

6. Where did he reside when he entered the service? Goshen N York

7. Is his statement supported by living witnesses, by documentary proof, by traditionary evidence, by incidental evidence, or by the rolls? By living witnesses

8. Are the papers defective as to form or authentication? and if so, in what respect? Correct

Certify that the foregoing statement and the answers agree with the evidence in the case above mentioned.

Richard Cutts Examining Clerk.

State of Ohio

5th September AD 1832

Huron County ss. Personally appeared before me Timothy Baker one of the Judges of Huron Com Pleas Captain John Brooks of Florence in said county & State aged Seventy six years and after being duly sworn did testify & declare That in the Revolutionary War he was well acquainted with Daniel Vanfleet now of Owasco in the County of Cayuga State of New York. That he always understood & has been informed that in the year AD 1776 said Vanfleet was drafted & served in the Militia of the State of New York at Fort Constitution & West point for six months & was assisting in building said Forts that the Deponent saw him at West point in October 1776 Serving in the service of the American cause. That he did not know his company or Regimental officers but he knew him to be in the Brigade commanded by Brigadier James Clinton that in the year 1777 the said Vanfleet served in the same Regiment with the Deponent for nine months at Fort Montgomery and Fort Clinton in the militia in a company commanded by Captain John Felton James Wilkins first Lieutenant William _____second Lieutenant Ensign John English. That he knew him to be at Hackensack at Ramapo at Paramus at the English Neighborhood. That said Vanfleet was in a Battle with the Indians up the Delaware river at Lackawaxen when the Deponent was with him & the Deponent was wounded in three places in his Body where Colonel Thurston was killed. That after the above the said Vanfleet remained on the Frontiers in Gumaugh Fort from AD 1781 to the close of the Revolutionary War he served under Jacob Dewitt Captain of the company Robert Cooke & Samuel King were first & second Lieutenants Timothy Coleman Ensign

John Brooks

State of Ohio

Norwalk Sept 5th 1832

Huron County ss. The within affidvit subscribed & sworn to by the within named John Brooks before me

Timothy Baker Judge of Huron Com Pleas

P.S. the word has - 7th Line, & the words, the deponent, 23rd Line interlined before signing

State of Ohio

Huron County ss. I David Gibbs Clerk of the Court of Common Pleas in said County do certify that Timothy Baker Esq. is an associate Judge of said Court & the above signature above purporting to be his is genuine. In testimony whereof I have hereunto subscribed my name & affixed the Seal of said Court at Norwalk this 5th day of September AD 1832.

D. Gibbs Clerk H. C. P.

State of New York

Cayuga County ss. On this 24th day of September AD 1832 personally appeared in open Court before the Court of Common Pleas in and for the County of Cayuga & State of New York now sitting Daniel Vanfleet of Owasco in the County of Cayuga in said State aged seventy seven years who being first duly sworn according to law doth on his oath make the following declaration in order to obtain the benefit of the act of Congress passed June 7th 1832. That he entered the service under the following named officers and served as herein stated That about the first of April AD 1776 then being a resident of the town of Goshen in the County of Orange & State of New York he entered the service of the United States but, (but the names of his company or field officers he does not now remember) that he marched to a place called fort Constitution on the Hudson River where he was stationed a term of three months that his services consisted of building Barracks & cleaning and making preparations to build fortifications and in performing the necessary duty of guard. He further declares that about the last of July in the same year he again entered said service in the company of militia commanded by Captain Tilford as he believes and Lieut Faulkenter for a term of three months that he marched with said company to fort Constitution at and near which place he continued to serve in said Company during the whole of said term. That he was employed in building fortifications and in keeping guard during said term. That he was employed part of the time in building fortifications on the West Point side of the Hudson River. That whilst at said place he saw frequently Gen James Clinton and thinks that he was under his command during said term. He further declares that very early in the spring of the year 1777 about the month of March as he believes he again entered said service in the company commanded by Captain John Felton or Faulkenter, James Wilkins first & William Rose second Lieutenant and John English Ensign that he marched ~~with said company~~ to fort Montgomery at which place and fort Clinton he continued to serve as Orderly Sargeant under the command of Gen George Clinton & James Clinton until fort Montgomery was taken by the enemy on the 17th October following. That fort Montgomery was commanded by Gen George Clinton and fort Clinton by Gen James Clinton during ~~said~~ most or all of said term. That when fort Montgomery was taken he escaped therefrom and returned home, spent two or three days there and returned to the Hudson River and entered the militia company of Captain Abraham Cudderback where he continued to serve as Sargeant over one month as he believes. That his whole term of service during this season was nine months according to his best recollection. That during said month the militia were all called out to watch the motion of the British shipping as it sailed up the the Hudson. That he was stationed during said month in the vicinity of Newburgh & New Windsor. That the General place of rendezvous was the house of Gen Geo. Clinton between Newburgh & New Windsor He further declares that early in the spring of the year 1778 he volunteered to hold himself in readiness to go out in said service upon the frontiers of Orange County during the year. That during the fore part of the season he was under the command of Captain Abraham Cudderback, and occasionally during the summer and fall under Captain Abraham Westfall. That during said term he was almost constantly in said service as a guard upon the frontiers, in scouting parties and as an Indian spy and that his whole time was spent as a guard upon the frontiers when not engaged in scouting parties and as an Indian spy. He further declares that immediately after Minisink was attacked & burned by the Indians and torys under the command of Brandt, a part of the militia of Orange County were collected and he among the number, to pursue them. That he entered the company of Captain Abraham Westfall in Col Tustons Regiment, and under Major Samuel Meeker and marched with said detachment to the mouth of the Lackawack or Lackawaxen River where they were attacked by the Indians and after an engagement were defeated with considerable loss. That Col Tuston was killed in said battle. That this was in the month of July 1779. He further declares that soon after this he was drafted and entered said service in the company of Captain Abraham Westfall as a guard upon the frontiers. That he continued to serve until the commencement of winter a term of about four months. He further declares that from this time until the close of the war he was frequently in said service as a guard upon the frontiers, in scouting parties and as an indian spy. That the country was kept in a constant state of alarm by the frequent incursions of the enemy, so that the inhabitants attended to no kind of ~~business~~ labour except so far as to provide for the actual necessaries of life. That they were accustomed to collect in the Piquet forts along the frontier for safety and from thence send out parties to range the woods and give notice of the approach of the enemy, and that he was frequently engaged in said excursions. He further declares that in the fall of 1776 as he believes he entered said service in the Regiment commanded by Major Philips and Col McClaughrey. That the name of his company officers he does not now remember. That he marched to Hackensack at & near which place he continued to serve until the following winter a term of about three months. That the enemy lay at Hackensack at this time, which place they evacuated about Christmass, from thence they retreated to the English neighbourhood and from thence to New York as he believes. That he did not return until after snow had falen. He further says that his memory fails him so that he is unable to state the names of his officers as correctly as he could wish, but that he has named them according to his best recollection. That he recollects the different tours distinctly as above stated and their length. ~~but~~ That he never received a written discharge from said service and has no documentary evidence thereof except the accompanying. That he saw Col Dubous with his regiment at Hackensack. That he saw Gen Putnam at fort Montgomery. He hereby relinquishes all claim to a pension except the present and declares that his name is not on the pension roll of any state or territory

Sworn & subscribed the day & year aforesaid
Asahel Fitch Judge

Daniel Vanfleet

State of New York

Cayuga County ss. On this 24th day of September AD 1832 personally appeared in open Court before the Court of Common Pleas in and for the County of Cayuga now sitting Martin Cuykendall of Owasco in said county who being duly sworn says that he is Sixty nine years of age, that he is well acquainted with Daniel Van Fleet of Owasco in said County and was well acquainted with him in the time of the Revolutionary War. That they both resided in the County of ~~Westchester~~ Orange in the town of Goshen & State of New York at that time & immediately on the frontiers of Orange County. That the said Daniel was engaged in service most part of the time during the war as he was informed and verily believes. That he resided on the frontiers and was engaged during the last years of the war frequently as a guard, in scouting parties, and as a spy to watch the motions of the Indians. That he judges of said services from having seen said Daniel frequently engaged, and from the situation of the country, it being exposed for about the distance of fifty miles, through the Countys of Orange and Ulster to the depredations of the Indians, and being guarded for the most of the time by the militia.

Sworn and subscribed before me date above
Asahel Fitch Judge

Martin Cuykendall

We Henry Van Etten & James Van Fleet do certify that we are well acquainted with Daniel Van Fleet mentioned in the foregoing affidavit & were well acquainted with him during the Revolutionary war that we have heard the said affidavit read and concur with the said Martin Cuykendall in the statement therein made

Sworn & subscribed in open Court the day & year aforsaid James Vanfleet
Asahel Fitch Judge

Henry Van Etten

And the Court certifies that the aforsaid witnesses are men of truth and veracity

Asahel Fitch Judge

We, Archibald McNiel a clergyman residing in Owasco in Cayuga County and Simeon Swartout residing in Owasco hereby certify, that we are well acquainted with Daniel Van Fleet who has subscribed and sworn to the preceding declaration; that we believe him to be 77 years of age; that he is reputed and believed, in the neighbourhood where he resides, to have been a Soldier of the Revolution, and that we concur in that opinion.

Sworn, and subscribed, the day and year aforesaid.
Asahel Fitch Judge

Archibald McNiel
Simeon Swartout

And the said Court do hereby declare their opinion, after the investigation of the matter, and after putting the interrogatories prescribed by the War Department, that the above named applicant was a Revolutionary Soldier, and served as he states. And the Court further certifies, that it appears to them, that Archibald McNiel who has signed the preceding certificate, is a Clergyman resident in Owasco and Simeon Swartout who has also signed the same, is a resident in Owasco and is a credible person, and that their statement is entitled to credit.

Asahel Fitch Judge

I, Michael S. Myers Clerk of the Court of Common Pleas in and for the County of Cayuga and State of New-York, do hereby certify, that the foregoing contains the original proceedings of the said Court in the matter of the application of Daniel Van Fleet for a pension. Witness my hand and the seal of said Court the 24th day of September AD 1832

M. S. Myers

State of New York

Cayuga County ss. Personally appeared before me the undersigned first Judge of the Court of Common Pleas in and for said County Daniel Van fleet who being duly sworn says that by reason of old age and the consequence of old age and ~~the~~ consequent loss of memory he cannot swear positively as to the precise length of his services but according to his best recollection he served not less than the periods mentioned below and in the following grade. For nine months I served as Sargeant, and for two years and five months I served as a private soldier and for such service I claim a pension. That in the spring of the year 1780 he was a soldier in the company of militia commanded by Captain Wilhelmus Westfall, first Lieut Elijah Van Auken & second Lieut Levi Van Etten, in the Regiment commanded by Col Hatfield, ~~that said~~ & Major John Decker that said company & Regiment lived immediately upon the frontiers and was left there to guard the same that he was a soldier in said company for upwards of two years and during the years 1780 & 1781 about eight months during the spring summer & fall of each year the said company & Regiment were kept constantly in service guarding said frontiers _____ scouting parties from said company & Regiment were kept constantly in service ranging the woods to give warning of the approach of the enemy. That he was frequently engaged in said scouting parties and when not so engaged he was stationed on the frontiers to guard the lines, and that his term of service during said two years would amount to sixteen months as he believes being eight months in each year He further says that he was born in the town of Goshen in the County of Orange & State of New York in the year 1756, and has no record of his age as he knows of. That when he was first called into said service he resided in the County of Ulster in said State of New York & continued to reside there until the year 1778 when he removed to Orange County in said State where he continued to reside until the year _____ when he removed to the County of Cayuga where he has resided ever since. That he was drafted, at each time that he entered said service as set forth in his declaration except the service performed during the years 1780 & 1781. That whilst in said service he saw Gen George Clinton, James Clinton Major Hatfield, Col Duboys and others that ~~he~~ when fort Montgomery was taken by the enemy he served in company with continental troops but what particular Regiments he does not now remember. That his whole term of service particularly set forth excepting the two years service in 1780 & 1781 amounts to nine months as a Sargeant and one year and one month as a private soldier

Sworn & subscribed the 13th day of April AD 1833
J. L. Richardson first Judge of Cayuga
Daniel Vanfleet

I certify my opinion that the said Daniel Van fleet is a man of truth and veracity

J. L. Richardson first Judge of Cayuga

State of New York
Cayuga County ss. I Michael S. Myers Clerk of the Court of Common Pleas in and for said County do certify that Joseph L Richardson before whom the annexed affidavit was made on the 13th day of April 1833 was on the day of the date of the same first Judge of said Court. That I am well acquainted with the hand writing of the said Judge and believe his signature to the said affidavit to be genuine Witness my hand and seal of office the 13th day of April 1833.

M. S. Myers Clerk

New York
Daniel Vanfleet of Owasco in the State of New York who was a Pr. & Sergt. in the compy commanded by Captain   of the Regt. commanded by Col. Tarrton in the New York line for 9 months Sergeant 15 months Private Inscribed on the Roll of New York at the rate of 95 Dollars 00 Cents per annum, to commence on the 4th day of March, 1831. Certificate of Pension issued the 22 day of July 1833 and sent A Gould Auburn N.Y.

Arrears to the 4th of Mar '33	$180.00
Semi-anl. allowance ending 4 Sep '33	47.50

$227.50

Recorded by Jno. _____ Clerk,
{Revolutionary Claim, _____ 1832.}
Book E Vol. 4 Page ___

State of New York

Cayuga County ss. On this twelfth day of May 1835 before me Joseph L Richardson first Judge of the Court of Common Pleas of said County appeared Daniel Vanfleet of Owasco in said County who being duly sworn doth on his oath make the following declaration in order to be reinstated upon the pension roll of the United States under the act of June 7th 1832. That about the month of May 1775 he entered the service of the United as a volunteer for a term of five months, that he marched to a place on the hudson river called fort Constitution, directly across the said river from West Point where he continued to serve during the whole of said term when he was dismissed. That he was engaged in clearing ~~away~~ the ground & fitting it for building barracks for soldiers and in performing duty as guard to protect the workmen whilst at work upon the said barracks. That when he arrived at said fort one building for a barrack had been raised and in a course of completion and when he left four barracks were in progress. That he does not now remember the names of any of his officers. That they were then strangers to him and their names he cannot now recall to mind. That it is his impression that there was stationed there at the same time one Capt Tilford and Major Hardin and that he was under their command at times but he is not certain

He further says that in the spring of the year one thousand ~~eight~~ seven hundred and seventy six he again entered said service as a volunteer as he believes in the company of militia commanded by Capt Abraham Cudaback for a term of three months, that he marched in said company to West point where he was stationed during the whole of said term when he was dismissed & returned home. He further says that he does not now remember with certainty, the names of any of his field officers but is of the impression that there was a Major Wood in command there at the time. That he was employed during said term in keeping garrison and building the fort at West point. That about the tenth of July 1776 he again entered said service as a volunteer for a term of five months in a company of militia but the names of his company officers he does not now remember in the Regiment commanded by Col McClaughery and Major Philips, that he marched first to Paramus about 15 miles from Hackensack at & near which place he was stationed until ~~the~~ about the month of December when said Regiment marched to Hackensack in order to attack the British stationed there. That the enemy fled upon our approach and we entered Hackensack. That in the morning after our arrival at Hackensack we were joined by the Regiment under the command of Col Duboys. The ~~same~~ same evening after our entry into Hackensack we returned to Paramus. That about the 25th of December we marched ~~to~~ towards the English neighbourhood to drive the enemy from that place. That when about five miles from that place being informed that the enemy had been reinforced we retreated back to Paramus where he remained until he was dismissed and returned home where he arrived on or about the 10th day of January following having served six months, one month longer than the time for which he entered service That early in the spring of the year 1777 he again entered said service as an orderly sargeant in the company of Militia commanded by Capt ~~Abraham Cudaback in Col McClaugherys Regiment~~ for a term of three months. That he marched to fort Mongomery where he was employed in keeping garrison and repairing said fort during said term That he does not remember the name of his Colonel but remembers that he was under the command of Gen James Clinton when in the fort That said term of service closed about the first of June when he returned home That he had been at home but a few days when he again entered said service in the company commanded by Capt Abraham Cudaback in the Regiment commanded by Col McClaughery, that he marched to fort Montgomery where he remained until said fort was taken by the enemy, when he in company with a number of others escaped from said fort and he went home, where he remained one day to have his clothes washed & prepared and then returned and joined his said company at New Windsor, from thence he marched to Murders creek from thence back to New Windsor and from thence to New Burgh, where he remained until the British returned from Esopus and until he was dismissed about the last of November following, having been in service about five months and fifteen days as he believes, that he served during this time as Sargeant. That this declaration is intended to be explanatory of one now on the files of the pension office under which he has received a pension. That his whole term of service as above stated he believes amounts to fourteen months as a private soldier and ~~and~~ eight months and a half as a Sargeant. That he has set forth his services agreeably to the requirements of the commissioner of pensions according to his best recollection and believes should any discrepancy appear between this and his former declaration that this is correct in as much as he has had more time to reflect upon his said service since making out his first declaration

Sworn this 12th day of May 1835 before me
J. L. Richardson first Judge of Cayuga

Daniel Vanfleet

I certify my opinion that said Daniel Vanfleet is a man of truth and veracity and that his foregoing statement may be relied upon

J. L. Richardson first Judge of Cayuga

Cayuga County Clerk's Office ss. I certify that Joseph L. Richardson Esquire is first Judge of the Court of Common Pleas in and for said county, duly commissioned and sworn, and that I am well acquainted with his hand writing, and verily believe that his signatures to the above affidavit and certificate are genuine. Given under my hand and the seal of said County the twelfth day of May A.D. 1835

M. S. Myers Clerk

Auburn N.Y. June 8th 1835

Sir

Enclosed I send another declaration of Mr Daniel Van Fleet explanatory of his supplemental affidavit agreeably to the direction contained in yours of the 28th May last He has also set forth 2½ months more service which he did not then remember

I hope the papers will now be considered satisfactory, to entitle him to be reinstated on the pension roll. Please send returns to me

Yours Respectfully
A Gould

Hon J L Edwards

Daniel Vanfleet Supplemental afft

State of New York

Cayuga County ss. I Daniel Vanfleet of Owasco in said County do hereby make the following declaration explanatory of the circumstances under which I served in companies other than that of Capt Abraham Cuddaback from the year 1776 to the ~~year~~ 20th of July 1779. In my tour of service to Hackensack and other places stated & set forth in my supplemental affidavit I served under Col McClaughery & Major Philips two of the field officers of the Regiment to which Capt Cuddabacks company belonged. At this time according to my best recollection a call was made for a certain number of men from said Regiment to meet which call said Regiment was called together, and all of the soldiers present who would were requested to volunteer to perform said service. And if a sufficient number did not volunteer the regular steps were taken to procure the requisite number of men. I volunteered in this service and I think that said Capt Cuddeback neither volunteered or was drafted at this time, but the reason of his staying at home I do not now remember. That Capt. Cuddebacks company was always as I believe at home and not on duty when I entered service in companies other than his. And when at home and having no call to active service in my own company I was accustomed to enter service as a volunteer in other companies as stated in my said supplemental affidavit. I dont remember that I was ever classed, but I entered service as a volunteer most comonly, when a call was made for soldiers. And this was the manner in which my services were performed. In the summer of the year 1779 I removed from the beat of said Capt Cuddebacks company a distance of about five miles from my former residence, into the beat of Capt Westfalls company in the same Regiment. Soon after my said removal I was called into service in said company and was stationed on the Neversink river about one mile above its mouth where I continued to serve two months and a half as a private soldier, and as a guard upon the lines. The last mentioned service was not stated and set forth in my said supplemental affidavit, having been omitted through forgetfulness. That Capt. Cuddebacks company was a company of militia and only bound to perform its regular tours of service as other militia companies in that secction, so that the soldiers of that company were much in the habit of performing service in other companies when not called upon to serve in their own either as substitutes ~~or vol~~ for others or as volunteers.

Sworn this 8th day of June 1835 before me
J. L. Richardson first Judge of Cayuga

Daniel Vanfleet

I certify my opinion that said Daniel Vanfleet is a man of truth and veracity
J. L. Richardson first Judge of Cayuga

State of New York

Cayuga County Clerks Office ss. I certify that Joseph L. Richardson Esquire is first Judge of said county; and that I am well acquainted with his hand writing and verily believe that the signatures subscribed to the foregoing affidavit & certificate, purporting to be his, are genuine. Given under my hand and the Seal of said county, this eighth day of June A.D. 1835

M. S. Myers Clerk

Auburn N.Y. Sept 30 1835

Sir

In pursuance of the suggestion contained in yours of the 17th June last on the subject of the claim of Daniel Vanfleet I have procured another afft from him which he says gives all the explanation which he is able to give. You will please examine it. I think there must have been some missapprehension upon your part in reference to the facts contained in his previous declarations. He has no recollection of having made the statements which you say he has in your letter He says that his three days service in 1776 of which you find evidence must have been upon a scout or in case of alarm but he has no recollection of it.

Will you please reexamine the matter and allow him so much pension as you think him entitled to He is aware that his age & forgetfulness will be the means of depriving him of the benefits of services which he knows he has performed but he must submit to the misfortune & be content with whatever you choose to give him

Very Respectfully Your Obt Servt
A Gould

Hon J L Edwards

State of New York

Cayuga County ss. Daniel Vanfleet being duly sworn makes the following declaration explanatory of his former declarations and statements in order to obtain the benefit of the act of June 7th 1832 and to be reinstated on the pension roll under said act. That he never was in a company of minute men commanded by Capt William Tilford, at any time neither has he any recollection of having ever stated that he was in a company of that kind. He thinks that in his former declarations he may have stated that he thought whilst stationed at Fort Constitution previous to his enrolment in Capt Cudebacks company he was part of the time under Capt Tilford. That he remembers seeing Capt Tilford who lived a distance of forty miles from the resident of him the said Vanfleet. He says he does not recollect distinctly that he ever was under the command of said Tilford That in the year 1776 previous to his commencing his service of about five months as heretofore stated by him he might have been out a short time in Captain Cuddebacks company but he has no distinct recollection of being out. That he has forgotten the name of the Captain and other company officers under whom he performed said service That they were strangers to him and did not as he believes belong to his company or Regiment whilst at home. That he volunteered about the month of July out of Capt Cuddebacks company, and as he believes was not embodied or enroled in any company until he arrived at Paramus, that he remembers the names of his field officers as stated in his last declaration He further says that he has no recollection that Capt Faulkner had command of a company of nine months men in 1777. That he has no distinct recollection of being in Capt

Faulkners company during this year 1777 but according to his best recollection he served about five months in said Capt Cuddebacks company during this season, altho the latter part of the time he was not under his immediate command. That his service continued until about two weeks after Burgoine was taken according to his best recollection.

Sworn this 19 day of September 1835 before me
John P Hulbert Justice of the Peace

Daniel Vanfleet

I certify my opinion that said Vanfleet is a man of truth and veracity
John P Hulbert Justice of the Peace

State of New York
Cayuga County Clerk's Office ss. I certify that John P Hulbert Esquire is a Justice of the Peace in & for said county; and I am well acquainted with his hand writing and verily believe that his signatures subscribed to the foregoing affidavit & certificate are genuine. Given under my hand & the Seal of said county this 19th day of September A.D. 1835

A. L. Miller Dep Clerk

Auburn N.Y. Oct 14 1835

Sir

Your communication of the 10th instant in relation to the claim of Daniel Vanfleet is received. In answer I would state that from all I can learn Mr Van fleet will not be able to furnish and further proof than he has done. He can furnish more of the same kind. If he is thought by the department to be entitled to any thing you will please issue his certificate for such amount as you think him entitled to

He thinks he has done all he can do He has attempted to give all the information which he was able. What he has done has been done honestly. If he has erred it has been from failure of memory

Please give an answer to this as soon as convenient

Yours Very Respectfully
A Gould

Hon J L Edwards

Washington June 11, 1836

Sir

I have the honor to transmit the enclosed, which please return with your answer as soon as convenient

I have the honor to be Respectfully your obt servt
U. F. Doubleday

Hon J. L. Edwards

Auburn June 4, 1836

Dear Sir

Mr Daniel Vanfleet of Owasco was placed on the pension roll under the act of 1832. In Feby 1835 his pension was suspended on account of some supposed defect in his original application, and further & explanatory affidavits were called for which were furnished. They were forwarded as will appear from the letter enclosed. The rolls of his company it is said do not agree with his statement and therefore suspicion is cast upon his claim altho his proof is very full The old gentleman is far advanced in life and his memory much impaired & his ideas confused on account of age and debility. The old man says that the discrepancy between his statement and the rolls is occasioned by the fact that there was another man of his name (Daniel Vanfleet) in the same company who served occasionally Will you have the goodness to examine the subject and call upon Mr Edwards & see what can be done. If you can bring about no arrangement for the allowance of the claim please obtain copies of all the declarations made by him, and send them to me in order that we may see where the difficulty is. I send the letters in order that you may understand the business, which I wish you to pursue & return to me

Please ascertain the precise difficulty in the case and see what can be done. M Cuykendall Esq of Owasco is one of his witnesses

Respectfully yours &c
A Gould

Hon U F Doubleday

State of New York

Cayuga County ss. Adam Branon of Owasco in said County being duly sworn saith that he is aged Seventy seven years on the second day of February last. That he is well acquainted with Daniel Van Fleet now of the town of Lysander in the County of Onodaga and State aforesaid, and that he knew the said Daniel Van Fleet in time of the Revolutionary war. That deponent was a private Soldier in Col. Dubois's Regiment during three years of the Revolutionary war. That in the year 1777 deponent was stationed in Fort Montgomery on the bank of the Hudson River from the fifth day of April (the day of his enlistment) untill about the fifth day of October of the same year, that is, untill the said Fort was captured by the British army That said Daniel Van Fleet served that year in Fort Montgomery in Col. McGlourghery's Regiment of Militia, deponent recollects that he saw the said Daniel Van Fleet about the time that Col. McGlourghery's Regiment arrived at the Fort which deponent thinks was in the month of May in the same year, that deponent then became acquainted with the said Daniel Van Fleet, and that the said Daniel Van Fleet remained ~~remained~~ there doing duty as a Militia man untill the Fort was captured, deponent saw him there on the day the Fort was captured, and further deponent saith not.

Subscribed and sworn to
January 5th 1837 before me
Cornelius Cuykendall Justice of the peace

his
Adam X Branon
mark

I Certify That I am well acquainted with Adam Branon, the above deponent, and that he as a witness is entitled to credit.

C. Cuykendall

Cayuga County Clerk's Office ss. I, B Franklin Hall, Deputy Clerk of said county do hereby certify that Cornelius Cuykendall Esquire before whom the within affidavit was taken, was at the time of taking such affidavit a Justice of the peace for said county, commissioned and sworn and duly authorized to take the same and further, that I am well acquainted with the hand writing of said Justice and verily believe his signature within to be genuine. Given under my hand and the Seal of the Court of Common Pleas of said county this 25th day of September 1837.

B Franklin Hall Deputy Clerk

State of New York
Orange County ss. William A. Cuddeback of the town of Deerpark in said

county, being duly sworn, saith that he is seventy seven years old, that he was well acquainted with Daniel Van Fleet, formerly of the town of Mamacating in ulster county, but now of the county of Onondaga and town of Lysander, and reccollects of the said Daniel Van Fleet starting out on two towers, to the forts on the north river near west point. But more particularly in the spring of the year of 1777, The said Daniel Van Fleet, started with his father Capt. Abraham Cuddeback to go to Fort Montgomery on the north river and acted as a serjent in his company, and reccollects that said Van Fleet was not at home again until after the fort was taken by the British in October, and then was home only a day or two, when he again started with others to the north river near Newburgh, that said Daniel Van Fleet's family lived but a short distance from his fathers house, and that said deponant frequently assisted Van Fleet's family when he was absent on duty, and that said Van Fleet was out frequently during the war in scouting parties on the frontiers, and also reccollects that said Daniel Van Fleet started with his father Capt. Abraham Cuddeback, in July 1779, in pursuit of the Indian warrior Brant, the day before the battle of Minisink, at Beaver Brook, and that after the battle he returned, and was out on duty he thinks most of the time, from that time to the close of the war, and further saith not.

Subscribed and sworn to October 2nd 1837 before me
William Cuddeback Justice of peace

William A Cuddeback

I certify that I am well acquainted with William A. Cuddeback the above deponant, and that he as a witness is entitled to credit

Charles C Boyd Jestus of peace

★

State of New York,

Orange County, ss. I, Lebbeus L. Vail, Clerk of the Court of Common Pleas, in and for the County of Orange, Do Hereby Certify, That William Cuddeback before whom the foregoing affidavit was subscribed and sworn, was at the time of taking the same, a Justice of the Peace in, and for said county: and further, that I am well acquainted with the hand writing of the said Justice and verily believe that the signature to the said affidavit is genuine. In Witness Whereof, I have hereunto set my Hand, and affixed the Seal of the said county, the 12th day of October 1837

Lebbeus L. Vail, Clerk

★

State of New York
Orange county ss. Jacobes Swartout of the Town of Deerpark in said County being Duly sworn saith that he is Eighty three years of age that he is was well acquainted with Daniel Van Fleet formerly of the town of Mamakating in Ulster County but now of the county Onondaga and town of Lysander and recolects of the said Daniel Van Fleet doing service on the frunteers in the town of Deerpark and on the Deleware river and at Cosehehten in the years of 1775 & 1776 and I and Daniel Van Fleet went on a tower as sagering to Fort Montgomery on the Hudson river in the Spring of 1777 I knew Daniel Van Fleet to go with the menitia on a tower above New Burgh a few Days after Fort Montgomery was taken how long he staid thare I do not know I and Daniel Van Fleet went a tower a sagering in the State of New Jersy we was thare severl months in the Latter part of the summer and winter of the year of 1778 I think but I may not be corect about the year and we left about the first of January and came home I hink our Captain was By the name of Falkner we was stacioned at paramus in New Jersy we Daniel Van Fleet and myself went to Fort Montgomery under Capt. Cuddeback
Subscribed and sworn October the 3rd 1837 before me Jacobus Swartwout
Charles C Boyd Jestus of the peace

I certify that I am well acquainted with Jacobes Swartwout the above deponent and that he as a witness entitled to credit

Charles C C Boyd Jestus of the peace

State of New York,
Orange County, ss. I, Lebbeus L. Vail, Clerk of the Court of Common Pleas, in and for the County of Orange, Do Hereby Certify, That Charles C Boyd before whom the foregoing affidavit was subscribed and sworn, was at the time of taking the same, a Justice of the Peace in, and for said county: and further, that I am well acquainted with the hand writing of the said Justice and verily believe that the signature to the said affidavit & Certificates is genuine. In Witness Whereof, I have hereunto set my Hand, and affixed the Seal of the said county, the 12th day of October 1837

Lebbeus L. Vail, Clerk

State of New York
Orange County ss. Thomas Van Etten of the Town of Deerpark in said county being duly sworn saith that he is sixty nine years ~~old~~ of age that he was well

acquainted with Daniel Van Fleete formerly of the Town of Minisink County of Orange but now of the County of Onondaga and Town of Lysander and recolects of the said Daniel Van Fleete duing duty as a minute Man in the fall of ~~seventy nin~~ _____ war in Seventeen hundred seventy nine for the space of three monts said Daniel Van Fleete was incampe in said deponants Mothers house and asisted in bilding a fort and the said Daniel Van Fleete lived on the este side of the Shawangunk Mountain at the time he asisted in bilding the fort

Subscribed and sworn Oct 3rd 1837 before me
Cornelius Gray Justice

Thomas Van Etten

I certify that I am well acquainted with Thomas Van Etten the above deponant and that he a witness intitled to credit.

Cornelius Gray Justice

State of New York,
Orange County, ss. I, Lebbeus L. Vail, Clerk of the Court of Common Pleas, in and for the County of Orange, Do Hereby Certify, That Cornelius Gray before whom the foregoing affidavit was subscribed and sworn, was, at the time of taking the same, a Justice of the Peace in, and for said county: and further that I am well acquainted with the hand writing of the said Justice and verily believe that the signature to the said affidavit & certificate is genuine. In Witness Whereof, I have hereunto set my Hand, and affixed the Seal of the said County, the 12th day of October 1837

Lebbeus L. Vail Clk

State of New York
Onondaga County ss. On this 24th day of January 1838 before me Myron L Mills one of the Justices of the Peace in and for said County comes William Mapes of the town of Marcellus County of Onondaga and State of New York aged Eighty four years who being duly sworn by and before me the said Justice doth on his oath depose and say that in the year 1777 he was well acquainted with Daniel Van Fleet who now resides in the town of Lysander in the County and State afforesaid and is an aplicant for a pension under the act of June 7th 1832 that this deponent then lived in the Town of Walkill in the County of Orange in said State and that said Daniel Van Fleet then lived in the town then called Mamacoten in the same County that this deponent was a soldier under

Capt Vail of the Militia and was at fort Montgomery from the Month of May untill the forepart of October or the surrender of the fort in the year 1777 that the said Daniel Van Fleet was a Serjeant in the company of Capt Abram Cuderback which was at fort Montgomery that some time in September this deponent was transfered to said Cuderbacks company and he well recollects that said Van Fleet was the Orderly Sergeant in said Company and drew their provisions this deponent cannot tell the time that said Van Fleet came to Fort Montgomery but he is certain he was there in Capt. Abram Cuderbacks company some time in the month of September and served untill after the surrender of the Fort which this deponent believes was on the 6th of October and that the said Company under Capt Cuderback were of the New York Militia and were drafted for the service as this deponent understood and believes

Sworn to and subscribed the day and year first above
writen and I certify that I am well acquainted with William
Mapes the above deponent and know him to be a credible
witness and his Statements are entitled to belief
Myron L Mills Justice of the Peace

his
William X Mapes
mark

I Certify that the reason of the above deponents making his Signature by mark is that he is old and quite infirm and unable to write his name so as to be Legible

Myron L Mills Justice of the Peace

State of New York

Onondaga County Clerks Office ss. I, William M. Clarke, Deputy Clerk of the said County do certify, that Myron L. Mills before whom the preceding affidavit purports to have been sworn, was at the date of the same one of the Justices of the Peace in and for the said County and that I am acquainted with his hand writing, and verily believe his name subscribed to the said affidavit, to be his proper hand writing and Signature. In testimony whereof I have here unto set my hand and affixed the Seal of the said County this 26th day of June 1838

W. M. Clarke Dep. Clerk

State of New York

Onondaga County ss. On the twelfth day of April 1838 personally appeared, before me, Squire Munro one of the Justices of the Peace of the town of Lysander in the county aforesaid, Benjamin Depuy of the said town of Lysander who being duly sworn did depose and say that he is now in the 74th year of his age, that before and at the time of the breaking out of the Revolutionary War he was personally acquainted with Daniel Van Fleet who was then a neighbour to him (the deponent) in a place called Peenpack, Present of Mamacating, Ulster County, State of New York and who now resides a neighbour to said deponent in the said Town of Lysander; That in the beginning of of the war the said Daniel Van Fleet was a militia man in Captain Abraham Cuddeback's company, of which King and Hardin were first and second lieutenants and Samuel Depuy Ensign; That said deponent lived about half a mile from said Daniel Van Fleet; That in the year 1775 or 1776, there was a call for some of the militia to go to the Hudson River and said deponent understood that said Daniel Van Fleet was gone, how long said deponent does not recollect; Said deponent believes that said Daniel Van Fleet turned out year after year to the Hudson River and in 1777 was at Fort Montgomery when taken and was in the battle; That said Daniel Van Fleet lived about half a mile from said deponent till about June 1778 when said Daniel moved south about five miles and was reputed to be a good and faithful soldier all the time he was in Captain Abraham Cuddeback's company which was about three years; In 1779 said Daniel Van Fleet was in Captain Abraham Westfall's company and remained there till the close of the war; When on the frontiers and when the Indians and Tories destroyed Minisink said Daniel Van Fleet turned out and pursued the Indians and Tories and was in the battle of Delaware river as said deponent understood and believes and was accounted a brave soldier in the service of his country and was so reputed by his fellow soldiers that served on that frontier; And said deponent understood and believes that said Daniel Van Fleet after he served and was in Capt. Abraham Westfall's company was the greatest part of his time in the neighbourhood called Minisink from 1778 to 1783 at the close of the war, and was on duty during that time as said deponent understood and believes.

Subscribed and sworn before me this 12th day of April 1838

Squire Munro Justice of the peace

Resworn before me this 31 day of May 1838 after altering the name "Van Vleet" whenever it occurs into "Van Fleet."

Squire Munro Justice of the peace

Benjamin De Puy

I certify that Benjamin Depuy is a resident of the Town of Lysander an is a man who is to be believed as a witness

Squire Munro

State of New York

Onondaga County Clerks Office ss. I, William M. Clarke Deputy Clerk of the said County, do certify, that Squire Munro, before whom the preceding affidavit purports to have been sworn was at the date of the same, one of the Justices of the Peace in and for said County and that I am acquainted with his hand writing, and verily believe his name subscribed to the said affidavit, to be his proper hand writing and signature. In testimony whereof I have hereunto set my hand and affixed the Seal of the said County this 26th day of June 1838

W. M. Clarke Dep. Clerk

State of New York

County of Ontario ss. On the twenty second day of May 1838, before me, Horace Pennell, Justice of the Peace in and for said county, came Asa Brown of the town of Bristol in Ontario county aged eighty three years, who, being duly sworn by & before the said Justice doth on his oath depose and say that he was well acquainted with Daniel Van Fleet in the Revolution war, who now resides in the town of Lysander county of Onondaga, and is an applicant for a pension under the act of June 7th 1832. That deponent then lived in the town of Minisink & county of Orange, & said Daniel Van Fleet then lived in the town of Mamakating, county of Orange, & said deponent says he knows that the said Daniel Van Fleet did duty as a militiaman under Capt. Abram Cuddeback in the years of 1775 & '76 on the frontiers as guarding the inhabitants and as one of scouting parties after the Indians & Tories, and in the year 1777 saw the said Daniel Van Fleet in Fort Montgomery in Capt. Cuddeback's company in the latter part of the month of September of that year before said fort was captured by the British army, and deponent says that he further knows that the said Daniel Van Fleet did duty under Captain Abram Westfall in the years of 1778 & '79, in guarding the inhabitants on the frontiers in the town of Minisink. And further knows that he saw the said Daniel Van Fleet on the morning of the day before the battle on the river Delaware in the town of Minisink at which time, he saw him, the said Daniel Van Fleet start in pursuit of the Indians and Tories under the command of Col. Tusten, and he further knows that the said Daniel Van Fleet did, and performed duty as a militia-man from the time last mentioned until the close of the war and deponent further says that he believes that the said Daniel Van Fleet served under the command of the officers above named to the amount of three years or upwards, and that the said Daniel Van Fleet was counted a faithful and true soldier in defence of his country and was so called by his fellow soldiers.

Sworn to and subscribed the day andyear first above written before me
Horace Pennell J. P.

his
Asa X Brown
mark

I certify that I am well acquainted with Asa Brown the above deponent and know him to be a credible witness and his statements are entitled to belief and the reason why he made his mark is because of his infirmity & blindness.

Horace Pennell J. P.

State of New York
County of Ontario ss. I Thomas Hall clerk of the common Pleas Courts of the county and State aforesaid do hereby certify that Horace Pennell is a Justice of the Peace in and for said county duly ~~commissioned~~ Elected and qualified that his term of Office commenced on the first day of January 1836 and will expire on the thirty first day of December 1839 and that his signature above written is Genuine. Given under my hand and the seal of said County this twenty third day of May A.D. 1838

Thomas Hall Clerk

Sir
I will thank you to give the enclosed papers an examination in connection with the case to which they belong so as to give me an answer before I leave for home

Respectfully yours
W. Taylor
House Representatives
July 3d 1838

Hon. J. L. Edwards

Rose New York
Nov 26th 1845

Sir,
Permit me Respectfully to request such information from your Office, as will

give me a knowledge of the objections, to the claim of Daniel Vanfleet whose claim for a pension was allowed, under act of 7th June 1832, but which has been withheld for some years past.

Please give me a statement of the services, as claimed to have been performed by him.

I, have the honor to be, very
respectfully, your humble Sevt
W. G. Wade

Hon J L Edwards
Comr of Pensions
Washington D.C.

House Representatives
Feby 7, 1848

Hon J. L. Edwards
Dear Sir:
A constituent writes me as follows, to wit:

"Daniel Van Fleet of Cayuga County N.Y. was a pensioner; his pension was probably suspended in 1833 & never restored. I ask the cause or reason of such suspension. Also a copy of the alledged service said to have been performed by him"

Please give me a reply to the above at your earliest convenience

I am respectfully yours
H. S. Conger

Washington March 7, 1850

Sir

You inform me that you do not feel at liberty to furnish me with copies of the alledged services of Peter Ball & Daniel Van Fleet, as requested by me, unless you know for what purpose copies are desired

I desire copies simply that the cases may be reviewed with a view of asking further action in them in case it shall be thought they show a good claim. I trust therefore you will see no objection to obliging me

I have the honor to be very respectfully yours
H. S. Conger

Col J. L. Edwards
Comr of Pensions

Sir

Washington May 20, 1850

Will you please do me the favor to comply with my requests in the following named cases.

1 Daniel Van Fleet of N.Y. was a pensioner but has been struck from the rolls. Why was he droped?

2 Ruth Branch of N.Y. widow of Samuel, is an applicant under law of 29 July 1848. What is the objection to granting her a pension. A copy of your report is asked.

3 David Rowland of N.Y. is a pensioner at about $32. He thinks he is entitled to increase. I would like a copy of his alleged service or what else that shall show whether he can assert a claim for increase

4 Peter Ball of N.Y. has been struck from the rolls. For what reason?

5 Asa Cheesebrough of N.Y. is an applicant under the law of 1818. What is the objection?

6 William Gillespie's widow of N.Y. married her second husband Jonathan Ellsworth. Pension certificates were issued to her, but were withheld. Why was this done & what is the objection to granting her a pension

7 Susanna Reigmond of N.Y. widow of Enoch late a pensioner is an applicant. What is the objection?

8 Timothy Windslow of N.Y. an applicant under the law of 1832. What is the objection

9 Olive Wooster widow of John of N.Y. Is she not entitled to pension? If the husband was a deserter was he not restored? It is said that his name does not appear in Boston as a deserter

10 William Foster of N.Y. What was the annual amount of his pension? Was he invalid or Revolutionary pensioner?

11 Joel Tiffany of N.Y. late a pensioner but struck form the rolls. If he ever deserted was he not restored? When the pension was granted was not his restoration admitted? Ought he not to be restored to the pension list, or if dead his widow or children

12 Samuel Morehouse of N.Y. an applicant under the act of 1832. What is the objection?

I have the honor to be yours &c
H S Conger

Col J. L. Edwards
Comr. of Pensions.

Auburn, N.Y. March 23rd 1853

To the 3rd Auditor of the Treasury

Sir

Please inform me as to what proofs of service were made in the case of Danl Van Fleet, a pensioner, formerly of Lysander, Onondaga Co. NY. under what law he drew a Pension, and what amount per year, and oblige

Yours truly
H H Bostwick

Auburn, N.Y. Nov 11th 1853

Honl L P Waldo

Comr of Pensions

Sir

On the 23rd of March I addressed a communication to the Third Auditor, asking information in the case of Daniel Van Fleet, which was referred to your office, and on the 29th of June a communication from you was addressed to me, relative to the matter, and saying that whenever I would file a Power of Attorney in the case, the desired information would be furnished. Enclosed I send the Power of Attorney required.

Yours truly
H H Bostwick

Power of Attorney and Certificate.

Know all men by these presents, That I, Sarah Van Auken of the County of Cayuga and State of New York do hereby constitute and appoint H. H. Bostwick, of Auburn, N.Y., my true and lawful attorney, irrevocable, for me in my behalf, to prosecute my claim for back pay, or unpaid pension on account of the military services of Daniel Van Fleet in the military service of the United States of America and to receive such certificate as may be issued, and to appoint one or more attorneys under him for the purposes aforesaid, hereby revoking any power of attorney which may have been heretofore given, and confirming all things that my attorney, the said H. H. Bostwick, may lawfully do in the premises. In testimony whereof, I have hereunto set my hand and seal this Fifteenth day of October A.D. 1853.

Sarah Vanaukn

Acknowledged before me the day and year last above written and I do hereby certify that the said Sarah Van Auken is one of the heirs of Daniel Van Fleet deceased

Justin Mason Justice of the Peace.

State of New York

County of Cayuga ss. I, E. B. Marvin clerk of the court of Record within and for the county and State aforesaid, do hereby certify that Justin Mason before whom the foregoing acknowledgment was taken, was at the date thereof an acting justice of the peace, duly elected and authorized to take the same. In testimony whereof, I have hereunto set my hand and affixed the seal of said court this 12 day of November A.D. 1853.

E. B. Marvin Clerk.

Auburn N.Y. Nov 24th 1853

Hon L P Waldo

Comr of Pensions

Sir

Yours of the 22d inst. relative to the case of Daniel Van Fleet, is recd. from which, together with the copy of letter from R Edwards to A T Wade Esq, I perceive that Danl. Van Fleet was a pensioner under the act of June 7th 1832. That his pension was suspended upon reexamination, and additional proofs required; that these proofs were furnished on the 9th of July 1838, and the result communicated to the Hon Wm Taylor.

It does not appear however when, or at what date, his pension was suspended nor how long it continued in suspension, nor whether in fact it was allowed on the reexamination, though I infer it was, from the fact that the requirement was complied with.

Please state, whether it was allowed upon reexamination. How long the suspension continued, whether the amount of pension he would otherwise have drawn, was made good to him for the time of suspension; the amt of pension per annum, originally allowed him; And whether it appears that there is anything due his heirs on his account.

Sarah van Aken, is a child of the said Daniel Van Fleet.

Please also furnish me with a copy of the letter to the Hon Wm Taylor, mentioned in the letter from R Edwards Commissioner to W T Wade Esq

Yours truly
H H Bostwick

Pension Office 10th Decr 1853

New York Rolls, $95 per annum, Act 7th June 1832. Daniel Van Fleet. To what time paid? When did he die?

To,

L P Waldo Comr.

Francis Burt Esqr. Third Auditor.

Daniel Vanfleet act 1832, $90 per annum. Paid to 4 Sept.1834, no date of death on the Books.

F. Burt Auditor

Auburn, N.Y. Dec 19th 1853

Hon L P Waldo
Comr of Pensions
Sir,

Yours of the 14th inst. in reply to mine of the 24th ult. is recd; together with Copy of a letter to the Honl Wm Taylor, bearing date Dec 22d 1837, relating to the suspended claim of Daniel Van Fleet.

I had also, on the 24th ult. recd yours of the 22nd ult. in reply to mine of the 11th ult. together with a copy of a letter to W S Wade Esq. bearing date Nov 3rd 1845, in the same case, alluding to a letter from Comr. Edwards to the Honl Wm Taylor, as communicating the result of a re-examination of said claim upon evidence filed on the 9th of July 1838

Allow me to call your attention to my letter of the 24th ult. above mentioned, and to the copies of the letters above referred to, from which you will perceive that the result communicated to the Honl Wm Taylor there (in my letter) alluded to, was upon a re-examination of the case, after the additional evidence required had been furnished; and that such additional evidence was filed on the 9th of July 1838. Subsequent to the date of the letter to the ~~letter~~ Honl Wm Taylor, (a copy of which accompanied your last letter,) and, that the letter to the Honl Wm Taylor alluded to in that of Comr. Edwards to W S Wade Esq., of Nov 3rd 1845, must have been of subsequent date to that of which a copy has been furnished me.

Please examine the case further, and reply more definitely to my letter of the 24th ult. Your early attention to the subject will much oblige

Yours truly
H H Bostwick

Decr 28 1853

Pension Office

Sir

Please state at what date the payment of the pension of the within named Danl. Van Fleet was suspended, whether payment was resumed, & if so, at what rate, and date of last payment.

Very Respectfully
S. Cole

F Burt Esq

3rd Audr &c

Daniel Vanfleet, Act 1832 $95 per annum. Paid to 4 Sept 1834, date of suspension not known; payment not resumed

F Burt Auditor

Auburn, NY Jany 4th 1854

Hon L P Waldo

Com of Pensions

Sir

Yours of the 31st ult is recd. In reply to mine of the 19th ult. relative to the suspended Pension of Daniel Van Fleet. From which it appears that there would be something due his ~~his~~ heirs upon an equitable adjustment of the case.

He died May 19th 1840, and his claim was suspended Feby 10th 1835, in effect from Sept 4th 1834.

Please state what course is necessary to be pursued to enable the heirs to realize the amount of whatever may be awarded them, and whether one of the heirs can draw the money for the whole, this would be desirable inasmuch as they are scattered widely apart, and residents of different states.

Yours truly
H H Bostwick

Auburn, N.Y. Jany 20th 1854

Honl L P Waldo,

Comr of Pensions,

Sir

In reply to your letter of the 11th inst., relative to the claim, by virtue of the revolutionary services of Daniel Van Fleet, I would earnestly urge a more liberal consideration of this case. An imputation or reflection of fraud should not be permitted to be made upon a Revolutionary Soldier because he could not prove the term of service claimed by him to have been performed. The difficulty of proving Revolutionary services, and the imperfection of the Records of the Revolutionary war, should induce a favorable construction of the declarations made by the soldier. It were better to grant pensions to ten who were not entitled to it, rather than exclude one who was deserving.

In the Copy of Letter to Honl H S Conger, accompanying your letter. I find reference again made to the Report made upon the additional evidence filed on the 9th of July 1838 and communicated to the Honl Wm. Taylor, a copy of which was furnished Mr Conger. This report has not been given me, (or a copy thereof) as requested in my former letters. A copy of one previous to that report was given, but not the one of subsequent date. If the one in question is on file, I would be glad to have a copy of it.

If a favorable view of the case as its proof present it, without embarrassing it with the consideration of a possibility of fraud, will admit of granting the claim after deducting the overpayment for the portion of term claimed but not subsequently proved, ~~will admit of its being~~ I trust you make a favorable adjustment, of the case.

Please let me hear further from you relative to it. And if expedient, I will see that a Declaration is duly made by an Administrator.

Very respectfully yours
H H Bostwick

Wauseon Ohio July 7 - '05

Department of the Interior,
Bureau of Pensions,
Washington D.C.

Kindly give me the Revolutionary War service of Daniel Van Fleet, New York men. Thanking you in advance

I am yours truly
(Miss) Lillian T. Strong
Wauseon Ohio

(stamped as received Jul 11 1905)

Revolutionary War Records Section.

Department of the Interior,
Bureau of Pensions.

Washington, D.C., , 19

In reply to your request of , received for a statement of the military history of Daniel Vanfleet a soldier of the Revolutionary War, you will find below the desired information as contained in his (or his widow's) application for pension on file in this Bureau.

DATES OF ENLISTMENT OR APPOINTMENT.	LENGTH OF SERVICE	RANK.	OFFICERS UNDER WHOM SERVICE WAS RENDERED.	OFFICERS UNDER WHOM SERVICE WAS RENDERED.	State
			CAPTAIN.	COLONEL.	
May 1775	served at various times 14 mos as Pvt 8 ½ mos as Sgt to close of war		Tilford	McClaughry	N.Y.
			John Fetter or Faulkner	Tuston	
			Abraham Cuddeback & Wilhelmus Westfall	Hatfield	
			Abraham Westfall		
			Jacob De Witt		
			Joseph Woods		

Battles engaged in, Ft Montgomery Minisink
Residence of soldier at enlistment, Goshen, Orange Co. N.Y.
Date of application for pension, Sept. 24, 1832. His cl was al
Residence at date of application, Owasco Cayuga Co. N.Y.
Age at date of application, born in 1756 at Goshen N.Y. died
May 19, 1840

Remarks: In 1853 Sarah Van Auken is referred to as a child of Daniel Vanfleet. There is no other data on file as to his family. It is also stated that soldier was a resident of Peenpack, Mamakating Precinct, Ulster Co. N.Y. during the war.

Heitman says Col. Tuston was killed at Minisink
Respectfully,
Commissioner.

Wauseon Ohio, May 28,1911

Department of the Interior,

Bureau of Pensions,
Washington D.C.
Sirs,

In 1905 I obtained from the Department of the Interior, the war record of my Great, Great Grand Father Daniel Van Fleet upon receipt of the same, and other important Data. I had hoped to join the D. A. R. but in looking over my material found two missing links, one being the date of this ancestors marriage, and the other, the wifes maiden name. As Daniel Van Fleet was a pensioner, thought perhaps you would have the date of his marriage & the wifes name on record. He was a resident of Ulster Co. N.Y. at time of enlistment, in May 1775 and at time of application for pension Sept 24, 1832 he resided at Owasco, Cayuga Co. N.Y. Perhaps this number will help you in searching his record, I. S. C. Inv. 14, 763 Rev. War. If you can give me this desired information I will be duly grateful, or if the date of marriage & the wifes name are not recorded, could you please tell me where I might be able to find same. Thanking you very kindly in advance for your trouble

I am Respectfully Yours
Miss Lillian Strong
Wauseon Ohio

Record Division.
Miss Lillian Strong,
Wauseon, Ohio.

June 5, 1911.

Madam:

In reply to your letter dated the 28th and received the 31st ultimo, you are advised that the papers in the pension claim of Daniel Vanfleet, Sur. File No. 14,763, fail to show the date of his marriage or the name of his wife.

It is possible that his marriage may be of record in Orange County, New York, where he resided from 1778 to 1810.

Very respectfully,
Commissioner.

Franklin K. Lane,
Secretary of the Interior,
Washington, D.C.

Alamosa, Colo., Feb. 10, 1918

Dear Sir:

Will you kindly give me the Revolutionary War standing of Marshall H. Van Fleet; whose father was Garrett B. Van Fleet, in the Civil War; whose father was Levy Van Fleet, in the war of 1812; whose father was Daniel Van Fleet, in the Revolutionary war from New York State, Orange County.

Yours truly,
M.H. Van Fleet

P.S. Also give the rank held in the War.

Invalid. File No. 14,763
Daniel Vanfleet Pri & Sergt Rev War
Act: June 7th 32
Index: - Vol. 1, Page 379
[Arrangement of 1870.]
Revolutionary War Pension and Bounty Land Warrant Application Files
Series M-804; Roll # 2445

Thomas Waters
New York (S.11,694)

State of New York

Orange County ss. On this 4th day of September personally appeared in open court before the Judges of the Court of Common Pleas in and for the County of Orange, being a court of record and now sitting, Thomas Waters a resident of the Town of Goshen, County and State aforesaid, aged Seventy years, who being first duly sworn according to law, doth on his oath make the following declaration in order to obtain the benefit of the act of Congress passed June 7th 1832.

That this deponent first entered the service of the United States as a private in a company commanded by Captain David Swezy at Goshen in the month of August 1776 as deponent thinks. There was an alarm and deponent volunteered to go to Fort Montgomery he cannot say how long he was there but thinks it was at least three weeks. Captain Jones of the militia was there Colonel Dubois commanded a regiment as deponent thinks of United States Troops.

And Deponent further saith that he was on several occasions at Fort Montgomery in the years 1776-77 under the command of the said David Swezy, that he cannot particularly state the various times and dates having no documents to refer to, but he feels confident that he was there under Captain Swezy at least three times and deponent thinks he was there once under Captain Jones Deponent has no hesitation in saying that he served at Fort Montgomery on three several occasions at least eight weeks Governor George Clinton and General James Clinton were there. Deponent thinks that at times there were ~~at times~~ two companies of United States Artillery. Capt Moody commanded one Lieutenant Oliver Lawrence of the United States Artillery was there.

And Deponent further saith that in the month of October 1777 Deponent went under Captain Swezy to Fort Montgomery when within a few miles of that place, information was brought that the fort was taken we were ordered to New Windsor thence up the North river to Kingston when on the way and near that place received information that it was burned by the enemy we were then ordered to Old Hurley a few miles west of Kingston, after remaining there a few days was ordered back to New Windsor. Deponent cannot recollect wether he went with Captain Swezy down the river or not but his strong impression is that he remained at New Windsor for several weeks under Col Hawthorn of the militia. Thinks Col Webbs regiment was there.

Deponent further saith that in the years 1777 & 78 he was out under the command of Captain Swezy on several occasions at Ramapo on the borders of New Jersey, an important pass and which was guarded during the greater part of the war. Deponent cannot state the different times he was out, and the periods he remained there in the service, but after bestowing his best reflections upon the subject he is confident he was in the service at Ramapo and in that neighbourhood seven or eight months Gen James Clinton was generally in command there, no United States troops there as deponent recollects Captain Jones, Captain Little and other officers were there.

And this Deponent further saith that in the years 1778, 1779 he was frequently out on the frontiers at Peenpack to defend them from the Savages it was on the west side of Shawngunk Mountain about twenty miles from Goshen. Deponent does not think he was out all the time Captain Swezy was there to whose company he belonged. Deponents recollection as to that service is indistinct he knows he was out on various occasions, and deponent thinks he can say with certainty, that all his service on the Peenpack frontiers, would amount to at least five months. Deponent was at a picket called fort Dewit, recollects Col. Newkirk was there, thinks Captain Little was there once or oftener. Capt Swezy was principally in command. There were frequent alarms and the militia called out on account of the Indians and deponent does not recollect one instance in which he did not turn out.

This Deponent further saith that about the middle of July 1779 he went out with the militia to the said Frontiers to meet the Indians who under Brandt had made incursions and killed some of the inhabitants. There was a general muster. Gen Hawthorn commanded. Col Thurston was there Captains Wood, Jones, Duncan, Sawyer and others not recollected. We followed the Indians from Minisink through the wilderness about twenty miles. On the east side of the Delaware river and opposite to the mouth of the Lackawanna we were attacked by the Indians and after ~~the~~ an engagement of four or five hours we were routed. Most of the officers were killed and we were compelled to retreat Col Tusten, Captain Tyler, Captain Jones, Capt Sawyer, Captain Duncan, Captain Vail were all killed. Deponent was out on that occasion about eight days.

And Deponent further saith that he was in the service on other occasions than those stated but his recollection is too imperfect to enable him to state them. Deponent verily believes that he served off and on in the revolutionary war more than two years.

And this Deponent further saith that he was in the service at Tappan on the North river during one winter and deponent thinks that it was in the winter 1776 & 1777 deponent went out under Captain Swezy after remaining there some time part of the forces were to return and part to remain there was a draft and deponent took the place of Albert Foster now deceased who was drafted to remain. Deponent cannot say how long he stayed on his own account and how long as a substitute for said Foster. But Deponent verily believes that he was there altogether at least four months. No United States troops there. Colonel Allison of the New York Militia was there Colonel Hathorn was also there thinks Captain Little was there. Deponent recollects on one occasion of marching from Tappan to the Three Pigeons in New Jersey three miles from Hoboken, having been informed that some of the enemy would be in that neighbourhood plundering, but we did not meet with the enemy and returned by the way of Hackensack.

Deponent thinks two regiments started on this excursion commanded by Colonels Hathorn Allison and Woodhull or Odle.

This Deponent further saith that he was born at Brooklyn Long Island in the year 1762. Has no record of his age lived at Goshen in the County of Orange when called into service and has lived there ever since.

This Deponent further saith that he first entered the service as a volunteer afterwards was ordered out by his officers was once a substitute as above stated never received a discharge. Gen James W Wilkin, David M Westcott, & George D Wickham reside in the neighbourhood who can testify to his veracity and their belief of his service in the revolutionary war.

This Deponent further saith he hath no documentary evidence of any kind whatsoever. That Captain David Swezy of Goshen aforesaid and Thomas Knight of the town of Walkill in the County of Orange aforesaid know of his services generally.

This Deponent here by relinquishes every claim whatever to a pension or annuity except the present and declares that his name is not on the Pension roll of the agency of any State.

Subscribed & Sworn the 4th September 1832
Asa Dunning Clerk

Thomas Waters

In the Matter of the Application of Thomas Waters for a Pension under the act of Congress passed June 7th 1832State of New York

Orange County ss. David Swezy being duly sworn doth depose and say that he knows the said Thomas Waters was in the service of the United States as a soldier in the revolution on various occasions.

Deponent recollects that said Waters was at Fort Montgomery under his command on several occasions, but deponent cannot state how often he was there nor how long he stayed. but should the said Waters was there three or four times. Deponent recollects that the said Waters was always ready and willing to turn out when the Company was ordered into service, and was a faithful soldier. Deponent does not recollect any instance in which he was not out. Deponent thinks that the said Waters served at Fort Montgomery altogether at least eight ~~months~~ weeks. Deponent recollects that said Waters was with him on his march to New Windsor & New Hurley up the North river on the way to Kingston shortly after Fort Montgomery was taken, and when Kingston was burnt but deponent cannot recollect wether the said Waters went with Deponent down the river after that period or not. The above service was in the years 1776 & 1777.

And this Deponent further says. That he recollects that in the winter of 1776 & 1777 the said Waters was in service at Tappan on the North river, now in the County of Rockland and State of New York. After deponent left there, said Waters remained behind, but Deponent does not know how long. Deponent was there about six weeks on one occasion. We marched to the three Pigeons in New Jersey, and back again by the way of Hackensack. There were there regiments commanded by Cols Hathorn, Allison, and Woodhull.

Deponent further says that in the years 1777 & 1778 the said Waters was in the service on several occasions at Ramapo under this Deponent, but deponents recollection will not enable him to state how long he was out. Deponent has no reason to believe but that the said Waters was at Ramapo as long as Deponent. which as he thinks was eight months. Deponent knows that the said Waters was always prompt to turn out on all occasions when required.

Deponent further says that he recollects that the said Waters was in the service at Peenpack on the Frontiers on several occasions in the years 1778 & 1779, the Indians had made frequent incursions and burnt some houses and killed several of the inhabitants. Deponent cannot state how often the said Waters was at Peenpack nor how long he served. but thinks he must have served there off and on at least five months. Deponents Company was often out and said Waters for ought this deponent was out as much as any of them.

And this Deponent further says that from his General recollection on the subject he has no doubt that from the commencement to the end of the war the said Waters was in the service more than two years.

Subscribed & Sworn in Open Court the 4th day of September 1832

Asa Dunning Clerk

David Swezy

And the said Court Do hereby declare their opinion after the investigation of the matter and after putting the interrogatories prescribed by the War Department that the above named applicant was a Revolutionary soldier and served as he states, and the Court further Certifies That it appears to them that David Swezy who has subscribed & Sworn to the foregoing affidavit is a credible person and that his statement is entitled to credit.

I Asa Dunning Clerk of the Court of Common pleas in & for the said County of Orange Do hereby Certify that the foregoing contains the original proceedings in the matter of the application of Thomas Waters for a pension. In Testimony Whereof I have hereunto Set my hand and Seal of Office this Fourth day of September 1832.

Asa Dunning

State of New York

City & County of New York ss. Personally appeared before me the undersigned James Hopson a Justice of the Peace in and for the City and County of New York, Thomas Waters the applicant within named who being duly sworn deposeth and saith that by reason of old age and the consequent loss of memory he cannot swear in all cases positively as to ~~all~~ the precise length of his service, but according to the best of his recollection and belief he served not less than the periods mentioned below in every case as a private. In 1776 he served three weeks at Fort Montgomery. During the residue of the year 1776 and in the year 1777 he served at the same place eight weeks. In October 1777 he served at New Windsor Hurley or Kingston about the time that place was burnt by the enemy four weeks. That he served at Ramapo in the years 1777 and 1778 seven months at least that in the year 1778 and 1779 he served at Peenpack Five months. That he was in the Minisink battle, and was in service Eight days about that time which was in July

1779. That he was at Tappan on the North river in the winter of 1776 & 1777, Four months that he served at various other times making altogether more than two years of service and for my service I claim a pension for the time specifically set forth in the above declaration and the said Thomas Waters further saith that when his former declaration was made he was informed that the certificate of a clergyman was not necessary as he had positive proof of service. That since his former declaration he moved to the city of New York where he now resides.

Sworn before me Novr 21st 1832

James Hopson Justice of the Peace

Thomas Waters

State of New-York,

City and County of New-York, ss. I, Abraham Asten, Clerk of the City and County of New-York, and Clerk of the Court of Common Pleas for the said City and County, do hereby certify, That James Hopson Esqr before whom the annexed affidavit was made was on the day when the same was made a Justice of the peace in & for the City & County of New York duly appointed & sworn & I further Certify that I am acquainted with the hand writing of the said Justice & verily believe that the signature to the Jurat of the said affidavit is genuine. In testimony whereof, I have hereunto set my Hand, and affixed the Seal of the said Court and County, the 21st day of November 1832.

Abm Asten Clerk

We Nathan Kingsbury a clergyman residing in the Town of Goshen County of Orange and State of New York, and Albert S Benton residing in the same place, hereby certify that we are well acquainted with Thomas Waters who has subscribed & sworn to the above declaration; that we believe him to be seventy years of age and that he is reputed and believed in the neighbourhood where he resides to have been a soldier of the Revolution, and that we concur in that opinion.

Sworn the 19 Novr 1832 before me

H W Elliott a Judge of Orange Com. Pleas

Nathan Kingsbury
A. S. Benton

State of New York

Orange County ss. I Horace W Elliott the Judge above named do certify that Nathan Kingsbury who has signed the above certificate is a clergyman & that he and the above named Albert S Benton are persons of credibility.

H W Elliott

Orange County ss. I Asa Dunning Clerk of the County of Orange aforesaid do certify that the certificate of Nathan Kingsbury and Albert S Benton endorsed on the annexed declaration of Thomas Waters & the certificate of Horace W Elliott a Judge &c thereunder written, were obtained since the date of my within certificate. I further certify that the said Horace W Elliott is one of the Judges of the Court of Common Pleas in and for the said County, and that full faith & credit should be given to his official acts. In witness whereof I have hereunto signed my name & affixed the seal of the said county this 19th of November 1832.

Asa Dunning Clerk

New York

Thomas Waters of New York in the State of N. York who was a Private in the Compy commanded by Captain Jones of the Regt. commanded by Col. Dubois in the New York line for 2 Years. Inscribed on the Roll of New York at the rate of 80 Dollars per annum, to commence on the 4th day of March, 1831. Certificate of Pension issued the 6th day of Decbr. 1832 and sent to Hon J. Wilkins

Arrears to the 4th of Septr 1832	$120
Semi-anl. allowance ending 4 Mar '33	40
	$160

Recorded by Geo. C. Stiles Clerk,
{Revolutionary Claim, Act June 7, 1832.}
Book D Vol. 5 Page 205

Invalid. File No. 11,694
Thomas Waters Priv Rev War
Act: June 7th 1832
Index: Vol. A, Page 894
[Arrangement of 1870.]

Revolutionary War Pension and Bounty Land Warrant Application Files
Series M-804; Roll # 2504

Benjamin Whitaker
New York (S.15,242)

Surgeon's CertificateThis may Certify that Benjamin Whitaker (a Private in Captain Coles Company in a Regiment of Militia of the County of Orange in the State of New York Commanded by Colonel William Ellison) was Wounded by a Rifle Ball which passed through his left Arm above the Elbow and tore the flesh and hurt the bone of his left breast, he was under my Care about Eight Weeks. The said Whitaker received said wound in an engagement with a party of Indians and Tories who made a descent upon the frontier part of the said County of Orange on the 22nd day of July in the year 1779 The Said Indians and Tories being in the service of his Britanic Majesty under the Command of Capt. Joseph Brant And the said Benjamin Whitaker when he received the said wound in the Actual Service of the United States under the command of the then Colonel now Brigadier General John Hathorn.

June 1811

Jonathn. Swezy Then Surgeon to Allisons Regiment

State of New York

Orange County ss. On this Seventh Day of September in the year 1832, Personally appeared Before me John Hallock Junr one of the Judges of the Court of Common Pleas, in and for said County of Orange Benjamin Whitaker, a resident of the Town of Minisink, County and State aforesaid, aged Eighty one years. Who being Duly sworn according to Law, doth on his Oath Depose, and make the following declaration in order to obtain the benefit of the, act, of Congress passed June the 7th 1832

Deponent saith that his first service as a private soldier in the Militia in the revolutionary war was at Fort Montgomery, under as he thinks the Command of Capt. David Swezy and thinks he served three weeks; and his Duty at that Tower of service was, preparing the fortifications, and in that service he recollects that Jamas Clinton was the General, and it was in the year as Deponent thinks to the best of his recollection 1776

During the same year Deponent was called at New Windsor under officers which he cannot now recollect their names, but thinks Daniel Everett was Capt. or Lieutenant; and served there ten Days or or more, to watch the Enemy on the Hudson River and then returned home

And further Deponent saith that he volunteered as a Militiaman in the year 1777 in Capt. Sawyers Company and marched to Fort Montgomery and there served two months or more. James Clinton was General & Harris was Col; at this time he was on fatiege, and then Returned home

And further Deponent saith that he was called upon the same year in the service at Tappan, on the Hudson River, as a guard against the Enemy, the company officers cannot recollect, William Allison was the Col. and there served two months.

And further Deponent saith that while belonging to Capt. Samuel Coles companey that he Deponent, Volunteered as a private to Defend the frontiers agains the Indians and Tories on what was then called the Minisink frontiers, and was there at that battle called the Minisink battle; wounded, and with said wound confined about two months as the annexed certificate of Dr. Jonathan Swezy will shew and further Deponent saith that he was in the aforesaid Indian battle in the year 1779 and on the 22 Day of July of the same year.

And further Deponent saith that he was called at different times into service in the Revolutionary war six months at Different places which deponent cannot now recollect to state time and place; neither his commanding officers as he was some times under one officer and some times under an other, some part of said service was rendered in scouting parties &c, and the Last part of his service as set forth is Exclusive of the foregoing.

And this Deponent saith and he verily believes that he was ~~that he wa~~s in the Revolutionary service as a Militia Private for the Term of one year one month and three Days or more

And Deponent further saith that he never had any written discharge, but that the common Practice was in the war to discharge by verbal order and in regard to proof of my character as a person of truth Integrity and veracity, and my services I refer to Fregift Cooley, Mary Van Tasell and Stepphen W. Fullerton Esq there Being no regular settled clergyman in this place that has an acquaintance with me for any Length of time. Deponent further that he is not on any Pension roll in any State.

Sworn and subscribed before me the 7th Day of Sept. 1832
John Hallock Junr

his
Benjamin X Whitaker
mark

In the Matter of Benjamin Whitaker

State of New York

Orange County ss. On tenth Day of September in the year 1832 Before me John Hallock Junior one of the Judges of the Court of Common Pleas in and for said County of Orange, Personally appeared, Mary Van Tasell, a resident of said county and being Duly sworn deposeth and saith, that she has been acquainted with the said Benjamin Whitaker forty years and upwards, and saith that she has never heard any ______ of any one that he was not a revolutionary soldier in Militia service. And further Deponent saith, that the said Benjamin Whitaker, is a person, of Integrity Honesty and Veracity an to be believed under Oath, and that is, the general oppinion of said Benjamin Whitaker where he has been acquainted.

And further this Deponent saith that she nursed the said Benjamin Whitaker when he was wounded in the battle called the Minisink battle, which conforms to the annexed certificate, of Dr. Jonathan Swezey. And further Deponent saith not

Sworn an subscribed the 10th
Day of Sept. 1832 before me John Hallock Junr

her
Mary X Van Tasell
mark

In the Matter of Benjamin Whitaker

State of New York

Orange County ss. On the tenth Day of September in the year 1832 Personally appeared, Before me John Hallock Junr. one of the Judges of the Court of Common Pleas in and said County of Orange Fregift Cooley and being Duly sworn according to law deposeth and saith, that he has been acquainted with the said Benjamin Whitaker forty years and upwards, has a portion of that time resided in the same neighbourhood and never heard any person have or set forth but what he was a soldier in the Revolutionary War, And this Deponent further saith, that said Benjamin Whitaker from his acquaintance with him has been reputed as a man of Integrity truth and veracity; and would have no hesitation of believing him under his Oath. And deponent further saith that he was acquainted with some of the services of the said Benjamin Whitaker as a Militia soldier in the Revolutionary War, and thinks at diferent times but cannot state precisely, Deponent was in the service himself and recollects of being in the service with said Benjamin Whitaker at Tappan on the Hudson river. And further Deponent saith not

Sworn by affirmation and Subscribed the
10th Day of Sept. in the year 1832 before me
John Hallock Junr

Freegift Cooley

In the Matter of Benjamin Whitaker

State of New York

Orange County ss. On the 10th Day of Sept. in the year 1832, Before me John Hallock Junr one of the Judges of of the Court of Com. Pleas in and for said County, Personally appeared Stephen W. Fullerton Esq and being Duly sworn according to Law, Deposeth and saith that he has been acquainted with the said Benjamin Whitaker for twenty years and upwards and as a man of truth Integrity and Veracity I concur in the Affidavits of Mary Van Tasell and Fregift Cooley.

And Deponent further saith that he is acquainted with the Infirmities of said Benjamin Whitaker, and believes what is set forth to be correct, and further saith not

Sworn and subscribed this 10th
Day of Sept. 1832 before me
John Hallock Jr

Stephen W. Fullerton

State of New York

Orange County ss. In the matter of Benjamin Whitaker a revolutionary soldier This may certify that I, have taken the foregoing Examination of Benjamin Whitaker who is making application as a pensioner under the act of Congress passed June the 7th 1832 And Propounded unto him, the questions laid down by the war department. And he the said Benjamin Whitaker has answered the Interrogatories unto him put, so far his recollection and memory is retained.

And I do further certify that the said Benjamin Whitaker is infirm, in body and has lost his sight, and believe him to be unable to appear at Open Court, a Distance of thirty miles from his residence, and that his recollection has much failed him.

And I do further certify that, I am acquainted with Mary Van Tasell, Fregift Cooley and Stephen W. Fullerton Esq whose affidavits are hereunto annexed, and believe them to be creditable, persons, and whose Oath would be taken in any Court of Record where they are known. And further I do concur in their

Testimony

Signed the 10th day of Sept. 1832

John Hallock Junr. A Judge of Orange Com. Pleas

State of New York

Orange County ss. I Asa Dunning Clerk of the County of Orange and Clerk of the Court of Common pleas in and for the said County of Orange Do hereby Certify That, John Hallock Junr. before whom the preceding Declaration of Benjamin Whitaker was made & Sworn & whose name appears to the preceding Certificate and also before whom the annexed affidavits of Mary Van Tassel, Fregift Cooley & Stephen W. Fullerton was sworn was at the time a Judge of the Court of Common Pleas in and for the said County of Orange which is a Court of record and that the name of John Hallock Junr Subscribed to the said affidavits and Certificate is his genuine signature. Witness my hand and Seal of Office this nineteenth day of September 1832

Asa Dunning

War Department,

Pension Office,

183

Sir:

The evidence in support of your claim, under the act of June 7, 1832, has been examined, and the papers are herewith returned. The following is a statement of your case in a tabular form. On comparing these papers with the following rules, and the subjoined notes, you will readily perceive that objections exist, which must be removed, before a pension can be allowed. The notes and the regulations will shew what is necessary to be done. Those points to which your attention is more particularly directed, you will find marked in the margin with a brace, (thus: }). You will, when you return your papers to this Department, send this printed letter with them; and you will, by complying with this request, greatly facilitate the investigation of your claim.

A Statement, shewing the Service of Benjn. Whitaker Orange Co. N.Y.

Period when service rendered	Duration of claimant's service			Rank of the Claimant	Names and Ranks of the Company Officers under whom he served	Names and Ranks of the Company Officers under whom he served
	Years	Months	Days			
Militia 1776			thinks 21	Private	Capt. David Swezy	General Clinton
Militia 1776			10			
vol 1777		2			Cap. Sawyer	Col. Harris
called 1777		2				Wm Allison Col.
vol 1779		2			Cap Saml Coles	
time not stated		6				
		13	31			

Battles in Which the applicant engaged	Country through which he marched	Place of abode when he entered service, and age at the present period	Evidence by which the declaration is supported
Minisink	Fort Mongomery	aged 81 Fort Mongomery	Freegift Cooley
	Hudson River		Mary van Tassell
			John Swezy Surgeon
	Tappan		Traditionary
	Minisink		

I am, respectfully,
Your obedient servant,
J. L. Edwards,
Commissioner of Pensions.

After the statement of service a copy of the regulations under the Act of June 7, 1832 is included but not reproduced here. An identical copy is included in the chapter entitled ***A Declaration of Service.***

Goshen May 23rd 1833

J L Edwards Esqr Comr of Pensions

I return you the papers with all the corrections that I can make the papers you will perceive has now been authenticated by the County Clerk as the regulations of the war office direct. The certificate of Judge Hallock as to the credibility of the witnesses you will find in the last paragraph of his certificate of the reason why the declarant could not appear in open Court

With the Highest considerations
I am yours &c. &c.
Chs. Morrell

New York

Benjamin Whitaker of Orange in the State of New York who was a private in the company commanded by Captain Everet of the Regt. commanded by Col Harris in the N.Y. State Troops line for 13 months Inscribed on the Roll of New York at the rate of 43 Dollars 33 Cents per annum, to commence on the 4th day of March, 1831. Certificate of Pension issued the 4 day of Septr. 1833 and sent to Chs. Morrell - Goshen N.Y.

Arrears to the 4th of Septr. 33	$108.32
Semi-anl. allowance ending 4 Mar 34	21.66
	$129.98

Recorded by ____ Franciose Clerk,
{Revolutionary Claim, Act June 7, 1832.}
Book E Vol. 4 Page 39

---

21 August 1919.
$1.00 enclosed

To the Honorable Commissioner of Pensions,
Washington D.C.
My dear Sir;

A pension was granted to Benjamin Whitaker, Private, in Revolutionary War somewhere between 1831 and 1833. He served from Orange County New York I very much want such particulars as he gave of himself and his family when he applied for his pension. This information is desired for genealogical purpose and will be very greatly appreciated. An old volume marked "U.S. Pensions Revolutionary War" p. 384 Vol. II reads as follows: "Orange County, New York, Benjamin Whitaker, Private New York Militia. Placed 4 Sept. 1833; Began 4 Mar. 1831. Age of pensioner 82." Everything relating to him will be of great interest to me.

Sincerely
Florence J. Gates
Mrs. Quincy A. Gates

236 South River Street
Wilkes-Barre Pa

Bureau of Pensions
Finance Division
Chief Revolutionary War Sec

Aug. 25 1919

Sir:

I have this day received from Mrs. Florence J. Gates draft No. Cash for $1.00 requesting post office addresses copy of record as per letter herewith

Very respectfully,
Wm. Campbell
Chief of Division.

August 26, 1919.

Rev. War Section.
Mrs. Q. A. Gates
236 South River ____
Wilkes-Barre, Penn

Madam:

In reply to your letter of the 21st instant, you are furnished the history of the Revolutionary War pensioner, Benjamin Whitaker of the New York Troops, for which there is no fee charged and therefore the dollar bill forwarded in your letter is herewith returned.

Very respectfully,
Commissioner.

Revolutionary War Records Section.
Department of the Interior,
Bureau of Pensions.
Washington, D.C., , 19

In reply to your request of , received for a statement of the military history of Benjamin Whitaker a soldier of the Revolutionary War, you will find below the desired information as contained in his (or his widow's) application for pension on file in this Bureau.

DATES OF ENLISTMENT OR APPOINTMENT	LENGTH OF SERVICE	RANK	OFFICERS UNDER WHOM SERVICE WAS RENDERED.	OFFICERS UNDER WHOM SERVICE WAS RENDERED	STATE
			CAPTAIN.	COLONEL.	
1776	3 weeks	Pvt.	David Swezy		NY
1776	10 day s		Daniel Everett		
1777	2 mos.		Sawyer	Harris	
1777	2 mos.			Wm Allison	
1779			Samuel Cole	Wm Allison	

6 mos. out at various times, officers not recollected

Battles engaged in, Minisink where wounded in left arm by rifle ball.
Residence of soldier at enlistment,
Date of application for pension, Sept. 7, 1832. His claim was al.
Residence at date of application, Minisink, Orange Co. N.Y.
Age at date of application, 81
Remarks: no data as to family
Respectfully,
Commissioner.

Invalid File No. 15,242
Benjamin Whitaker Priv. Rev. War
Act: June 7th 1832
Index: - Vol. 1, Page 405
[Arrangement of 1870.]

Revolutionary War Pension and Bounty Land Warrant Application Files
Series M-804; Roll # 2549

Hezekiah Woodward
New York (S.11,865)

State of New York

Orange County ss. On this fourth day of September 1832 Personally appeared in open Court before Samuel S Seward John Hallock Jr Gilbert O Fowler Nathaniel P. Hill and Horace W. Elliott, Judges of the Court of Common pleas in and for the County of Orange in said State being a Court of record now sitting Hezekiah Woodward a resident of the Town of Calhoun in the County of Orange and State of New York aged seventy seven years who being first duly sworn according to law doth on his oath make the following declaration in order to obtain the benefit of the Act of Congress passed June 7th 1832.

That he entered the service of the United States under the following named Officers and served as herein stated, he enlisted as a minute man for six months at Deerpark Orange County State of New York under Capt Burnet as a private in the year one thousand seven hundred and seventy five cannot tell the Capt first name and on the fifteenth day of the following June 1776 he again enlisted as a volunteer under Capt William Faulkner for six months in Col Hardenburgh's Regiment served in Gen Scotts Brigade he was marched from Orange County to New York and from thence to Green Wich and lay there till two British Ships the Phenix and Rose came up the North River he stayed until the fifteenth of the following December and was then discharged and returned home and stayed nine days and was called out again under Capt William Dennison in the Militia and went to Paramus Ramapo and Hackensack in New Jersey and stayed one month and was then transferred to Capt Falkeners Company and stayed three months more at another time he was called out and helped build Fort Montgomery was there two weeks and he was then sent out by Capt Benjamin Vail to Cochecton after a deserter Daniel Skinner and took him he was at West point two weeks under Captain Cuddeback and one month at Jacob Gumaers in Peenpack protecting his house which was a Garrison against the Indians and one month at Westbrooks in Ulster Co. under Capt Owen whose house was also a Garrison. After New York was taken by the British he went one third part of the time for three years in the Militia took turns one third went out and the rest stayed he says he always went his third part of the time and at every alarm during his turn at home he was called out at one time he was called on the night and sent to warn the company by Capt Vail at an alarm of the Indians, he was in the Indian Battle at the Delaware when our men were mostly killed by Brants men only six of his company returned he went under Capt Benjn. Vail who was killed in the Battle he was a Sergeant in the Militia the whole time (except the six months he enlisted as a volunteer) he was born in Stonington New London County State of Connecticut, in 1755, he has a record of his age in his Bible, he moved from Con. before the Revolution with his Father and has ever since lived where he now lives he has no written or printed evidence of his service he kept an account of his services but has lost it he has looked for it without effect he knows only three men by whom he can prove any part of it Amos Wheat and Thomas Oliver and Daniel Skinner both of Orange County. He hereby relinquishes every claim to a pension or annuity except the present and declares that his name is not on the pension roll of the agency of any State he says he was in the Battle when New York was taken by the British in Col Hardenburghs Regiment and he was on the Picket Guard when the British came to take white planes he was one sent by Capt Tyler to Scout for Indians gone one week, he was at New Windsor and Newburgh one month at the time the British burnt Esopus, that he is acquainted with nearly half of the people of the County where he lives

Subscribed & Sworn the 4th day of September 1832
Asa Dunning Clerk

Hezekiah Woodward

⁂

Orange County ss. Daniel Skinner of Calhoun in the said county being sworn in open Court says he is well acquainted Hezekiah Woodward who has subscribed and sworn to the above declaration and that he was a Soldier in the Revolutionary war he this deponent was with him one Tour of service against the Indians in Capt Owens Company he has heard of his being out in the service a great many times and believes he served as stated above and believes he is of the age above stated, he has been acquainted with him ever since he knew of his going after Daniel Skinner uncle to this deponent a deserter in the war

Subscribed & Sworn the 4th day of September 1832
Asa Dunning Clerk

Danl Skinner

⁂

And the said Court do hereby Certify and declare their opinion after the investigation of the matter and after putting the interrogatories prescribed by the war department the above named applicant was a Revolutionary Soldier and served as he states and they further Certify that Daniel Skinner of Calhoun of our said county who has subscribed and sworn to the above affidavit is a credible person and that his statement is entitled to credit.

⁂

Asa Dunning Clerk of the Court of Common Pleas in and for the County of Orange Do hereby Certify that the foregoing contains the original proceedings in the matter of the application of Hezekiah Woodward for a pension. In Testimony Whereof I have hereunto Set my hand and Seal of office this ~~third~~ Fourth day of September 1832

Asa Dunning

Orange County ss. Thomas Oliver of Wallkill in the county aforesaid being duly sworn saith that he is well acquainted with Hezekiah Woodward of Calhoun in said county who is about making application for a Pension under the Act of Congress of June 7th 1832 That he this deponent was acquainted with him during the Revolutionary war and knew that he the said Hezekiah Woodward was a Soldier of the Revolution, that this deponent served with the said Hezekiah Woodward nearly one year a part of the time in the same company, six months was in Col. Hardenburghs Regiment and about four months in Col. McGlaughans Regiment and this deponent has been acquainted with the said Hezekiah Woodward ever since the Revolutionary war and knows he is person of good character.

Sworn and subscribed the 12th day of Sept 1832 before me

Alex Bell Justice of the Peace

Thomas Oliver

The within Affidavit was not obtained until after the Court adjourned and there will be no more Court till December, Mr Oliver who has sworn to the within Affidavit is a credible person and the Court will certify the same if necessary he is aged & infirm.

J. Chattle Counsellor at Law

I Alexander Bell before whom the within Affidavit of Thomas Oliver was taken on the twelfth day of September 1832 Do Certify that I am acquainted with the said Thomas Oliver. That he is a resident of the Town of Wallkill in the County of Orange and that he is a credible person and that his statement is entitled to credit and that he was unable to attend Court when the said Affidavit was taken by reason of age and infirmity.

March 19th 1833

Alex Bell Justice of the Peace

State of New York

Orange County ss. I Asa Dunning Clerk of the County of Orange and Clerk of the Court of Common Pleas in and for the said County of Orange Do hereby Certify That Alexander Bell Esq before whom the preceding affidavit purports to have been Sworn was at the date of the same a Justice of the peace in and for the said County of Orange and that I am acquainted with his handwriting verily believe his name subscribed to the said affidavit is his proper handwriting and usual signature. In Testimony Whereof I have hereunto Set my hand & Seal of office this 21st day of September 1832

Asa Dunning

Sir

Mount Hope Orange County N.Y. March 25th 1833

I herein return to your Office the papers of Hezekiah Woodward, he has made an additional declaration wherein he has answered all the objections noted in the Brief, and the Certificate of the Justice to the credibility of the second witness is procured and the proceedings duly certified.

I have twelve other cases in your Office a part of which were sent last Sept. Please not to permit them to lose their turn & direct the result to me as soon as examined.

Yours Respectfully
Joseph Chattle

Personally appeared before me the undersigned a Justice of the peace in the Town of Wallkill in the County of Orange and State of New York Hezekiah Woodward who being duly sworn deposeth and saith that by reason of old age and the consequent loss of memory he cannot swear positively as to the precise length of his service, but according to the best of his recollection he served not less than the periods mentioned below and in the following grades: For six months as a private under Captain Burnet at Deerpark Orange County State of New York protecting the frontiers against the Indians I served as a Sergeant as a Sergeant for six months under Captain William Faulkner. I served one month under Captain William Dennison. I served three months in Captain William Faulkners Company I served two weeks at Fort Montgomery. I served under Captain Benjamin Vail one week after a deserter and under Captain Vail at the indian battle three ~~weeks~~ days. I served under Captain Cuddeback at West point two weeks I served one month at Peenpack protecting the inhabitants against the indians under Capt. Cuddeback. and one month under Captain Owen at Westbrooks in Ulster County I served one third part of the time for three years in the Militia under different Captains a part of the time under Captain Vail part under Capt Cuddeback part under Captain Owen. I served two years and two months as a Sergeant in the different companies he has above mentioned and six months as a private and for ____ ____ service I claim a pension. And I further state that there is no settled Clergyman in the immediate neighborhood and never has been, there have been Preachers that have only stayed a few years at a time, and then left the place. I was a Sergeant all the time mentioned in my declaration annexed except six months and that all my services were in the Militia in regular organized companies the names of some of the Captains I cannot now recollect. I cannot state the particular periods of my service with more accuracy than I have stated them but am certain as to the length of the service as stated.

Sworn and Subscribed the 19th Day of March 1833 Before me
Alex Bell Justice of the Peace

Hezekiah Woodward

State of New York

Orange County ss. I Asa Dunning Clerk of the County of Orange & Clerk of the Court of Common pleas in & for the said County of Orange Do Certify that Alexander Bell Esq before whom the above affidavit was sworn and whose name appears to the preceding Certificate was at date of the same a Justice of the peace in & for the said county, and that I am acquainted with his handwriting & verily believe his name subscribed to the said affidavit & Certificate is his genuine signature Witness my hand & Seal of Office this 25th day of March 1833

Asa Dunning

New York

Hezekiah Woodward of Orange in the State of N. York who was a Private in the Company commanded by Captain Burnet of the Regt commanded by Col Hardenburg in the N York Militia for 2 years Inscribed on the Roll of New York at the rate of 80 Dollars Cents per annum, to commence on the 4th day of

March, 1831. Certificate of Pension issued the 11th day of June 1833 and sent Jos Chattle Mt Hope Orange Co

Arrears to 4th of Mar 1833	$160.00
Semi-anl. allowance ending 4 Sep 33	40.00
	$200.00

Recorded by Geo. B. Whiting Clerk,
{Revolutionary Claim, Act June 7, 1832.}
Book E Vol. 4 Page 37

Invalid File No. 11,865
Hezekiah Woodward Priv. Rev. War
Act: June 7th 1832
Index: Vol. A, Page 415
[Arrangement of 1870.]
Revolutionary War Pension and Bounty Land Warrant Application Files
Series M-804; Roll # 2639

Chapter 6
The Jerseymen

In the Summer previous, we had a fight with the Indians at Sheholi, under Major Samuel Meeker & Col John Seward. Capt. Hill was in that engagement. Daniel Talmage was killed in that fight, & I was shot in the fleshy part of the thigh, the wound is yet visible.

Isaac Bedell
Sussex County Militiaman at Battle at Minisink

The following pensions are from those enlisted men and officers who served in New Jersey units and are listed alphabetically. The notation at the end of the pension identifies the source of the material and if the application was rejected.

Isaac Bedell
New Jersey (S.15,322)

Declaration &c of Isaac Bedell
Declaration, In Order To Obtain
The Benefit Of The Act Of Congress, Passed June 7, 1832.

State of New-Jersey,
County of Morris ss. On this Twenty seventh day of July personally appeared, in open court, before the court of of Common Pleas now sitting Isaac Bedell a resident of the Township of Chatham in the County of Morris and State of New-Jersey, aged Sixty eight years, who being first duly sworn according to law, doth, on his oath, make the following declaration, in order to obtain the benefit of the Act of Congress, passed June 7, 1832.

That he entered the service of the United States, under the following named Officers, and served as herein stated:

1 - He says he was Born in the County of Essex and his father removed to Long hill in Morris County he believes in 1764, that he was born in the year 1763, that he has a Record of his age in an old bible. He lived at Long hill when he first went into the Service, he volunteerd and Sometimes served as a Substitute, he has lived most of the time since the War in Morris County, and that he served in the Militia of the State of New Jersey half a month in July of 1776 as a fifer ~~partly under Capt. Peter Layton~~ under Captain Stephen Day at Paulus Hook. 2d - served half a month as fifer partly under Captain Peter Layton and partly under Captain John Lindsley. 3d - went from Long hill to Elizabeth Town on an alarm, out four days

4th - Served one month as a Substitute for his father John Bedell in August 1777 at Elizabeth Town and Newark under Captain David Bates.

5th - In March and April 1778 served under Captain Israel Ward of Chatham one month at Elizabeth Town.

In April 1778 His Father removed to Sussex County and there this applicant in June was out as volunteer across the Delaware River after the Indians seven Days, in July was an Indian alarm and he volunteerd again and was out eight days and in august another and he was out 7 days. Then in September same year he served as a Substitute for his Father one month and in October was an alarm and he was out nine days. In these several Tours since April he was under the command of Captain Peter Westbrook, John Westbrook 1st Lieut, Gilbert Smith 2d Lieut

In June 1779 was an Indian alarm in which he served 9 days and again in august 6 days and was this time wounded by a musket shot in the thigh, and in October an Indian alarm in which he was out 5 days. In april 1780 an alarm again ~~of~~ in which he was out five days Jacob Bockover was our Captain in 1779

In ~~April 178~~ May and June 1780 he served one month under Coonrod Countryman along the Delaware River, and in June and July served one month under Lieut Benoni Ashton as a Substitute for Henry Bedell an uncle, and immediately after went out half a month on a Scouting Party to keep off the Indians and guard the Farmers in their harvesting and in which he was wounded in the head by an Indian bayonet in a single encounter, one other good soldier coming to his assistance they two took the Indian Prisoner

Never had a written discharge and knows William Parrot and a Mr Johnson to prove part of his service if he can obtain them.

He hereby relinquishes every claim whatever to a pension or annuity except the present, and declares that his name is not on the Pension Roll of the Agency of any State, is only on that of the Agency of the State of New Jersey.

Sworn to, and subscribed, the day and year aforesaid.

William Brittin

Isaac Bedell

State of New Jersey

Morris County ss. William Parrot appeard in open Court and being sworn on his oath says that he served in the Militia of the State of New Jersey in the time of the Revolutionary war as a soldier with Isaac Bedell and that said Bedell performed a months duty in March and April 1777 as he believes (possibly it may have been in 1778) in Captain Israel Wards Company, at Elizabeth Town in the County of Essex

Sworn and Subscribed this 27th day July 1832

William Brittin

William Parrot

We William Britton of Morris County & John Littell residing in the County of Essex hereby certify, that we are well acquainted with Isaac Bedell who has subscribed and sworn to the above declaration, that we believe him to be Sixty eight years of age; that he is reputed and believed, in the neighborhood where he resides, to have been a soldier of the revolution, and that we concur in that opinion.

Sworn and subscribed, the day and year aforesaid.

Thomas Dickerson

William Brittin
John Littell

And the said court do hereby declare their opinion, after the investigation of the matter, and after putting the interrogatories prescribed by the War Department, that the above named applicant was a Revolutionary Soldier, and served as he states. And the court further certifies, that it appears to them, that William Britton who has signed the preceding certificate, is resident in the County of Morris aforesaid, and that John Littell who has also signed the same, is a resident in the County of Essex and are credible persons, and that their statement is entitled to credit.

Thomas Dickerson
Stephen Vail
Silas Lindsey

I David Day Clerk of the Court of Common Pleas, do hereby certify that the foregoing contains the original proceedings of the said court in the matter of the application of Isaac Bedell for a pension. In Testimony Whereof, I have hereunto set my hand and seal of office, this twenty seventh day of July in the year of our lord one thousand eight hundred and thirty two.

David Day Clk

State of New Jersey

Morris County ss. Be it known that on this eighth day of June 1833, before the subscriber one of the Justices of the peace in & for said county personally came John Johnson well known to me as a man of truth & veracity, who being duly sworn, on his oath saith; that he was well acquainted with Isaac Bedell, now an applicant for a Pension under the act of June 7th 1832, in the war of the Revolution and knows that the said Isaac Bedell was engaged in active service in the militia of the state of New Jersey at various times & places as hereinafter stated, viz. that in the month of January 1777, while this deponent was doing militia duty in the Company commanded by Captain Thomas Dickerson at _______ near Pluckamin New Jersey, he often saw and was personally well acquainted with the said Isaac Bedell who was in the Company under Captain Peter Layton & ~~knows that~~ knew him at that place for the space of one month. That after this, in the month of August following this deponent was again in service at Elizabeth town, he thinks in a Company under Lieutenant Elisha Drake, (the Captain being sick) he again saw the said Isaac Bedell doing militia duty in a Company commanded by Capt David Bates for the space of 2 weeks, when both companies were marched to Newark where they ~~assisted~~ were employed two weeks more when both Companies were verbally dismissed. That in the month of October of the same year this deponent was again in service at Newark in the Company of Joseph Luse, and ~~was~~ again saw the said Isaac Bedell in the Company commanded by Capt. Stanbury, that both Companies together with other troops under Colonel Samuel Potter & General Winds, were marched through Hackensack up the North River to near Newburg, when the company in which Bedell was, was with others ordered back to Newark, this deponent with a portion of the troops going on to Newburg, & at the end of the month returning to Newark again saw the said Isaac Bedell & the Companies in which they both were, were verbally dismissed & went home. That in the month of March 1778 this deponent was again in service in Capt. Robert Youngs Company, & again saw the said Isaac Bedell in Capt. Israel Wards Company at ~~Newark~~ Elizabeth town & knows that they both served for the space of one month. And further this deponent saith not.

Sworn & subscribed before me the day & year last above written.
Charles Freeman J. Peace

his
John X Johnson
mark

I hereby certify that John Johnson, the within Deponent, is a man of good standing, & of fair character for veracity in Morris County. And that Charles Freeman Esqr. before whom the deposition was taken, is a Justice of Peace in the said County, & that the Signature of his name thereto is genuine.

Morris town, N.J. June 8th 1833.

Lewis Condict, late a Repr.
from N. Jersey

Isaac Bedell's Amended Declaration

State of New Jersey,

Morris County ss. Be it known that on this 8th day of June, 1833 before the subscriber a Justice of the Peace in and for said county, personally appeared Isaac Bedell well known to me as a man of truth and veracity, who being duly sworn saith, that he is the same Isaac Bedell who on the 27th day of July, 1832 made and subscribed a Declaration in order to obtain the benefit of the Act of Congress of the 7th June 1832, before the Judges of the Inferior Court of Common Pleas of said county. That at the time of making said declaration he was not aware of the necessity of detailing minutely particularly all the ~~rendered~~ services rendered by him in the War of the Revolution, that having since that time bestowed more thought upon the subject & refreshend his memory, he desires that this his amended declaration may be received by the Secretary of War & Commissioner of Pensions as setting forth with more clearness & more in detail the nature, times & places of the service done by him in the ~~War~~ Revolutionary war. That at the beginning of the War he was living with his father at Long Hill, Morris County, N. Jersey. That in July 1776 he enter'd the Militia of the State of New Jersey in the Company under the command of Captain Stephen Day, Uzal Johnson Lieutenant, Jacob Drake being Colonel of the Regiment. That he volunteered as a Fifer, the company was muster'd at Chatam and was marched from thence through Springfield & Newark to Poules Hook where the Company was stationed for 2 weeks upon guard duty preventing the enemy from plundering the inhabitants & the surrounding country, the company was then verbally dismissed & went home. In the month of October following, his father was called upon to do duty, but the Company being in want of a Fifer, this deponent volunteered, and enter'd the Compy of Capt. Peter Layton, & Colonel Drake at the head of the Regiment, & was stationed at Elizabeth town upon guard duty as before for the space of one month, acting as Fifer to the Company. That in November he served 2 weeks, the first week as Fifer in the Company of Capt. Layton, the other week as Fifer in the company of Capt. John Lindly, both companies being attached to the regiment of Colonel Jacob Ford. It was at the time when Washington retreated through N Jersey upon what was called the Mudrounds; with his company, went nearly to the Delaware river, having joined the main army in their retreat, at Elizabeth town, & then came by way of Pluckamin & Baskinridge to Morris town Again in January following, viz. '77, he went in place of his father John Bedell, to Van Muleners, near Quibbletown while the enemy lay at New Brunswick & its vicinity, and did guard duty as a private soldier for one month, under Capt Layton, & Lieutenant Hugh Caldwell, General Winds being in command of the militia. In August following, he again went out in his fathers place, to Elizabeth town as a private where he was stationed 2 weeks, the company was then marched to Newark & staid 2 weeks more; their duty was to guard the lines & protect the country from plundering parties of the enemy. Capt. David Bates, & Lieutenant Uzal Johnson were the Company officers. Major Seely commanding the whole, there being no Colonel present. In October following, he went out in place of a cousin Abraham Bedell, with whom he then lived, as a private under Capt. Stanberry, & Colonel Samuel Potter and General Winds; the company mustered at Newark, was marched through Aquacanoc to the North River, up nearly to Newburg, when he with a part of his Company went back to Newark & finished the month & was verbally dismissed at that place. In the year '78 on the 7th day of March he went out in place of his father to Elizabeth town & was stationed there 1 month upon guard duty as before doing duty as a private soldier. The officers of the company were Capt. Israel Ward, Benjamin Carter 1st Lieutenant, Phineas Farrand 2nd Lieutenant, David Provost Ensign, Colonel Seely & General Winds. In the latter part of the month of April his father removed to Sussex County N. Jersey, near the Delaware River. And in the month of June an alarm was given on account of the Indians, and he went as a volunteer under the command of Capt Peter Westbrook, Colonel Rosegrant & did guard duty 2 weeks, protecting the country from Indian depredations. Again he in October following was stationed under the same officers as on the preceding tour at Rosegrants fort on the Delaware & upon the same duty as before, there was at this time a large body of militia collected under the command of General Winds, an attack from the Indians being apprehended. Upon this tour he did duty as a private for one month. In the month of March '79 he went to live with his uncle at Pepper Cotton, in the same county Sussex, and in the month of August upon an alarm from the Indians, he went out in the Company of Capt. Jacob Bockover, Colonel ~~Seely~~ John Seward, & Major Samuel Meeker, the troops crossed the Delaware at Smiths ferry & followed the Indians as far as Scheholy, where a severe skirmish was had with them. Major Meeker was wounded by a musket shot in the shoulder. In this engagement this deponent was wounded by a bayonet on the right side of the head, by a musket ball in the right thigh & by a buck shot in the right leg, near the ankle, the scars of these wounds are still remaining. Many were killed on both sides, and of the American force the heaviest loss was sustained by the N. York troops. He did duty for 15 days as a private upon this occasion. The next year 1780 on the 2nd day of May he went out as Sergeant in a Company commanded by Capt. Conrad Countryman & Major Samuel Westbrook, and was stationed at Westbrook fort for the space of 1 month. He was elected Sergeant by the Company at a training in April preceding & received a commission from the Captain as Sergeant: He was dismissed on the 2nd day of June and immediately took the place of his uncle Henry Bedell & continued to do duty as Sergeant for one month more at the same place under Lieutenant Benoni Ashton, Capt Hill being sick. There was during this month a skirmish with the Indians in Pennsylvania when a Capt Westbrook & Lieutenant Daniel Amis was killed & in which this deponent was personally engaged. In the month of October following, he did another months duty as Sergeant at Chambers fort on the Delaware under Capt. Thomas Hill upon guard as before. In November, the next month he was called out by Ensign Adams of the Company to which he belonged to guard the public stores at Newton Sussex County & was upon this duty for 2 weeks as Sergeant of a guard of thirty men. Beside the several monthly tours above recited, this deponent was often called out upon alarms. When early in the war the British came over from Staten Island he was out 4 days under Lord Stirling, & in Capt Peter Laytons company, this was in June '77. When the Pensylvania troops revolted on the 1st of January '81, he was out 3 days with his Company at Woodbridge & its vicinity. Beside these there were many other alarms when he was out but is unable to recite them particularly, amounting as he knows to more than one month of active duty as a private. He has recited two months duty done as a Fifer; seven months duty as a private & three & a half months as Sergeant & for this service he claims a pension. This deponent received as before stated a Commission as Sergeant which has been Long since lost, but never received a written discharge. The clergyman at Mendham Rev Mr Johnson, where this deponent now lives has been settled there but a short time, is a young man & of course has no knowledge of the Revolutionary service of this deponent.

Sworn & subscribed the day & year above named before me
Charles Freeman Justice Peace

Isaac Bedell

I hereby certify, that Charles Freeman Esqr. before whom the forgoing Deposition was taken, is a Justice of peace in the County of Morris & State of New Jersey & that the signature of his name therto is genuine.

Morris town New Jersey 8th June, 1833.

Lewis Condict late a
Repr. from New Jersey

State of New Jersy

Morris County ss. Before the subscriber, a Justice of peace in said County, personally appears Isaac Bedell, the within applicant for a pension, well known to me as a man of credibility & good standing for truth, & being duly sworn, saith, that in his original declaration, it is not stated particularly, that he served in the Militia, in various capacities of private, fifer, & Sergeant, altho' he actually performed duties in each character & grade, & so stated to the person, (Mr. Littell), who drew up his papers, but was answered by Mr. Littill that the distinction was unnecessary, & would make no difference; & that it was because Mr. Little so informed this Deponent, that he did not insist on stating the fact, that he actually performed the Services, in each grade which are specified in his amended Declaration, which he prays may be recd. & admitted. He also declares under oath, that in the haste, in which his first Declaration was made, time was not allowed to recollect all his Services, but that he very soon remembered very distinctly, other monthly tours of Militia Service, which were actually performed, & which are distinctly set forth in his amended Declaration, which he prays may be recd. & for which he claims a pension.

Sworn & subscribed before me at Morris town, this 11th July, 1833.

Henry Stevens Justice of the Peace

Isaac Bedell

I hereby certify to the Secretary of War, & Commissioner of pensions, that Henry Stevens Esqr. before whom the foregoing Deposition was taken, is a Justice of peace in & for the County of Morris, & that the signature of his name thereto, is genuine.

Morris town, N.J. July 11th 1833.

Lewis Condict late a Repr. from N. Jersey

Ohio

Isaac Bedell of Chatham in the State of N.J. who was a Pri in the Com commanded by Captain of the Reg commanded by Col in the N.J. Militia 6 mo Musi. 2 mo Sergt. 3 ½ mo Inscribed on the Roll of NJ at the rate of 44 Dollars 83 Cents per annum, to commence on the 4th day of March, 1831. Certificate of Pension issued the 5 day of Augt 1833 and sent Hon L Condict

Arrears to the 4th of March ___	$89.66
Semi-anl. allowance ending Sep 4 ___	22.44
	$113.10

Recorded by E D Bullock Clerk,
{Revolutionary Claim, Act June 7, 1832.}
Book E Vol. 3 Page 102

Application for transfer to Ohio agency

State of New Jersey, Morris County ss:

On this fifth day of September, Domini 1842, before me, the subscriber, a Justice of the peace for the said County of Morris, personally appeared Isaac Bedell, who on his oath declares, that he is the same person who formerly belonged to the company commanded by Captain Peter Layton, of Morris County Militia, in the Regiment commanded by Colonel Jacob Ford & Col: Seely, in the service of the United States in the Revolutionary war: that he now is in the State of New Jersey, where he has lived all his days, but is on the point of removal with his family to the State of Ohio, where he intends & expects to remain, & wishes his pension to be there payable in future.

The following are his reasons for removing from Morris County to Ohio. His son Isaac Bedell Junr. owned the place or little farm on which we have all lived together as one family. In March last, he Sold said place to James B. Millen, & is now almost ready to start with his family for Ohio, & I expect to go with him. It is my expectation & intention, to reside in Lodi township, Athens County Ohio; & it is my desire, that the Commissioner of Pensions, will enclose my new Certificate for my pension, to my address, under his parole, at Shade Post Office, Lodi township Athens Co. Ohio, making it payable at the nearest Agency from Athens, as he may judge shall be most convenient & least expensive to me. My name was placed on the Pension roll of the State of New Jersey, & my certificate of pension, dated August 5th 1833, which has been regularly paid to me, is herewith returned, for exchange or transfer to Ohio.

Sworn & Subscribed to, before me, the day & year aforesaid.

Silas L. Condict J P

Isaac Bedell

Morris County, State of New Jersey ss:

Before the Subscriber, a Justice of peace in said County, personally appeared Lewis Condict known to me as a man of veracity, who being by me duly sworn according to law, saith:

I have been personally well acquainted with Isaac Bedell who has taken the above oath for ten years or more, & know him to be the same person described in the affidavit.

Sworn & Subscribed, September 5th 1842, before me,

Silas L. Condict J P

Lewis Condict

And I also certify, that I am well acquainted personally with said Isaac Bedell, & know him to be the same person described in the above affidavit, signed by him.

Silas L. Condict J P

State of New Jersey Morris County ss:

I, David B. Hurd, Clerk of Morris County aforesaid, hereby certify, that Silas L. Condict, the Magistrate before whom the within affidavits of Isaac Bedell & of Lewis Condict were taken & Subscribed, is a Justice of the peace in & for said County, duly commissioned & authorized, & that the signature of his name thereto, "Silas L. Condict," is genuine. Witness my signature, & Seal of my office at Morris town, this 6th day of September, 1842.

David B Hurd Clerk

Morris Town, N.J. Sept. 26th, 1842

Dear Sir,

Enclosed is the Pension Certificate of Isaac Bedell who desires for reasons assigned in his Deposition which accompanies it, to be transferred to an Agency in Ohio, for which State he has started, never expecting to return. In his behalf, I will request that his future payments may be made at whichsoever Agency in Ohio may be nearest his residence in Athens Co. And to save him postage, will you be kind enough to enclose the certificate to his address, Lodi township Athens Co. Ohio, Shade Post Office, under your parole?

Very respectfully Your obt. Servt.
Lewis Condict.

Col: J. L. Edwards
Commr. Of Pensions

Trenton Bank
octr 24, 1842.

Sir,

Isaac Bedell a pensioner under the Act of June 7, 1832 at the rate of 44 83/100 per annum was paid up to the 4th Sep 1842.

Your instruction as to his transfer to Ohio is noted & our books regulated accordingly

Very Respectly yr ob ser

John Dickinson Pension
Agent N. Jersey

J. L. Edwards Esqr
Commnr of Pensions

Thomas H. Schooley
Attorney
293 Avery Ave. Ph. Grand 5286
Detroit

Octo. 7th, 1918

Commissioner of Pensions,
Washington, D.C.

Dear Sir:

In a volume named "Senate Documents, Pension Roll, 1st Sess, 23d Cong." under the head "New Jersey, Morris Co." on page 614, or page 68 of the New Jersey pensions is the following: Isaac Bedell, Pri. Mus. Serg't, New Jersey Militia. Placed on Pension Roll 1833, age 70.

Will you be kind enough to send me a certificate showing the facts as stated in his petition for a pension?

If the petition states names of his parents, his brothers or sisters, where he was born and when, when and whom he married, names of his children, where he died and when, if pension was allowed to his widow if she survived him.

You will have our sincere thanks for this favor.

Respectfully,
Ths Schooley

C.O.S.

Revolutionary War Records Section.
Department of the Interior,
Bureau of Pensions

Washington, D.C., , 19

In reply to your request of , received for a statement of the military history of Isaac Bedell a soldier of the Revolutionary War, you will find below the desired information as contained in his (or his widow's) application for pension on file in this Bureau.

DATES OF ENLISTMENT OR APPOINTMENT.	LENGTH OF SERVICE	RANK.	OFFICERS UNDER WHOM SERVICE WAS RENDERED.	OFFICERS UNDER WHOM SERVICE WAS RENDERED.	STATE
			CAPTAIN.	COLONEL.	
July 1776	served at various times until 1781	Fifer & Sergt.	Stephen Day	Samuel Potter	N.J.
			Peter Layton	Jacob Drake	
			John Lindsley David Bates	Jacob Ford	
			Israel Ward Peter Westbrook	John Seward	
			Jacob Bockover, Coonrod Countryman Stanbury	Seely	

Battles engaged in, was wounded in thigh, leg & head while out in alarms with Indians.

Residence of soldier at enlistment, Long hill, Morris Co. N.J.

Date of application for pension, July 27, 1832. His cl. was al.

Residence at date of application, Chatham, Morris Co. N.J.

Age at date of application, born in 1763 in Essex Co. N.J.

Remarks: Soldier was the son of John Bedell. In 1842 he left New Jersey with his son Isaac Bedell Jr. and went to Lodi Township Athens Co. Ohio, to live. It is not stated whether there were other children and there is no mention of his wife, mother, brothers or sisters, or the date & place of his death. Uncle Henry Bedell

Respectfully,
Commissioner.

Invalid. File No. 15,322
Isaac Bedell Pri Rev War
Act: June 7th 32
Index: Vol. 1, Page 31
[Arrangement of 1870.]
Revolutionary War Pension and Bounty Land Warrant Application Files
Series M-804; Roll # 198

Solomon Cartwright
New Jersey (S.4,993)

BRIEF in the case of Solomon Cartwright of Tompkins county in the State of New York

(Act 7th June, 1832.)

1. Was the declaration made before a Court or a Judge? court
2. If before a Judge, does it appear that the applicant is disabled by bodily infirmity?
3. How old is he? 80
4. State his service, as directed in the form annexed.

Period.	Duration of Service			Rank.	Names of General and Field Officers under whom he served.
	Years	Months	Days		
Volunteer in 1776		1		Private	Capt. Harper, Col Martin
Volunteer June 1776		5		Private	Capt. Harper, Col Martin
Volunteer 1777 & 78		2		Private	Capt. Cole and Capt. Coykendoll
Enlisted 1781		9		Private	Capt. Bunnell, Col Martin

5. In what battles was he engaged?
6. Where did he reside when he entered the service? Sussex county, New Jersey
7. Is his statement supported by living witnesses, by documentary proof, by traditionary evidence, by incidental evidence, or by the rolls? living witness, tradition

8. Are the papers defective as to form or authentication? and if so, in what respect? Not ______ as required

Certify that the foregoing statement and the answers agree with the evidence in the case above mentioned.

______ Bradford Examining Clerk.

Declaration in order to obtain the benefit of the act of Congress, passed June 7th, 1832.

State of New York

Tompkins County ss. On this 3rd day of October personally appeared in open court before the Judges of the Court of Common Pleas of said county now sitting, Solomon Cortwright a resident of the town of Danby in the county of Tompkins and state of New York, aged Eighty years, who being first duly sworn according to law, doth on his oath make the following declaration in order to obtain the benefit of the act of Congress, passed June 7th, 1832. That he entered the service of the United States, under the following named officers and served as herein stated.

That he first volunteered into said service for one month at the town of Wantage in Sussex County, New Jersey where he then resided, in the month of March 1776, in Capt. Joseph Harpers company and Col. Martins Regiment of N. Jersey Militia, immedeatily after his enlistment he march with his said company to New York & there continued in said service during the month when he was discharged & he returned home about the first of April 1776. That afterwards and about the first of June 1776 he volunteered a private into the same service, at the town of Wantage aforesaid for the term of five months in Captain Sowards company in said Col. Ephraim Martins Regiment of N.J. Militia, that immedeatily or soon after his said enlistment for five months he joined his said Regiment at Morristown N.J. about the twentieth of June 1776 and marched from thence under Captain Bond in the aforesaid Regiment to New York, thence to Long-Island when the British Army landed on Long-Island. They were forced by the British back to New York and from thence to Fort Washington near Kings Bridge. From thence this declarant with said regiment crossed the Hudson River to Fort Lee and from thence they marched to Hackinsack New Jersey where after a short time their term of service expired and this declarant was discharged on the last day of November 1776 and then returned to his place of residence in the town of Wantage, aforesaid.

That during the years 1777 & '78 he volunteered with his company into the monthly service at first for one month in Capt. Cole's company and afterwards for one month in Capt. Coykendolls company in said Col. Martins Regt. and during the said two months service by this declarant such service was rendered by standing as a guard at Morristown but the precise time of that sercice he cannot now recollect, that after that & during the years 1778, 79 & 80 he was several times called into actual service for a few weeks at a time, but this declarant cannot now state the precise times nor the names of his officers.

That on the first day of March 1781 at the said town of Wantage, county & state aforesaid this declarant enlisted into the same service for the term of nine months in Capt. James Bunnels company in which Caleb Hopkins was first Lieutenant and in Col. ~~Isaac~~ Ephraim Martin's Regiment of New Jersey Militia as aforesaid in which said Regt. John Welling was Quarter Master. That under his said last mentioned enlistment he served as second Sergeant in the aforesaid Capt. Bunnels company for nine months and during the term of such enlistment at and in the immediate vicinity of a place called the Block House in the town of Montague in said County of Sussex & state of New Jersey, that during that service he was employed in guarding the banks of the Delaware against innovations by the British and indians. That he was discharged from said service at the Block House about the first day of December 1781 and immedeatily returned to his parents with whom he resided at the town of Wantage aforesaid, that during his said service he was in the Battle at Flatbush on Long Island, at Tide Mills, at Elizabethtown N. Jersey, at Aquehanock, Lackawax at a Battle against the Indians & tories.

That he has no documentary evidence and that he knows of no person whose testimony he can procure who can testify to his service other than Moses Brink and Solomon Middough whose affidavits are hereunto annexed.

He hereby relinquishes every claim whatever to a pension or annuity except the present, and declares that his name is not on the pension roll of the agency of any state.

Sworn to and subscribed, the day and year aforesaid.
S. Love Clk

his
Solomon X Cortwright
mark

The following questions were propounded by the Court to the applicant.

1. Where and in what year were you born?

Answer. in Wantage Sussex County N.J. in 1752 the 8th of July

2. Have you any record of your age, and if so where is it?

Answer. I have not here but there is a record in New Jersey where I was baptized

3. Where were you living when called into service; where have you lived since the Revolutionary war; and where do you now live?

Answer. When I went into the service I lived in Wantage N. Jersey - has since lived at at same place till about 25 years ago he moved to Danby, where he now lives

4. How were you called into service; were you drafted, did you volunteer, or were you a substitute; and if a substitute, for whom?

Answer. I volunteered

5. State the names of some of the regular officers who were with the troops when you served; such continental and militia regiments as you can recollect, and the general circumstances of your service.

Answer. Genl. Washington, Genl. Putnam, has seen Genl. Lee at Elizabethtown Col. Martin, Capt. Cole & Coykendoll in the militia His service is stated in the body of his declaration

6. Did you ever receive a discharge from the service, and if so by whom was it given, and what has become of it?

Answer. I never received any written discharge

7. State the names of persons to whom you are known in your present neighborhood, and who can testify as to your character for veracity and their belief of your services as a soldier of the Revolution.

Answer. Judge Jennings Hudson Jennings Stephen ________ Hezekiah Clark Chester W Lord

---

We, Solomon ________ a Clergyman residing in the town of Danby and county of Tompkins and State of New York, and Benjn. Jennings residing in the same place, hereby certify that we are well acquainted with Solomon Cortwright who has subscribed and sworn to the above declaration, and that we believe him to be Eighty years of age, that he is reputed and believed in the neighbourhood where he resides, to have been a soldier of the Revolution, and that we concur in that opinion.

Sworn and subscribed, the day and year aforesaid,
S. Love Clk

Solomon ________
Benjn. Jennings

And the said Court do hereby declare their opinion after the investigation of the matter, and after putting the interrogations prescribed by the War Department, that the above named applicant was a Revolutionary Soldier, and served as he states, and the Court further certifies that it appears to them that Solomon ________ Clergyman & Benjn. Jennings who have signed the preceding certificate are credible citizens resident in the town of Danby and county of Tompkins and are creditable persons, and that their statements are entitled to credit.

State of New York

Tompkins County ss. Moses Brink of the town of Candor in the County of Tioga in the state aforesaid being duly sworn doth on his oath depose and say that he is sixty nine years of age on the tenth day of March next that he was born and brought up in the town of Montague in the County of Sussex and State of New Jersey and that he resided there until about twenty years after the close of the revolutionary war. And this Deponant further saith that he was well acquainted with Solomon Cortwright who has subscribed and sworn to the annexed Declaration previous to the Revolutionary war, that he first became acquainted with the said Solomon Cortwright when he was but a child about eight or nine years old, that at that time the said Solomon resided in the town of Wantage in the County of Sussex & State of New Jersey about nine miles from the residence of this deponant. That he has been well acquainted with the said Solomon Cortwright from that time to the present. That he saw the said Solomon Cortwright actually engaged as a soldier in the American Revolution in a company of militia of the State of New Jersey commanded by Capt James Bunnel at the Block house in the aforesaid town of Montague & County of Sussex & state aforesaid in early part of the year 1781. That he saw him the said Solomon at and near that place almost daily during the spring, summer and fall months of the year 1781 engaged as aforesaid in said Capt Bunnel's company & Col. Isaac Martins Regiment, that the Block house where the said Solomon was in the aforesaid service during the spring summer and fall aforesaid was situate about three quarters of a mile from where this Deponant then resided with his parents. That during the aforesaid period of about nine months while he was knowing of the said Solomons service as aforesaid he this deponant was induced to visit the Block house where the said company were stationed from his previous auquaintance with the said Solomon Cortwright and others who served with him in the aforesaid company commanded by Capt Bunnel.

Subscribed & sworn before me this 3rd day of October 1832

Benjn. Jennings one of the Judges of Common Pleas for said County

his
Moses X Brink
mark

Tompkins County ss. I, Benjamin Jennings, one of the Judges of the Court of Common Pleas in and for the County of Tompkins Do hereby certify that I am well acquainted with Moses Brink and Solomon Middough who have subscribed & sworn to the two several affidavits hereunto annexed and that their reputation for truth is good and in my best Judgement & from my acquaintance with them in their respective neighborhoods they are entitled to fully credit. Given under my hand this 4th day of October 1832

Benjn. Jennings

In the matter of the application of Solomon Cortwright
for a Pension
Declaration &c.

State of New York

Tompkins County ss. Solomon Middough of the town of Caroline in the county and state aforesaid being first duly sworn in open court before the Judges of the Court of Common Pleas in & for the said County of Tompkins doth on his oath depose and say that he was well acquainted with Solomon Cortwright who has subscribed and sworn to the preceding declaration before and during the Revolutionary war, that this Deponant was born and brought up in the same town with the said Solomon Cortwright viz in the town of Wantage, in the County of Sussex and State of New Jersey. That during the Revolution and in the month of March 1776 this deponant and the said Solomon Cortwright served a campaign of one month, together in Capt Joseph Harpers company in Col Martins Regiment in the service of the United States in the City of New York where they were employed in repairing batteries at that place. That immediately or soon after the said Solomon Cortwright returned from the aforesaid service he the said Solomon Cortwright again enlisted into the said service at Wantage N.J. for the term of five months, in Capt. Sowards company and in Col. Ephraim Martins Regiment of the Militia of said State, that the said Solomon Cortwright was absent from home during the summer of 1776 and this deponant has always understood & believed & still believes the said Solomon Cortwright was during that time engaged in said Corps in the aforesaid service, that this Deponant was present at the time of his said last mentioned enlistment at Wantage. That during the month of January 1777 the Solomon Cortwright was in the service aforesaid for one month at Scotch plains N. Jersey in the same company with this Deponant, that at that campaign this deponant in the absence of Capt. Cole commanded said company as Lieutenant Commandant, in Col. Ephraim Martins Regiment and afterwards commanded by Col. Hankison.

That in the summer of 1778 the said Solomon Cortwright was engaged as a soldier for one month in the aforesaid service on the Delaware River in the town of Minisink in a company then under the command of this deponant as first Lieutenant, in Col Hankisons Regiment then commanded by Lieutenant Col Soward, and were engaged during that time in constructing a fort and scouting at that place and along the Delaware River.

And this Deponant further saith that the said Solomon Cortwright was in said service in the aforesaid company under the command of Capt. Cole & this Deponant as Lieutenant for a short tour on the Delaware River at Lackawax among the indians in the month of July 1779.

That he then understood and believed and still verily believes that during the years 1780 & '81 the said Solomon Cortwright was engaged in said service mostly within the State of New Jersey under an enlistment for nine months, during and for that period. That such was the report in the neighborhood at that time where he lived and this Deponant knows that during a part of that period the said Solomon Cortwright was absent from home. And further saith not.

Subscribed & sworn in open Court aforesaid before the Judges thereof this 4 day of Oct. 1832

Saml. Love Clk

Sollomon Middaugh

I Samuel Love Clerk of the Court of Common Pleas for the County of Tompkins do hereby certify that the foregoing contains the original proceedings of the said Court in the matter of the application of Solomon Cortwright for a pension. In Testimony Whereof, I have hereunto set my hand and seal of office, this 4th day of October 1832

Saml. Love Clk

BRIEF objections to the admission of Pension Claims, of which those answered in the negative, in red ink, will apply to the claim of Solomon Cartwright of Tompkins co New York which require further proof or explanation.

I. Does the declaration show where the applicant resides? yes

II. Did the declarant make the declaration in the county where he resides? yes

III. If the applicant did not make the declaration in the county where he resides, is there any reason assigned for his not doing so?

IV. Has the declarant mentioned the period or periods of the war when he served? yes

V. Has he mentioned with precision the length of his service, and the different grades in which he served, in language so definite as to enable the department to determine to what amount of pension he is entitled? yes

VI. Has he given the names of the officers under whom he served, in conformity with the regulations? yes

VII. Has he made a relinquishment of every claim to any other pension than the one under the Act of June 7, 1832? yes

VIII. Has the Court given their opinion? yes

IX. Has the Clerk given his certificate? yes

X. Is the Clerk's seal affixed? And if so, has it a device or inscription by which it can be distinguished from any other seal? yes

XI. Has the applicant obtained the evidence of a clergyman and another respectable citizen as to their belief respecting his age and the general belief in his neighborhood relative to his revolutionary service? yes

XII. If the clergyman's affidavit has not been produced, is there any reason assigned for not obtaining it?

XIII. If the applicant has no documentary evidence, and has not obtained the testimony of at least one living witness, has he stated in his declaration that such proof cannot be had?

XIV. If the applicant's name is not found on the rolls of military service, has he produced a witness or witnesses as to his service?

If he has produced direct proof as to service, has the magistrate who administered the oath, certified to the credibility of the witness or witnesses? yes

Has the proper officer certified to the official character and signature of the magistrate as the regulations direct? yes

XV. Does it appear from any of the papers that the witness was in such a situation, or of such an age, as to have a personal knowledge of the applicant's service? yes

XVI. Are the papers authenticated as the regulations direct? No

XVII. Are the seven interrogatories prescribed by the War Department answered satisfactorily? And, if not, which of them is not so answered? all answered

EXPLANATORY NOTES.

QUERE V. In a case where the applicant cannot, by reason of the loss of memory, state precisely how long he served, he should amend his declaration by making an affidavit in the following words: "Personally appeared before me, the undersigned, a Justice of the Peace, &c. A. B. who, being duly sworn, deposeth and saith, that, by reason of old age, and the consequent loss of memory, he cannot swear positively as to the precise length of his service, but according to the best of his recollection he served not less than the periods mentioned below, and in the following grades: For year months, and days, I served as a For months and days, I served as a ; and for such service I claim a pension."

It is important in all cases to determine with precision the period for which each applicant served, and the particular rank he held, as the law directs the pension to be paid according to the grade of the pensioner and the length of his service. The use of the phrase about three or four months is too indefinite, and all such qualifying expressions are objectionable. Some persons who apply for pensions merely state they served two years in the militia, &c. without specifying the tours; the names of the officers, and other particulars respecting their service. This form of a declaration is highly objectionable. It must in every case be clearly shewn under what officers the applicant served; the duration of each term of engagement; the particular place or places where the service was performed; that the applicant served with an embodied corps called into service by competent authority; that he was either in the field or in garrison; and for the time during which the service was performed, he was not employed in any civil pursuit.

QUERE XIV. Proof of Service.

In a case where the name of the applicant is not found on the records of the Department, he must prove his service by two credible witnesses, who are required to set forth in their affidavits the time of the claimants entering the service, and the time and manner of his leaving the same, as well as the regiment, company, and line to which he belonged. The magistrate who may administer the oaths must certify to the credibility of the witnesses, and the official character and signature of the magistrate must be certified by the proper officer, under his seal of office.

QUERE XVI. Mode of authenticating papers.

In every instance where the certificate of the certifying officer who authenticates the papers is not written on the same sheet of paper which contains the affidavit, or other paper authenticated, the certificate must be attached thereto by a piece of tape or narrow ribbon, the ends of which must pass under the seal of office of the certifying officer, so as to prevent any paper from being improperly attached to the certificate.

* In every case where the clerk of the court, or other certifying officer, has no public seal of office, the certificate of a member of congress, proving the official character and signature of the certifying officer, should accompany the papers.

~~New York~~ Ohio

Solomon Cartwright of Tompkins Co. in the State of N. York who was a private in the Comp. commanded by Captain Harper of the Regt. commanded by Col. Martin in the New Jersey line for 17 mos. Inscribed on the Roll of N.

York at the rate of 56 Dollars 66 Cents per annum, to commence on the 4th day of March, 1831. Certificate of Pension issued the 12 day of Sept. 1833 and sent to S. Crittenden Ithaca

Arrears to the 4th of Sept. '33	$113.33
	28.33
Semi-anl. allowance ending 4 Sep '33	28.33
	$169.99

Recorded by ~~Geo C. Stiles~~ W. L. Williams Clerk,
{Revolutionary Claim, Act June 7, 1832.}
Book E Vol. 4 Page 52

In the matter of the application of Solomon Cortwright
for a Pension
Declaration &c.

State of New York

Tompkins County Clerks Office I Samuel Love Clerk of the Court of Common Pleas for the County of Tompkins Do hereby certify that the following Papers, hereunto annexed contain the original proceedings of the said Court in the matter of the application of Solomon Cortwright for a Pension. In Testimony whereof I have hereunto set my hand and seal of office this 16 day of May A.D. 1838

Saml. Love Clerk Tomp. County

The State of Ohio

Delaware County ss. On this 8th day of February A.D. 1839, before me the subscriber a justice of the peace for the said county, personally appeared Solomon Cartwright who being duly sworn upon his oath declares that he is the identical person to whom a pension Certificate was issued in conformity with the law of the United States of the 7th of June 1832, by which he the said Solomon Cartwright late a private in the army of the Revolution was inscribed on the pension list, Roll of the New York agency, at the rate of fifty six dollars and 66 cents per annum during his natural life commencing on the 4th of March 1831, and bearing date at the War Office of the United States the 12th day of September 1833, and numbered 20417. That his name was placed on the pension roll of the State of New York from whence he has lately removed; that he now resides in the State of Ohio, where he intends to remain, and wishes his pension to be there paid in future. The following are his reasons, for removing from the state of New York to the State of Ohio, to wit; That his children & friends were residing in the County of Delaware in the State of Ohio, and it was his wishes to reside with or near them, and therefore removed to the state of Ohio.

Sworn & subscribed to the day & year above written. (The words "of Ohio" interlined before _______) Before me Joel Z Mendenhall Justice of the peace

Signed & sworn to in the presence of
J. W. Powell
John Ward

his
Solomon X Cartwright
mark

Also at the same time & place personally came Ira Finch and Solomon Cartwright junr. who being duly sworn upon their oaths say that they are personally well acquainted with the within named Solomon Cartwright and also appeared before the said justice and signed, & was sworn to, the foregoing affidavit, that they personally know that he is the person to whom was issued the pension Certificate, and the identicle pensioner described in the said affidavit & certificate and know the facts in the said affidavit are true, and further say not.

Ira Finch
Solomon Courtright Jr

Sworn & subscribed to before me on this 8th day of February 1839, and I certify that I am well acquainted with the above named Ira Finch & Solomon Cartwright junr. that they are persons of good character for truth and entitled to full belief.

Joel Z Mendenhall Justice of the Peace

State of Ohio

Delaware County ss. I Thomas Reynolds Clerk of the Court of Common pleas of the County & State aforesaid, do hereby certify, that Joel Z Mendenhall Esqr. is a Justice of the peace in & for said County duly commissioned & qualified; that his commission was dated on the 18th day of Decr. 1838 & will expire on the 8th day of Decr. 1841, & that his signatures before written, I believe to be genuine. Given under my hand & the seal of said County, this 8th day of Feby. AD 1839.

Thos. Reynolds Clerk
by Wm. D. Heim Dep.

⋆

The State of Ohio
Delaware County ss. On this 4th day of May in the year 1839 before me the subscriber a justice of the peace for the said County of Delaware personally appeared Solomon Cartwright late of the state of New York who on his oath, declares that he is the same person who formerly belonged to the company commanded by Captain William Bond in the Regiment commanded by Colonel Ephraim Martin in the New York State Militia in the Revolutionary War and in the service of the United States; that his name was placed on the pension roll of the State of New York from whence he has lately removed; that he now resides in the state of Ohio, where he intends to remain, and wishes his pension to be there payable in future. The following are his reasons for removing from New York to the state of Ohio, to wit; that his children had previously removed to Delaware County in the state of Ohio, and that it was his wishes to reside with them and therefore removed to this county.

Subscribed and sworn to on the day & year above written before me.
Joel Z Mendenhall Justice of the peace

his
Solomon X Cartwright
mark

⋆

And also at the same time and place came Elias Vansickle who being duly sworn upon his oath says that he has been well acquainted with the above named Solomon Cartwright who appeared and made oath to the above affidavit and signed the same, for the last thirty years, and personally knows him to be the identicle person to whom a pension was granted as he claims in his above affidavit, and further say not.

Elias Vansickle

Subscribed & sworn to the day & year first above written before me and I certify that I am personally acquainted with the said Elias Vansickle and that he is a person of good character for truth and his statements are entitled to full belief.

Joel Z Medenhall Justice of the peace

⋆

State of Ohio
Delaware County ss. I William D Heim Clerk of the Court of Common

Pleas in & for said County, do hereby certify that Joel Z Mendenhall Esqr. is a Justice of the Peace in & for said County, duly commissioned & qualified; that his commission was dated on the 18th day of Decr. AD 1838 & will expire on the 18th day of Decr. AD 1841 & that his signature above written I believe to be genuine. Given under my hand & the seal of said County this 6th day of May AD 1839.

W D Heim Clerk

Delaware(Ohio) May 6, 1839.

Sir,

At the request of Solomon Cartwright I forward the enclosed vouchers &c for the purpose of transferring the payment of his pension from New York to this state. The certificate of transfer you may forward to my care at this place.

J. W. Powell

Honl. J. L. Edwards Comr. of Pensions

Mechanics' & Farmer's Bank,
Albany
June 10, 1839

James L. Edwards, Esquire,
Pension Agency,
Commissioner of Pensions.

Sir,

In reply to your letter of the 7th inst. I have to state, that Solomon Cartwright a Pensioner under the Act of 7 June '32 at 56.66 Dollars per annum was last paid to the 4 day of Sept. 1838.

I am, Sir,
Very respectfully,
Your ob't serv't,
T. W. Olcutt
Pres't and Pension Agent.

The Pension Dept.

Bowling Green O. Feb 26 - 31

I am writing for information regarding a pensioner of the Revolutionary war, Solomon Cortwright of the Sussex Co. N.J. Militia Would like to know his marriage date and residence after removing to Ohio.Also any information will be gratefully received.

Resp
Mrs. W. W. Waugh
330 S. Grove St.
Bowling Green O

Pension Dept.

Bowling Green O. Mar 11 - 31.

My ancestor Solomon Cortwright a soldier of the Revolutionary war was a pensioner up to the time of his death in 1844 could you give me any information as to his place of residence & anything concerning his wife Hannah Ayers Cortwright.

Resp Yours
Mrs. W. W. Waugh
330 S. Grove St.
Bowling Green O.

April 6, 1931
Rev. & 1812 Wars Section
MLB

Mrs. W. W. Waugh
330 South Grove Street
Bowling Green, Ohio
Dear Madam:

You are advised that it appears from the papers in the Revolutionary War pension claim, S.4993, that Samuel Cartwright or Cortwright was born July 8, 1752 in Wantage, Sussex County, New Jersey.

While residing in Wantage, New Jersey, he volunteered in March, 1776 and served with the New Jersey troops, as follows: - one month in Captain Joseph Harker's company, Colonel Martin's regiment; from June, 1776, five months as private in Captains Soward's and Bond's companies in Colonel Ephraim Martin's regiment and was discharged November 30, 1776. During 1777 and 1778 he served two months as guard at Morristown, in Captains Cole's and Coykindall's companies in Colonel Martin's regiment. In 1778, 1779 and 1780 he served at various times, a few weeks at a time, dates of service and names of officers not given. He enlisted March 1, 1781 and served nine months as second sergeant in Captain James Bunnell's company, Colonel Ephraim Martin's regiment and was stationed at the Block House, in Montague, Sussex County, New Jersey, where he was discharged about December 1, 1781. During his service he was in the battles of Flatbush, Tide Mills, Elizabethtown, Aquahanock and Lackawax.

He was allowed pension on his application executed October 3, 1832, while a resident of Danby, Tompkins County, New York, where he had lived for twenty-five years.

In 1839 the soldier was living in Delaware County, Ohio, where he stated he had removed to be with his children.

There are no data on file in regard to soldier's marriage and no record of his death.

Very truly yours
E. W. Morgan
Acting Commissioner

Charles C. Pinkerton, M. D.
435 Second National Bldg.
Akron, Ohio
November 21, 1933

Commissioner of Pensions,
Department of the Interior,
Washington, D.C.
Dear Sir:

I should like a statement of the military history of Solomon Kortryk (Kortreyght Cortreight or Cortrite) a soldier of the American Revolution. I am not certain as to the spelling of his name but he is said to have been a member of the third Battalion of Sussex County, New Jersey Militia. This information is sought for membership in the Daughters of the American Revolution.

With sincere thanks, I am
Very truly yours,
Chas. C. Pinkerton

MP:HVD

Record Division

Revolutionary War Records Section
Department of the Interior,
Bureau of Pensions.

Washington, D.C., , 19

In reply to your request of , received for a statement of the military history of Solomon Cartwright/Cortwright a soldier of the Revolutionary War, you will find below the desired information as contained in his (or his widow's) application for pension on file in this Bureau.

DATES OF ENLISTMENT OR APPOINTMENT.	LENGTH OF SERVICE	RANK.	OFFICERS UNDER WHOM SERVICE WAS RENDERED.	OFFICERS UNDER WHOM SERVICE WAS RENDERED.	STATE
			CAPTAIN.	COLONEL.	
Mar 1776	and served 17 mos at various times until Dec. 1781	Pvt. & 2nd Sgt.	Joseph Harp(k)er	Ephraim Martin	N.J.
			Soward		
			Bond		
			Cole		
			Coykendall		
			James Bunnell		

Battles engaged in, Flatbush, Tide Mills, Elizabethtown Aquahanock & Laxkawax

Residence of soldier at enlistment, Wantage, N.J.

Date of application for pension, Oct. 3, 1832 His cl. was allowed

Residence at date of application, Danby Tompkins Co NY

Age at date of application, b July 8, 1752 in Wantage, Sussex Co. N.J.

Remarks: In 1839 he was living in Delaware Co. Ohio with his children. Date of death not on record.

Respectfully,
Commissioner.

December 1, 1933.
BA-J/ILL
Solomon Cartwright
S.4993

Dr. Charles C. Pinkerton
35 Secon National Building
Akron, Ohio.
Dear Sir:

Reference is made to your letter in which you request the record of Solomon Kortryk (variously spelled), a soldier of Sussex County, New Jersey, in the Revolutionary War.

The record of Solomon Cartwright, as his name is borne on the pension rolls, is furnished herein as found in the papers on file in his claim for pension, S.4993, based upon his service in that war.

Solomon Cartwright was born July 8, 1752 in Wantage, Sussex County, New Jersey, and resided there with his parents during the Revolutionary War; the names of his parents are not stated.

He volunteered and served in the New Jersey troops as follow: from March 1, 1776, one month in Captain Joseph Harker's company, Colonel Ephrain Martin's regiment; from about June 1, 1776, in Captains Seward's and Bond's companies, Colonel Ephraim Martin's regiment, was in the battle of Flat Bush on Long Island and was discharged at the expiration of service of five months; in January, 1777, one month in Captain Cole's company, Colonels Ephraim Martin and Hankinson's regiment; in the summer, 1778, one month, Captain Kirkendall's company, Lieutenant Colonel Seward and Hankinson's regiment; in July, 1779, one month in Captain Cole's company at Lackawaxon and was in an engagement with the Indians; a short tour in 1780, no officers named; from March, 1, 1781, nine months as second sergeant in Captain James Bonnel's company, Colonel Isaac Martin's regiment, on duty at the Block House in Montague, Sussex County, New Jersey. He was in engagements at Tide Mills, Elizabethtown, and at Aquahanock.

After the war, the soldier resided in Wantage, New Jersey, then moved to Danby, Tompkins County, New York.

Solomon Cartwright was allowed pension on his application executed October 3, 1832, at which time he resided in Danby, Tompkins County, New York; he had lived there for about twenty-five years.

In February, 1839, the soldier was a resident of Delaware County, Ohio; he stated then that he had moved to that county recently from New York State because his children resided in Delaware County, Ohio; he did not give the names of said children, nor is the name of his wife shown in the papers in this pension claim. Solomon Courtright, Junior, as he signed his name, made affidavit in behalf of the soldier in Delaware County, Ohio, in 1839.

In May 1839, one Elias Vansickle stated in Delaware County, Ohio, that he had been acquainted with the soldier for thirty years.

Very truly yours,

A. D. Hiller,
Assistant to the Administrator.

December 8, 1933.
BA-J/_____/ILL
Solomon Cartwright
S.4993

Mrs. W. W. Waugh
330 South Grove Street
Bowling Green, Ohio.
Dear Madam:

This letter has reference to a recent reexamination of pension claim, S.4993, based upon service of Solomon Cartwright, whose record you requested in March 1931. It is noted that in the record furnished you of that soldier on April 6, 1931, the name was inadvertently written Samuel, instead of Solomon. A corrected record is furnished you herewith to be used in place of the one dated April 6, 1931.

The record of Solomon Cartwright, as his name is borne on the pension rolls, is given below as shown in the papers on file in his claim for pension, S.4993, based upon service in the Revolutionary War.

Solomon Cartwright was born July 8, 1752 in Wantage, Sussex County, New Jersey, and resided there with his parents during the Revolutionary War; the names of his parents are not shown.

He volunteered and served in the New Jersey troops as follow: from March 1, 1776, one month in Captain Joseph Harker's company, Colonel Ephraim Martin's regiment; from about June 1, 1776, in Captains Seward's and Bond's companies, Colonel Ephraim Martin's regiment, was in the battle of Flat Bush on Long Island and was discharged at the expiration of service of of five months; in January, 1777, one month in Captain Cole's company, Colonels Ephraim Martin and Hankinson's regiment; in the summer, 1778, one month, Captain Kirkendall's company, Lieutenant Colonel Seward and Colonel Hankinson's regiment; in July 1779, one month in Captain Cole's company at Lackawaxon and was in an engagement with the Indians; a short tour in 1780, no officers named; from March 1, 1781, nine months as second sergeant in Captain James Bonnel's company, Colonel Isaac Martin's regiment, on duty at the Block House in Montague, Sussex County, New Jersey. He was in engagements at Tide Mills, Elizabethtown, and at Aquahanock.

After the war, the soldier resided in Wantage, New Jersey, later moved to Danby, Tompkins County, New York.

Solomon Cartwright was allowed pension on his application executed October 3, 1832, at which time he resided in Danby, Tompkins County, New York; he had lived there about twenty-five years.

In February, 1839, the soldier was a resident of Delaware County, Ohio; he stated then that he had moved to that county recently from New York State because his children resided in Delaware County, Ohio; he did not give the names of said children, nor is the name of his wife shown in the papers in this pension claim. In 1839, Solomon Courtright, Junior, as he signed his name, made affidavit in behalf of the soldier in Delaware County, Ohio.

In May 1838, one Elias Vansickle stated in Delaware County, Ohio, that he had been acquainted with the soldier for thirty years.

Very truly yours,
A. D. Hiller,
Assistant to the Administrator.

(Dec 8 - 1933 Note - I drew this claim at request of Dr. Chas. C. Pinkerton, Akron, Ohio - reexamined, corrected and mailed on Dec. 1 - 1933. The former record, furnished April 6 - 1931 - MLB -\as Samuel/ Mrs. Finch requested me to send the corrected record.) _____

Invalid. File No. 4,993
Solomon Cartwright Privt Rev War
Act: June 7th 1832
Index: - Vol. M, Page 57
[Arrangement of 1870.]
Revolutionary War Pension and Bounty Land Warrant Application Files
Series M-804; Roll # 488

William De Shay
New Jersey (S.28,709)

State of New York,

Erie County, ss. On this 3d day of November 1832 personally appeared ~~in open Court~~, at Buffalo in said county before the Hon James Stryker one of the Judges of the Court of Common Pleas, ~~now sitting at Buffalo~~, in said County, William De Shay a resident of Buffalo in the County of Erie and State of New York, aforesaid, aged 69 years, who being first duly sworn, according to law, doth on his oath, make the following declaration, in order to obtain the benefit of the

Act of Congress, passed June 7, 1832, That he entered the service of the United States under the following named officers, and served as herein stated was born in the Town of Nolton in the State of New Jersey in the year 1763. Has no Record of his age his Family was lost a number of years since, was living in Sandustown when called into Service. has lived since the revolutionary war a few years in Canada but has lived in Buffalo the last 36 years where he now lives.

Entered the Service as a Volunteer at Sandustown aforesaid under Captain Westbrook in Colonel Rosecrants Regiment. Went out on Scouting parties against the Indians & was out guarding the country until the Fort at Sandustown was built & then continued guarding the country & Sandustown Fort for the Term of 4 years

Captain Westbrook & the Lieutenant of his company were killed in a Battle with the Indians about the first part of his service Has never received a discharge from the Service other than a Verbal one

And further States that he is unable to obtain a clergyman to certify for him as there is none residing in his neighborhood And this Deponant says that Cyrenius Chapin and James McKnight are persons to whom he is Known in his present neighbourhood And whom he offers as Witnesses as to his truth and veracity and their belief as to his services in the Revolution

He hereby relinquishes every claim, whatever, to a Pension, or annuity, except the present, and declares that his name is not on the Pension Roll of the Agency of any State,

Sworn and subscribed this third day of November 1832
James StrykerJudge of Erie County Courts

his
William De X Shay
mark

We, James McKnight residing in the town of Buffalo and County of Erie and Cyrenius Chapin residing in the same place, hereby certify, that we are well acquainted with William De Shay who has subscribed and sworn to the foregoing declaration; that we believe him to be 69 years of age; that he is reputed and believed in the neighborhood where he resides to have been a Soldier of the Revolution, and that we concur in that opinion.

Subscribed and Sworn, this third day of November 1832
James Stryker Judge of Erie County Courts

Cyrenius Chapin
Jas M Knight

And I the said Judge do hereby certify & declare my opinion after the investigation of the matter and after putting the interrogatories prescribed by the War Department that the above named applicant William Deshay was a Revolutionary soldier and served as he states. And I the said Judge do further certify that it appears to me that Cyrenius Chapin & James McKnight who have signed the foregoing certificate are residents of the Town of Buffalo in said County and are credible persons and that _____ & each of their Statements are entitled to credit. And I do further certify that, the foregoing contains the original proceedings before me in the matter of the application of the said William De Shay for a Pension. And I the said Judge do further Certify that it appears satisfactorily to me that the Said William Deshay could not without too much inconvenience procure the attendance of a Clergyman before me to testify to his character & services or make the certificate required by the War Department

And I the said Judge do hereby certify that in consequence of the Prevalence of the cholera in the City of Buffalo in August last when a Term of the Court of Common Pleas of said County should have been held, the said Court was not held and that on a representation of the peculiar circumstances which rendered it proper the War Department deemed it expedient to permit me the Said Judge to receive pension applications and take testimony thereon until the next Term of the Said Court in December next and no longer which said permission or authority of the War Department was communicated to me the said Judge by a letter under the date of September 21st AD 1832

And I the Said Judge do hereby further certify that the foregoing proceedings have been _____ taken and allowed before me under and by virtue of the aforesaid permission & authority of the War Department. Given under my hand & Seal at Buffalo this 3rd day of November 1832.

James Stryker Judge of Erie County Courts

Erie County

Clerks Office ss. I, Noah P. Sprague Clerk of the County of Erie aforesaid do hereby certify that the Hon. James Stryker before whom the foregoing proceedings in the matter of the application of William De Shay for a pension _____ had was on the day of the date thereof and still is a Judge of the County Courts of the County of Erie duly appointed and sworn and that the signatures of the said Judge the jurat of the deposition or declaration of the said William DeShay and to the jurat of the joint certificate or affidavit of Cyrenius Chapin and James McKnight, and to the certificate of the said Judge following the same are all the genuine signatures of the said Judge. In testimony whereof I have hereunto set my hand and seal of office this 3d day of November 1832

Noah P. Sprague Clerk of Erie County

State of New York

County of Erie ss. Ezekiel Lane now of Cleveland in the State of Ohio formerly of Buffalo in the County of Erie in the State of New York being duly sworn deposes & Says that he is of the age of 79 years & upwards and is well acquainted with Mr. William De Shay now of said Buffalo who is nearly two years younger than this Deponent,& that he was acquainted with him & they were boys together in the same neighborhood before the Revolutionary War and at the commencement of the War this Deponent lived in Yankeetown on the Pensylvania side of the State Line opposite Sandytown in the State of New Jersey and soon after the ______ War commenced they were driven off to the Jersey side that the second year of the War this Deponent & the said William De Shay both entered the Service of the United States by joining a company under Captain Peter Westbrook of Sandytown aforesaid about the middle of April of that year that the other officers of Company were Lieut James Westfall & Ensign Cornelius Westfall that they were engaged in Scouting & guarding fording places & also in little Garrisons which were kept up on & near the fording places on the Delaware river that soon after they entered the Service they crossed over in to Pennsylvania & while there & soon after they had a skirmish with Indians & Tories in which Capt Westbrook was killed that some time after John Westbrook the Uncle of the deceased Capt Peter Westbrook was elected Captain & took the Command & Cornelius Westbrook was elected Lieutenant & under these Officers this Deponent & the said William De Shay continued in the Service until winter doing active Duty. The next Spring & he thinks in April & about the first of that month they again both entered the service under the Same officers & both done the same duty as they had done the Season before or nearly the same & continued in active service until the approach of winter, that again the next Spring they both entered the Service at the early part of the Spring & he thinks about the first of April under the same officers & again done nearly the same duty until winter of that year. That during all that time they were both private soldiers and were either in the field or in some little Garrison and were employed in no civil pursuit whatever. That their duty was generally about the same as before stated & they were called out and joind only with one two or more Companys together most usually but sometimes ______ three Companys and with other Companies they were under the command of Col Rosecrantz. Saml Westbrook was also the Major of the Regiment After the above service was performed this Deponent moved into the State of New York so that he cannot tell how much more the said William De Shay served.

Sworn & subscribed before me this 22nd day of July 1833
Harry Slade J. P.

his
Ezekiel X Lane
mark

I certify that it satisfactorily appears to me that the above named Ezekiel Lane is a credible person & that his Statement is entitled to credit

Harry Slade J P

State of New York
Erie County Clerks Office ss. I Noah J Sprague Clerk of the County of Erie do hereby certify that Harry Slade before whom the above affidavit was sworn was on the day of the date thereof and still is a Justice of the peace in and for said County duly elected and sworn, and authorized to act as such, and I further certify that I am well acquainted with the hand writing and signature of the said Justice and that the signature of said Justice to the jurat of the said affidavit and also to the above certificate are the genuine signatures of said Justice. In testimony whereof I have hereunto set my hand and ~~affixed~~ and seal of office this 23rd day of August 1833.

Noah P Sprague Clerk of Erie Co

State of New York
County of Erie ss. On this 23rd day of August in the year 1833, personally appeared before me a justice of the peace in and for the County of Erie William Deshay of Buffalo in the County of Erie aforesaid, and being by me duly sworn deposes and says: that he was born in the Town of Nolton in the State of New Jersey in the year 1763: he has a record of his age at home in his bible, which he has found since his former declaration hereunto annexed was made, that the statement in the annexed declaration that "the family was lost a number of years since," is incorrect and should read, that "the family record was lost a number of years since." And he further says that he was living in Sandustown in Sussex County in New Jersey, and continued to make it his home there during the War, and until some years after the War was ended when he moved into Canada and lived a few years and then into the Town of Buffalo in Erie County where he now resides.

And this deponent further says that in the year 1776 or 1777 and he thinks in 1776, and he thinks about the first of April he volunteered into the Militia of the State of New Jersey at ~~Nolton~~ Sandustown aforesaid, that he joined his company at Sandustown, and he thinks the company joined the regiment at the same place, that he volunteered into Captain Westbrooks Company in Colonnel Rosecrants regiment, that he cannot recollect the name of their General, that one Westbrook a brother of the Captain was Major, that James Westfall was Lieutenant of the company, that he thinks a man by the name of Westbrook was Orderly Sergeant, that one Cornelius Westfall a brother of James the Lieutenant, was the Ensign, that he cannot recollect the names of any more of the Officers of the company or regiment: that when he joined his company and regiment as aforesaid which was about the first of April as near as he can recollect, they took possession of the place and built a fort that he was employed in standing guard and working on the fort and was out on scouting parties against the Indians a number of times, that he cannot recollect the names of any other regiments that came there during this season, that other regiments came there often and staid a few days but he cannot recollect their names, that he was not in any battles during this season, that he conntinued doing duty as above stated until about the middle of October following when he was discharged by his Colonel: that he was actually out in the service at this time and doing duty not less than the term of Six months that his Captain was killed some time in the month of April as near as he can recollect in a skirmish with some Indians about three miles from the fort, that after his Captain was killed as above stated, one John Westbrook took his place and was elected Captain, that during this time he was out ~~and~~ he was doing duty all the time and was not engaged in any civil pursuit: that he was in the skirmish at the time his Captain was killed.

And this deponent further says that the next season he volunteered again into the Militia of New Jersey at the same place and about the same time, that he joined his company there and regiment, that he served this term under the same officers as the other as near as he can recollect, that he was engaged in the same business very nearly viz: standing guard & was out on scouting parties &c. and was discharged about the middle of October as near as he can recollect, by his Captain, that he was out _______ during this term and actually engaged in doing duty and was not employed in any civil business the full term of Six months: he was not in any battles that there were other companys came there during the season but he cannot recollect who commanded them. And this deponent further says that the next season after, he volunteered again into the Militia as aforesaid at the same place, and under the same officers, and about the first of April, that he joined his company there and served the time out there and about there that he was employed in standing guard and was out in scouting parties during the time: that he was discharged about the middle of October as near as he can now recollect, that he was out and actually doing duty this season and not engaged in any civil occupation not less than Six months: was not in any battles; had some skirmishes with the Indians

And this deponent further says that he again volunteered the next season at the same place and under the same officers; about the first of April and joined the company there, & continued to do duty and was out and not engaged in any civil pursuit until about the middle of October following when he was again discharged by his Captain, that he served at this time the full period of Six months as aforesaid, that he was employed this season in standing guard &c. and guarding the Country: that he served all of the above times in the capacity of a private soldier; that he never received a written discharge from the service, nothing more than a verbal dismissal: that this deponent and Ezekiel Lane whose affidavit is hereunto annexed served the first two terms above enumerated by this deponent in the same company, that then the said Ezekiel moved from there into the State of New York; that he does not know of any other person except Mr Lane, now living by whom he can prove any part of his services by; that he has made frequent and very extensive inquiries and searches and has not been able to find any one: and this deponent further says that the reason why his services were all performed in the summer and about the same time in the year was that in the summer the settlements in that part were harrassed by the Indians and it was necessary to have the militia or a part of them under arms and ready at a moments call and in the winter the Indians were kept at home by the cold: and this deponent says that he stood as a minute man the greatest part of the time during the winters but never happened to be called out excepting in the summers as is above stated.

Sworn before me this day 23rd day of August 1833
J. Clary J. P.

his
William X De Shay
mark

I certify that I am well acquainted with William De Shay who has subscribed and sworn to the foregoing affidavit and he is a credible person and a man of truth and veracity and his statements are entitled to full credit.

J. Clary J. P.

Buffalo Nov 19, 1833

J L Edwards Esq
Comr of Pensions
Sir
We send Enclosed herewith the Pension papers of Mr Wm. De Shay with

certificate of Mr Clary signed as required which we suppose renders them perfect
He will wish to be placed on the roll of the Buffalo agency

Very Respty Your obt servt
Fillmon & Hall

State of New York

Erie County Clerks Office ss. I Noah P. Sprague Clerk of the County of Erie do hereby certify that Joseph Clary before whom the foregoing affidavit or deposition was sworn was on the day of the date thereof and still is a justice of the peace in and for said County of Erie duly elected and sworn and authorized to act as such and that I am acquainted with the handwriting and signature of the said Justice that the signature of the said Justice to the jurat of the said deposition _____ affidavit and also to the above certificate are the given signatures of the said Justice In testimony whereof I have hereto set my hand and Seal of Office this 23rd day of Augt. 1833.

Noah P. Sprague Clerk of Erie County

War Department,
Pension Office,
May 27 - Nov 4th 1833

Sir:

The evidence in support of your claim, under the act of June 7, 1832, has been examined, and the papers are herewith returned. The following is a statement of your case in a tabular form. On comparing these papers with the following rules, and the subjoined notes, you will readily perceive that objections exist, which must be removed, before a pension can be allowed. The notes and the regulations will shew what is necessary to be done. Those points to which your attention is more particularly directed, you will find marked in the margin with a brace, (thus: }). You will, when you return your papers to this Department, send this printed letter with them; and you will, by complying with this request, greatly facilitate the investigation of your claim.

A Statement, shewing the Service of William De Shay, Erie co. N.Y.

Period when service rendered	Duration of claimant's service			Rank of the Claimant	Names and Ranks of the Company Officers under whom he served	Names and Ranks of the Company Officers under whom he served
	Years	Months	Days			
Volunteer!	4			private	Capt Westbrook	Col. ________

Battles in Which the applicant engaged	Country through which he marched	Place of abode when he entered service, and age at the present period	Evidence by which the declaration is supported
		69 Sandustown	tradition

I am, respectfully,
Your obedient servant,
J. L. Edwards,
Commissioner of Pensions.

After the statement of service a copy of the regulations under the Act of June 7, 1832 is included but not reproduced here. An identical copy is included in the chapter entitled ***A Declaration of Service****. In the middle of the regulation is the following: 2. Have you any record of your age, and if so, where is it? Family lost a number of years!!!*

Buffalo

William De Shay of Erie Co in the State of New York who was a private in the company commanded by Captain Westbrook of the regt. commanded by Col. R   in the N.J. malitia line for 29 ms. Inscribed on the Roll of Buffalo at the rate of 80 Dollars 00 Cents per annum, to commence on the 4th day of March, 1831. Certificate of Pension issued the 16 day of Jay 1834 and sent to Fillman & Hall Buffalo

Arrears to the 4th of Sept 1833	$200.00
Semi-anl. allowance ending 4 March 3	$40.00
	$240.00

Recorded by W. L. Williams Clerk,
{Revolutionary Claim, Act June 7, 1832.}
Book E   Vol. 4   Page 9

Wm. De Shay 80 pr. yr Payments to be made at Buffalo by Prest US Br: Bank

Agent for paying Pensioners, in the Agency of Buffalo Recorded in the Pension Office, in Book E Vol. 4 Page 9 by W. L. Williams Clerk.

Hon S. N. Clark or Hon M. Fillmore
Members of Congress
Washington City

Silver Creek Post Office July 28 1842

Hon S. N. Clark,

Sir I enclose you herewith papers to obtain a restoration of the pension of William De Shay, requesting that you lay them before the proper branch of the government at your earliest convenience, Mr De Shay has drawn a pension for several years until last March when he was suspended for some defect in the statement and evidence on which it was originally granted. His being a resident of your county suggests that I should enclose the papers to you, and his extreme indigence together with the simplicity and honest worth of the poor old man who seems to have preserved through the _____ and privations of four score years the virtues of temperance and laborious industry will no doubt interest you as they have me warmly in his behalf, and relieve me from the necessity of further apology as well as excuse the benefit of your influence in obtaining as early action upon his case as the circumstances of business will admit

very respectfully yours
Elisha Ward

N. B. To guard against possible contingency so near the close of the Session I have _____ the authority of any acquaintance directed in the _____ to the Hon M. Fillmore

E Ward

House of Reps
Augt 1st 1842

J. L. Edwards Esq

Dear Sir,

I enclose herein the declaration of William De Shay a Revolutionary Soldier asking the restoration of his pension

Yours respectfully
S. N. Clarke

State of New York
Cataraugus County ss. On this fourth day of July in the year one thousand eight hundred and forty two personally appeared before the subscriber a Judge of the court of common pleas of said county of cataraugus in the State of New York William De Shay a resident of the town of New Albion in the county of cataraugus and State of New York, aforesaid aged eighty years who being first duly Sworn according to law, doth on his oath make the following declaration in order to obtain the benefit of the act of congress, passed June 7th 1832

That he entered the service of the United States under the following named officers and served as herein stated, that the Majors name under whom he entered the service was Westbrook, his first or christian name he does not now recollect. That the captain of the company to which he belonged at the time of entering the service was Peter Westbrook and a brother to the Major that he served in the company of Captain Peter Westbrook, that some afterwards and according to his best recollection during the second summer after he entered the service Captain Peter Westbrook was killed in skirmish with the Indians. After that event John Westbrook became Captain of the company and continued so during the remainder of the time that he was engaged in the service and he served under him and in the same company, until he finally left the service.

That he entered the service in the State of New Jersey as a volunteer that he cant tell the month or year when he entered the service nor the year when he left it, but that he entered the service in the spring of the year when he was between sixteen and seventeen years old, and according to his present impression after the taking of Burguoyne that he was stationed principally on Jersey side of the Delaware river at a place called the Minisink where was a small fort which erected during the first year or summer of his services he recollects being employed in erecting it the fort was about a quarter of a mile from the Dlaware river and that during the whole time that he was in the service after the fort was completed his principal service and duty was about the fort either in keeping guard or in guarding the fords of the river with occasional excursions that the first season he entered the service in the spring and continued to serve until near the setting in of winter when no further danger was apprehended from the enemy in consequence of the winter season, when he went home to his fathers who in the winter resided about three miles from the fort, though his fathers family most times in the summer when the most danger was apprehended from the Indians lived in the fort, that the next spring soon after the snow was off he returned to the fort and served as before through the summer and fall, and then left as before, and remained unemployed until the opening of spring when he again returned to the service and continued in service during the season until near the approach of winter, when he again left and continued out of the service until the spring when he returned again for the fourth season and served as before until near winter when he again left the service and does not now remember whether he ever returned or not, that he served four seasons and each year from the opening of spring until late in the fall that he cant say the precise time that he served in each year further than above, that he thinks he did not serve less than six months each year that he never received any written discharge but left with the consent of the officers that during the whole four years he was engaged in no other business that the publick service and was only absent in consequence of no danger being apprehended in the winter season, that his fathers farm was on the Pennsylvania side of the river which he was compelled to leave and take up a residence on the other side of the river to which he did not return until after hostilities had ceased.

That he was twice engaged in battle with the indians during the time that he was in the service, that the first time he belonged to a party of about thirty under the command of Captain Peter Westbrook who had an engagement with the indians on the Pennsylvania side of the river and about three miles from the river, the Indians killed three of the party to which he belonged and forced the others to retreat among the killed was Captain Peter Westbrook. This skirmish to place early in the spring in the forest, and according to his best recollection during the second season that he was in the service, that some time afterwards but in what year he does not recollect he belonged to a party of near four hundred who were in pursuit of indians near the Delaware river and as he thinks within the bounds of the State of New York, when after having followed the indians nearly all day with the intention of surprising them at night, and the troops being much fatigued they were towards night suddenly attacked by a large number of Indians and being surrounded the whole party with the exception of some eight or ten were killed and the few who escaped saved themselves by running. The indians were supposed to be commanded by Brant, their numbers were unknown, this happened in the summer time some miles from the white settlements that none of the officers who commanded the party escaped, that he does not now know who commanded the party on the part of the americans, if he ever knew none of the officers which belonged to his company in which he usually served were there, as the party was hastily formed for the occasion, that he does not know the name of the Colonel under whom he served but thinks his name was Rosencrants, that he, the applicant, was born at Knowlton in the State of New Jersey that he never enlisted but served as a volunteer, that he has no record of his age and does not of any, that he was eighty years old in January last according to his best information and belief, that he did not receive a written discharge that he served as a private during the whole time he was in service that he has never been on any pension Roll except on the roll from which he has been suspended and asks to be restored, that there is no living witness by whom he can prove his services, that at the time his pension was first granted which was granted on the sixteenth day of January 1834 one Ezekiel Lane was a witness as he believes, that the witness Lane is now dead having died about two years ago, and he further states that he is unable to write his name never having learned to write, and he hereby relinquishes every claim to a pension or annuity except the present and the arrears thereof and declares that his name is not on the pension Roll of any state except that of the State of New York, as above stated

Sworn to and subscribed the day and year aforesaid before me Thos J Wheeler a Judge of Catts Co.

his
William X De Shay
mark
Witness Levi Steel

We Jonas Glazier a clergyman residing in the town of Napoli county of Cataraugus and state of New York and Bennedick Russel residing in the same place hereby certify that we are well acquainted with William De Shay who has subscribed and sworn to the above declaration, that we believe him to be eighty years of age, that he is reputed and believed in the neighbourhood where he resides to have been a soldier of the revolution and that we concur in that opinion

Sworn and subscribed the day and year aforesaid before me

Thos J Wheeler a Judge of Catts Co

Jonas
Glazier
his
Benedick X Russel
mark
Witness Levi Steel

And I Thomas J Wheeler a Judge of the court of common pleas of the county of cataraugus do hereby declare my opinion after the investigation of the matter and after putting the interrogatories prescribed by the war department, that the above named applicant was a revolutionary soldier and served as he states, and I do further certify that the said applicant by reason of bodily infirmity is unable to appear in open court, and I do further certify that it appears to me that Jonas Glazier who has signed the preceding certificate is a clergy man resident in the town of Napoli in the county of cataraugus and that Benedick Russel is a credible person and that their statement is entitled to credit

Thomas J Wheeler a Judge of Catts Com Pleas

State of New York

Cattaraugus County clerk's Office ss. I George W. Gillet clerk of said county Do hereby certify that Thomas J Wheeler Esqr whose name is subscribed to the foregoing Certificate was at the day of the dates of the said certificates a Judge of the court of Common Pleas of Cattaraugus county in said state, the same being a court of Record, and that the signatures annexed are his genuine signatures In Testimony Whereof I have hereunto set my hand and affixed the seal of said court and county this 25th day of July 1842

G. W. Gillet Clerk

Elisha Ward Applies for restoration of William Deshay to the pension roll.

Hon John C. Spencer,

Silver Creek Chau. Co. NY, Sept 19, 1842

Sir, In the case of William De Shay of the State of New York, a revolutionary pensioner who was suspended in March last, application was made by me, through the Hon S. N. Clark for a restoration of his pension founded on a supplemental statement or declaration of his services. The commissioner of pensions in answer to Mr Clark on the 5th of August addressed a note to Mr Clark, saying that the case had been examined and afforded confirmation of the views continued in his letter of the 5th Feb, 1834, a copy of which was enclosed and which purports to be of a letter of that date to Mr Fillmore stating that the claim of William De Shay had been admitted.

From that communication it was infered that his pension had been restored, until it was founded that he continues suspended at the agency, I am therefore under necessity of again bringing the case before the department, to enquire whether he has been restored and if so to apply for such proceedings in his case as shall autherize the payment of his pension at the agency.

very respectfully yours &c
Elisha Ward

Conewango Cattaraugus Co N.Y.
March 17th 1843

Sir

Permit me to trouble you with an Enquiry in the case of William De Shay By your letter to the Hon S. N. Clarke of Aug 3rd 1842 I observe that one reason of his suspension is that he is too young to have performed the duties of a soldier at the of the commencement of his alleged term of service. The old man says there must have been an error in drawing up his declaration which escaped his observation at the time, as he says that he is now 81 years of age which would make the time of his birth 1762 consequently he was 14 years old in 1776 he also states that it was not until some time after the commencement of the war that his first term of service commenced, that he was then 16 years of age. He also says that one Ezekiel Lane was in the service with him. Now if the testimony of Lane can be procured to prove his actual service as a soldier together with his own affidavit correcting his age will it be satisfactory to the Department? It will not be worth while for the old man's friends to incur the expense of finding Lane, if he is still living, unless his evidence will be sufficient to reinstate De Shay on the Pension List.

You will confer a favor on a poor old man by answering this at your Earliest convenience

Respectfully Your Obedt Servt
Thos J Wheeler

Hon. J. L. Edwards

Pension Office,
April 29th 1843.

Leon Mills P Office
Cattaraugus County
Sir,

In answer to yours of 17th inst. in the case of William De Shay, I would remark that the statements in your letter respecting his age conflict with his statement upon Oath. He swore that he was born in 1763 and entered the service in 1776.

With respect to the evidence of Lane no opinion can be formed until it is submitted. The claimant should make a brief statement of the period, length and grade of each term of his service, names of Officers, stations &c. and obtain the certificate of the Secretary of New Jersey shewing how far his records either confirm or discredit each.

I am very respectfully, Your Ob Servant
J L Edwards

Thomas J. Wheeler, Esq.
Conewango
Cattaraugus County
New York

Hs. of Reps.
Washingn. 19th Feby. '44

To/James L. Edwards Esqr. Comr. &c

Sir

I beg leave to enclose you the Pension Certif. of William De Shay & a letter in explanation which accompanied the same & to assure you the writer Genl. Burt is one of the most respectable & _____ citizens of Buffalo. Your early attention to the same will oblige

Your obedt. servt.
Wm. A. Moseley

Buffalo February 12 1844

Dear Sir

You know the Old saying is if you have friends they should be willing to be used A little to prove their sincerity and as such friend I ask the request of you to attend to A Little Business for an Old Revolutionary Soldier & has been Pensioner untill some 2 years since I enclose to you his Original Certificate and A Letter to Thomas J Wheeler Esq in which Mr Edwards says that some Statements in Respects to his age are the Reasons of Suspending him But I apprehend there is no intentional Error in Relation to his case I have known him almost since the time I first came to Buffalo which is about 28 years he then Lived at Black Rock or did when I first collected his pension for him To be sure he is very ignorant man and may have made Errors by foregetfullness as he is an Old Man, I have no earthly interest about this Business and would not interfere for him if I did not believe him honest and really deserving, he is also very Poor and his Son in Law came to me and requested me to write to Mr Edwards for him and informs me that he is 81 or 82 years of age the reason of his coming to me is I suppose in consequence of my previous procuring his pension money for him when he resided at Black Rock some years since he removed to Cattaraugus County and I knew nothing of him untill Last week or that is of his being suspended my knowledge of his poverty and Belief of his having been A Soldier of the Revolution and honestly Entitled to his pension is my reason for writing if Mr Edwards desires to know who were his Officers that he served under he says John Rosecrants Col Samuel Westbrook Maj Peter Westbrook Capt James Westfall Lieut Cornelius Westfall Ensign says he served as minute man in the State of New Jersey I believe Mr Edwards will on reveiwing his case reinstate him from Sept 4th 1841 I have

Enclosed his pension certificate which you can Hand or Send to Mr Edwards and if his application is successfull _____ if Otherwise let me know the particulars and I will endeavour to have all Explained

Yours &c Most Respectfully
David Burt

Hon W. A. Mosely
In Congress
Washington City

War Department,
Revolutionary Claim.

I certify that, in conformity with the law of the United States of the 7th June, 1832, William De Shay of the State of New York who was a private in the revolutionary war is entitled to receive eighty dollars and cents per annum, during his natural life, commencing on the 4th of March, 1831, and payable semiannually on the 4th of March, and 4th of September in every year. Given at the War Office of the United States, this Sixteenth day of January one thousand eight hundred and thirty-four.

Lew Cass
Secretary of War.

Examined and Countersigned.]
J. L. Edwards
Commissioner of Pensions.

Seelysburgh NY July 8th 1846

To W. G. Wade Esq
Dear Sir

I have another case of a Revolutionary Soldier by the name of William Deshay who served under Capt Westbrook in Col Westfalls regiment of New Jersey served 4 years as a minute man said Deshay drew a pension of Eighty dollars per annum til between 4 and 5 years since when it was stopped for what cause I am unable to say, the old man died last January and has left a widow who is entitled to pension, if she can make the necessary proof but she tells me that she does not know of any person living that knew of her marriage at the time and there is no record of it to her knowledge

I address this _____ rather as an enquiry with a request that you will inform me what if any testamony will enable her to obtain her pension aside from what

I stated above as beyond our reach also please consider yourself as employed in obtaining the old mans back pension and the widows if it can be done

In haste Yours respectfully
Lewis P. Thorp

---

20 St Johns Place
Buffalo N.Y.

To the Commissioner of Pensions
Washington D.C.

Will you kindly send me the record of William Desha of Erie Co N.Y. in Pension List 1831 Do not mention the D.A.R. and when report is recieved kindly send to Mrs Chas P Lytle

20 St Johns Place
Buffalo N.Y.

Revolutionary War Records Section.Department of the Interior,
Bureau of Pensions.
Washington, D.C., , 19

In reply to your request of , received for a statement of the military history of William De Shay a soldier of the Revolutionary War, you will find below the desired information as contained in his (~~or his widow's~~) application for pension on file in this Bureau.

DATES OF ENLISTMENT OR APPOINTMENT.	LENGTH OF SERVICE	RANK.	OFFICERS UNDER WHOM SERVICE WAS RENDERED.		STATE
			CAPTAIN.	COLONEL.	

Soldier stated that he served from Apl to Oct for four seasons (~~amounting to 2 years~~) as Priv. in Capt. Peter Westbrook's & John Westbrook's co. Col. Rosekranz's regt. of N.J. troops, & that he was in two skirmishes with Indians, in the first of which Capt Peter Westbrook was killed.

Battles engaged in, see above.
Residence of soldier at enlistment, Sandiston, sussex co, N.J.
Date of application for pension, nov 3, 1832 Allowed.
Residence at date of application, Buffalo, Erie Co, N.Y.
Age at date of application, Born in Jany 1763 in Knowlton, N.J.
Remarks: sol. died Jany 1846 leaving widow - name not stated

Invalid. File No. 28,709
William De Shay Pri Rev War
Act: June 7th 32
Index: - Vol. 2, Page 106
[Arrangement of 1870.]

Revolutionary War Pension and Bounty Land Warrant Application Files
Series M-804; Roll # 801

Moses Dewitt
New Jersey (R.2,921)

State of New York
Tompkins County ss. Solomon Middaugh of the town of Danby in the County of Tompkins and State of New York being duly sworn says, that he was seventy nine years old the twentieth day of August Eighteen hundred and thirty four. That he was born in the town of Wantage in the County of Sussex and State of New Jersey, and resided there untill he was of the age of Fifty years. That while he resided in Wantage as above stated he became acquainted with Moses Dewitt who was then and still resides there and knew him at Wantage for the space and period of Forty years. That this deponent was a soldier in the Army of the Revolution, and served as such from the commencement of said War untill peace was declared in the year Seventeen hundred and Eighty three. That this deponent was a Lieutenant in Captain Joseph Harkars Company, Captain Samuel Meekers Company, and Captain John Davis company in the Second New Jersey Regiment. That he knew said Moses Dewitt in the Revolutionary war and he was a private in Captain Josiah Coe's Company in the Second New Jersey Regiment, under the command of Colonel John Seward That he frequently saw him doing duty in said company, and said Moses Dewitt was always with said Company whenever it was called into service. That this deponent on the twenty second day of July 1779 was in the Battle of Lackawack, and saw said Moses Dewitt in said Battle. That he can not precisely state the length of time said Dewitt was in the service, from the failure of his memory and the time that has elapsed since said service
Subscribed & sworn October 3, 1834 before me
Frederick G. Stanley a Commissioner of deeds & c

Solomon Middaugh

I do hereby certify that I am well acquainted with Solomon Middaugh above named who made the above affidavit, that he is a man of good moral character and full faith and credit may and ought to be given to his statement

Frederick G. Stanley a Commissioner of deeds & c

State of New York

Tompkins County Clerks office I certify that Frederick G. Stanley Esqr before whom the foregoing affidavit purports to have been Sworn was at the date of the same a commissioner of deeds in & for Said county duly qualified and that his Signatures ~~are genuine~~ above written are genuine In testimony whereof I have hereunto set my hand and affixed my official Seal at Ithaca the 3d day of October 1834

Saml. Love Clerk of Tompkins Co.

State of New Jersey

Sussex County ss. On this twenty seventh day of November in the year of our LORD Eighteen hundred and thirty four, personally appeared in open Court before the Judges of the Inferior Court of Common pleas now sitting in and for the County of Sussex aforesaid, Moses Dewitt, a resident of Wantage Township in the County of Sussex and State of New Jersey aforesaid aged seventy three years who being first duly sworn according to law doth on his Oath make the following Declaration in order to obtain the benefit of the provision made by the act of Congress passed June 7th 1832. That he entered the Service of the United States and served under the following named Officers Towit in the 2d New Jersey Regiment comd. by colonel John Seward, in the company of capt Josiah Cole, Lieutenant Solomon Middaugh, Ensign Joseph Elston, 1st Sergeant Samuel Van-Vleet that he entered as a volunteer in the year 1779 thinks in the spring of that year, that he served in said Company upon the western or north western frontiers of the State ~~upon~~ principally upon the Delaware and Lackawack Rivers, that he was at the Battle of Lackawack on the 22nd July 1779, that he was out after the Indians when they burnt Peepack, and also when they killed Job Jobs and Lewis and Solomon Westbrooks, that he was also stationed at Ten Mile River, that he was Stationed at various places on the sd. Delaware Lackawack and Navesink Rivers, at which places, and throughout that Section of Country he served and done duty in the said Company and under the Officers aforesaid from the said time he entered until the close of the war in 1783, he says that section of country was subject to frequent and almost continual invasions and depredations of the Indians throughout the War, and that it was necessary to have an armed force constantly in action against them, he says that he was born in the sd. Township of Wantage in sd. County of Sussex the 23rd of October in the year 1761, that there is a record of his age in his father's Bible that at the time he entered the Service he was residing at Wantage where he has lived since the Revolution and where he now lives, that he volunteered as aforesaid, that he never received any written discharge, and has no documentary ~~discharge~~ evidence of any kind, that he knows the said Lieu Solomon Middaugh, and sd. Samuel VanVleet two of the said Officers under whom he served who can testify as to his service, that he is acquainted with, the Revd. Abraham Ketcham and William Cole Esquire who can testify to his character for veracity and their belief of his services as a soldier of the revolution, he hereby relinquishes every claim whatever to a pension or annuity except the present and declares, that his name is not on the pension roll of the agency of any State.

Sworn to and subscribed the day and year aforesaid.

Jos. Y. Miller Presiding Judge &c

Moses Dewitt

We Abraham Ketcham a clergyman residing in the neighbourhood of the sd. Moses Dewitt, and William Cole residing in the same do hereby certify that we are well acquainted with Moses Dewitt who subscribed and Sworn to the above Declaration that we believe him to be 73 years of age; that he is reputed and be lieved in the neighborhood where he resides to have been a soldier of the Revolution and that we concur in that opinion.

Sworn and subscribed the day andyear above written in open Court

Jos. Y. Miller Presiding Judge

Abm. Ketcham
William Cole

And the said Court do hereby declare their Opinion after putting the interrogatories prescribed by the War Department, that the above named applicant was a revolutionary soldier and Served as he States, and the Court further certifies that it appears to them that Abraham Ketcham who has signed the preceeding certificate is a clergyman resident in the neighbourhood of said Moses Dewitt, and that William Cole Esquire who has also subscribed the same is a resident in the same neighbourhood, and is a credible person, and that their statement is entitled to full credit. Done the day above written

Jos. Y. Miller Presiding Judge & c

On the same day and year in Open Court before the same Court personally appeared Samuel Van Vleet aged 75 years and being duly sworn according to law on his Oath doth depose and say, that he this deponent was sergeant in the Company comd. by Capt Josiah Cole in the 2d New Jersey Regiment comd by Colonel John Seward in the service of the United States during the revolutionary war, and that he is acquainted ~~well~~ with Moses Dewitt who has subscribed and sworn to the above Declaration, that he was acquainted with him during said war, and knows that the sd. Moses Dewitt, was a member of said Company and served as he states

Sworn and subscribed the day and year aforesaid.

Jos. Y. Miller Presiding Judge & c

Samuel Vanfleet

The Court hereby certifies that Samuel Van Vleet who has subscribed to the above affidavit is a credible person and that his Statement is entitled to full credit.

Jos. Y. Miller Judge & c

I Joseph E Edsall Clerk of the Court of common pleas in and for the County of Sussex and State of New Jersey do hereby certify that the foregoing ~~the~~ contains the Original proceedings of the said Court in the matter of the application of Moses Dewitt for a pension. In Testimony whereof I have hereunto Set my hand and Seal of Office at Newton this twenty seventh day of november AD eighteen hundred and thirty four

J E Edsall Clk

War Det
P O
Decr 11 Decr 1839

Sir

The claim of Samuel Van Vleet to have served ~~two years~~ from 1778 to 1783 as a sergeant in the 2d New Jersey Regiment cannot be allowed. ~~upon the facts~~ stated Constant service of such duration was not performed by any corps excepting such as belonged to the continental army, in which it appears by the rolls the claimant did not ~~belong~~ serve. The ~~present~~ papers are placed on file. ~~If the claimant~~

The claim of Moses Dewitt who alleges that he was in constant service from 1779 to the close of the war in the same corps is in a similar condition. such service was not performed by any corps raised in New Jersey for the service of that State. The papers remain on file

Hon Samuel Fowler
HR

Newton Sussex County NJ

Dr sir

I forward you the annexed Depositions in the case of Moses Dewitt for a pension he made application to the department in the year 1834, the papers were presentd by the Hon. Samuel Fowler then member of Congress from our County, for me, the objection to the allowance of the claim then was that the evidence was too general & indefinite, not showing the precise time of service, and the papers were retained and placed on file as appears, by your answer to the Hon S. Fowler

Mr. Middaughs residence is in Tompkins County N.Y., and Mr. Dewitt being a very old & infirm man & without means could not obtain from Mr. Middaugh any further deposition, until now, Mr. Middaugh has visited this County, and thus enabled Mr. Dewitt to avail himself of the deposition of Mr. Middaugh, here to annexed, their service was in the militia, and altho not strictly in actual service during the years spoken of, yet there was a continual repetition of service of short periods at a time, throughout the years spoken of, they where what were called minute-men, and almost in constant service having kept no record or account of the particular times, & length of time, it is imposible for them at this late day to undertake to state, the different times or length of time precisely they were in actual service but living on the borders of the State than a wilderness, where the indians were making constant invasions, they were almost constantly on duty.

P.S. W. Vanvleet, who made application at the same time & who was a witness for Mr. Dewitt is now dead.

Yours Respectfully
Robt. Hamilton

Hon James L. Edwards

State of New Jersey
Sussex County ss. Personally came before me the undersigned, a Justice of

the peace in & for said County, Moses Dewitt aged seventy nine years who being duly sworn, deposeth and saith that, by reason of old age, and the consequent loss of memory, he cannot swear positively as to the precise length of service, but according to the best of his recollection, he served not less than the period and time mentioned below, as a private, during the years 1779. 80. 81. & 82, in actual service, at least one year, for which service I claim a pension.

Sworn to and subscribed before me this 11th day of July 1840
Rob Hm Carter Justice of the peace

Moses Dewitt

State of New Jersey
Sussex County ss Personally came before me the undersigned a Justice of the peace in & for said County, Solomon Middaugh who being duly sworn upon his oath deposeth and saith, that he is a pensioner of the United States & served as lieutenant in the Company of Capt. Josiah Cole, under Colonel John Seward, that he is acquainted with Moses Dewitt of said County who has sworn to the foregoing affidavit, and well recollects that the said Moses Dewitt was a private in said Company under the deponent, that the said Moses served in the same during the years 1779, 80, 81, & 82 whenever the said Company was out, that the said Company, rendered services on the frontiers of the State of New Jersey on the delaware & Lackawack Rivers, and that it was out a great deal during the years spoken of, but the deponent never kept any record of the different times, or length of time and from old age & consequent loss of memory cannot state the precise times or length of time, the said Company was out in actual service, but the deponent is confident that the said Moses Dewitt served in said Company, during said years at least one year in actual service.

Sworn to & subscribed before me this 11th day of July 1840 -
the word "Company" in the 5th line from bottom first interlined
Rob Hm Carter Justice of the peace

Solomon Middaugh

I certify that Moses Dewitt, and Solomon Middaugh who have subscribed the foregoing depositions before me are credible persons and that there statements are entitled to belief.

Rob Hm Carter Justice of the peace

Pension Office
11 Aug 1840

Sir

The additional evidence in support of the claim of Moses Dewitt enclosed in your letter of the 14th ulto has been examined. This is a claim for service as a minuteman in the New Jersey Militia estimated at 12 months service: In order to further investigation the duration of each term of actual Service must be specified as well as the period ~~in every instance~~ in which the Service was rendered

Robert Hamilton Esq
Newton Sussex Co.
New Jersey

Old War Desk
Department of the Interior.
Pension Office.
Washington, D.C. August 13 1878.

Sir:

In reply to your communication requesting copies of the declarations of Peter Sitts, Joseph Carpenter, Solomon Middaugh Moses DeWitt, and John McKinley, Revolutionary pensioners, I have to inform you that neither the original nor a copy of any essential paper filed in a claim before this office, can be furnished except upon the call of an officer of the Government, or of a court in which it is required for purposes of litigation. It is proper however to add that if the facts desired, be stated, you will be furnished the information, and if necessary, extracts from the ~~data as afforded by the~~ records of this office.

Lyman C Draper
Madison Wis

1459 West Second St., Ottumwa, Ia., May 14, 1929.
U.S. Pention Dept.,
Washington, D.C.

Dear Sirs;

Will you please send me the records pertaining to Moses DeWitt, who applied for a Revolutionary Pension about 1853 or 1854, supposedly from Ohio, as it was through Daniel D.T. Benedict, "Agent for Pensions," of Gallipolis, Ohio.

Very respectfully,
Mrs. A. L. Towns.

August 23, 1929.

Rev. and 1812 Wars Section
MEE/HMM
Mrs. A. L. Towns
1459 West Second Street
Ottumwa Iowa.

Dear Madam:

I advise you from the papers in the Revolutionary War pension claim, R.2921, it appears that Moses Dewitt was born October 23, 1761 in Wantage Township, Sussex County, New Jersey.

He applied for pension November 27, 1834, at which time he was a resident of Wantage Township, New Jersey, where he had always lived, and alleged that he enlisted in the spring of 1779 and served at various times until the close of the war in 1783, amounting to about one year in all, as a private in Captain Josiah Cole's Company, Colonel John Seward's New Jersey Regiment, defending the New Jersey and Pennsylvania frontiers against the Indians, and was in the battle of Lackawaxen.

His claim was not allowed as he failed to furnish proof of six months military service as required by the pension law.

The above ~~named~~ noted Moses Dewitt is the only soldier by that name found on the Revolutionary War records of this Bureau.

Very truly yours,
Earl D. Church
Commissioner.

Mrs. Joseph D. Kinney
1733 Virginia Park,
Detroit, Michigan
March 18 - 1938.

Pension Bureau
Washington D.C.

Sirs -

Could you give me proof of service at the Battle of Ticonderoga of a Moses DeWitt? If so, what was his rank, did he get a pension, or any other related facts?

I do not know his residence. I desire this information for Genealogical Family History and would appreciate an early reply

Very Truly

Ethel S. Kinney
(Mrs. J. D. K.)
above address

April 5, 1938
Moses Dewitt
R.2921
BA - J/MLB

Mrs. Joseph D. Kinney
1733 Virginia Park
Detroit, Michigan
Dear Madam:

Reference is made to your letter in which you request the Revolutionary War record of Moses Dewitt, and desire to know whether he served at the battle of Ticonderoga.

You are furnished herein the record of the only Moses Dewitt found in the Revolutionary War pension records of this office. The data therein were obtained from the papers on file in claim for pension, R.2921, based upon his service in the Revolutionary War.

Moses Dewitt was born October 23, 1761, in the Township of Wantage, Sussex County, New Jersey. The names of his parents were not given.

Moses Dewitt, while still living in Wantage Township, New Jersey, where he had always lived, applied for pension November 27, 1834, and alleged that while a resident of Wantage Township, New Jersey, he volunteered in the spring of 1779, served at various times until the close of the Revolution in 1783, amounting to about one year in all, as a private in Captain Josiah Cole's company, Colonel John Seward's New Jersey regiment engaged in defending the frontiers of New Jersey and Pennsylvania against the Indians, that he was in the battle of Lackawaxen. It is not stated that he was in the battle of Ticonderoga.

His claim for pension was not allowed as he failed to furnish proof of his six months' service in accordance with the requirements of the pension law.

There are no family data in the claim.

Very truly yours
A. D. Hiller
Executive Assistant
to the Administrator

Rejected

Revolutionary War Pension and Bounty Land Warrant Application Files
Series M-804; Roll # 805

Thomas and Charity Hill
New Jersey (W.935)

Declaration to obtain the benefit of the 3rd Section of the Act of Congress, of the 4th July, 1836.

State of New Jersey

Morris County ss. On this 4th day of October, 1836, personally appeared before the Court of Common Pleas of the County of Morris & State of New Jersey, Charity Hill, at present, residing in Morris town, in the County of Morris & State of New Jersey aged 82 years on the 23d day of February last (1836), who being first duly sworn according to Law doth on her oath make the following Declaration in order to obtain the benefit of the provision made by the act of Congress, passed July 4th 1836. That she is the widow of Captain Thomas Hill decd. who was, as she verily believes a Captain of a Militia company in Hardiston township, Sussex County, in Colonel Hankinson's Regiment, in the beginning of the Revolutionary war. She is the daughter of John Jerolman, was born in Bergen Co. N.J. near where the Town of Paterson now is, her father moved from there to Sparta, in Hardiston, Sussex Co. when she was about 12 years old. She was married at that place by Revd. Mr. Cox, a Baptist Minister in the year 1773 on the 20th day of December, to, Thomas Hill, who then lived in Sparta, & with whom she lived in the bonds of marriage until his death, which was on the 18th August, 1814, according to the record of births, marriages & deaths of her family, as contained in the family bible, kept by her husband& entered by him, which bible is now in her possession, & from which the date of her own age & marriage, & the time of her husbands death, are now taken. She is now his widow, & has remained so, having never married since the death of her said husband Captain Thomas Hill, as she trusts may more fully appear by reference to the proof hereto annexed. The troubles of the Revolutionary war began soon after her marriage, trainings of the Militia were often held at her husbands place of residence, viz: in Sparta, Sussex County & he was chosen Captain of a company, either just before or immediately after the first blood was said to be spilled at Lexington. The Indian frontier was along the Delaware river, distant some 15 or 20 miles from her,

& her husband with his company of Militia, was often called out on alarms there, to drive back the Indians & to disperse the Tories & Refugees, who were very troublesome. She believes, that before she had been married three years, her husband had been called out by Col. Hankinson & Col. Seward, with his company, to defend the frontier settlements along the Delaware river & the Minisink creek, not less than five or six times, often for a month, & sometimes for ten & twelve days at once, & making within these three years, not less than four months of active Service, as she believes. These Services against the Indians were in the year immediately before the war, the latter part of that year, & in the early part of the first year of the war. Sometimes a part of his company, joined with part of one or two neighbouring companies. Vincent Dey, & one Randolph or Randall, & Wm. Johnson, were Lieutenants & Ensign under Capt. Hill. They were often with him at his house, & she knew them well. Her husband, Captain Hill, was out in Service every year, more or less, from the beginning to the end of the war. His absence when on duty, was usually, one month at a time, but there were many occasions in which he was on duty with his company six weeks, & two months when the enemy were very strong & the country was overrun. He frequently volunteered his services, & She believes he never omitted nor neglected to go, when he was ordered by his Colonel, Hankinson. She recollects he was out with his company in the summer season, when the battle of Long Island was fought, & as she understood was stationed somewhere near Elizabeth town. He was from home on Militia Service, the most part of that summer. He owned a mill, & a tannery, & she, this deponent had the charge & care of both in his absence. She has herself personally attended the mill, & ground Wheat & Corn for weeks, & months in his absence, & superintended the Tannery. He also owned a Shoemaker's shop & carried on that business, which in his absence she drove on as well as she was able, altho' it suffered more or less in his absence. Her husband was out with his Company, when Genl. Washington retreated through N. Jersey into Pennsylvania, & sometime in the winter which followed, he was also present as she understood him to say & as she believes, at a fight with the enemy near Springfield, & aided in taking a number of Hessian prisoners who were sent to Morris town, & toward the spring, or early in the spring season, he lay with his company near Quibble town a month or near, under command of General Winds, & a large body of Militia. He often was with Genl. Winds in the neighbourhood of Elizabeth town & Springfield & Second river, & Wardsesson, now called Bloomfield & Connecticut farms. He also was in the Service with his company as she has heard him declare, at Woodbridge, & Amboy & Rahway & Connecticut farms & Hackensack & she verily believes he has actually performed tours of Militia service at each & all of these places, of not less duration than a month each, & frequently of six weeks or longer, when the enemy was strong & the Militia was all out. Beside his skirmishes with the Indians on the frontiers, she has often heard him tell of being engaged in the battle of Monmouth, and Brandywine, and the the Short hills & German town & Springfield & Red bank. She has heard him often tell of the murder of Mrs. Caldwell, after a fight with the enemy near where she lived, in which he was engaged, about the time of the burning of the Churches & towns of Springfield & Connecticut farms. Her brother, John Jerolman often drove the baggage waggon of Capt. Hill on his tours of duty, & when out one Summer with Capt. Hill near Aquahononk & Second river, John was taken very Sick of fever; & her father then living near her, & one of her Sisters, went after him, taking a bed in a waggon, & brought him home to Sparta, leaving Capt. Hill with his company, under Genl. Winds, as she understood. This Deponent has no Documentary evidence of the services of her husband Captain Hill, except what has been accidentally found in the office of Secretary of State of N. Jersey, & of which she never knew it's existence until after her own narrative had been given of his Services. It gives but a few instances, of his many tours of military duty, & speaks of his commission as being dated May 16th 1777. She well knows he performed many tours against the Indians as Captain, many months before that period, & several tours along the frontiers near New York, at Elizth. town & elsewhere, in the Summer & fall & winter of 1776. This she hopes to prove, altho' she has found but few survivors who served with him. The records at Trenton, shew him on duty as Captain, prior to May 16th 77, viz: on the 12th April 77, under Col. Hankinson. She believes his former Commission as Captain (probably first granted by the Crown) had been directed to be sent in, & a new one to be issued under State authority. He could not have been of the "newly elected," as spoken of in Col. Hankinson's letter, because paymt. had been made to him for services rendered as Captain, a month previous. She does not know with certainty, whether or not Capt. Hill was ever "drafted" into the service, nor does she know, whether or not his Services were performed in strict military language, as a "volunteer," all she knows on that point is, he always went into service, voluntarily & readily, whenever he learned from his superior officers, or others, that his services were wanted. His heart seemed devoted to the cause of the revolution, & he greatly neglected his own private business by his long & repeated absence from home, at the head of his Company. She cannot say, how long he was actually in service, but certainly not less than two years & a half, & she verily believes, as much as three full years, including Indian campaigns, always under the orders & command of his superior officers, in active & constant duty, always as Captain at head of his company, in full dress, with Sword, hat, epaulettes, Sash &c. The officers under whom Captain Hill served, were, Col. Hankinson, Col. Seward, Genl. Winds, Genl. Heard & Lord Stirling. She thinks she has often heard him speak of serving with Col. Ford, Col. Drake, Col. Seely, Col Frelinghuysen & Genl. Maxwell. He lived in Sparta, Sussex Co. when his first Service was performed. He removed from Sparta to Baskingridge in Somerset Co. at which place he died, in August,

18th day 1814, & she has never since married. She does not think she can particularise further, his services, but hopes to find some living witnesses. She believes his Services in the latter years of the war, including those on the Indian frontier, averaged 4 months per year, & more than 4, in the 1st & 2nd & 3rd years thereof. I do not know that I ever saw his commission. It was most probably passed over into the hands of his Administrators or others, & most likely was lost.

Sworn & subscribed in open Court, October 4th 1836 before me
Silas Cook

her
Charity X Hill
mark

State of N. Jersey, Morris County ss.

Morris Common Pleas. Personally appears in open Court, Isaac Bedell of said County, well known to some of the Judges as a man of fair character & good standing for veracity, & having been first duly sworn, according to law, saith on his oath, as follows, viz: I am in the 74th year of my age. I was born in the month of August, 21st day 1763, & was a fifer in Capt. Layton's company of Militia in 1776, & was in his company stationed on guard duty at Elizabeth town point, on guard duty & building breast works, in the Summer & fall of 1776. We all lay under cloth tents, & I believe under Col. Ford's command. I then & there saw Captain Thomas Hill, performing Militia duty one month commanding a company of Militia from Sussex County. In the year 1780, I had gone to live with an uncle in Sussex County, & was then 17 years old, & was under Captain Thomas Hills command, on a tour of a month against the Indians on the Delaware frontier. We were stationed at Fort Chambers, on the river, in Sussex County one month. Aaron Rolefson, Henry Westbrook, Peter Smith, & Henry Bemer (now all dead) were with us at the fort. Rosekrantz, I believe was our Colonel, Saml. Westbrook, Major, & Hill was our Captain. He lived in Sparta, & lived near the river. In the Summer previous, we had a fight with the Indians at Sheholi, under Major Samuel Meeker & Col John Seward. Capt. Hill was in that engagement. Daniel Talmage was killed in that fight, & I was shot in the fleshy part of the thigh, the wound is yet visible. Wilson Johnson was Hill's Lieutenant & Wilhelmus Westbrook Ensign. Caleb Kemble & David Shay, belonged to his company. We continued on the station & performed duty a month & were discharged.

Sworn & subscribed in open Court, this 4th day of October, 1836.Silas Cook
Isaac Bedell

State of New Jersey, Morris County ss. I, Joseph Dalrimple Esqr Clerk of Morris County hereby certify, that the foregoing are the original proceedings in said Court, in the matter of Mrs. Charity Hill, widow of Captain Thomas Hill, an applicant for a pension from the United States, on account of the Revolutionary services of her said husband. And I do hereby further certify, that Silas Cook Esqr. before whom the Declaration of said Charity Hill was made, & before whom also the deposition of Isaac Bedell was taken, is a Judge of said Court, & that the signature of his name thereto is genuine. I also certify, that Elijah Ward Esqr. the Majistrate before whom the respective depositions of Rachel Toan, Stephen Lyon, Job Loree, & Robert Young are taken; which are hereunto annexed, is a Justice of peace in said County, duly commissioned & authorized, & that the signature of his name thereto, "Elijah Ward," is genuine. In testimony whereof, I hereunto subscribe my name, & affix my seal of office at Morris town, August 7th 1837.

Jos. Dalrimple Clerk

★

State of New Jersey, Morris County ss. Be it known that on this 18th day of October 1836, before the subscriber, a Notary Public, duly commissioned, & qualified by the Governor of said State, personally appears Mrs. Rachel Tone widow of William Tone decd. residing in Somerset, & occasionally being with her friends in Morris town for several weeks together, & having first been by me duly sworn according to law, doth on her oath declare as follows, viz: I am in the 75th year of my age, & have spent my life principally in Baskingridge Somerset County. Capt. Goin McKoy of Somerset, was my father, & commanded a Militia company in the Revolutionary war. I have been well acquainted with Mrs. Charity Hill, the widow of Captain Thomas Hill, for Several years years past. Captain Hill with his family, removed from Sussex County into Somerset near my fathers dwelling, some time after the war ended, & died in that neighbourhood some 12 or 15 years since. I have known Mrs. Charity Hill his widow ever since their removal from Sussex, till now. She now lives in the house with one of my sons, on the farm adjoining my own residence. I am confident she has never married Since the death of Capt. Hill, but that she is now his widow. I know that she is poor & destitute, entirely dependent on the kindness of friends for a home, & has in a considerable degree lost her sight, & cannot use her needle at all in consequence of her blindness.

Sworn & subscribed, the date above written, before me
Lewis Condict Noty. Public

Rachel Toan

★

State of New Jersey, Morris County ss. The above named Deponent Mrs. Rachel Tone, believed by me to be a woman of truth & respectable character, comes before me & makes oath in due form of law, that the facts & Statements set forth in the above deposition respecting the widow Mrs. Charity Hill, _____ of Captain Thomas Hill decd. are strictly true, to the best of her knowledge & beliefs.

Sworn & subscribed, at Morris town July 28th 1837 before me
Elijah Ward J. P.

Rachel Toan

State of New Jersey, Morris County ss. Before the subscriber, a Justice of peace in said County, personally appeared Stephen Lyon, of Somerset County in N. Jersey aforesaid, believed by me to be a man of truth & of good character, & being by me duly Sworn according to law, saith as follows, viz: Mrs. Charity Hill, widow of the late Capt. Thomas Hill, & now an applicant for a pension under the act of Congress of July 4th 1836, is the mother of my wife, & is now living with her daughter my wife, in my family near Baskinridge in Somerset. The Bible from which two leaves have been taken in my presence this day, is the family bible, which was kept & owned by my father in law, Capt. Thomas Hill, & the leaves so cut out from said bible, contain the family record, which was kept by said Thomas Hill, & are known by me to be entries made therein, by said Thomas Hill himself, in his own hand writing. I was married to his daughter Nancy Hill, my present wife 3 or 4 years before her fathers death, & am well acquainted with his hand writing. The said family record & bible leaves, are attached to this deposition, & accompany the pension papers of said Charity Hill.

Sworn & subscribed before me at Morris town, N. Jersey
this 12th of July, Domini 1837.
Elijah Ward J. P.

Stephen Lyon

State of New Jersey, Morris County ss. Before the subscriber, a Justice of peace in & for said County, personally appeared Job Loree, aged 77 years, known to me as a man of truth & fair character, & being first duly sworn according to law, on his oath saith as follows, viz: I am now in my 78th year, since the 16th Sept. last, was a Militia Soldier in the revolutionary war, & was acquainted with Captain Thomas Hill of Sparta in Sussex County, who commanded a Militia company & whom I often saw at the head of his company performing Militia duties along the lines, & in every year of the war. In the summer of 76, I was performing a months Militia duty at Elizth. town point, in Capt. Pierson's company, & Col. Jacob Ford's regiment, opposite Staten Island, building breast works. Capt. Thomas Hill was also there a month, at the head of his company, engaged in the same duties. I stood Sentry several nights under Capt. Hills orders. He was a man who made a good appearance as an officer, & after this tour, I always knew him at once, wherever I met him. I remember him in the same fall again at Elizth. town, when Genl. Washington retreated thro' Jersey. We were then performing guard duty at the same place, under the same officers, & when the american army passed through, our Militia followed in the rear, so far as N. Brunswick, & Capt. Hill & his company also. We returned home through Bound brook, Pluckemin & Vealtown. Col. Ford, with Capt. Hill & most of the Morris & Sussex Militia were out all this fall, the whole British army pursuing as far as Brunswick & Amboy, when they wintered. The march is known as the "mud-rounds" march, & the Militia were in the field not less than 6 weeks. Early in Decemr 76, there was a battle at Springfield, in which the Militia of all the neighbouring Counties were ordered out. I saw Capt. Hill there with his company, under command of General Heard, Colonels Ford, Drake & others. A number of Hessians were taken prisoners, & the Militia stationed a month, in the neighbourhood of Springfield, Elizth. town & Staten Island. A Militia Encampment was made near Brunswick & Amboy this winter; under Genl. Winds & Genl. Dickenson, to guard the country against the enemy, which was kept up till the British left Brunswick, about the middle of April. I was there as much as 3 months, & Capt. Hill & his company were there also the whole time. Col. Ford & Col. Drake commanded the Morris Militia, & there were many skirmishes with the enemy. A party of the enemy came out of a wood near Quibble town & Capt. Hill & his compy. with others, engaged them. Some time after this, I was stationed on a months duty under Capt. Pierson at Elizth. town, in Price's barn in the fall season of 77, & remember Capt. Hill with his company was there also all the time. I remember Capt. Hill & his company in an engagement with the enemy at Aquakanonk, under Genl. Winds, in the fall season. We were stationed in that neighbourhood & near Newark a month. Col. Seely, Major Hays & Genl. Winds commanding. I saw him & his company at Connecticut farms battle, when a Mrs. Caldwell was killed, in the spring of 80, all the Militia in our part of the State were out on this occasion, as they were again at the Springfield battle in June following, where I also saw Capt. Hill & his company, & on both occasions remained on Militia duty in that quarter of the State, to protect the people against invasion by the enemy for a month at each time. Genl. Winds had command of the Militia, & there was two or three regiments of U.S. troops also engaged at Springfield. In the summer previous to taking of Cornwallis in August 1781, I saw Capt. Hill & his company perform a months tour of Militia duty at Elizth. town. He was frequently seen by me on horseback this month. Col. Seely commanded the Regiment.

Sworn & subscribed before me, at Morris town, June 15th 1837
Elijah Ward J. P.

Job Loree

State of New Jersey, Morris County ss. Be it known, that before the subscriber, a Justice of peace in & for said County of Morris, personally appeared Robert Young of said County, aged 84 years, a Soldier of the revolution, & well known to me personally as a man of truth, & fully entitled to belief, who being by me duly sworn according to law, on his oath saith as follows.

I am in my 85th year, was a Soldier in the war of the revolution, & performed military service throughout the whole war. I was well acquainted with Thomas Hill, of Sussex County, who lived in Sparta. He was Captain of a Militia Company in the beginning of the revolution in Col. Hankinson's, afterward Colonel Seward's regiment. I was frequently in Sparta & became acquainted there with Captain Hill before the war began, & frequently Saw him at the head of his Company as Captain, during the war, performing Militia duty in defence of his Country. The first time I recollect him on duty, was at the first battle of Springfield, in December 1776, under orders of Col Ford & General Heard. The enemy were principally Hessians & Refugees, & a number of Hessians were made prisoners. He was stationed at Connecticut farms & Springfield a month. In January following He was stationed near New Brunswick, when the British army occupied that town & Amboy, under command of Genl. Winds & Genl. Dickinson. There was a large Militia body encamped there, until the enemy left N. Brunswick & Amboy, which was late in the spring, & the Militia was not discharged till sometime in April. Captain Hill & his company were on this duty not less than three months, & there was much Skirmishing with the enemy. I was with the Morris Militia there, under Col Ford, Col. Drake & Genl. Winds. I recollect Captain Hill on Militia duty with his company at Rahway, one month near Trembly's point, in the Summer of 1777. I saw him & his company near Hackensack early in the fall of that year, under Genl. Winds, Col. Seely & Col. Hays, where we had a skirmish with the Tories & Refugees, after which we were marched under Genl. Winds to the neighbourhood of Pompton, & were out a month, Capt. Hill also. Late in the Same fall, I recollect him a month at Elizabeth town, stationed at one Price's. In a skirmish with the enemy at Aquakanonk bridge. I also remember Capt. Hill & his company, under Genl. Winds & Col. Seely a month there & in the neighbourhood of Newark. One Jabez Bell was accidentally shot through his head & was killed at this place. It was in the fall season. I also recollect Captain Hill on a months tour at Minisink, against the Indians, in the Summer Season, under Genl. Winds & Col. Seward. Captain Hill I saw on a tour of duty of a month near Connecticut farms in the spring of 1780 & in the battle at that place, under General Winds. I saw him on another tour of a months duty at Springfield in June 80, in that battle, when Springfield was burnt. Genl. Winds commanded. After the battle, the enemy retreated to Staten Island, & Capt. Hill & the other Militia performed guard duties a month between Elizabeth town & the landing opposite Staten Island.

Sworn & subscribed before me at Morris town June 22d 1837
Elijah Ward J. P.

Robert Young

State of New York, County of Kings, viz: Be it known, that on the 19th day of July Domini, 1837, before Samuel Smith Esqr. one of the Judges of the Court of Common Pleas in & for the County & State aforesaid, personally appeared Teunis Joralemon of the City of Brooklyn in said County, well known to me to be a man of very respectable character & fully entitled to belief, who being by me duly sworn according to Law, on his oath saith as follows, viz: I am now in the 77th year of my age, now a resident of Brooklyn City aforesaid, & in the earlier part of my life a resident of Sussex County New Jersey. I was well acquainted with Captain Thomas Hill, an officer in the Militia service, in Sussex County New Jersey in time of the Revolutionary war, whose widow, Charity Hill, is now applicant for a pension. The said Charity Hill, was formerly, & before her marriage, Charity Jeroleman, my own Sister, living before her marriage at my father's home in Sparta, Sussex County, where Captain Hill also lived. I knew him, before he was married to my said Sister, & until his death. They were married at my father's house in Sparta, on the 20th day of December, in the year of our Lord 1773, by a Baptist Clergyman whose name was Cox. I was present, & saw them married. They had several children before the war ended, & some before it began. I lived neighbour to Captain Hill, during the greater part of the revolutionary war, & performed Several tours of Militia duty under his command as Captain during said war, some of which are yet within my recollection. I was with him & under his command at Redbank, called Fort Mifflin, & Mud Island when it was surrendered to the enemy, & the Company performed Militia duty in that part of Jersey, at Haddonfield & Morris town, more than a month, under Colonel Seely, & General I served another month under Captain Hill at the battle of Springfield, in June 1780, when that village & Church were burnt by the enemy, after which they retreated to Staten Island & the Militia pursued them, & guarded the shores along Staten Island, & in the neighbourhood of Elizabeth town & Springfield a month or more.

This deponent is unable to remember what General officers were at Fort Mifflin or who commanded the Regiment at that time. My brother John Joralemon _____ was very often called out on tours of duty very often as I believe with Captain Hill, but the particular times & places I am not able to specify in detail, but know that Captain Hill was in service a great portion of the time which he cannot particularise, as my father removed to Somerset County in the spring of 1780 & did not of course know so well the different tours & duty in the order in which they were performed

Sworn & subscribed before me the day &
year above mentioned at the City of Brooklyn
Samuel Smith one of the Judges of the Court of Common
Pleas in And for the County of Kings in the State of New York

Teunis Joralemon

★

Abraham Vanderveer Clerk of the Court of Common Pleas for Kings County, in the State of New York do certify that Samuel Smith Esq. before whom the foregoing deposition has been taken is a Judge of the Court of Common Pleas duly commissioned to administer oaths & that the signature of his name therefor is genuine. In witness whereof I subscribe my name & affix my seal of office this 19 day of July 1837

Abm. Vanderveer

★

State of New Jersey Sussex County ss. Be it known that before me the subscriber a justice of the peace in and for said county Personally appeared William Perry of said county aged 76 years a Soldier of the revolution & known to me personally as a man of truth & fully entitled to belief who being by me duly sworn according to law on his oath saith as follows. I am in my 76th year, was a soldier in the war of the revolution and performed military service at different times during the war. I was well acquainted with Thomas Hill of Sparta in this county. He was a Captain of a Militia Company in the war of the revolution. I was first under the command of Capt. Thomas Hill at Elizabeth Town N.J. in the summer of 1778 on a tour of one month, we were then under the command of Genl. Winds & Col Seward. I was at Pompton on a tour of one month, under Capt. Hill, in the year 1777, was one of the guard of the public stores. I was with Capt. Hill on a months tour of duty against the Indians at Minisink, on the Delaware river, in the summer season: the year I do not now recollect. Genl. Winds & Col Seward commanded.

Subscribed and sworn before me the 26th day July 1837
Matthias H Ogden J. P.

his
William X Perry
mark

State of New Jersey
Sussex County ss. I Joseph E Edsall clerk of the Court of Common pleas in & for said County do hereby certify, that Matthias H Ogden Esquire before whom the foregoing depositions appear to have been taken at the dates thereof respectively was and now is a Justice of the peace in and for said County duly commissioned and sworn, and that his Signatures thereto written are genuine. Witness my hand and Seal of Office this 27 day of July 1837

J E Edsall Clk

J. L. Edwards Esq.
Dear Sir,
I have a good deal of _____ in the case of Charity Hill & have no doubt she has rights. Would it be improper for you to send to me her papers, that I might examine them in my Room?

I am respectfully &c.
Saml. L. Southard
Washn. 9 April 1840

New Jersey

Charity Hill, widow of Thomas Hill, decd. who died on the 18th August 1814 of Morris Co. in the State of N. J. who was a Sgt & Capt in the compy. commanded by himself of the Regt commanded by Colo Hankison in the New Jersey line for 4 months & 2 days Captain 2 mos Sergeant Inscribed on the Roll of New Jersey at the rate of 91 Dollars 33 Cents per annum, to commence on the 4th day of March, 1831. Certificate of Pension issued the 3d day of Sept 1841 and sent to S. L. Southard, Senate U.S.

Arrears to the 4th of March 1841	$913.30
Semi-anl. allowance ending Sept 1841	45.66
	$958.96

Recorded by D. D. Addison Clerk,
{Revolutionary Claim, Act July 4, 1836. Section the 3d}
Book ___ Vol. 1st Page ___

Morristown N.J. May 1851

Hon: the Secry of the Departmt. of the Interior

Sir,

The children & heirs of Mrs. Charity Hill, decd. widow of Capt. Thos. Hill, & a Revolutionary Pensioner of the U. States, being much dissatisfied with the annuity allowed said widow, respectfully solicit through you, a re-examination of her claim, believing that she was entitled equitably & legally, to the full pay of a widow of a Captain of Calvary.

By the original papers on file, it will appear that I was her Atty & Agent & so continued to her death, which occurred as I believe some time in 1848. Her children desire me in their behalf, to present to you their claim for Increase.

The Declaration of the widow (being very infirm both of body & mind) unaided by any memoranda, of course could present but a very imperfect detail of his duties, altho' she was conscious of his long & repeated absences from home, on public service.

His companions were mostly dead or dispersed, & the proofs were slowly furnished. Yet finally, the testimony of many very credible eye witnesses, established a long career of public faithful duty, at the head of his troop of Dragoons, along the valley of the Delaware (then an Indian frontier) as well as at various other places, & in several engagements.

It is believed that Documentary proof of his rank, will be found on file. Extracts from N.J. records duly certified, will be forwarded in a few days, shewing as well his rank, as the various places & dates of service, & monies paid him. The precise date of his commission is not known, but all his service is believed to have been rendered as Capt. of Cavalry, in which character an Increase is claimed.

If desired, proof of the time of the widow's death & number of her children will be furnished.

Very respectfully
Your obt. servt. &c.
Lewis Condict

Dear Sir

Your favour of the 3rd Inst is at hand and I would reply that Mr Thomas Hill decd. lived and died in Somerset County State of New Jersey. The Pension was Issued to his widdow many years after his death who live in the same county and state. His heirs have been informed that there is a considerable amount due to Cap Hill, (or that there should be) from the documents which were presented to enable his widdow to receive what she did, but that she did not receive all that was due.

I am obliged for your attention to the matter and trust you will have the goodness to investigate the matter by a reference to the documents and if any thing is due inform me the amount.

With much regard yours &c.

Thos. G. Talmage

New York 10th June 1851

⁎

Claim of Charity Hill
Act of 4th July 1836

State of New Jersey

County Mercer ss. Be it known, That on this Twenty first day of June A.D., eighteen hundred and fifty one, before the undersigned, a ~~Justice~~ Judge of the ~~Peace~~ pleas in and for the County and State aforesaid, personally appeared James J. Hill a resident of Somerset ~~in said~~ County, and made oath according to law that he is the son of Charity Hill who was a Revolutionary Pensioner of the New Jersey agency at the rate of ninety one 33/100 Dollars per annum; that he has been informed that the said Pension ought to be increased; that he is directly interested as a claimant in said Pension, and makes this affidavit to be filed with such additional evidence or arguments as my Agent may use in prosecuting said claim.

Sworn to and subscribed, the day and year aforesaid, before me

Chas. Burroughs ~~J. Peace~~

James J. Hill

⁎

I, Charles Burroughs ~~Justice~~ Judge of the ~~Peace~~ pleas in and for said County do hereby certify that James J. Hill who executed the foregoing paper is a respectable resident of ~~this~~ Somerset County, and is, I believe, directly interested in said claim as therein set forth.

Chas. Burroughs ~~J. Peace~~ Judge

⁎

Power of Attorney- Revolutionary Pension Claims

Know all Men by these Presents, That I, James J. Hill of Somerset County N Jersey hereby constitute and appoint William S. Allison of Washington City my true and lawful Agent and Attorney, to prosecute the Claim of Charity Hill for any amount of Revolutionary Pension, or Increase of Pension, that may be due; and I hereby authorize my said Agent to examine all the papers and documents in relation to said claim, on file in the Departments at Washington City, or elsewhere; to file additional evidence or arguments; and to receive the Certificate which may be issued for said Claim, which Certificate I wish made payable to myself as admistrator of said Charity Hill to appoint one or more substitutes under him for the purposes herein expressed; and to do all things that I might or could do were I personally present. Hereby ratifying and confirming all that my said Attorney and Agent shall lawfully do in the premises. Witness my hand and Seal this 21st day June A.D., 1851

Signed and Sealed in the presence of Chas. Burroughs

James J. Hill

Chas. Burroughs

State of New Jersey

County of Mercer ss. On this Twenty first day of June A.D. 1851 before me, the subscriber, a ~~Justice~~ Judge of the ~~Peace~~ pleas in and for the County aforesaid, personally appeared James J. Hill and acknowledged the foregoing Power, of Attorney to be his act and deed, for the purposes therein mentioned. In Testimony Whereof, I hereunto set my hand, the day and year aforesaid.

Chas. Burroughs ~~J. Peace.~~ Judge

Washington
Jany. 27, 1852

J. E. Heath Esqr
Comr of Pensions
Sir,

In the case of Charity Hill widow of Capt. Thomas Hill, late a pensioner of the New Jersey roll under act July 36.

I have recd from Doctr. Lewis Condict of New Jersey (the former agent) the enclosed certified extracts from the Sect. of State of New Jersey and upon which an increase of her pension is claimed.

Not being prepared to say upon evidence the claim was granted, yet it is hoped this evidence in connection with the prior on file, will justify the Department in granting an augmentation of the stipend.

very respectfully
I remain your most Obt Serv
Alex. Ray

The proof now exhibited shows Capt. Hill
was in the service for the mos. of Ap. & Nov.
1777 & July & Oct 1778 making four months.
Capt Thomas Hill Col Hankinsons regt. Sussex County N. Jersey

				days
185.12	1777	12 April	32. 9.0 Sussex	6
375.06	1777	19 Nov.	141.15.0 Red Bank	28
375.06	1778	29 May	141.15.0	
87.78	1778	14 July	33.15.0 Delaware	7
20.12	1778	July	7.16.0 Sussex	2
50.54	1778	5 Octob	3. 7.0 Sussex	1
50.54	1778	5 Octob	15. 9.6 Aquacknack	3
24.00	1778	18 Octob	9. 1.0 Delaware	2
52.54	1778	8 July	19.18.0 Eliz Town	4
114.38	1778	2 Nov	42.19.4 Frontiers	9
			448. 4.10	90
			deduct Red Bank	28
				62

Adrian Lyon,
Counsellor-at-Law,
Perth Amboy, N.J.
April 6th, 1897

The Pension Department,
Washington, D.C.
Gentlemen:

Captain Thomas Hill of the Sussex County, New Jersey militia served in the Revolutionary War. After his death his widow received a pension from the government. She was at that time living at Lyons Station near Basking Ridge, New Jersey. When the application was made for this pension, as a part of the evidence of her right thereto, a copy of the family Bible was sent to the department. It was either the original record taken from the Bible or a transcript thereof.

I desire to get a copy of this record from this family Bible, and if such is on file in your office, will you please send it to me.

Yours truly,
Adrian Lyon
(Dictated)

L. W. B.
May 26, 1897

Sir:

Replying to your communication of recent date, wherein you request a copy of a bible leaf on file in the pension claim of Charity Hill, widow of Thomas Hill, who was a soldier of the Revolutionary War, you are advised that such a copy, or the original if desired, will be furnished you upon receipt of an affidavit showing that you are the eldest surviving lineal descendant of the soldier, Thomas Hill. If you cannot comply with these requirements, such an affidavit from the eldest surviving lineal descendant of the soldier, with an endorsement permitting you to receive the leaf will be accepted.

Very respectfully,
Commissioner.

Mr. Adrian Lyon,
Perth Amboy, N.J.

Adrian Lyon,
Counsellor-at-Law,
Perth Amboy, N.J.
June 2nd, 1897

Department of the Interior,
Bureau of Pensions, Washington, D.C.
Dear Sir:

In reply to yours of the 26th ult. in reference to the Bible leaf in the pension claim of Charity Hill widow of Thomas Hill who was a soldier in the Revolutionary War, I enclose you an affidavit from the eldest, surviving, lineal descendant of the soldier Thomas Hill with also an endorsement permitting me to receive the Bible leaf.

Will you please mail the original to me, and oblige
Yours truly,

Adrian Lyon
(Dictated)

State of New York,
County of Schoharie, ss. Rev. David Lyon, of the town of Sloansville, in said County and State being duly sworn says that he is the grandson of Charity Hill, widow of Thomas Hill, who was a soldier of the Revolutionary war; that said Thomas Hill and Charity Hill were the father and mother of this deponent's mother; that at the time of securing a pension for said Charity Hill she lived at deponent's fathers house and a leaf was taken from the Bible of said Thomas Hill and sent to the Pension Department at Washington; that this deponent is eighty five years of age and is the eldest, surviving, lineal descendant of said soldier, Thomas Hill.

Subscribed and sworn to before me, this 31 day: of May A.D
1897
Delwin H. Gordon Notary Public

David Lyon

I, David Lyon, of the town of Sloansville, New York, being the eldest, surviving, lineal descendant of Thomas Hill, a soldier in the Revolutionary War, whose widow was Charity Hill, do hereby permit my nephew, Adrian Lyon, of the City of Perth Amboy, New Jersey, to receive the Bible leaf from the Bible of said Thomas Hill, on file in the Pension Department at Washington, and hereby request that the same be delivered to him.

Dated May 31st, 1897.

David Lyon

O. W. & N. Div.
Revolutionary.

Department of the Interior,
Bureau of Pensions,
Washington, D.C.,
June 25, 1897.

Sir:
In compliance with your request I enclose, herewith, the Bible Leaf filed in the

claim (Revolutionary war) of Charity, widow of Thomas Hill. Kindly acknowledge receipt by an endorsement on the reverse side of this sheet, returning same by using enclosed penalty envelope.

Very respectfully,
Commissioner.

Mr. Adrian Lyon,
Perth Amboy, N.J.

Thomas Hill was born in Elizabeth Town, East Jersey, in the year of 1747, the 31st May. The son of John and Margaret Hill, and married the 20th December, 1773.

October 4, 1777, was born, John, the son of Thomas and Charity Hill, the County Sussex, East Jersey, and deceased in August 1779.

At 4 o'clock in the morn, Sunday October 24th, in the year 1779, was born John, the son of Thomas and Charity Hill, in the County of Sussex and deceased 28th March, 1782.

September 11th, 1781, was born Elizabeth, the daughter of Thomas and Charity Hill, the same place and time of day as above, and deceased the 31st of May, 1782.

March 12, 1783, Mary the daughter of Thomas and Charity Hill was born it being Thursday near 3 o'clock in the afternoon.

April 1st, 1785, was born Margret, the daughter of Thomas and Charity Hill, on Friday near 2 in the morn.

Saturday, June 28th, 1788, was born Elizabeth, the daughter of Thomas and Charity Hill, near six in the morn, and deceased 11th February, 1790.

Wednesday, July 21st, 1790, was born Hannah, the daughter of Thomas and Charity Hill, near 6 o'clock in the morn.

Friday, February 17th, 1792, was born Nancy, the daughter of Thomas and Charity Hill, near 4 in the morn.

Wednesday, February 19th, 1794, was born Thomas, the son of Thomas and Charity Hill, near five in the morn.

Saturday, May 14th, 1796, was born James, the son of Thomas and Charity Hill, near 4 o'clock in the afternoon.

June 26th, 1798, was born Edward, the son of Thomas and Charity Hill, in the county of Sussex, East Jersey.

I certify that the above is a true and correct copy of a Bible Leaf filed in the claim of Charity, widow of Thomas Hill, (a soldier of the Revolutionary war), the same being returned this day to Adrian Lyon, Perth Amboy, N.J., in accordance with the attached affidavits.

June 25, 1897.

N B Prentice
actg Chief, O. W. & N. Division.

Copied by F. W. B.
Compared by A. D. S. and F W B

Sir:

I have yours of the 25th instant Enclosing the Bible leaf in matter of the claim of Charity Hill (Revolutionary War), for which please accept my thanks

Yours truly,
Adrian Lyon

6/29/97

Record Division
July 9, 1909

Mrs. Henry Young,
#310 Court Street,
Portsmouth, Va.

Madam:

In reply to your personal inquiry of the 6" instant, you are advised that from the papers in pension claim Wid. File No. 935, it appears that Thomas Hill the son of John and Margaret Hill, was born May 31, 1747 at Elizabeth Town, East Jersey, and married at Sparta, Sussex County, New Jersey, December 20, 1773 Charity Jerolman, daughter of John Jerolman, who was born February 23, 1754 in Bergen County, New Jersey; that he resided at Sparta, New Jersey when he was appointed a Captain of the Sussex County Militia under Colonels Hankinson and Seward, and served at least four months every year of the War, amounting to nearly three years' service; that he was in many skirmishes with the Indians along the Delaware river, and in the battles of Short Hills, Brandywine, Germantown, Red Bank, Fort Mifflin, Monmouth, Connecticut Farms and Springfield.

Thomas Hill died August 18, 1814, at Baskingridge, New Jersey, and his widow Charity was allowed pension on her application executed October 4, 1836, while a resident of Morristown, New Jersey.

Their children were:

John	born October 4, 1777,	died August - - 1779.
John	born October 24, 1779,	died March 28, 1782.
Elizabeth	born September 11, 1781,	died May 31, 1782.

Mary	born March 12, 1783,	
Margaret	born April 1, 1785,	
Elizabeth	born June 28, 1788,	died February 11, 1790.
Hannah	born July 21, 1790,	
Nancy	born February 17, 1792,	married Stephen Lyon.
Thomas	born February 19, 1794,	
James	born May 14, 1796,	
Edward	born June 26, 1798,	

In 1837 the widow's brother, Teunis Joralemon of Brooklyn, New York, testified in the case.

For the date of the widow's death and name of the person to whom the arrears of her pension were paid, apply to the Auditor for the Interior Department, U.S. Treasury Department, citing the following data: "Charity Hill, widow of Thomas Hill, Certificate No. 3,598, issued September 3, 1841, under the Act of July 4, 1836 at the New Jersey Agency."

Very respectfully,
Actg. Commissioner.

Rejected

Widow, &c. File No. 935
Charity Hill, widow
Thomas Hill dec'd Sergt. & Capt. Rev. War
Act: July 4, 1836.
Index: - Vol. A, Page 333
[Arrangement of 1870.]
Revolutionary War Pension and Bounty Land Warrant Application Files
Series M-804; Roll # 1279

Sollomon or Solomon Middaugh or Madoh
New Jersey (S.23,323)

THE STATE OF NEW-JERSEY,
TO Solomon Madoh Gent. GREETING:

WHEREAS it hath been certified to His Excellency WILLIAM LIVINGSTON, Esquire, Governor and Commander in Chief of this State that you have been duly chosen by the   Company of Militia in   in the County of Sussex being the Company in the Second Rigment of the said County, whereof Aaron Hankison Esqr. is Colonel to be of the said Company; you are therefore to take the said Company of Militia into your Charge and Care as First Leutenant thereof, and duly to exercise both Officers and Soldiers of the said Company in Arms: And as they are hereby directed to obey you as their First Leut. you are likewise to obey and follow such Orders and Directions, from Time to Time, as you shall receive from your Colonel or other your superior Officer or Officers; and for your so doing this shall be your Commission. IN TESTIMONY whereof the Great Seal of the said State is hereunto affixed: Witness WILLIAM LIVINGSTON, Esquire, Governor, Captain-General and Commander in Chief in and over the State of New-Jersey and Territories thereunto belonging, Chancellor and Ordinary in the same, at Haddinfield the Sixtenth Day of May in the Year of our Lord One Thousand Seven Hundred and Seventy Seaven

Excellency's Command,
Chas Pettit Secty

Wil. Livingston

State of New York

County of Tompkins ss. Solomon Cortwright of the town of Danby in the county & state aforesaid being first duly sworn in open court doth on his oath depose, and say that he is Eighty years of age and upwards, that he was born & brought up in the town of Wantage, County of Sussex & State of New Jersey. That he was well acquainted with Solomon Middough who has signed the preceding declaration who then resided in the same town with and about 3 miles from this Deponant before the Revolutionary War. That he this Deponant was present with the said Solomon Middough in the said town of Wantage and saw him enlist as a soldier in the Revolutionary service in the spring of 1776, in Capt Joseph Harkers Company in Col Ephraim Martin's Regiment and that he this deponant served in the same company with said Solomon Middough for one month. That during the remaining part of the year 1776 this Deponant was in the service in New York & on Long Island but at that time he understood & believed & now believes the said Solomon Middough was in said service during that time in the State of New Jersey.

That in the spring & summer of 1777 this Deponant saw at several different times _____ the said Solomon Middough in said service at the Scotch Plaines in the State of New Jersey in Capt Meekers Company and Col Martin's Regiment of New Jersey State troops, that this deponant served about one month in the same company with said Solomon Middough and when the deponant left the aforesaid service at the Scotch planes he left the said Solomon Middough in the service.

That in the month of June 1777 he again saw the said Simeon Middough in the revolutionary service at Pumpton planes in the said State of New Jersey under Capt Benjamin Coykendoll, & in Col Frelinghousen's Regiment.

That this deponant was in the aforesaid service in the same company with the said Solomon Middough for about one month at ~~Germantown~~ Minisink in the ~~State of Pensylvania under Capt.~~ New York. That during the said last mentioned service the said Solomon Middough commanded the company as Lieutenant in the absence of Capt Cole who was the Capt of said company under Col. Aaron Hankerson, that said last mentioned service was in the summer of 1778.

And this Deponant further saith that he often saw the said Solomon Middough in actual service as a Lieutenant in said Capt Coles Company under Col. Hankerson at Minisink in the State of New York at Montague in the State of New Jersey, and at various other places along the Delaware river in the State of New Jersey during the years 1778, 1779, & 1780 and until near the close of the revolutionary war in 17783, and that he verily believes that the said Solomon Middough was actually engaged as Lieutenant in said service during that time. And further saith not.

Subscribed & sworn in open court this 3rd day of October 1832 before the Judges thereof Saml Love Clerk

his
Solomon X Cortwright
mark

I further certify that I am acquainted with the above named Solomon Cortright, that he is a person of credibility and that his statement is entitled to credit

Saml. Love Clk Tompkins County

Declaration in order to obtain the benefit of the act of Congress, passed June 7th, 1832.

State of New York,

Tompkins County, ss. On this 4th day of October personally appeared in open court before the Judges of the Court of Common Pleas in & for said county now sitting, Solomon Middough a resident of the town of Caroline in the county of Tompkins and state of New York, aged seventy seven years, who being first duly sworn according to law, doth on his oath make the following declaration in order to obtain the benefit of the act of Congress, passed June 7th, 1832. That he entered the service of the United States, under the following named officers and served as herein stated.

That he volunteered into said service for one month in March 1776 in Capt Harkers Company of New Jersey Militia in a Regiment under the command of Col. Ephraim Martin, that at that time he resided in the town of Wantage, County of Sussex and State of New Jersey where he volunteered as aforesaid. That during the aforesaid month of March he served on a tour to New York and was engaged in building and repairing batteries in the City of New York. That in the month of August in the same year he again volunteered into said service for one month and was called to Perth Amboy in the State of New Jersey where he served for one month in a company commanded by Samuel Meeker in Battallion under the command of Major Samuel Westbrook in the New Jersey Militia under Gen. Robedeaux. That in the month of October in the same year he again volunteered to serve for one month in a company commanded by William McLennen in Col Brodirck's Regiment of New Jersey Militia and was called to South Amboy where he performed the last aforesaid months service during which time he was chosen by the voice of said company to serve as Ensign and he served accordingly as Ensign and was employed in guarding the Shores. That in December 1776 the said Samuel Meeker was promoted to the office of Major in the Regiment commanded by Col. Hankison and this declarant was then elected by the said company as first Lieutenant in said company. That in January 1777 he again volunteered with said company to serve for two months under Col. Hankison, Lieut. Col. John Seward and Major Meeker. That served with said company as Lieut Commandant during the said two months & until in the month of March 1777, and was engaged in said last mentioned service as Lieut Commandant as aforesaid at Scotch Plains in the State of New Jersey in guarding the lines & in scouting in that neighborhood. That during the last mentioned service at Scotch Plains he saw a Brigade of U. S. Troops under General Sullivan. That in the month of May he received his commission from William Livingston Esq. Govenor of the State of New Jersey which said commission is hereunto annexed.

That he with ~~he with~~ said company again volunteered into said service for about five weeks during the months of May & June 1777 and served for that term in said company then commanded by Capt Coykendoll, as first Lieutenant under his aforesaid commission as such hereunto annexed, at Pumpton Plains and at Paramus in the State of New Jersey in said Col. Hankisons Regiment then under the command of Lieut. Col. Seward and Major Gordon and General Hurd commanded the Brigade to which said Corps was attached and belonged.

That in February and March 1778 he again volunteered with said company then under John Holbert in which the applicant served as first Lieutenant as aforesaid for about six weeks in the Regiment under Col. Smith & Col. Wines at Elizabethtown N.J. during which time they were there stationed as a guard & served as such. That in the months of May & June 1778 he was again in actual service, in the Revolution as a volunteer with the aforesaid company under his command as Lieut Commandant for one month at Minisink in the State of New York & was employed in building a fort at that place. That in September & October 1778 he was again actually employed with said company in the aforesaid service as a volunteer for one month at Germantown in the State of Pennsylvania , that during that month he commanded said company in the absence of Capt. Cole under Col. Hankison & belonged to the Brigade under Genl. Furman & was present there & engaged in the battle of Germantown where General Washington commanded the American forces in person aided by Genl. Green. That during the fall & winter of 1778 he was engaged with his said company as volunteers for two months in said service in which he served as Lieut. in the town of Montague & along the Delaware river in N.J. in the aforesaid Regiment. That from 1779 to the close of the Revolutionary War in 1783 he was actually engaged in said service as the Commandant of said company under his aforesaid commission in the States of Pennsylvania, N. Jersey & N. York, was present at the battle at Lackawax on the Delaware on the 22nd of July 1779, at which Col. John Haythorn commanded. That he has no documentary evidence except the commission hereto annexed and that he knows of no person whose testimony he can procure who can testify to his service except those whose affidavits are hereto annexed. He hereby relinquishes every claim whatever to a pension or annuity except the present, and declares that his name is not on the pension roll of the agency of any state.

Sworn to and subscribed, the day and year aforesaid
in open Court before the Judges aforesaid.
Saml. Love Clk

Sollomon Middaugh

The following questions were propounded by the Court to the applicant.

1. Where and in what year were you born?

Answer. At Wantage in the County of Sussex, New Jersey, August 20, 1755

2. Have you any record of your age, and if so where is it?

Answer. I have in my Bible at home

3. Where were you living when called into service; where have you lived since the Revolutionary war; and where do you now live?

Answer. In Wantage New Jersey I was first called into service, after Revolution lived in N. Jersey in the County of Sussex till 1806, then moved to Tompkins County in N.Y. & _____ now (in the Town of Caroline)

4. How were you called into service; were you drafted, did you volunteer, or were you a substitute, and if a substitute, for whom?

Answer. I volunteered, I never was drafted, nor never will be

5. State the names of some of the regular officers who were with the troops when you served; such continental and militia regiments as you can recollect, and the general circumstances of your service. Answer. Generals Washington, Genl. Sullivan, Gen. Maxfield & Genl. Putnam, Col. Spencer, in the Militia I knew Generals Hurd, Furman, Robedoux, Colonels Hankison, Seward.

6. Did you ever receive a discharge from the service, and if so by whom was it given, and what has become of it? and Did you ever Receive a Commission & from whom.

Answer. I never received any written discharge. I received a Lieut's Commission from Gen. William Livingston, Chas. Pettit which is hereto annexed

7. State the names of persons to whom you are known in your present neighborhood, and who can testify as to your character for veracity and their belief of your services as a soldier of the Revolution. Answer. John King, David Quigy David Woodcock Esq. Joseph Chambers, Gorham Hanford Bartholomew Grim, Peter Lounsbery, Levi Stater, Aaron Bull, Silas Hutchinson.

We, David Woodcock residing in the town of Ithaca and county of Tompkins and State of New York, and John King & David Quigy residing in the same place, hereby certify that we are well acquainted with Solomon Middough who has subscribed and sworn to the above declaration, and that we believe him to be seventy seven years of age, that he is reputed and believed in the neighbourhood where he resides, to have been a soldier of the Revolution, and that we concur in that opinion.

Sworn and subscribed, the day and year aforesaid,
Saml. Love

D Woodcock
John King
David Quigy

And the said Court do hereby declare their opinion after the investigation of the matter, and after putting the interrogations prescribed by the War Department, that the above named applicant was a Revolutionary Soldier, and served as he states, and the Court further certifies that it appears to them that David Woodcock, John King & David Quigy who have signed the preceding certificate are citizens resident in the town of Ithaca and county of Tompkins and are creditable persons, and that their statements are entitled to credit.

I Samuel Love Clerk of the Court of Common Pleas for the County of Tompkins do hereby certify that the foregoing contains the original proceedings of the said Court in the matter of the application of Solomon Middough for a pension. In Testimony Whereof, I have hereunto set my hand and seal of office, this 4th day of October 1832.

Saml. Love Clk

State of New York

Tompkins County ss. Personally appeared before me the undersigned a Majistrate in and for the said County of Tompkins Solomon Middough who being by me duly sworn deposeth and saith that he is the same identical Solomon Middough named on and who has subscribed and sworn to the annexed Declaration on his application for a Pension for services in the Revolutionary War. And further this Deponent on his oath aforesaid saith that by reason of old age and the consequent loss of memory he cannot swear positively as to the precise length of his service in the revolutionary war, nor to the precise time or times of his entering such service under his various enlistments, but according to the best of his recollection he served not less than the periods mentioned before and in the following grades to wit:

For one month I served as a Private in Capt Joseph Harkers company of N. Jersey Militia at New York.

For one month I served as a Private in Capt Samuel Meekers company of N. Jersey Militia at Perth Amboy, N. Jersey.

For one month I served as Ensign in Capt. William McLennins company of Militia aforesaid at South Amboy N.J.

For two months I served as 1st Lieutenant, commandant, in Capt John Davis' company of the aforesaid Militia at the Scotch Plaines N. Jersey, the said Capt not being present in service.

For one month & 15 days I served as 1st Lieutenant in Capt Holbert's compy of the aforesaid Militia at Elizabethtown N. Jersey.

For one month I served as 1st Lieutenant Commandant in Capt Cole's compy his said Capt not being in service, at Minisink in N. York State.

For one month I served as 1st Lieutenant commandant in the absence of Capt Cole in his company at Germantown in Pennsylvania

For two months I served as 1st Lieutenant & as such commanded said Capt Coles company at Montague in the State of N. Jersey & during which time I had the command of the main guard.

For three years and 3 months I served constantly as 1st Lieutenant in Col Hankesons Regiment in the States of Pennsylvania, New Jersey & New York, to wit from the summer of 1779 to the close of the war.

That my whole service as a Private was not less than 2 months

That my whole service as Ensign was not less than 1 month

That my whole service as 1st Lieutenant was not less than 3 yrs 10 ½ months

And further saith not.

Subscribed & sworn before me this 31st day of May 1833

Arthur S. Johnson JP

Sollomon Middaugh

I Arthur S. Johnson the above named Justice of the Peace certify that I am well acquainted with the above named Solomon Middough who has before me subscribed & sworn to the above afft and that he is a person of good reputation for truth & veracity & his statement is intitled to full credit.

May 31st 1833

Arthur S. Johnson J Peace

State of New York

Tompkins County Clerks Office I Samuel Love Clerk of said county do hereby certify that Arthur S. Johnson Esq. before whom the within affidavit purports to have been sworn and also by whom the within certificate purports to have been given, was at the date of the same a Justice of the peace in and for said county and that I am well acquainted with his hand writing and verily believe that his name subscribed to said affidavit and to the said certificate is genuine. I further certify that the annexed contains the original proceedings of the Court of Common Pleas of said county in the matter of of the application of Solomon Middaugh for a Pension. In testimony whereof I have hereunto set my hand and affixed my Seal of office at Ithaca

May 31, 1833.

Saml. Love Clk Tompkins County

War Department,
Pension Office,

July 25 1833

Sir:

The evidence in support of your claim, under the act of June 7, 1832, has been examined, and the papers are herewith returned. The following is a statement of your case in a tabular form. On comparing these papers with the following rules, and the subjoined notes, you will readily perceive that objections exist, which must be removed, before a pension can be allowed. The notes and the regulations will shew what is necessary to be done. Those points to which your attention is more particularly directed, you will find marked in the margin with a brace, (thus: }). You will, when you return your papers to this Department, send this printed letter with them; and you will, by complying with this request, greatly facilitate the investigation of your claim.

A Statement, shewing the Service of Solomon Middough

Period.	Duration of Service			Rank.	Names of General and Field Officers under whom he served.	Age at present, and place of abode when he entered the service.	Proof by which the declaration is supported.
	Years	Months	Days				

I am respectfully
Your obedient servant
J. E. Edwards
Commissioner of Pension

After the statement of service a copy of the regulations under the Act of June 7, 1832 is included but not reproduced here. An identical copy is included in the chapter entitled ***A Declaration of Service****.*

In the Matter of the Application of Solomon Middough
for a Pension
Declaration

State of New York

Tompkins County ss. Personally appeared before me the undersigned a Justice of the Peace in & for the said County of Tompkins Solomon Middaugh who being duly sworn deposeth and saith that he is the identical Solomon Middaugh named in and who subscribed & sworn to the annexed Declaration and affidavit.

And this Deponant further saith that by reason of old age and the consequent loss of memory, he cannot swear positively as to the precise length of his service but according to the best of his recollection he served not less than the periods mentioned below & in the following grades.

For one month I served as a private in Capt. Harkers company & Col Martins Regt of N. Jersey Militia during the months of March & April 1776 at the City of New York.

For one month I served as a private in Capt. Meeker's company & Maj. S. Westbrooks Battalion of N.J. Militia during the month of August 1776 at Perth Amboy N. Jersey.

For one month I served as Ensign Elect in Capt. McLennin's company & Col. Brodircks Regt. of N.J. Militia in the months of October & November 1776 at South Amboy N.J.

For two months I served as 1st Lieutenant in Capt Davis company & Col Hankison's Regt. of N.J. Militia during the months of January February & March 1777, at Scotch Plains N.J. during which I commanded said company in the absence of the Capt.

For one & an half months I served as 1st Lieutenant, commanded in Capt Holbert's company & Col Wynes Regt. in the months of Feby & March 1777 at Elizabethtown N. Jersey.

For five weeks I served as 1st Lieutenant in Capt Coykendolls company & Col Hankisons Regt. of N.J. Militia at Pumpton Plains & at Paramus during the months of May & June 1777.

For one month I served as 1st Lieutenant in Capt Coles company & Col Seward Regt. at Minisink N.Y. during the months of May & June 1778.

For one month I served as 1st Lieutenant commandant in the months of Sept & October 1778 in said Capt Cole's company under Col Hankson & Lt. Col Seward at Germantown in the State of Pennsylvania.

For two months, I served as 1st Lieutenant Commandant during the months of October November & December & immediately after the last above mentioned service in said Capt Coles company at Montague on the Delaware in the State of N. Jersey and during which time I had the command of the guard.

For three years and three months I served as first Lieutenant in said Capt Cole's company in Col Sewards Regt. from the summer of 1779 to the fall of 1782, during the last above mentioned service I was enlisted in said corps as a minute man, & served as such for & during the war, that during that service I was stationed at & near Montague on the Delaware River & was there in pursuance of the orders of Maj Westbrook who had the immediate command of the minute men in that vicinity on the banks of the Delaware, that during the said last mentioned term of service I was engaged in several engagements with and scouting parties among the British & Indians on and along the banks of the Delaware, that in that service

I was engaged in the Battle of Lackawax at which seven men were killed from said Capt Coles company, that my principal engagement ~~engagement~~ during the said term of service was to guard the Banks of the Delaware and keep the aforesaid Capt. Coles company at all times in readiness to march at a moments warning, that I often and some considerable part of _____ under the _____ mustering said company after which the alarms would often die away until again repeated. That to keep said company in readiness at all times for a march as ordered to do by my said officers required my whole time till in the fall of 1782 after which I was called out only for a few days at a time & then would return to my own business till the summer or spring of 1783 when they were discharged from further service. That during the whole of my service as Lieutenant as aforesaid I served under & by virtue of a commission signed by Wil. Livingston Governor & Commander in Chief of the State of New Jersey, which said commission is hereunto annexed with the original application of this Deponant for a pension.

Subscribed & sworn before me this 19 day of August 1833

Chauncey G. Heath J. Peace

Sollomon Middaugh

I certify that I am acquainted with the above named Solomon Middaugh and verily believe he is a person of truth and veracity, and that his Statement above contained is entitled to credit.

Chauncey G. Heath J. Peace

Tompkins County Clerk's Office ss. I do hereby certify that Chancey G Heath Esquire before whom the foregoing affidavit purports to have been sworn, was at the date of the same one of the Justices of the Peace in and for said county, and that I am well acquainted with his hand writing and verily believe his name subscribed to said affidavit to be his own proper hand writing & signature Witness my hand and official seal this 19th day of August 1833

Samuel Love Clerk

Pr Geo. G. Freer Deputy Clerk

Ithaca Dec 30th 1833

Sir

Enclosed I send you some affidavits &c in furtherance of the application of Capt. Solomon Middough, of Caroline for a Pension. By a letter from the Coms-sr. of Pensions some time since I was informed that his case was suspended & his papers placed on file. The difficulty was not I believe that his service was not proven, but that the precise time or times of his service was not particularly specified; and the precise day of his entering the service as well as the day of his leaving such service in his various campaigns was not pointed out. Now upon an examination of his papers it will appear that during that part of his service but about which there can be any doubt he was an officer on the Indian frontiers on & near the Delaware River, that for some 3 or 4 years he was there actively engaged in the service, that he as an officer was compelled to turn his whole time & attention to such service while the privates who were divided into classes only devoted a portion of their time to the service. It will be perceived that to require Capt. M. to specify particularly the day, week or month that he was not actually at the head of his company or platoon, after a lapse of so many years, during that 3 or 4 years service, will only tend to exclude him absolutely from the provisions of the act of Congress, providing for those who served in the Revolutionary war which I believe would be unjust.

The enclosed affts. will probably furnish to the Department under the circumstances all the proof that reasonably can be required. Will you be good enough to submit the papers to Mr Edwards in person & in case he wishes any further proof as to his respectability I presume you are acquainted with Mr. M. the father of _____ Middough and can give him a statement which will be satisfactory. The case of James Smith an applicant, the papers have been a long time before the Department. Please enquire in relation to this application also, and let me hear from both as early as may be. In conclusion permit me to wish you the compliments of the season

Yours very Respectfully
Saml. Crittenden Jr

Hon N. Halsey
House of Reps. Washington

State of New York
County of Tioga ss. On this 18th day of Dec 1833 personally appeared before me the undersigned one of the Justices of the peace of the County of Tioga Capt Benjamin Coykindoll a resident of the Town of Vetran in the County aforesaid and State of New York aged eighty four years who being duly sworn according to law doth on his oath make the following declaration, and saith that he was well acquainted with First Lieutenant Sollomon Middaugh in seventeen hundred and seventy six and afterwards during the Revolutionary War. And resided within six miles of him in the Town of Wantage County of Sussex and State of New Jer-

sey. And further more this deponant saith that he personally knew that Lieu Middaugh served under Capt Josias Cole and under the Command of Col. Seward Maj. Samuel Westbrook on the Indian frontiers on the Delaware River and he further deposes and saith that Lieut Middaugh lived within eight miles of the frontier and That he was there on duty the greater part of the time from the commencement of the Indian war untill General Sullivan drove them off, which this deponant verily believes to be at least three years. This deponant further saith that Lieut Middaugh was a faithful officer in the discharge of his duty to the law and to his country and he was always ready and willing to volunteer his services when called on. And this deponant further saith that Capt. Coles Company and his (the Deponant's) lay the nearest the Frontier and were very frequently invaded by the Indians and they always had to be in readiness by the frequent alarms which was occasioned by scouting parties the immediate vicinity of the neighbourhoods

Subscribed and sworn before me this 18th day of Dec 1833

S. Conkling J P

Benj Coykindall

Tioga County ss. I do hereby Certify that I am well acquainted with the above named Capt Benjamin Cuykindall and know him to be a man of good Morel Caractor & of truth and varasity and an Exelant mind for one of his age

S Conkling J P

State of New York

Tioga County ss. I Green M Tuthill Clerk of said county do certify that Stoddard Conkling Esq before whom the within affidavit purports to have been sworn & also by whom the certificate of identity is subscribed was on the day the same purports to have been sworn a Justice of the peace in & for said county. And that I am acquainted with his hand writing & verily believe his name subscribed thereto to be his proper hand writing and signature Witness my hand & seal of said county this 21st day of December 1833

Green M Tuthill Clk

State of New York

Tioga County ss. Be it remembered that on the 20th day of December 1833 Personally appeared before me Francis Armstrong a Justice of the peace of said county Jonathan Catterlin who being duly sworn deposeth & saith that he is a resident of the Town of Tioga County & State aforesaid & that he was seventy nine years of age last January that he was well acquainted with Solomon Middaugh when he resided in the Town of Wantage County Sussex & State of New Jersey that this deponent then resided within two miles of the said Middaugh which was in the year 1775 & up to the year 1783, and this deponant further saith that Solomon Middaugh served as first Lieutenant in Captain Josias Coles Company of Malitia & that this deponant was a private in sd company, that Said Middaugh served as Lieut. & this deponant as private in sd company aforesaid on the Indian frontiers on the Dileware River commencing as this deponant fully believes in the year 1778 & continuing untill some time in the year 1781, or untill they were drove off by Geonl. Sulivans army, that they each resided within Eight miles of the frontiers that sd company consisted of about fifty privates that they were divided into classes & that one of those classes were constantly on duty & that the Officers tower were more frequent then the privates & that I never knew Lieut. Middaugh to fail of serving his tower & this deponant further saith that he & Lieutenant Middaugh was in the Battle at Jerman town, that he does not reccollect the name of the Captain who commanded in the engagement but that he fully reccollects that Geoneral Washington was commander in chief & this deponant further saith that said Lieut Middaugh was a faithful Officer & brave soldier & that he frequently volenteerd his services during the war, & this deponant further saith that haveing never learnt to write he is obliged to make his mark for his signature

Subscribed & sworn to the day & year first above written before me Francis Armstrong J. P.

his
Jonathan X Catterlin
mark

Tioga County ss. I do hereby certify that I am well acquainted with the within named Jonathan Catterlin that I fully believe him to be a man of good moral character & that he is a person of truth and veracity & that for a man of his age his mental faculties are full more than common

Owego Dec 20, 1833

Francis Armstrong J. P.

State of New York

Tioga County ss. I Green M Tuthill Clerk of said county do certify that Francis Armstrong Esq before whom the preceding affidavit purports to have been sworn whose name is subscribed to the preceding affidavit and certificate of identity was on the day of the date of the same one of the Justices of the peace in & for said court. And further that I am acquainted with the hand writing of said Justice and verily believe his name subscribed to said affidavit & certificate to be his proper hand writing and signature Witness my hand & seal of Office this 21 day of December AD 1833

Green M Tuthill Clk

State of New York

Tompkins County ss. Samuel Crittenden Junr. of Ithaca in said county being duly sworn deposeth and saith that Solomon Middough named in the annexed & preceding affidavits is the same identical Solomon Middough named in the an original declaration & papers accompanying the same, made on his application for a pension and forwarded by this deponant to the Pension office at Washington: which said declaration &c. this Deponant is informed by letter from J. L. Edwards Esq Comr. of Pensions is placed on file in said Pension office. And further this Deponant saith not.

Subscribed & sworn Before me this 28 day of December 1833

Frederick G Stanley a Commissioner of Deeds for Tompkins County

Saml. Crittenden Jr.

State of New York

Tompkins County Clerks Office ss. I Samuel Love Clerk of the county courts in & for the County of Tompkins & of said county Do hereby certify that Frederick G Stanley Esq before whom the annexed affidavit of Saml Crittenden Jr purports to have been sworn was on the day of the date thereof a Commissioner of Deeds &c duley authorized by law to take affidavits & to administer in ~~& for~~ said county. And that I am well acquainted with the hand writing of said Commissioner & verily believe his signature to the Jurat to be his own proper hand writing & signature. In witness whereof I have hereunto set my hand and affixed my official seal this 28th day of December AD 1833.

Saml. Love Clk

New York

Solomon Middaugh of Tompkins Co. in the State of New York who was a Pr.

Ensign & Lt. in the company commanded by Captain Harker of the Regt. commanded by Col.

in the N. Jersey Troops for Private 2 months		$6.66
	Ensign 1 month	10.00
	Lieut. 14 months & 20 days	195.54
		$212.20

Inscribed on the Roll of New York at the rate of 212 Dollars 20 Cents per annum, to commence on the 4th day of March, 1831. Certificate of Pension issued the 27 day of Jany. 1834 and sent to Hon. N. Halsey H. R.

Arrears to the 4th of Sep. 1833	530.50
Semi-anl. allowance ending 4 March '34	106.10
	$636.60

Recorded by D. Brown Clerk,
{Revolutionary Claim, Act June 7, 1832.}
Book E Vol. 4 Page 106

Hon J L Edwards
Commissioner of Pensions
Sir

I have appointed Arad Joy of Ovid New York agent to act in all thins for me in procuring or investigating Solomon Middaughs claim to a pension

Dated July 1846

Jane Paige

To Hon J L Edwards Coms. of Pensions
Sir

I have appointed in be half of all of the children of Solomon Middaugh, Arad Joy of Ovid our agent to act for us in investigating our claim to an increase or back pension

Dated Tompkins Co. March 30 '47

Simeon W Middaugh

Power of Atty case of Solomon Middaugh

Ovid N.Y. April 12 '47

Dear Sir

I enclose my agency in the case of Solomon Middaugh, who was a pensioner of the State of New York. I am informed that he died 23rd Feby. 1843. If I have a right to do so, I beg leave respectfully to ask if there is not an increase of pension due to him, since he only recd. $220 and served as I understand as a Lieut. I may be allowed to say, that I supposed that the pension should have been $320 of this of course you can at a glance determine.

Most Respectfully Your Obt Servt
Arad Joy

Hon J. L. Edwards
Comr of Pensions
Washington DC

I assign my agency to N. Leake Esq
Arad Joy

Ithaca Tompkins Co N York
March 25, 1850

J. L. Edwards Esq Comr. of Pensions
Sir,
Solomon Middaugh formerly of this county, in his life time drew a Pension of $212.20 per annum for his services in the Revolution under the law of 1832. He died Feb 20, 1843. His pay was regularly drawn up to the time of his death.

Since his death has there been a certificate of Pension issued from your Department to any one allowing to his heirs or any other person any arrearages or back pay? If so to whom and when? Your answer will much oblige

Your Obt Huml Servt
S Crittenden Jr.

Power of Attorney case Solomon Middaugh New York 32 ad
Hon L P Waldo,

Washington July 18th 1854

I enclose herewith, a power of attorney from the heirs of Solomon Middaugh of NY whose claim was allowed under act 1832. I present the claim for an increase of Pension believing that Mr Middaugh should have been pensioned for 2 years service as a Lieut. You will find parol evidence sufficient to establish the claim I respectfully ask a reexamination.

Your Obt Svt
A H Markland

State of New York

County of Tompkins Be it known, that on this Sixth day of July A.D. 1854 before me, the undersigned, a Justice of the Peace, in and for the County and State aforesaid, personally appeared Simeon W. Middaugh resident of Ithaca in said County, and made oath according to law that he is the son of the identical Solomon Middaugh whose name, under Act of 7th June 1832 has been placed on the New York Pension Roll at the rate of $212.20 Dollars per annum.

That he is credibly informed that the said Pension may and ought to be increased from its commencement and during its continuance; that he is directly interested as a claimant in said increase, and makes this affidavit to be filed with such additional evidence as my Agent may use in prosecuting said claim.

Countermanding and revoking all Powers of Attorney and authority heretofore given, I hereby constitute and appoint A. H. Markland of Washington City, my true and lawful Attorney to prosecute the claim of Solomon Middaugh for any amount of Revolutionary Pension, or Increase of Pension that may be due; and for the purpose of establishing additional service by the soldier for whose military duty in the Revolution the aforesaid Pension was granted, I hereby authorize my said Attorney to apply for and obtain from State archives or other sources, certified proofs of such service. To examine all papers or documents in relation to said service, on file in the Departments at Washington City or elsewhere; to file additional evidence or arguments, and to receive for my benefit the Certificate for Increase which may be issued in satisfaction of said claim, which certificate I wish made payable at the Pension Agency of this State; to appoint one or more substitutes under him for the purposes herein expressed, and to do all things that I might or could do were I personally present. Hereby ratifying and confirming all that my said Agent and Attorney shall lawfully do in the premises. A. H. Markland interlined before signing

Simeon W. Middaugh

Sworn to, Subscribed and Acknowledged before me the day and year above written, and I hereby certify that I believe the said Simeon W. Middaugh to be directly interested in said claim as set forth, and to be the identical party therein alleged, in whose claim I am in no wise interested.

C. G. Heath J. P.

State of New York

County of Tompkins   It is hereby Certified, that satisfactory evidence has been exhibited before me, Clerk of the Court of Tompkins County that Solomon Middaugh was a Revolutionary Pensioner of the United States, at the rate of $212.20 per annum, and that he died on the Twentieth day of February Eighteen Hundred and forty three leaving surviving the following named (widow and) children and none other, to wit: Jedediah Middaugh Simeon W. Middaugh Jane Paige Richard Middaugh

It is further certified that Chauncey G. Heath Esq. before whom the preceding affidavit and Power of Attorney were made and acknowledged, was at the time of so doing a duly authorized Justice of the Peace in and for said county, and that the signature purporting to be his _____ genuine; and further, that in the within claim for Increase of Pension I am in no degree interested. In Testimony Whereof, I have hereunto set my hand and affixed my seal of Office, this Sixth day of July 1854

Ezra Weaver Clerk

State of New Jersey,
Department of State,
Trenton, July 24 1854

A H Markland Esq

Sir

I find there was a Solomon Middaugh of Sussex County, in Service, a Lieutenant, and as it seems, commanding a company. I send herewith ~~an abstr~~ a copy, of payments made to him, which is all the reference I can find of his name, upon the Slender records in this office

Respectfully
Thos S Allison

Fees two dollars

Payments made by Joseph Lewis Pay Master to the Militia of
the Counties of Morris Sussex & Bergen

Cash paid Major Meeker on the following Pay Rolls.

No. 309 For Lt. Solomon Middaugh's Compy. of 22 men of Col Hankinsons Regt. which served at Delaware from 7th to 9th June 1778.

£11. 9.0

No. 310 For Lt. Solomon Middaugh's Compy. of 21 men from Sep. 24 to Oct

24 64.18.5

No. 311 For Lt. Solomon Middaugh's Compy. of 42 men 23d Oct to 21st Nov. 123.16.6

State of New Jersey I Thomas S Allison Secretary of State of the State of New Jersey do hereby certify that the foregoing is a true extract from a Book headed as above, as taken from and compared with the original on file in my Office. In Testimony whereof I have hereunto set my hand and affixed my seal of Office at Trenton this twenty fourth day of July Eighteen hundred and fifty four.

Thos. S. Allison Secretary of State

Washington City July 26th 1854

Hon L P Waldo,

Commr of Pensions

I acknowledge the receipt of your letter of the 20th Inst. respecting the case of Solomon Middaugh of NY, act 1832, admitted. I call your especial attention to the statement of Middaugh made on the 19th August 1833, which states That "for three years & three months I served as 1st Lieut in said Capt Coles company in Col Sewards Regt. from the summer of '79 to the fall of '82, during the last above mentioned service I was enlisted in said corps as a minute man & served as such for and during the war, that during that service I was stationed at & near Montague on the Delaware river & was there in pursuance of the orders of Maj Westbrook, who had the immediate command of the minute men in that vicinity &c" the statement was made under oath at the town of Caroline County of Tompkins State of New York on the day above mentioned.

On the 17th January 1833 John Swartwood made a declaration at the town of Cayuta, County of Tioga, State of NY, he made the following statement "In April 1779 the company under Capt Hoover to which the deponent was Lieut was ordered to the Delaware Indian frontier to act as Rangers along the frontier for the term of seven months, the deponent served that time as Lieut in the said company. This company composed a part of a battalion assigned to that service under command of Maj Saml Westbrook. From the expiration of this term which was in the month of Nov 1779, the company in which the deponent acted as Lieut aforesaid under Capt Hoover was assigned to guard the said frontier in the capacity of Rangers until the close of the war."

You will perceive that the above persons who were residing in different parts of the country and who did not know of the existence of the other, made statements under oath almost precisely similar respecting the service on the Delaware river. Mr Swartwood was pensioned for two years service on that frontier. Middaugh is only pensioned for 14 months. Maj Saml Westbrook under whom the service was performed is pensioned for two years service on that frontier. I ask sir, that Middaugh shall be pensioned for two years service also. I further refer you to the cases of Solomon Cartwright, Ohio 1832 admitted & Benj Cuykendall NY '32 ad

Your Obt Svt
A H Markland

State of New Jersey,
Department of State,
Trenton, Aug 5 1854

A H Markland Esq
Sir

Our books of accounts &c for revolutionary service, all stop in 1778. There was a Captain named Cole, in the service of Sussex County. There was a Colonol ~~Seward~~ Seward, also, and Major Westbrook was a distinguished officer, and had charge of the frontier on the Delaware. The books of Mr Lewis, refer to Seward's regiment and Hankinson's regiment as under the command of Westbrook on the Delaware. If these facts will be of any service to you I can furnish a certificate

Yours &c
Thos S. Allison

Treasury Department,
Third Auditors Office
Sept 16, 1854

Sir

I certify that it appears from the files and records of this Office that Solomon Middaugh, was a pensioner of New York, under act 1832, at $212.20 per annum, that he died 23 Feb 1843, leaving 4 children only whose names are Jedediah Middaugh Simeon W Middaugh, Jane Paige & Richard Middaugh & that the arrears of pension due at the time of the pensioner's death were paid to his Executors for the benefit of the above named children

____
Third Auditor

Case of Solomon Middaugh - N.Y. admd. 1832
Application for increase

Solomon Middaugh declared in 1833 That he served as follows:
In Mar & Aprl 1776 in Harker's Co. Martins N.J. Rt. 1 mo
In Aug 76 in Meeker's Co. Westbrook's battaln 1 mo
In Oct & Nov 76 - Ensign McLennins Co. Brodericks Rt. 1 mo
Lieut 2 m In Jany & Feb 77 - 1st Lt Davis' Co. Hankinsons Rt (commdt) 2 mo
Lieut 1 m 15 ds Feb & Mar 78 1st Lt Holbert's Co Wynes Rt 1 mo 15 ds
Lieut 1 m 05 ds May & June 77 - Coykendalls Co Hankinsons Rt 1 mo 5 ds
Lieut 1 m May & June 78 - Coles Co Sewards Rt (Lt Col) 1 mo
Lieut 1 m Sept & Oct 78 - Coles Co Hankinsons Rt 1 mo
Lieut 2 m Oct Nov & Dec 78 - Coles Co Hankinsons Rt 2 mo
8 m 20 ds From summer of 79 to fall of 82-3 minute man in almost 3 yrs.
(& 3yrs 10 ½ m constant service on the Delaware Lt Coles Co. Westbrooks
3 8 20 estimated by battalion He was Lt Comdt. Hankinson was Col & Seward Lt Col. himself)

Capt Benjn Coykendall - Middaugh served under Capt Josias Cole, Col Seward & Maj Westbrook, for at least 3 yrs as dept believes, as Lt. Cole's & depts Co's were nearest the frontier & almost constantly in service.

Jonathan Catterlin - Resided within 2 miles of Middaugh from 1775 to 1783, dept was private in Coles Co in which Middaugh was Lt in service for some 3 yrs from 78 to 81 The company was 50 men, in classes & one class constantly on duty. The officers were on duty more constantly than the men. Never knew Middaugh to fail his tour, and he frequently volunteered.

On this evidence the claim was admitted for the 2 mos as private, 1 mo Ensign & for 14 mos & 20 das as Lt. Probably adding the 6 mos as Lt for the service after 1778. The records of New Jersey show that he was a Lt of militia in 1778 and must have performed considerable service, but the payments having been made for both Officers & men, it is impossible to determine the duration of the service from the accountts.

Benjamin Coykendall of N. York - admd 32 No. 20,375 who declared for 23 ½ mos as Capt & 1 ½ as Lt was first pensioned for 23 ½ mos Capt & ½ mos ~~Ensign~~ Lt. - his service extended from 1775 to 1779, inclusive. Afterwards his pension was reduced by leaving off his 6 mos alleged service in 1779. His was militia service It is not quite clear upon what grounds this reduction was made, but it apparently was because the militia accounts of N.J. which are exceedingly imperfect did not show him in service later than 1778. The parol proof was ample

---

Jona Catterlin was pensioned (N.Y. 1832 admd No. 14,657) for 2 yrs service, one year of which was as a regular and the other year was estimated for 2 yrs service in the militia under Lt Middaugh, sworn by Middaugh.

John Swartwood ~~was~~ of Cayuta N.Y. was pensioned, N.Y. Act 1832, No. 8310, for 2 yrs service as Lieut. in the same Major Westbrook's battalion extending from 1779 to 1782-3, alleged by Middaugh, which service Swartwood estimated at 34 months in all. His witnesses were James and Daniel Swartwood

---

Baranardus Swartwood, Pa Pike Co. admd act 1832 No 11,101 for 2 yrs service mostly performed in the same Indian wars on the Delaware, in which Middaugh served, up to near close of war, served mostly in Pa militia.

Henry C. T. Middaugh, Pike Co Pa admd act 32 No 26,338 for 2 yrs service, almost precisely the same as B. Swartwood's.

On the 5th June 1777 an act was passed authorizing the Governor & privy council to station or employ guards ~~for~~ on any frontier of the state, relievable at discretion Under this ~~ord~~ act guards were to be stationed ~~at Mon~~ in Monmouth

&c, relieveable monthly and to serve during the whole war.

These claimants, Middaugh &c allege that they engaged for the war and served until in the spring of 1783. From all these facts it is apparent that Middaugh must have served more than two years as Lieutenant

---

Exd. Proof in cases of Jon. Catterlin & Benjn. Coykendall N.Y. admd 32 also David Depew admd NY 32 (widow Mary susp act 53) and Cornelia Ann Dean NY admd 2d Feby '48. Confirm but do not augment

---

Brief in the case of Solomon Middaugh, deceased
Tompkins County and State of New York
act June 7, 1832

Claim, (~~"original," or~~ "for increase.")

Proof exhibited, (if original.)

Is it documentary, traditionary, or supported by rolls? If either, state the substance.

Solomon Middaugh was pensioned under act of 1832 at 212 $ per annum for 14 mos 20 das Lieut 1 mo Ensign & 2 mos private. Died 23d Feby 1843, leaving surviving the four children now claiming arrears as appears by certif of 3d Auditor.

(If for increase.)

Has additional evidence been filed since the admission of the claim? If so, what?

The additional evidence of service consists of a large number of admitted cases, mostly from New York, admitted for the same and similar service, from a careful examination of which it is apparent that, in the deliberate and repeated judgements and decision of this office, well sustained by proof, the service in which he was engaged as Lieut must have exceeded 2 years. Administration and heirship certified by court. See brief of evidence.

Name and residence of Agent

A. H. Markland Present.

S. G. Dodge Examining Clerk

The People of the State of New York by the grace of God Free and Independent
To all to whom these presents shall come or may

concern send greeting

Know ye That at the town of Ithaca in the County of Tompkins on the 24 day of April in the year of our Lord one thousand eight hundred and forty three before George G. Freer Surrogate of our said County the last will and testament of Solomon Middaugh deceased a copy whereof is hereunto annexed was proved and is now approved and allowed of by us & the said deceased having whilst he lived and at the time of his death goods chattels ~~and~~ or credits within this state by means whereof the the proving and registering the said will and granting administration of all and singular the said goods chattels and credits and also the auditing _____ and final discharging the account thereof doth belong unto us the administration of all and singular the goods chattels and credits of the said deceased and any way concerning will is granted unto Samuel A. Dean and Richard W. Middaugh Executors in the said will named they having first taken and subscribed an oath before the said Surrogate faithfully & honestly to discharge the duties of such executors hereby requiring you the said Samuel H. Dean and Richard W. Middaugh to make or cause to be made a true and perfect inventory of all and singular the goods chattels and credits of the said deceased which have or shall come to your hands possession or knowledge as also to make or cause to be made duplicates of such inventory and cause the same to be signed by the appraisers and the same so made and signed that you make return thereof to the Surrogate of the said County within three months from the date thereof In testimony whereof we have caused the seal of office of our said Surrogate to be hereunto affixed Witness George G. Freer Esquire Surrogate of the said County at the town of Ithaca in the said County on the 24 day of April in the year of our Lord one thousand Eight hundred and forty three

Geo G Freer Surrogate

I certify the above to be a true record.

Gco C Frccr Surrogatc

State of New York

Tompkins County Surrogates Court ss. I do hereby certify that I have compared the foregoing copy of the Letters Testamentary on the last Will & Testament of Solomon Middaugh of said County deceased & of the endorsements therein with the original records of the same remaining in this office & that the same are correct transcripts therefrom and of the whole of said original records & that the same still remains in full force & in no way revoked or canceled In testimony whereof I have caused the seal of the said Surrogates Court to be hereunto affixed this eleventh day of September in the year 1854

D Boardman Tompkins County Judge & Ex officio Surrogate

Solomon Middaugh deceased
New York

Act June 7, 1832 Admitted, for increase. Lieut of Infy 2 years, at $320 per annum from March 4, 1831 to Feby 23, 1843, when he died, deducting former payts at $212.20 per annum. Certificate payable to Samuel H. Dean and Richard W. Middaugh Execrs. of Solomon Middaugh, decd for benefit of Jedediah Middaugh Simeon W. Middaugh Jane Paige and Richard Middaugh, his only surviving children

Atty A H Markland Present

New York Albany

Solomon Middaugh decd. of in the State of who was a Lt. of Inf in the commanded by Captain of the commanded by in the Revolution for 2 years. Increased from $212.20 Inscribed on the Roll of Albany at the rate of 320 Dollars 00 Cents per annum, to commence on the 4th day of March, 1831, & end February 23d 1843. Certificate of Pension issued the 18th day of September 1854 and Sent to A. H. Markland Present.

Recorded by G. T. Getty Clerk,

{Revolutionary Claim, Act June 7, 1832.}

Book H Vol. 1 Page 205

Invalid. File No. 23,323
Solomon Middaugh Lieut Rev War
Act: June 7th 32
Index:- Vol.2, Page 275
[Arrangement of 1870.]
Revolutionary War Pension and Bounty Land Warrant Application Files
Series M-804; Roll # 1719

THOMAS AND MARGARET SCHOONOVER
NEW JERSEY (R.9,270)

State of Pennsylvania
Wayne County On the 18th day of June AD 1838 personally appeared before me a Justice of the Peace in and for said County Margaret Schoonover a resident of Berlin Township in said County and state aforesaid aged Eighty three years who being duly sworn according to Law doth on her oath make the following declaration in order to obtain the benefit of the provision of the Act of Congress passed July 4th 1836. That she is the widow of Thomas Schoonover who served as a private in the Revolutionary War at several times - deponent states that she was married to said Thomas in December 1780. That she had no acquaintance with said Thomas till one year before their marriage. That she is quite aged and her memory much impaired. That at the time of her marriage and subsequent she has heard him tell of his being in the service as a Private in the Revolution at one time nine months and at another time three months. That she heard others at that time speak of the same. That this deponent ~~knows~~ believes that in June 1782 he enlisted under Captain Josiah Cole and went with him from Wantage in Sussex County into Sulivan County N York, where at the mouth of Half Way Brook the Company of Capt Cole had an Engagement with the Indians in which Capt Coles company was defeated. That he was absent in this tour part of the time along the Delaware River three months. That her husband was out at other times susequent to the marriage the length of time and under what officers deponent does not reccollect. That her husband the said Thomas died in August 1821. That there is no documentary evidence of the several facts set forth as this deponent knows of. That she has remained a widow ever since that period as will appear by reference to the proff hereto annexed.

Sworn and Subscribed on the day and year above written
David Tarbox J. Peace

hir
Margaret X Schoonover
mark

Wayne County ss. I David Tarbox the Justice of the Peace before whom the foregoing deposition was sworn and subscribed do hereby certify that Mrs Margaret Schoonover is a woman of good character and entitled to credence and do further certify that from bodily infirmity is unable to attend the Court for the purpose of making her declaration

David Tarbox J. Peace

Pennsylvania

Wayne Co ss. I Paul S Preston Prothonotary and Clerk of the several Courts in and for the County of Wayne do hereby certify that David Tarbox is an acting Justice of the Peace in and for said County duly commissioned. that his Commission bears date the Eighteenth day of March AD 1836 and that the signature attached to the foregoing certificate and deposition purporting to be his are true and genuine. In witness whereof I have hereunto set my hand and seal of office this Sixth day of August 1838.

Paul S Preston Prot.

Pennsylvania

Wayne County ss. Personally appeared before me David Tarbox Esquire a Justice of the Peace in and for said County Betsey Cole of the Township of Texas County and state aforesaid, who being duly sworn according to Law doth depose and say, That she is ~~the~~ aged Seventy one years That she is the sister of Thomas Schoonover decd whose widow is now applying for a Pension. That during the Revolutionary War her father mother and brother resided in the Township of Wantage in Sussex County and State of New Jersey. That she reccollects that one occasion, when she was about twelve years old deponent thinks in June 1779 that Captain Josiah Cole who resided in that Township called at her fathers house and told her mother that her brother Thomas ~~was~~ had enlisted in the army. That her mother being unwell was much affected and that a few days after her brother Thomas was marched down into the interior of Jersey That she does not reccollect under what officer he was. That he was absent three months, as a private. That when he came back she reccollects distinctly that he was dressed in military dress. That in July 1782 she reccollects her brother Thomas enlisted under Captain Josiah Cole the length of time she does not reccollect That during that month Captain Cole marched his Company from Wantage in Sussex County to the mouth of Half Way Brook in Sulivan County N York where his Company had an Engagement with the Indians in which Capt Cole's Company was defeated, and many of deponents neighbors killed and wounded. That deponent does not know how long said Thomas was absent in this Expedition with Capt Cole. That she reccollects of her brother being absent in the service at other times, the particular periods or officers under whom he served she does not reccollect. That she was present at the marriage between her brother Thomas and the present applicant That it took place in said Sussex County in December 1780 by the Revd John

Coree decd That this deponent thinks ~~the~~ her brother the husband of the present applicant died in August 1821 in Wayne County Penna. That said applicant was never married but once

Sworn and Subscribed before me June 18th 1838

NB In the 12th line from the bottom on the 2nd page the word "enlisted" interlined before signing Also the last line interlined before signing

David Tarbox J. Peace

her
Betsey X Cole
mark

Wayne County ss. I David Tarbox Esquire the Justice of the Peace whose name is superscribed above do hereby certify that Mrs Betsey Cole who made the foregoing deposition is a woman of integrity and entitled to credence Witness my hand this 18th June 1838.

David Tarbox J. Peace

Pennsylvania

Wayne County ss. I Paul S Preston Prothonotary and Clerk of the several Courts in and for the said County do hereby certify that David Tarbox is and acting Justice of the Peace for said County duly commissioned and that his Commission bears date the Eighteenth day of March AD 1836 and that the signatures attached to the foregoing affidavit and certificate purporting to be his are true and genuine. In witness whereof I have hereto set my hand and Seal of Office this Sixth day of August 1838.

Paul S Preston Prot.

Bethany, Wayne Co. Penna.
June 18, 1842

Dear Sir,

I am informed by the friends of Margaret Schoonover widow of Thomas Schoonover of this county that her declaration and other proofs were forwarded to the Pension Office some years ago, and that no answer has ever been received by any person here in relation to them. Will you have the goodness to address me on the Subject. Mrs. Schoonover yet lives.

I am very respectfully your obt. Servant
Earl Wheeler

J. L. Edwards Esq.
Com. of Pensions

Rejected

Revolutionary War Pension and Bounty Land Warrant Application Files
Series M-804; Roll # 2132

Samuel Vanfleet or Van Vleet
New Jersey (R.10,900)

State of New York

Tompkins County ss. Solomon Middaugh of the town of Danby in the County of Tompkins in the State of New York being duly Sworn deposes and says, that he was Seventy nine years old on the twentieth day of August 1834. That this deponent was born in the town of Wantage in the County of Sussex in the State of New Jersey and resided there untill he was of the age of Fifty years. That while he resided in Wantage as aforesaid he became acquainted with Samuel Van Vleet, who then resided at Wantage and knew him then about fifty years. That this deponent was in the Army of the Revolution from the Commencement of hostilities to the declaration of peace in the year 1783, and was while in the Service a Lieutenant in Captain Joseph Harkar's Company, also in Captain Samuel Meeker's Company and in Captain John Davis Company, in the Second New Jersey Regiment. That while he was in the Service he frequently saw said Samuel Van Vleet in the Service doing duty as a Sergeant, in Captain John Davis Company, in which this deponent was a Lieutenant, and he afterwards held the same post under Captain Josias Cole, in which Company this deponent was also a Lieutenant and served as such untill the close of the war. That said service of the said Samuel Van Vleet commenced in the year 1778 and continued untill the close of the war in the year 1783, and was performed in the Second New Jersey Regiment under Colonel John Seward.

Subscribed and Sworn before me October 4, 1834
Frederick G Stanley a Commissioner of deeds &c

Solomon Middaugh

I do certify that I am well acquainted with Solomon Middaugh above named, who made the above affidavit, that he is a man of good Moral Character, and full faith and credit ought to be, and should be given to his Statements

Frederick G. Stanley a Commissioner of deeds &c

State of New York

Tompkins County Clerks Office I certify that Frederick G. Stanley Esqr. before whom the foregoing affidavit purports to have been sworn was at the date of the Same a commissioner of deeds in and for the said county duly qualified and that his signatures above written are genuine. In testimony whereof I have hereunto set my hand and affixed my official Seal at Ithaca October 4, 1834.

Samuel Love Clerk of Tompkins County

State of New York

Tioga County ss. Elias Clark of the Town of Vetran in the county of Tioga and State of New York being duly Sworn deposes and says that he was Eighty years old on the Thirtieth day of January 1834 That this deponent was born in the town of Morris town County of Morris in the State of New Jersey and that at the age of about fifteen years he moved in the town of Wantage County of Sussex and State of New Jersey That while he resided in the town of Wantage aforesaid he became acquainted with Samuel Van Vleet who then resided at Wantage and knew him there untill six or seven years after the revolutionary war That this deponent at the commencement of Hostilities enlisted for the term of five months and went to Long Island and after his term of service expired he returned to the town of Wantage aforesaid and that shortly afterwards went in the service under Captain Samuel Meeker with the said Samuel Van Vleet and that during the term of service the said Samuel Van Vleet served as Sergant which said Company was attached to the Second New Jersey Regiment and that after the promotion of Captain Samuel Meeker to the Office of Major of the aforesaid regiment Captain John Davis was elected in his place and that this deponent served under the said Captain John Davis for the space of two years or there abouts when the said Capt John Davis removed from the beat of the said Company and that during the aforesaid term of two years the said Samuel Van Vleet served as Sergant in said company and that after the said Captain John Davis removed as aforesaid Captain Josias Cole was elected in his place and that this deponent and the said

Samuel Van Vleet served in the Company of the said Captain Josias Cole during the term of the war and that the said Samuel Van Vleet served as Sergant under the said Capt Josias Cole during the whole term which said service was performed in the Seccond New Jersey Regiment under Colonel John Seward

Subscribed and Sworn before me Oct 6th 1834

John W Wisner J P

his
Elias X Clark
mark

I do certify that I am well acquainted with Elias Clark above named who made the above affidavit. That he is a man of good Moral Character and full faith and credit ought & should be given to his Statements.

John W Wisner JP

State of New York

Tioga County Clerks Office ss. I Certify that John W Wisner Esq before whom the foregoing Affidavit purports to have been sworn was at the date of of the same a Justice of the Peace in and for said county duly commissioned and Sworn and that his Signatures above written are genuine. In Testimony whereof I have hereunto set my hand and affixed the Seal of said county this 8th day of October AD 1834.

Edward Suffern Dep Clk in the absence of Green M Tuthill Clk

Samuel Van Vleet Application

State of New Jersey

Sussex County ss. On this twenty Seventh day of November in the year of our Lord Eighteen hundred and thirty four, personally appeared in Open Court before the Judges of the Inferior Court of Common Pleas in and for the County of Sussex aforesaid Samuel Van Vleet, a resident of the Township of Wantage in said County and State of New Jersey aforesaid aged Seventy five years who being first duly sworn according to law doth on his Oath make the following Declaration, in Order to Obtain the benefit of the provision made by the act of Congress, passed June 7th 1832. That he entered the Army of the United States in the year 1778 in the Second New Jersey Regiment, under the following named Officers, and served as hereinafter Stated. Colonel John Seward, in the Company of Capt. Samuel Meeker who was soon promoted to Major of sd. Regiment, and Captain John Davis was appointed in his place. Solomon Middaugh was Lieutenant, Joseph Elston Ensign, and this declarant, 1st Sergeant. That he resided in the Township of Wantage aforesaid at the time he entered the Service. That he entered as a volunteer and served as Sergeant in the said Company under sd. Capt. Meeker and under the sd. Captain Davis for the Space of two years. That the said Capt. Davis then moved out of the State and out of the bounds of said Company, and Capt. Josias Cole was elected in his place - ~~and~~ that he continued to serve in sd. Company as Sergeant under the sd. Capt. Cole and the aforesaid Officers until the close of the war in 1783. That he marched under the sd. Capt Davis from Wantage Sussex County aforesaid, to Newark, Elizabeth Town point, Scotch Plains, Westfield, and other places on the eastern border of said State of New Jersey, at which places he was stationed and served most of the Term under Said Capt. Davis, and in the said Regiment comd. by Col. Seward as aforesaid. That he also served upon the north western frontiers of the State upon the borders of Delaware and Lackawack Rivers. That he marched with his company under Capt. Cole towards Lackawack at the time of the Battle of that place, but became so lame as to be unable to proceed on, to the place of the Battle. That he was out after the Indians when they burnt Peepack, and also at another time when Job Jobs and Lewis and Solomon Westbrooks were killed by the Indians - he says that the frontiers of the State were almost constantly invaded by the Indians and that it was necessary to keep an armed force in that section of the Country continually to prevent their Savage depredations - he says that he was born in the Township of Wantage aforesaid, on the 5th of January 1759. That there is a record of his age in his family bible which was transcribed from his Grand Fathers Bible. That at the time he entered the service he resided in Wantage as before stated where he has lived since the Revolutionary War. That he entered as a volunteer, that he never received a Commission, and has no documentary evidence, but that he knows the Lieutenant of the sd. Company in which he served, to wit Solomon Middaugh, and also Elias Clark and Moses Dewitt who were members of said Company and who can testify to his Service - he says he is acquainted with the Revd. Abraham Ketcham and William Cole Esquire ~~and~~ who can Testify as to his character for veracity and their belief of his Services as a Revolutionary Soldier - he hereby relinquishes every claim whatever to a pension or annuity except the present and declares that his name is not on the pension roll of the Agency of any State.

Sworn to and Subscribed the day & year aforesaid
Jos. Y. Miller Presiding Judge &c

Samuel Vanfleet

We Abraham Ketcham a clergyman residing in the neighbourhood of the said Samuel Van Vleet, and William Cole Esquire residing in the same do hereby certify that we are well acquainted with Samuel Van Vleet who has subscribed and sworn to the above declaration. That we believe him to be 75 years of age. That he is reputed and believed in the neighbourhood where he resides, to have been a soldier of the Revolution and that we concur in that opinion

Sworn and Subscribed the day and year aforesaid in Open Court.Jos. Y. Miller Judge &c

Abm Ketcham
William Cole

On the same day and year first aforesaid personally appeared in Open Court before the same Court Moses Dewitt aged 73 years, of Wantage in the County of Sussex aforesaid who being first duly sworn on his oath doth depose and say that he is and has been acquainted with, the said Samuel Van Vleet who has sworn to the above Declaration, for Sixty years past. That he knows the said Samuel Van Vleet was Sergeant in the Company comd. by Capt. Josias Cole as by him Stated in his declaration, and that he served in sd. Company under sd. Capt. Cole in the Service of the United States as by him Stated in his Declaration, and further this deponent saith not

Sworn and Subscribed in Open Court the day first above written.
Jos. Y. Miller Judge &c

Moses Dewitt

The Court hereby certifies that Moses Dewitt who has signed the above affidavit is a credible person and that his Statement is entitled to credit.

done as above. Jos. Y. Miller Judge &c

And the said Court do hereby declare their opinion after the investigation of the matter and after putting the interrogatories prescribed by the War Department, that the above named applicant was a revolutionary soldier and served as he states. And the Court further certifies, that it appears to them that Abraham Ketcham who has signed the preceeding certificate is a clergyman resident in the nieghbourhood of said Samuel Vanvleet and that William Cole Esquire who has also signed the same is a resident in the same neighbourhood and is a credible

person and that their Statements is entitled to credit.
done the day aforesaid
Jos. Y. Miller Presiding Judge &c

I Joseph E Edsall Clerk of the Court of Common Pleas in and for the County of Sussex and State of New Jersey do hereby certify that the foregoing contains the original proceedings of the said Court in the matter of the application of Samuel Vanvleet for a pension. In Testimony whereof I have hereunto set my hand and Seal of Office at Newton this twenty seventh day of November AD Eighteen Hundred and thirty four AD 1834.

J E Edsall Clk

Rejected

Revolutionary War Pension and Bounty Land Warrant Application Files Series M-804; Roll # 2445

SAMUEL VANGORDER OR VANGORDEN
NEW JERSEY/PENNSYLVANIA (S.9,503)

BRIEF in the case of Samuel Vangorden of the County of Tioga in the State of Penna.

(Act 7th June, 1832.)

1. Was the declaration made before a Court or a Judge? Court
2. If before a Judge, does it appear that the applicant is disabled by bodily infirmity?
3. How old is he? 69 years
4. State his service, as directed in the form annexed.

Period.	Duration Of Service			Rank.	Names of General and Field Officers under whom he served.
	Years	Months	Days		
1779		1		Volunr.	Maj Saml Westbrook, Capt Peter Westbrook

1779		1		Volunr	Capt. Hoover, same Regt.
1780		7		Private	Col. Stroud, Capt. Vannatter

5. In what battles was he engaged? Niscopac Falls with the Indians

6. Where did he reside when he entered the service? Sussex County N. Jersey

7. Is his statement supported by living witnesses, by documentary proof, by traditionary evidence, by incidental evidence, or by the rolls? nothing

8. Are the papers defective as to form or authentication? and if so, in what respect? Regular

Certify that the foregoing statement and the answers agree with the evidence in the case above mentioned.

__________ Examining Clerk.

In the matter of Samuel Vangorder a Pensioner
Declaration in order to obtain the benefit of the act of Congress
passed June 7th 1832

State of Pennsylvania

County of Tioga ss. On this twenty fourth day of September A.D. One thousand Eight hundred and thirty two Personally appeared in Open Court before the honorable Edward Herrick President and Ira Kilburn and Samuel W Morris Esquires his associate Judges of the court of common pleas in and for the county of Tioga being a court of record now sitting in and for said county Samuel Vangorder a resident of the township of Middlebury in said county of Tioga and State of Pennsylvania aged sixty nine years and upwards who being first ~~sworn~~ duly sworn according to Law doth on his oath make the following declaration in order to obtain the benefit of the act of Congress passed June 7th 1832.

That he entered the service of the United States under the following named officers and served as herein after stated that is to say That when he was about sixteen years old about the year 1779 he volunteered in a militia company commanded by Captain Peter Westbrook in Sussex County New Jersey the number of the regiment he does not now recollect but recollects that Samuel Westbrook was Major and one Cahoun or Calhoun was the quarter master stationed at the Continental Store in said county of Sussex. He thinks the name of the Colonel was Rosecrantz. He volunteered for and served one month in the army of the United States as a soldier in what was then called the State troops, that during this time he was in the battle of Lacawack which commenced at Beaver Creek. After serving out this month I volunteered as a soldier in a company commanded by Captain Henry Hoover Lieutenant Daniel DePuy in the same regiment above named in the county of Sussex State of New Jersey the same year as above stated. I volunteered for and served in the Army of the United States in this company for One Month. I received my pay for the above services from one Cahoun or Calhoun at the rate of Eight Dollars per month.

The next year after I moved into Easton in Pennsylvania State on the Delaware river I there enlisted in a company Commanded by Captain John Vannatter Lieut Wm. Fish in a Regiment commanded by Colonel Jacob Stroud for the term of seven months. I served as a soldier the whole time I enlisted for and was discharged. My discharge I lost about twenty five years ago. I lost my pocket Book at the same time. This company and regiment I understood were of the Continental line of troops. during this term I was engaged in a Battle with the Indians near the Niscopac Falls on the Susquehanna river in which engagement thirteen of the company in which I enlisted were killed the name of the Commissary of this department was Derrick Van Vleck. I do not now know that there is any person living that was engaged with me in the above service or knows of my services.

I hereby relinquish every claim whatever to a pension or annuity except the present and I do declare that my name is not on the Pension roll of the agency of any state

Sworn to & subscribed the day & year aforesaid.
Jonah Brewster Prothy

his
Samuel X Vangorder
mark

In answer to the interrogatories put by the court the said
Samuel Vangorder says -

He resided in Sussex county when I volunteered the two first times mentioned in his declaration. When I enlisted for seven months I resided twenty five miles above Easton on the Delaware river in the State of Pennsylvania. About 4 years after I was discharged I moved to Wyoming on the susquehanna river two miles below Wilkesbarre. I resided there about two years I then moved to Fishing creek on the susquehanna about 30 miles below Wyoming I resided there two years I then moved back to Wyoming I resided there about one year & Eight months I then moved to near the Owego on the north branch of the susquehanna river in the county of Tioga ~~county~~ New York State I resided there about three years I then moved to the town of Chemung in said county I resided there about six years I then moved into the town of Sullivan Tioga county Pa. I resided there about seven years I then moved into the town of Lawrence in said county of Tioga I resided there about five years I then moved into the township of Middlebury and have resided there ever since. I was born in Sussex county New Jersey in the year 1763 & resided there until I first volunteered. I never served as a substitute for any person I have been personally acquainted with the Honorable James Ford member of congress and the Hon. Ira Kilburn associate Judge for upwards of seven years. I am generally acquainted in the county of Tioga Pa. I had the record of my age in my house but some years ago it became destroyed by children. I have none now

Subscribed & sworn to as above
Jonah Brewster Prothy

his
Samuel X Vangorder
mark

⁂

We Thomas Kinney a farmer & deacon, residing in the township of Middlebury in the county of Tioga and State of Pennsylvania and Samuel Gronolt a minister of the gospel residing in the same county hereby certify that we are well acquainted with Samuel Vangorder who has subscribed & sworn to the above declaration. that we believe him to be sixty nine years of age & upwards that he is reputed and believed in the nieghbourhood where he resides to have been a soldier of the revolution and that we concur in that opinion and that his character for truth and veracity is good.

Sworn & subscribed the day
& year aforesaid in open court
Jonah Brewster Prothy

Thomas Kinney
Samuel Gronolt

⁂

And the said court do hereby declare their opinion that the above named applicant was a revolutionary soldier and served as he states

I Jonah Brewster Clerk of the ~~circuit court~~ court of common pleas in and for the county of Tioga in the State of Pennsylvania do hereby certify that the foregoing contains the original proceedings of the said court in the matter of the application of Samuel Vangorder for a Pension In testimony whereof I have hereunto set my hand and Seal of Office this twenty fourth day of September A.D. 1832.

Jonah Brewster Prothy

⁂

BRIEF objections to the admission of Pension Claims, of which those answered in the negative, in red ink, will apply to the claim of Samuel Vangorden which require further proof or explanation.

I. Does the declaration show where the applicant resides?

II. Did the declarant make the declaration in the county where he resides?

III. If the applicant did not make the declaration in the county where he resides, is there any reason assigned for his not doing so?

IV. Has the declarant mentioned the period or periods of the war when he served?

V. Has he mentioned with precision the length of his service, and the different grades in which he served, in language so definite as to enable the department to determine to what amount of pension he is entitled?

VI. Has he given the names of the officers under whom he served, in conformity with the regulations?

VII. Has he made a relinquishment of every claim to any other pension than the one under the Act of June 7, 1832?

VIII. Has the Court given their opinion?

IX. Has the Clerk given his certificate?

X. Is the Clerk's seal affixed? And if so, has it a device or inscription by which it can be distinguished from any other seal?

XI. Has the applicant obtained the evidence of a clergyman and another respectable citizen as to their belief respecting his age and the general belief in his neighborhood relative to his revolutionary service?

XII. If the clergyman's affidavit has not been produced, is there any reason assigned for not obtaining it?

XIII. If the applicant has no documentary evidence, and has not obtained the testimony of at least one living witness, has he stated in his declaration that such proof cannot be had?

XIV. If the applicant has produced a witness or witnesses as to his service, has the magistrate who administered the oath certified that he, she, or they are persons of credibility? Has the official character and signature of the magistrate been certified by the proper officer under his seal of office?

XV. Does it appear from any of the papers that the witness was in such a situation, or of such an age, as to have a personal knowledge of the applicant's service?

XVI. Are the papers authenticated as the regulations direct? No. See Quere 16.

XVII. Are the seven interrogatories prescribed by the War Department answered satisfactorily? And, if not, which of them is not so answered?

EXPLANATORY NOTES.

QUERE V. In a case where the applicant cannot, by reason of the loss of memory, state precisely how long he served, he should amend his declaration by making an affidavit in the following words: "Personally appeared before me, the undersigned, a Justice of the Peace, &c. A. B. who, being duly sworn, deposeth and saith, that, by reason of old age, and the consequent loss of memory, he cannot swear positively as to the precise length of his service, but according to the best of his recollection he served not less than the periods mentioned below, and in the following grades: For year months and days, I served as a . For months and days, I served as a ; and for such service I claim a pension."

It is important in all cases to determine with precision the period for which the applicant served, and the particular rank he held, as the law directs the pension to be paid according to the grade of the pensioner and the length of his service. The use of the phrase about three or four months is too indefinite, and all such qualifying expressions are objectionable.

QUERE XIV. Proof of Service.

In a case where the name of the applicant is not found on the records of the Department, he must prove his service by two credible witnesses, who are required to set forth in their affidavits the time of the claimants entering the service, and the time and manner of his leaving the same, as well as the regiment, company, and line to which he belonged. The magistrate who may administer the oaths must certify to the credibility of the witnesses, and the official character and signature of the magistrate must be certified by the proper officer, under his seal of office.

QUERE XVI. Mode of authenticating papers.

In every instance where the certificate of the certifying officer who authenticated the papers is not written on the same sheet of paper which contains the affidavit, or other paper authenticated, the certificate must be attached thereto by a piece of tape or narrow ribbon, the end of which must pass under the seal of office of the certifying officer, so as to prevent any paper from being improperly attached to the certificate.

Philada Oct 17, 1837.

J L Edwards Esq
Comr of Pensions Washington City
Dear Sir

I have just presented the enclosed Pension Drafts to the Agent for paying Pensions in Pa he declines paying for the reason that there is more than Twelve Months pay due. As I am a Stranger in the City and waiting Wholly and Solely to receive the amt due on them you will very much oblige me by giving your Order on the Agent at as early a period as possible to pay them. As that is all that is wanting

I am Dr Sir Yours Very Respy
Wm Herrick

please address me at Dunlaps City Hotel
N 3d. St Philada

War Department
Pension Office
Sir
19 Oct 1837

The Pension Agent at Philad. has this day been instructed to pay the amount due ~~William~~ Manassah Powers and Samuel Vangorden, pensioners under the Act of the 7th June 1832. The vouchers are herewith enclosed.

I am & c
William Herrick Esq
Dunlaps City Hotel
North 3d Street
Philad

---

34 N. Main st.,
Cortland, N.Y.,
Jan. 28, 1905.

Commissioner of Pensions,
Washington, D.C.
Dear Sir;

My great, great grandfather, Samuel Van Gorder was in the Revolutionary War. He served as a private in Captain Johannes Van Etten's company (Northampton Co.) Pennsylvania Volunteers. He was a pensioner. The Military Secretary, War Department, Washington, D.C. thought you would be able to give me a full record of the military service upon which his pension was granted, together with such other information concerning him, that is not obtainable from other sources.

Will you kindly tell me everything with regard to him - what he swore to when he received his pension - who he married - ect

Very truly yours,
Mary C. Van Gorden.

April 25, 1905.

The Commissioner of Pensions,
Washington, D.C
Dear Sir:

I desire information as to the records in your department relating to the pension of Samuel Van Gorder or Van Gorden for services in the Revolutionary War. I am not sure what state he enlisted from but I think from Pennsylvania. There is, I find, in the pension list of 1840 of soldiers in Pennsylvania, a Samuel Van Gorden residing with Chancy Terry at Charleston, Tioga County, Pa. His age was seventy-six years.

If there are any charges for searching the records I will gladly remit. I want to find out what I can about Samuel Van Gorder or Van Gorden. We are informed that he drew a pension for services in the Revolutionary War.

Yours very truly,
J. W. Cole

CB

---

34 N. Main st.,
Cortland, N.Y.
April 21, 1906.

Commissioner of Pensions,
Washington, D.C.
Dear Sir:

I wrote you some time ago with regard to the military history of Samuel Van Gorder, a soldier of the Revolutionary War. You sent me a statement as to when and where he enlisted, battles engaged in, etc. He applied for his pension Sept. 24, 1832. Can you tell me where and in what year, day and month he was born and where he died and in what year, month and day? I want to join the D.A.R.

Very truly yours,
Casendana Van Gorder.

Copy.

Surv. File #9503.
Rev. War.

L.M.H.

May 2, 1906.

Mrs. Casendana Van Gorder,
No. 34 North Main Street,
Cortland, New York.
Madam:

In reply to your letter dated April 21, 1906 and received April 24, 1906, you are advised that the papers on file in the above cited claim show that Samuel Van Gorder was born in Sussex county, New Jersey in 1763, day and month not stated. Date of death is not given; for date of last payment of pension apply to the Auditor for the Interior Department, Treasury Department, citing the following data; "certificate 6,073, issued February 23, 1833, act of June 7, 1832, Pennsylvania Agency."

Very respectfully,
Commissioner.

December 21, 1931.

Commissioner of Pensions,
Department of the Interior,
Washington, D.C.
Dear Sir:

I am enclosing copy of a paper received from the Pension Bureau many years ago by a member of the Van Gorder family and showing a synopsis of the service of Samuel Van Gorder, a solder of the Revolutionary War.

I would like to obtain a certified photostatic copy of the original papers on file in your Bureau covering the application of the soldier for pension in 1832.

Will you kindly advise me if the photostatic copy referred to can be furnished and if so the cost of same and to whom remittance should be made. The copy of the papers is wanted solely for the purpose of filing with some family records.

Thanking you in advance for an early reply,
Very truly yours,
Frederick H. Wells

January 19, 1932.

Mr. Frederick H. Wells,
105 West 40th Street,
New York City.

BA-J/MMHF

Dear Sir:

Reference is made to your letter in regard to securing photostatic copies of original papers on file in Revolutionary War pension claim, S.9503, Samuel Vangorder. He was pensioned as Vangorden.

The number of pages to be copied is five.

The cost per page is fifteen cents. The additional cost for certification is twenty-five cents for the set.

The proper amount should be forwarded to this office by money order drawn to the order of the Treasurer of the United States.

Very truly yours,
A. D. Hiller
Assistant to Administrator.

January 20, 1932.

Mr. A. D. Hiller,
Assistant to Administrator,
Veterans Administration,
Washington, D.C.

Dear Sir:

I am in receipt of your letter of January 19th reference BA-J/MMHF. In accordance therewith I am enclosing postal money order payable to the Treasurer of the United States for $1.00 for which kindly send me certified photostatic copies of original papers on file in Revolutionary War pension claim, S.9503 Samuel Vangorder, pensioned as Vangorden.

I shall appreciate it if you will send the copies to me as promptly as possible.

Very truly yours,
Frederick H. Wells
406 Tilden Building,
105 West 40th Street,
New York, N.Y.

February 9, 1932

United States Veteran's Bureau,
\Veterans' Administration,
Washington, D.C.

Gentlemen:

Attention: Mr. J. B. Schommer.
Disbursing Clerk

I will greatly appreciate any efforts you can make in the way of forwarding the copy of Pension Record to me covered by my remittance of January 20, 1932, your receipt No. 33692.

The copy has not as yet been received.

FHW/H

Yours very truly,
Frederick H. Wells

Veterans Administration

Date FEB 27 1932
DAC-B

Memorandum
From Chief, Accounting Division
To Chief Clerk
Subject Official Receipt No. 33,692.
Attention: Mrs. Finch

There is forwarded herewith for any action deemed necessary letter dated February 9, 1932 signed by Frederick H. Wells, wherein inquiry is made concerning the forwarding of copy of pension record which had previously been requested by him.

The records of the Accounting Division show the amount of $1.00 was received for which Official Receipt No. 33,692, dated January 22, 1932, was issued. The salmon copy of this receipt was forwarded to your Division for appropriate action under date of January 26, 1932 and returned under date of February 9, 1932. The amount thereof has been credited to Miscellaneous Receipts.

1 Enc.

Wm. H. Holmes

Veterans Administration
Washington DC

Dear sir: I wrote you some time ago regarding the war record of one Abram Fulkerson but had insufficient data. You asked me in replying to refer to BA-J/MMHF:MLB. I now have a copy of his commission in the Confederate Army, dated Feb. 12, 1864 Richmond, in which Abram Fulkerson was commissioned Colonel in the 63rd Tennessee Regiment, to rank as such from the 9th of Nov. 1863 to report for duty to Gen. Longstreet.

Abram Fulkerson lived, I think in Staunton Virginia. In the commission he was instructed to report his age, residence, when appointed, and the state in which he was born. I'd like to have, if possible, all the data concerning this man.

Also, I'd like data concerning William Leeper Fulkerson of Lawrence Co. Pennsylvania who also was in the Civil War, Northern army. I think he deserted and came home to his children whose mother had died in 1861.

Very truly yours,
(Mrs. F. E.) Lucille Van Gorder Sturgeon
121 First St. Aspinwall, Pittsburgh 15, Pa.
Mar. 16, 1938.

also, please, information about Samuel Van Gorder in Revolutionary war, from Sussex Co. N.J. Send to Mrs Finch Navy R1851

☆

March 25, 1938
BAHB

Mrs. Lucille Van Gorder Sturgeon,
121 First Street, Aspinwall,
Pittsburgh 15, Pennsylvania.

Dear Madam:

Receipt is acknowledged of your communication of March 16, 1938, requesting information concerning the Civil War records of Abram Fulkerson and William L. Fulkerson and the Revolutionary War record of Samuel Van Gorder.

The United States Government has not granted pensions based on service in the Confederate Army; therefore, this office does not have a record for Abram Fulkerson.

The attached report contains information concerning the Civil War service of William L. Fulkerson as shown on his widow's application for a pension.

Your inquiry concerning Samuel Van Gorder will be answered under separate cover.

By direction,
W. C. Black,
Chief Clerk.

Enc.

☆

Mrs. Lucille Van Gorder Sturgeon
121 First Street, Aspinwall
April 8, 1938.
Pittsburgh 15, Pennsylvania.

BA-J/ILL
Samuel Vangorder
S. 9503

Dear Madam:

Further reference is made to your letter of March 16, 1938, in which you request information in regard to Samuel Van Gorder of Sussex County, New Jersey, a Revolutionary War soldier. The data which follow are found in pension claim, S.9503, based upon service of Samuel Van Gorder (Vangorden) in that war.

Samuel Van Gorder (Van Gorden) was born in the year 1763, in Sussex County, New Jersey; the day of his birth and the names of his parents are not shown.

While a resident of Sussex County, New Jersey, he enlisted about the year 1779, and served one month as a private in Captain Peter Westbrook's company, Colonel Rosekrans' New Jersey regiment, during which he was in an engagement on Beaver Creek; he enlisted after that and served one month as a private in Captain Henry Hoover's (Hover's) company in that same regiment. He moved the next year and lived about twenty-five miles above Easton, on the Delaware River, in Pennsylvania; he enlisted there and served seven months as a private in Captain John Van Etten's company, Colonel Jacob Stroud's Pennsylvania regiment, during which he was in a severe engagement with the Indians at Nescopeck Falls.

About four years after his discharge from service, the soldier moved to Wyoming, near Wilkes-Barre, Pennsylvania, where he resided about two years, then moved to Fishing Creek on the Susquehanna River and remained two years, returned to Wyoming and remained one year and eight months, after which he moved to Owego, Tioga County, New York, and resided about three years, thence to Chemung in that same county and resided about six years, then moved to Sullivan, Tioga County, Pennsylvania, and remained about seven years, moved from there to Lawrence in that same county where he resided about five years, from which place he moved to Middleburg Township, Tioga County, Pennsylvania. ~~His name was borne on the pension roll as Samuel Van Gorden.~~

There are no data pertaining to the family of this soldier.

In order to obtain the date of last payment of pension, name of person paid, and possibly the date of death of this pensioner, you should apply to the Comptroller General, General Accounting Office, Records Division, this city, and furnish the following: Samuel Van Gorden (as his name was borne on the pen. roll)

Certificate #6073
Issued February 23, 1833
Rate $30 per annum
Commenced March 4, 1831.
Act of June 7, 1832.
Pennsylvania Agency.

Very truly yours,

A. D. Hiller,
Executive Assistant
to the Administrator.

Invalid. File No. 9,503
Samuel Van Gorden Prvt Rev War
Act: June 7th 1832
Index: - Vol. 1, Page 388
[Arrangement of 1870.]
Revolutionary War Pension and Bounty Land Warrant Application Files
Series M-804; Roll # 2446

Chapter 7
The Pennsylvanian

That this action was fought according to the best of his reccellection some time in the month of July and that about three hundred militiamen were killed, that among the officers killed were Capt. Veal Captain William Tyler and Colonel Trawson was shot from his large brown horse and fell into the brook which he was endeavoring to cross, that after the battle when the indians and tories had retired they were sent back to the scene of action for the purpose of burying the dead, and that the bodies smelt so offensively that it was impossible to perform the burial.

Reuben Place
Pennsylvania Militiaman at Battle at Minisink

The Pennsylvania Militiamen

The following pension is from the only enlisted man who served in a Pennsylvania unit. The notation at the end of the pension identifies the source of the material and if the application was rejected.

Reuben Place
Pennsylvania (R.8,279)

Reuben Place's Pension Application
under Act of Congress passed June 7th 1832
Presented by his Attorney Wm. B. Jarvis Columbus Ohio

Know all men by these presents that I the subscriber have appointed and do hereby appoint William B. Jarvis of Columbus Ohio my true and lawful attorney for me, and in my name to receive from the Secretary of War Paymaster General

Second or Third Auditor of the Treasury, or any officer or officers authorized to pay the same, all moneys of every kind, nature and name due me for extra pay bounty, clothing money, withheld pay, commutation money, pension money, or any other dues of every kind nature and name due me from the United States, whether paid by any of the before mentioned or any other officers of the United States, and I do hereby request said officers, or any or either of them, to pay said ~~to pay said~~ amounts to the said Wm B. Jarvis, and I do hereby give the said Wm B. Jarvis full power to endorse all drafts pay certificates or any other papers in reference to said business, and to execute any and all receipts or discharges, and to do any other act, which I could do if present in proper person. Witness my hand and seal this 18th day of September A.D. 1851

In presence of N. Campbell
John Brocklehurst

his
Reuben X Place
mark

State of Ohio

County of Delaware ss. On this 18th day of September A.D. 1851 personally appeared Reuben Place before me a Judge of the Court of Common Pleas in and for the County and State aforesaid, aged eighty eight years (and concerning whom I certify that I believe him prevented from bodily infirmity and extreme old age from attending in open court) a resident of said County and State, who being first duly sworn according to law doth on his oath make the following declaration in order to obtain the benefit of the act of Congress passed June 7th A.D. 1832 That he entered the service ~~service~~ of the United States under the following named officers and served as herein named. That in or about the month of July AD 1777 or 78, the indians being very troublesome, and continually committing depredations upon the settlements, ruthlessly destroying the houses and property of the settlers and murdering the women and children, the people of his neighborhood capable of bearing arms were called en masse into service, to protect the settlements. That he was attached to a Company of Pennsylvania Rangers, who were dressed in a Dark Green uniform, that he resided at the time of entering the service as aforesaid about two miles west of the Delaware river in the State of Pennsylvania with his father Philip Place, that he does not now recollect the name of the County in which his father lived, that the said company of Rangers was under the command of Captain Henry Shoemaker, Major Nelson and Colonel Jacob Strouds; That from the time he was called into the service until the close of the war, he was actively engaged in performing the various duties of a frontier

Ranger that he was dismissed each fall, remaining at home during the winter months and generally resuming the duties of a Ranger about the first of May of the following spring, and that he continued to serve as aforesaid until the close of the war. That he cannot positively state the events which transpired during the time of his service, in the exact order in which such events were transacted, but the following are among the principal of such said events. That about the first service he rendered was to help build Captain Henry Shoemaker's fort, which was situated on or very near the Delaware river and on Capt Shoemakers farm. ~~that~~ That he helped to build Captain Jean's fort which was situated about 7 miles west of the Delaware River, that he helped to build one wing of Bill Smiths fort which was situated on the Bushkill and that the said wing which he helped to build was made of pitch pine logs, that he was at this time about 15 years of age. That he was at Peanpack and was in a battle fought there, with the tories and indians commanded by one Butler (one of whose parents as he thinks was a negro, the other an indian) it took place as he verily believes during the month of July, and the whites drove back the indians. That immediately after this action he was stationed at Col Jacob Stroud's fort and remained there some six weeks, the fort being commanded if he rightly remembers by Major Nelson, and that from this fort he went home and remained till the following spring when he was called into service on or about the first of May, that Captain Dewit's company served about the same time and was occasionally in company with Capt Shoemakers command, and that the said Capt. Dewits company was engaged as he verily believes in service of the same nature as that in which this affiant was engaged viz. as Rangers employed in guarding the frontiers.

He further states that he was present and engaged in the battle fought at Beaver Brook State of Pennsylvania between the tories and indians commanded by Butler and Brant and the Rangers and New York Militia commanded as he thinks by Colonel Trawson. That this action was fought according to the best of his reccellection some time in the month of July and that about three hundred militiamen were killed, that among the officers killed were Capt. Veal Captain William Tyler and Colonel Trawson was shot from his large brown horse and fell into the brook which he was endeavoring to cross, that after the battle when the indians and tories had retired they were sent back to the scene of action for the purpose of burying the dead, and that the bodies smelt so offensively that it was impossible to perform the burial. That they then went to Captain Shoemakers fort where they staid one night and were then ordered to Colonel Jacob Strouds' fort for the purpose of protecting the women and children who were assembled therein from the surrounding country, and that they remained here some six weeks during which time they were kept in a continual state of alarm and anxiety by the continual depredations of the Indians and were daily expecting to be attacked by a large body of Indians that from this place he was ordered to Capt. Henry

Shoemaker's fort where he remained some 30 days and then, it being late in the fall he went home and remained during the winter. That while he was stationed at Colonel Stroud's fort the last above mentioned time the fort was frequently visited by many officers who were elsewhere stationed, for the purpose of consultation and that among them were the following to wit; Captain _______ Captain Chambers, Captain Benjamins

That he thinks the next service rendered by him, after leaving as aforesaid Capt Henry Shoemakers fort was on the Ten mile River from which place all the women and children who had escaped the tomahawk and knife of the Indians were collected and taken into Captain Dewitts fort, for protection and defence.

He further states that while in performance of his duty as a Ranger at or near Bill Smiths fort the were attacked by a superior force of Indians and retreated through the fields of rye (which was ripe and sown already harvested) and were hotly ~~hotly~~ pursuit by the Indians but accomplished the retreat safely.

That these (the aforesaid) are the principal events connected with his services, and that he continued to serve every summer as aforesaid from the time of entering the service as aforesaid, until the close of the war, that he never received any written discharge, and that he never received one cent of pay as compensation for his services. That the following persons were among others this affiants fellow soldiers and messmates in said war and service to wit, James Place, who if now living would be ninety six years of age and Peleg Place who if now living would be ninety two years of age, but who departed this life the said James at the State of New York about twenty years ago and the said Peleg Place who was a regular pensioner of the United States on the Revolutionary Roll on account of the services performed at the same place and time and under the same officers as is herein claimed by this affiant, that said Peleg drew a pension of ninety six dollars a year up to the time of his death, which took place on or about the 15th day of August A.D. 1838, at Delaware County Ohio that the widow of the said Peleg drew also a pension in right of the said services of the said Peleg from the said period of his death until she also departed this live at the County of Allen and State of Ohio some four or six years ago.

This affiant begs leave to refer the department, in support of this his application to the proofs filed, and produced by the said Peleg, although he does not know that said proofs would directly aid him therein, but the general facts stated must be the same as herein set forth if true, that he knows of no person now living whose testimony he can procure to prove his said service neither has he any documentary evidence to offer. That this affiants ~~services~~ swears most positively, that he was in actual service under the officers herein before mentioned, and in performance of the services herein before mentioned from the time of his entering the service as aforesaid, until the close of the war.

He further declares that he resided at the time of entering the service, with his father, in ~~Luzern County~~ the State of Pennsylvania, that after the war he continued to reside in said County for the period of ___ years that from thence he removed to the North ~~fork~~ Branch of the Susquehannah River, that he does not recollect the name of the County but thinks it was Luzern, and that he continued to reside at said North Branch until December A.D. 1818, when he removed to Delaware County and State of Ohio, where he has ever since continued to reside.

That the following named persons knew him while he lived on the North Branch of the Susquehannah River, but does not now know whether they are living or dead, to wit Daniel Starling John Starling, John ______, William Janes, Solomon Bunnel, John Bunnel, Cornelius Woodcock

That the following named persons to wit, Cornelius Woodcock, Thomas Carney, Elijah Carney, Jacob C Rosacrance, Samuel Landen, Joseph Patrick Richard H. Hodgsdon, are acquainted with him in the County where he resides. He hereby relinquishes every claim what ever to a pension or annuity, except the present; and declares that his name is not ~~found~~ on the pension roll of the agency of any state, ~~or if any, only on that of the agency of the State of Ohio~~.

Sworn to and subscribed the day and year first above written

Almon Stark Judge as aforesaid

Claimants signature
his
Reuben X Place
mark

And I the said Judge further certify that upon putting the following questions to the said Reuben Place the following answers, as annexed were given.

1st Ques. Where and in what year, were you born?

Ans. I was born in Duchess County New York on the 30th day of January A.D. 1763

2d Ques. Have you any record of your age and if so where is it?

Ans. There is a record which is in my daughters Abigail Roberts possession, and who resides in Union County Ohio

3rd Ques. Where were you living when called into service, where have you lived since the Revolutionary War and where do you now live?

Ans. At the time of entering the service I resided in ~~Luzern County~~ State of Pennsylvania, with my father from whence ~~he~~ I removed shortly after the war to the North Branch of the Susquehannah River where I ~~have~~ continued to reside until A.D. 1818 when ~~he~~ I removed to Delaware County State of Ohio, where ~~he~~ I have ever since resided.

4th Ques. How were you called into service, were you drafted, did you volunteer, or were you a substitute and if a substitute, for whom?

Ans. I was pressed or called into service as stated in the foregoing declaration.

5th Ques. State the names of some of the Regular officers who were with the troops where you served, such continental and militia regiments as you can recollect, and the general circumstances of your service.

Ans. The only answers which I can give to this question are embodied in the above declaration.

Ques. 6th Did you ever receive a written discharge from service, and if so, by whom was it given and what has become of it?

Ans. I never received a written discharge.

Ques. 7th State the names of persons to whom you are known in your present neighborhood and who can testify as to your character for veracity, and their belief of your services as a soldier of the Revolution.

Ans. Cornelius Woodcock, Jacob C. Rosacrance Charles Hodgsdon, Benjamin Benedict, Richard H. Hodgsdon, Joseph Patrick Samuel Landen, Ira Stephens, William Shaffer.

Sworn to and subscribed the day and year first above written and I certify that all interlineations and erasures were made prior to the signing of this my certificate

Almon Stark Judge as aforesaid

State of Ohio

County of Delaware ss. On this 18th day of September AD 1851 before me a Judge of the Court of Common Pleas, in and for said County and State personally came Cornelius Woodcock a respectable and reliable citizen of this county and state who being duly sworn according to law on his oath declares as follows. That he has been acquainted with Reuben Place now a resident of Delaware County Ohio for a period of 45 or 50 years. That when he first knew him he Place resided in Luzerne County and State of Pennsylvania, and that afterwards for a period of 30 or 35 years the said Place resided in Delaware County Ohio. That during all this period the said Place has represented himself to have been a soldier of the Revolution, and that as far as his knowledge extends this has been his general reputation, and he states that this reputation has never been disputed to his knowledge. That said Reuben Place is in my opinion a man of veracity and is generally reputed such in the neighborhood where he resides. He further states that he has no interest whatever in this matter being only influenced in thus stating the fact by a conviction that said Reuben Place is fully entitled to the benefit of the Revolutionary Pension laws of the United States.

Sworn to and subscribed the day and year first above written, and I hereby certify that the said Cornelius Woodcock is a man of truth and veracity and worthy of all credit.

Almon Stark Judge as aforesaid

his
Cornelius X Woodcock
mark

⁂

State of Ohio

County of Delaware ss. On this 18th day of September A.D. 1851 personally appeared before me a ~~Justice of the Peace~~ Judge of the Common Pleas Court in and for said County and State the several persons whose names are hereunto annexed, and whom I certify to be respectable and reliable citizens of this vicinity, and who being duly sworn according to law on their oaths make the following declaration. That they have been severally well acquainted with Reuben Place of Delaware County and State of Ohio for the periods set opposite their names that during that time he has been universally reputed to have been in the Revolutionary War, that this reputation has never been doubted to their knowledge and that they fully believe such to be the fact. They further declare that the said Reuben Place is in our opinion a man of truth and veracity, that such is his general reputation and they declare that they are severally disinterested in this matter, being only influenced in thus stating the fact, by a conviction that said Reuben Place is fully entitled to the benefit of the Revolutionary Pension laws of the United States

Ira Stevens — has known him 7 years
his
William X Shaver — has known him 15 years
mark
Richard H Hodgsdon — has known him 35 years
Sarah hodgsdon — has known him 35 years
Jacob C Rosecrans — has known him 30 years

Sworn to and subscribed the day and year aforesaid and I certify that the said affiants are all ~~men~~ persons of truth and veracity and worthy of all credit.

Almon Stark ~~Justice of the Peace~~
Judge as aforesaid

⁂

We Ira Chase, a clergyman residing in Porter township Delaware County State of Ohio, and Richard H. Hodgsdon a resident of Kingston township in the aforesaid county, hereby certify, that we are well acquainted with Reuben Place who has subscribed and sworn to the above declaration, that we believe him to be 88 years of age, that he is reputed and believed, in the neighborhood where he resides to have been in the Revolutionary war, and that we concur in that opinion.

Sworn to and subscribed the day and year first above written

Almon Stark Judge as aforesaid

Ira Chase

Richard H. Hodgsdon

★

And I the said Judge do hereby declare it to be my opinion after a thorough investigation of the matter and after putting the interrogatories prescribed by the War Department, that the above named applicant was a Revolutionary soldier and served as he states, and that he is prevented by reason of bodily infirmity and extreme old age from attending in open Court. And I do further certify that it appears to me that Ira Chase who has signed the preceding certificate is a clergyman resident in Porter township Delaware County State of Ohio and that Richard H. Hodgsdon is a credible person and that their statement is entitled to credit.

Almon Stark Judge as aforesaid

★

State of Ohio

County of Delaware ss. On this day of September A.D. 1851, before me a Judge Court Common Pleas in and for said County and State personally appeared Reuben Place and acknowledged that he had executed the foregoing instrument, and I certify that I know him to be the person described in the same Given at Kingston township the day and year aforesaid

Almon Stark Judge as aforesaid

★

United States of America.

The State of Ohio,

Secretary of State's Office. I, Henry W. King, Secretary of State of the State of Ohio, do hereby certify, That Almon Stark whose attestation is hereunto annexed, was, on the day of the date thereof, an acting Associate Judge, within and for the County of Delaware in the State of Ohio, duly Commissioned as such; that he is the proper officer to take such attestation; that his attestation is in due form; and that his official acts are entitled to full faith and credit, as well in Courts of Justice as thereout. In Testimony Whereof, I have hereunto subscribed my name, and affixed the Great Seal of the State of Ohio, at Columbus, the 22d day of September, in the year of our Lord one thousand eight hundred and fifty one

Henry W. King Secretary of State.

Rejected

Revolutionary War Pension and Bounty Land Warrant Application Files
Series M-804; Roll # 1941

Epilogue

Posterity who are to reap the blessings will scarcely be able to conceive the hardships and sufferings of their ancestors.
Abigail Adams

The debate will continue long into the future about what exactly happened on that hillside at Minisink Ford along the Delaware River on July 22, 1779. Surely there will be other historians who look into this battle that had such important consequences for those living in the frontier settlements at the time.

As one drives along the river one can see no evidence of the battle save a few historic markers and directional signage. Unlike so many battlefields with their museums, restored zigzag fencing, and cleared fields, this battleground is now covered by a dense second growth forest dotted with housing. Bluestone quarries are scattered all across the landscape, and one, in particular, near Hospital Rock, caused concerned citizens to lobby for the Last Stand Hill's purchase by the Minisink Valley Historical Society.

Today a beautiful park with trails and exhibits is maintained in good condition by the Sullivan County Department of Public Works, at the place where the battle ended. Commemorative ceremonies are held on the site every year on the anniversary of the battle. The Sullivan County Historical Society has long coordinated the ceremonies and has read aloud the names of the men who perished there.

We will never know for certain who all the participants were on that memorable day, although we probably now know as much as can be known with the resources that are available. We also cannot know what went through the minds of Hathorn, Tusten and Meeker, who watched the catastrophe unfold before their eyes, although Hathorn's after action report gives us a great deal of insight into what he felt. One cannot help but wonder what they were thinking as they saw neighbors, friends and colleagues killed or disappearing into the woods during the final moments.

We cannot recreate the fear, the raw bravery and courage, and perhaps some cowardice and poor leadership, of the men on the hillside that day. If not for the accidental firing of a rifle, it all might have ended differently. Anyone who has been in combat will understand how all those factors play out in the course of battles and warfare in general. From the accounts that have been passed down, however, we know that there was a great deal of courage shown that day as well as tremendous bravery. There had to be profound sadness and guilt as friends, neighbors and relatives were left behind, either dead or severely wounded.

Today we are simply left with the memories on paper in the form of pension applications, affidavits, bills, and receipts. We also have traditions passed down through families which are probably not as accurate as we would like. The *Draper Manuscripts*, an interesting source of materials related to the battle, have also led historians astray with some of the material included therein.

There is certainly some archaeological evidence left on the site, and every year after the battle's memorial service is conducted, we have all walked across the battleground late in the day knowing that the battle had probably ended at about that time. There was most certainly the smell and feel of death on that hill and probably a stillness that one can now sense if he sits on one of the boulders that the militiamen may have hidden behind.

In closing, it is our sincere hope as the reader examines some of the new information presented in this work, and as we remember those who fought and shed their blood for the founding of our country, that we remember them with a greater appreciation of what they went through to establish this nation. Let us also remember the wives and children who endured great privation and danger while their husbands, brothers and fathers struggled by force of arms to gain our independence.

Mark Hendrickson
Jon Inners
Peter Osborne
March 31, 2010

May the Souls of The Valiant Rest In Peace

Epilogue

Bibliography

Over the course of creating this book we have used a variety of sources and, because of our professional backgrounds have used three distinctive styles in citing them. We all like what Vernon Leslie said in his foreword in *The Battle of Minisink: A Revolutionary War Engagement in the Upper Delaware Valley* with regard to sourcing. He wrote "Where there are deviations (of style) and there are several in regard to footnotes, bibliographic citations, references, etc., they arise from personal preferences on the part of the author which he has every right to indulge." And so have we.

With that said, the bibliographic material in Chapters 1 and 3 are comprehensive and reflect all of the sources that we have used. The reader should look at Vernon Leslie's bibliography because he has a wide array of resources that he used for his book.

We might also add another quote from his foreword: "It is hoped that this volume will be judged not only on its own merits, but also against a background of what has previously appeared on the subject."

Frontispiece

We were about one third Tories: McCullough, David, *John Adams,* New York, Simon and Schuster, 2001, p. 78.

Introduction

To the Heroes of 1776: Osborne, Peter, *Our Day of Deliverance: The 11th Annual Reading of the Declaration of Independence,* Delivered at Orange Square, Port Jervis, New York, July 4, 2009.

That on the nineteenth day: Wilhelmus Westfall Pension Application, Revolutionary War Pension Record S.15322, New York, NARA, Revolutionary War Pension and Bounty Land Warrant Application Files, Series M-804, Roll # 2538.

But to return to the western: Jacob Aldrich Pension Application, Revolutionary War Pension Record R.80, New York, Revolutionary War Pension and Bounty Land Warrant Application Files, Series M-804; Roll # 26

Chapter 1
Blood and Mayhem At The Minisink

...[He] has endeavoured to bring: Osborne, Peter, *Our Day of Deliverance: The 11th Annual Reading of the Declaration of Independence,* Delivered at Orange Square, Port Jervis, New York, July 4, 2009.

Donehoo, G. P., 1928, A history of Indian villages and place names in Pennsylvania: Harrisburg, PA, 290 p. (Reprinted by Wennawoods Publishing, Lewisburg, PA, in 1998.)

Fischer, Joseph R. 1997, A well-executed failure - the Sullivan Campaign against the Iroquois, July-September 1779: Columbia, SC, University of South Carolina Press, 265 p.

Kelsey, Isabel T., 1984, Joseph Brant 1743-1807 - man of two worlds: Syracuse, NY, Syracuse University Press, 248 p.

Leslie, Vernon, 1976, The battle of Minisink: (2nd ed.): Middletown, NY, T. Emmett Henderson, Publisher, 248 p.

Sir Frederick Haldimand: Unpublished Papers and Correspondence 1758-1784; B.100; 21,760; 173-17

Stickney, Charles E., 1867, A history of the Minisink region: Middletown, NY, Coe Finch and I. F. Guiwits, Publishers, 211 p. (Reprinted by the Minisink Valley Historical Society in 1989.)

Stone, William L., 1865, Life of Joseph Brant (Thayendanegea), vol.1: Albany, NY, J. Munsell, 500 p.

Williams, Glenn F., 2005, Year of the Hangman - George Washington's campaign against the Iroquois: Yardley, PA, Westholme Publishing, LLC, 355 p.

Wilson, James R., 1822, Funeral Eulogium delivered in Goshen on the 22nd of July, 1822, at the internment of the bones of those who fell at the Battle of Minisink on the 22nd of July, 1779: Orange Farmer [Goshen, NY]. Monday, August 19, 1822. (Reprinted in Leslie, 1976, p. 193-200.)

Witte, R. W., 2001, Late Wisconsinan deglaciation and post-glacial history of Minisink Valley: Delaware Water Gap to Port Jervis, New York, in Inners, J. D., and Fleeger, G. M., eds., 2001 - a Delaware River Odyssey: Guidebook, 66th Annual Field Conference of Pennsylvania Geologists, Shawnee-on-Delaware, PA, p. 99-118.

Chapter 2
The Loss of So Many Brave Men

We have burnt all the: Sir Frederick Haldimand: Unpublished Papers and Cor-

respondence 1758-1784; Reel #42; Add. MSS. 21,760; 173-174.
. . . altho we Suffered by the Loss: Draper Manuscripts; 20F35

Chapter 3
The Patriot's Deed is Honored

Our duty is now done: Johnston, John W., and Albert Stage, *Centennial Celebration of the Minisink Battle on the Actual Battlefield, July 22d, 1879: With the Prayer, Oration and Speeches Delivered on the Occasion : and the Incidents Connected with the Celebration*, John Waller, Monticello, N.Y., 1879.

Books:

Beers, F. W., *County Atlas of Orange, New York*, Andreas, Baskin and Burr, Chicago, Illinois, 1875.

Bevier, Abraham G., *The Indians: or Narratives of Massacres and Depredations on the Frontier, in Wawasink and it's Vicinity, During the American Revolution,* originally published in Rondout, N.Y., 1846; reprinted by the Town of Wawarsing American Revolution Bicentennial Commission, Wawarsing, N.Y., 1975.

Blanco, Richard L. and Paul J. Sanborn, *The American Revolution 1775-1783 an Encyclopedia, 2 volumes*, Garland Publishing, Inc., New York & London, 1993.

Bockstruck, Lloyd De Witt, *Revolutionary War Bounty Land Grants Awarded by State Governments*, Genealogical Publishing Co, Baltimore, Maryland, 1996.

Bowden, A. J., *The Unpublished Revolutionary Papers of Major-General Edward Hand of Pennsylvania, 1777-1784*, George H. Richmond, New York, 1907.

Brick, John, *The Raid*, Farrar, Straus and Young, New York City, 1951.

Buffalo Historical Society, *Publications of the Buffalo Historical Society, volume 7*, published by Bigelow Brothers, Buffalo, N.Y., 1904.

Burbank, James W., *Cushetunk 1754-1784 The First White Settlement in the Upper Delaware River Valley*, Sullivan County Democrat, Callicoon, N.Y., 1975.

Cartwright, Conway Edward, *Life and Letters of the Late Hon. Richard Cartwright, Member of Legislative Council in the First Parliament of Upper Canada*, Belford Brothers, Toronto, Canada, 1876.

Clark, Murtie June, *Index to U. S. Invalid Pension Records 1801-1815*, Genealogical Publishing Co., Baltimore, Maryland, 1991.

Clark, Murtie June, *The Pension Lists of 1792-1795 With Other Revolutionary War Pension Records*, Genealogical Publishing Co., Baltimore, Maryland, 1991.

Clearfield Company, *Rejected or Suspended Applications for Revolutionary War Pensions*, Washington, D.C., 1852; reprinted with an added Index to States, Genealogical Publishing Co., 1969; reprinted by Genealogical Publishing Co., Baltimore, Maryland, 1991.

Clearfield Company, *The Pension Roll of 1835 Indexed Edition*, 4 volumes, originally published in 1835 in three volumes as *Senate Document 514, Serial*

Nos. 249-51 (23rd Cong., 1st Sess.) under the title *Report from the Secretary of War in Obedience to Resolutions of the Senate of the 5th and 30th of June, 1834, and the 3rd of March, 1835, in Relation to the Pension Establishment of the United States*; rearranged and reprinted in four volumes by Genealogical Publishing Co., 1968; reprinted with a new index by Genealogical Publishing Co., 1992; reprinted by Genealogical Publishing Co., Baltimore, Maryland, 2002.

Clearwater, Alphonso T., *The History of Ulster County New York*, W. J. Van Deusen, Kingston, N.Y., 1907.

Coleman, Charles C., *The Early Records of the First Presbyterian Church at Goshen, New York From 1767 to 1885*, originally published in 1934; reprinted by Willow Bend Books, Westminster, Maryland, 2006.

Cook, Frederick, *Journals of the Military Expedition of Major General John Sullivan against the Six Nations of Indians in 1779 with Records of Centennial Celebrations*, Knapp, Peck, & Thomson, Auburn, N.Y., 1887.

Eager, Samuel W., *An Outline History of Orange County, With an Enumeration of the Names of its Towns, Villages, Rivers, Creeks, Lakes, Ponds, Mountains, Hills and Other Known Localities, and Their Etymologies or Historical Reasons Therefor; Together With Local Traditions and Short Biographical Sketches of Early Settlers, Etc.*, S. T. Callahan, Newburgh, N.Y., 1846-7.

Eckert, Allan W., *The Wilderness War*, Bantam Books, New York, 1982.

Elliott, James, *If Ponies Rode Men: the Journey of Robert Land, 1777 - 1791*, Barrie Press, Stoney Creek, Ontario, 1999.

Empire State Society, *Register of the Empire State Society of the Sons of the American Revolution*, Sons of the American Revolution, published by the Society, 1899.

Fernow, Berthold, *Calendar of Wills on File and Recorded in the Offices of the Clerk of the Court of Appeals, of the County Clerk at Albany, and of the Secretary of State 1626-1836*, originally published New York, 1896; reprinted by Genealogical Publishing Co., Baltimore, Maryland, 1967.

Fernow, Berthold, *New York in the Revolution*, Weed, Parsons and Company, Albany, N.Y., 1887.

Fraser, Alexander, *United Empire Loyalists: Enquiry into the Losses and Services in Consequence of Their Loyalty, Evidence in the Canadian Claims, Second Report of the Bureau of Archives for the Province of Ontario,* 2 volumes, originally published as Second Report of the Bureau of Archives for the Province of Canada, 1905; reprinted by Genealogical Publishing Co., Baltimore, Maryland, 1994.

Gardner, Sue, and Gary Randall, Kathy Randall, and the Town of Warwick Historical Society, compiled by, *Days Gone By: A History In Pictures Town of Warwick, New York,* Keene Publishing,Warwick, N.Y.. 2006.

Graymont, Barbara, *The Iroquois in the American Revolution*, Syracuse University Press, Syracuse, N.Y., 1972.

Gumaer, Harry T., *A Minisink Fragment, Early American Frontier Life in*

Southeastern New York, with a Brief Gumaer Family History, published by H. T. Gumaer, Oakland, N.J., 1981.

Gumaer, Peter E., *A History of Deerpark in Orange County, N.Y.*, Minisink Valley Historical Society, Port Jervis, N.Y., 1890.

Halsey, Francis W., *A Tour of the Hudson, the Mohawk, the Susquehanna, and the Delaware in 1769, Being the Journal of Richard Smith of Burlington, New Jersey, originally published as A Tour of Four Great Rivers, the Hudson, Mohawk, Susquehanna and Delaware*, Scribner's, New York, 1906; reprinted by Ira J. Friedman, Inc., Port Washington, N.Y., 1964; reprinted by Purple Mountain Press, Ltd., Fleischmanns, N.Y., 1989.

Harper, Josephine L., *Guide to the Draper Manuscripts*, State Historical Society of Wisconsin, Madison, Wisconsin, 1983.

Harvey, Oscar Jewell and Ernest Gray Smith, *A History of Wilkes-Barré, Luzerne County, Pennsylvania: From its First Beginnings to the Present Time, Including Chapters of Newly-Discovered Early Wyoming Valley History, Together With Many Biographical Sketches and Much Genealogical Material*, 3 volumes, Raeder Press, Wilkes-Barre, Pennsylvania, 1909.

Hazard, Samuel, *Pennsylvania Archives Selected and Arranged from Original Documents in the Office of the Secretary of the Commonwealth, Conformably to Acts of the General Assembly, February 15, 1851, & March 1, 1852*, 10 volumes, Joseph Severns & Co., Philadelphia, Pennsylvania, 1853.

Hubbard, John Niles, *Sketches of Border Adventures in the Life and Times of Major Moses Van Campen*, J.S. Minard, Fillmore, N.Y., 1893.

Johnston, John W., and Albert Stage, *Centennial Celebration of the Minisink Battle on the Actual Battlefield, July 22d, 1879: With the Prayer, Oration and Speeches Delivered on the Occasion : and the Incidents Connected with the Celebration*, John Waller, Monticello, N.Y., 1879.

Judd, Jacob, *The Revolutionary War Memoir and Selected Correspondence of Philip Van Cortlandt*, Sleepy Hollow Restorations, Tarrytown, New York, 1976.

Kelsey, Isabel Thompson, *Joseph Brant 1743-1807: Man of Two Worlds*, Syracuse University Press, Syracuse, N.Y., 1984.

Leslie, Vernon, *The Battle of Minisink: A Revolutionary War Engagement in the Upper Delaware Valley*, T. Emmett Henderson, Middletown, N.Y., 1976.

Leslie, Vernon, *The Tom Quick Legends*, T. Emmett Henderson, Middletown, N.Y., 1977.

Lewis, Edgar Donald, *Minisink Kingdom Forbidding Loyalty*, Exposition Press, New York, N.Y., 1949.

Lossing, Benson J., *The Pictorial Field-Book of the Revolution; or Illustrations, by Pen and Pencil, of the History, Biography, Scenery, Relics, and Traditions of the War for Independence*, 2 volumes, Harper & Brothers, New York, 1852.

Mathews, Alfred, *History of Wayne, Pike and Monroe Counties, Pennsylvania*, R.

T. Peck and Company, Philadelphia, 1886.

Montgomery, Thomas Lynch, *Pennsylvania Archives: Fifth Series,* 8 volumes, Harrisburg Publishing Co., Harrisburg, Pennsylvania, 1906.

Nelson, William, *Documents Relating to the Revolutionary History of the State of New Jersey: Second Series,* 5 volumes, John L. Murphy Publishing Co., Trenton, N.J., 1906.

New Jersey Historical Society, *Proceedings of the New Jersey Historical Society, volume 7,* Newark, N.J., 1922.

New York State, *The Public Papers of George Clinton, First Governor of New York 1777-1795, 1801-1804,* 10 volumes, Wynkoop Hallenbeck Crawford Co., New York and Albany, 1899-1914.

New York State Historical Association, *Proceedings of the New York State Historical Association. The Eighth Annual Meeting, with Constitution, By-laws and List of Members, volume 7,* Glens Falls Publishing Company, 1907.

Ogden, Lucile G., *The Journal of the Records of Peter E. Gumaer, 1771-1869,* published by Lucile G. Ogden, Middletown, N.Y., 1983.

Osborne, Peter, *A Guide to the Historic Markers of the City of Port Jervis and the Town of Deerpark, New York,* Minisink Valley Historical Society, Port Jervis, N.Y., 2002.

Osborne, Peter, *The New York - New Jersey Boundary Line: While New Jersey Dozed, New York Was Wide Awake,* Minisink Valley Historical Society, Port Jervis, N.Y., 1992.

Osborne, Peter, *Perseverance & Vigilance, a History of Fort Decker, the Revolutionary War Stone House that Survived Fires, Raids, and Nearly Three Centuries of Public and Private Service to Become a Museum of Local History,* Minisink Press, Port Jervis, N.Y., 2007.

Osborne, Peter, *The Port Jervis Area Heritage Commission Salutes the Decker Stone House 1793 - 1993,* Port Jervis Heritage Commission, Port Jervis, N.Y., 1993.

Palmer, Gregory, *Biographical Sketches of Loyalists of the American Revolution,* Meckler Publishing, Westport, Connecticut, 1984.

Priest, Josiah, *The Captivity and Sufferings of Gen. Freegift Patchin,* originally published by Packard, Hoffman and White, Albany, N.Y., 1833; reprinted by Garland Publishing, New York, 1977.

Quinlan, James Eldridge, *History of Sullivan County: Embracing an Account of Its Geology, Climate, Aborigines, Early Settlement, Organization; the Formation of Its Towns With Biographical Sketches of Prominent Residents, Etc., Etc.,* G. M. Beebe and W. T. Morgans, Liberty, N.Y. 1873.

Quinlan, James E., *Tom Quick the Indian Slayer and the Pioneers of Minisink and Wawarsink,* DeVoe & Quinlan, Monticello, N.Y., 1851.

Richards, Mark V., *The Sesquicentennial of the Battle of Minisink, A Story of the*

Commemoration Held on the battlefield at Minisink Ford, Sullivan County, N.Y., July 22, 1929, published by the Republican Watchman, Monticello, N.Y., 1929.

Ruttenber, E. M. and L. H. Clark, *History of Orange County, New York, with Illustrations and Biographical Sketches of Many of its Pioneers and Prominent Men*, Everts and Peck, Philadelphia, 1881.

Sivertsen, Barbara J. and Barbara L. Covey, *The Legend of Cushetunk: The Nathan Skinner Manuscript and the Early History of Cochecton*, Heritage Books, Inc., Bowie, Maryland, 1993.

Smith, Philip H., *Legends of the Shawangunk and its Environs, including Historical Sketches, Biographical Notices, and Thrilling Border Incidents and Adventures Relating to Those Portions of the Counties of Orange, Ulster and Sullivan Lying in the Shawangunk Region*, Smith and Company, Pawling, N.Y., 1887.

Snell, James, *History of Sussex and Warren Counties, New Jersey, with illustrations and biographical sketches of its prominent men and pioneers*, Everts & Peck, Philadelphia, 1881.

Sparks, Jared, *The Writings of George Washington; Being his Correspondence, Addresses, Messages, and Other Papers, Official and Private, Selected and Published from the Original Manuscripts; With a Life of the Author, Notes, and Illustrations*, 12 volumes, American Stationer's Co., John B. Russell, Boston, Massachusetts, 1834-1837.

State of Pennsylvania, *Colonial Records of Pennsylvania*, 16 volumes, Joseph Severns & Co., Philadelphia, 1852.

Stickney, Charles E., *A History of the Minisink Region*, originally published by Coe Finch and I. F. Guiwits, Middletown, N.Y., 1867; reprinted by the Minisink Valley Historical Society, Port Jervis, N.Y., 1989.

Stickney, Charles E., *Woodburn, a descriptive and historical poem of the part taken in the Battle of Minisink by Jerseymen*, Wantage Recorder Office, Deckertown, N.J., 1897.

Stone, William Leete, *Life of Joseph Brant - Thayendanegea: Including the Border Wars of the American Revolution, and Sketches of the Indian Campaigns of Generals Harmar, St. Clair, and Wayne, and Other Matters Connected With the Indian Relations of the United States and Great Britain, From the Peace of 1783 to the Indian Peace of 1795*, 2 volumes, Alexander V. Blake, New York, 1838; reprinted by Scholarly Press, St. Clair Shores, Michigan, 1970.

Stryker, William S., *Official Register of the Officers and Men of New Jersey in the Revolutionary War*, William T. Nicholson and Co., Trenton, N.J., 1872.

Swiggett, Howard, *War Out of Niagara: Walter Butler and the Tory Rangers*, Columbia University Press, New York City, 1933.

Sylvester, Nathaniel Bartlett, *History of Ulster County, New York: With Illustrations and Biographical Sketches of Its Prominent Men and Pioneers*, Everts & Peck, Philadelphia,1880.

Sypher, F. J., *The Journal of William Scudder*, Scholars' Facsimiles & Reprints, Ann Arbor, 2005.

Szymanski, Leszek, *Casimir Pulaski, A Hero of the American Revolution*, Hippocrene Books, New York, 1994.

Terwilliger, Katherine T., *Wawarsing, Where the Streams Wind: Historical Glimpses of the Town*, Rondout Valley Publishing Company, Inc., Ellenville, N.Y., 1977.

Twichell, Horace E., *History of the Minisink Country*, Schilling Press, New York City, 1912.

University of the State of New York, *The Sullivan-Clinton Campaign in 1779 Chronology and Selected Documents*, prepared by the Division of Archives and History, Albany, N.Y., 1929.

Van Buren, Augustus H., *A History of Ulster County Under the Dominion of the Dutch*, originally published Kingston, N.Y., 1923; reprinted J. C. & A. L. Fawcett, Inc., Astoria, N.Y., 1989.

Wallace, Paul A. W., *Indian Paths of Pennsylvania*, Pennsylvania Historical & Museum Commission, Harrisburg, Pennsylvania, 1998.

White, Virgil D., *Genealogical Abstracts of Revolutionary War Pension Files,* 4 volumes, National Historical Publishing Company, Waynesboro, Tennessee, 1990-1992.

White, Virgil D., *Index to Revolutionary War Service Records,* 4 volumes, National Historical Publishing Company, Waynesboro, Tennessee, 1995.

Williams, Glenn F., *Year of the Hangman: George Washington's Campaign Against the Iroquois*, Westholme Publishing, Yardley, Pa., 2005.

Wilson, Thomas B., *Notices From New Jersey Newspapers 1781-1790*, Hunterdon House, Lambertville, N.J., 1988.

Newspapers

The Connecticut Courant and the Weekly Intelligencer, Hartford, Ct., August 3, 1779.

The Continental Journal, Poughkeepsie, N.Y., May 20, 1779; September 2, 1779.

The Daily Argus, Middletown, N.Y., August 6, 1879.

The Daily Register, Newburgh, N.Y., February 2, 1892.

The Delaware Courier, September 26, 1863.

The Delaware Gazette, Delhi, N.Y., January 7, 1874; January 21, 1874; October 21, 1874; October 21, 1879.

The Evening Gazette, Port Jervis, N.Y., January 8, 1878.

The Goshen Democrat, Goshen, N.Y., July 24, 1879.

The Goshen Independent Republican, Goshen, N.Y., March 24, 1828; September 12, 1877; September 19, 1877; September 26, 1877; October 10, 1877; October 17, 1877; January 9, 1878; January 16, 1878; January 30, 1878; February 6, 1878; May 14, 1879; July 16, 1879; July 23, 1879; July 30, 1879; August 22, 1879.

The Mercury, Middletown, N.Y., August 1, 1879.

The New York Gazette and the Weekly Mercury, New York, N.Y., April 12, 1779; August 2, 1779; August 9, 1779; August 23, 1779; August 30, 1779; October 4, 1779.

The New York Spectator, Warwick, N.Y., April 4, 1828.

The Orange County Press, Middletown, N.Y., May 16, 1879; July 18, 1879; July 25, 1879.

The Pennsylvania Evening Post, Philadelphia, Pa., August 28, 1779.

The Pennsylvania Gazette and Weekly Advertiser, Philadelphia, Pa., August 11, 1779.

The Pennsylvania Packet, Lancaster, Pa., August 3, 1779; August 7, 1779; August 10, 1779; August 19, 1779.

The Port Jervis Weekly Gazette, Port Jervis, N.Y., April 26, 1878; July 18, 1879; July 25, 1879; January 16, 1885.

The Royal Gazette, New York, N. Y., July 31, 1779; August 14, 1779; August 18, 1779.

The Tri-States Union, Port Jervis, N.Y., July 18, 1879; July 23, 1879; July25, 1879.

The Weekly Register, Newburgh, N.Y., July 23, 1879.

The World, New York, N.Y., July 16, 1879.

Microfilm

Audit Office Series 12 (AO12) *(American Loyalists Claims: Series I: 1776-1831),* 146 volumes, David Library of the American Revolution, originals in the Public Records Office, London, England.

Audit Office Series 13 (AO13) *(American Loyalists Claims: Series II: 1780-1835),* 141 bundles, David Library of the American Revolution, originals in the Public Records Office, London, England.

British Headquarters (Sir Guy Carleton) Papers, 107 volumes, Duke University, originals at the Public Record Office, London, England.

Joseph Brant Papers, Series F, 22 Volumes, *Draper Manuscript Collection*, Duke University, originals at the State Historical Society of Wisconsin.

Department of Defense Adjutant General's Office (Revolutionary War) Numbered Manuscripts, 1770s-1890s, Control Number SDEA1001, 30 rolls, New Jersey Division of Archives & Records Management, Trenton, New Jersey.

Edward Hand Papers, 1 roll, David Library of the American Revolution, originals at the Division of Archives and Manuscripts (Pennsylvania State Archives), Pennsylvania Historical and Museum Commission, Harrisburg, Pennsylvania.

National Archives Microfilm Publication M247 (Papers of the Continental Congress 1774-1789), 204 rolls, Record Group 360, Records of the Continental and Confederation Congresses and the Constitutional Convention, National Archives Building, Washington, D.C.

National Archives Microfilm Publication M804 *(Revolutionary War Pension and Bounty Land Warrant Application Files)*, 2,670 rolls, Record Group 15, Records of the Veterans Administration, State Library of North Carolina, originals at the National Archives Building, Washington, D.C.

National Archives Microfilm Publication M881 (Compiled Service Records of Soldiers Who Served in the American Army During the Revolutionary War), 1,096 rolls, Record Group 93, War Department Collection of Revolutionary War Records, State Library of North Carolina, originals at the National Archives Building, Washington, D.C.

New York State Comptroller's Office Series Number A0200 (Revolutionary War Accounts and Claims (1775-1808)), 42 rolls, New York State Archives, Albany, New York.

Papers of Major-General William Alexander Lord Stirling, 2 volumes, David Library of the American Revolution, originals at The New York Historical Society, New York City.

Sir Frederick Haldimand: Unpublished Papers and Correspondence 1758-1784, 232 volumes, Duke University, originals at the British Library, London, England.

Sussex County Court of Common Pleas/Court of General Quarter Sessions Rough [Pamphlet] Minutes, 1776-1852, Control Number CSUCP005, 14 boxes, New Jersey Division of Archives & Records Management, Trenton, New Jersey.

Websites

Diary of Ensign John Barr 1781, Ancestry.com, transcribed by Arlene Goodwin, rootsweb.ancestry.com/~nygreen2/diary_of_john_barr.htm, July 13, 2009.

George Washington Papers 1741-1799, Manuscript Division, Library of Congress, lcweb2.loc.gov/ammem/gwhtml/gwhome.html, May 4, 2009.

Poppino/Popenoe/Popnoe & Allied Families, by Oliver Popenoe, popenoe.com/index.html, July 20, 2009.

Revolutionary War pensions, Albert Wisner Public Library, Warwick, New York, albertwisnerlibrary.org, November 8, 2008.

Revolutionary War pensions, Footnote.com, November 19, 2008.

Revolutionary War pensions, Heritagequestonline.com, August 14, 2008.

Revolutionary War pensions, Morrison's Pensions, by James F. Morrison and Berry Enterprises, morrisonspensions.org, December 5, 2008.

Sir Frederick Haldimand: Unpublished Papers and Correspondence 1758-1784, by Mario Lemoine and Alain M. Tremblay, haldimand-collection.ca, January 11, 2009.

Miscellaneous

Transcriptions of The Mamakating Committee of Safety, Revolutionary War Collection, Minisink Valley Historical Society, Port Jervis, New York.

Centennial of Burial of Victims of Minisink Battle is Observed, undated (October 7, 1922?) and unidentified newspaper clipping, Battle of Minisink Collection, Minisink Valley Historical Society, Port Jervis, New York

Chapter 4
A Declaration of Service

Every claimant must state: Pension of Jacob Aldrich, Revolutionary War Pension and Bounty Land Warrant Application Files, Series M-804; Roll # 26.

Chapter 5
The Yorkers

But to return to: Pension of Jacob Aldrich, Revolutionary War Pension and Bounty Land Warrant Application Files, Series M-804; Roll # 26.

The source for each veteran's pension application is shown at the end of each application.

Chapter 6
The Jerseymen

In the Summer previous: Pension of Isaac Bedell, Revolutionary War Pension and Bounty Land Warrant Application Files, Series M-804; Roll # 198.

The source for each veteran's pension application is shown at the end of each application.

Chapter 7
The Pennsylvanian

That this action was fought: Pension of Reuben Place , Revolutionary War Pension and Bounty Land Warrant Application Files, Series M-804; Roll # 1941.

The source for each veteran's pension application is shown at the end of each application.

Epilogue

We were about one third Tories: McCullough, David, *John Adams,* New York, Simon and Schuster, 2001, p. 169.

Index

THE FOLLOWING INDEX uses the same format that we have used in the bibliography and again, we rely on Vernon Leslie's adage about an author's indulgence. Because of the nature of the material in the book we have limited our index to place names used during the American Revolution, current place names as they relate to the Revolutionary War era, the battle itself and the Minisink region in general. We have included all of the battle participants, and their wives, officers and leaders of the time, and any other citation we deemed important. We have not included place names where veterans may have been living at the time of their pension application in the nineteenth century. We also have not included the names of clerks or witnesses to the pension applications unless they were witnesses to the Revolutionary War events described. Most of that information is not relevant to the battle. Finally, we have maintained the spelling and capitalization of words as they appear in the contemporary documents.

C

O

Acknowledgments

This book is the result of the efforts of a number of people and organizations. People have assisted us and opened their institutions' doors to us with open arms. This present effort is five years in the making. For Mark, it was a dream of the last twenty years to publish the results of his research.

We are grateful to the National Archives and Records Administration (NARA) in Washington, D.C. NARA is the nation's record keeper, or really the people's archives. Contained within the walls of its numerous buildings are documents, photographs, films, reports, and a mind-boggling collection of our nation's most important historic manuscripts. All of the original pension records that have been reproduced here reside at NARA.

The Center for Legislative Archives, a section within NARA, provided help in tracking down records of men known or suspected to have been in the battle or who were from the area at that time.

A. J. Berry, who has been doing her own work on Revolutionary War pensions, was kind enough to allow her own transcriptions to be copied and incorporated into this work. She very graciously offered unlimited access to everything she has worked on over the years and was quick to offer advice when her help was sought. Her help and friendship has been invaluable and has contributed greatly to this project.

Paula Radwanski also provided a great deal of assistance. She was quick to share her own research and offer advice. She sent numerous photocopies and always had words of encouragement when dedication to this work would falter.

Cheryl McLean and the entire staff at the North Carolina State Library in Raleigh, North Carolina went out of their way to help to help us go through countless reels of microfilm. They were frequently sent running back and forth for reel after reel of microfilm, ten or twelve at a time, and were quick to help whenever there were the inevitable problems with the microfilm machines, or there was difficulty locating certain records. You would be hard-pressed to find a nicer group of people.

The staff at the Perkins-Bostock Library of Duke University in Durham, North Carolina, and in particular Robbin Ernest, were very helpful in locating and working with the British Headquarters (Carleton) Papers, the Draper Manuscripts, and the Haldimand Papers among their collection of microfilm.

Bette M. Epstein at the New Jersey State Archives was extremely helpful when contacted. There was knowledge of the existence of one document in particular and in the course of searching for it, she located a number more that were never known about and she brought them to our attention. Bette located, copied and mailed just about everything that is credited with coming from records at the New Jersey State Archives. Without her help, many of the records credited to that location would be missing from this work.

The staff at the D. H. Hill Library at North Carolina State University in Raleigh, North Carolina helped us navigate around their huge collection of books on the Revolutionary War, and the biographies of the men who took part in it.

The staff at the Olivia Raney Local History Library in Raleigh, North Carolina helped when problems came up using Heritage Quest through the online services provided by the Wake County Library System. They were also instrumental in helping to track down and borrow microfilm records from the State of New York.

Katherine Ludwig, of the David Library of the American Revolution in Washington Crossing, Pennsylvania, helped us navigate through the Loyalist Claims and other microfilm records at their facility. The staff at the Goshen Library in Orange County, New York helped with research on Captain John Wood. We are grateful to Gary Fleeger of the Pennsylvania Geological Survey, who created the maps that allowed the reader to understand how the battle unfolded.

The Minisink Valley Historical Society, which is now the repository for all of the photocopies of the pensions that were used in this book, has been an important repository of archival materials relating to the battle for more than one hundred years. We are grateful to volunteers and members past and present who have saved so many of the archival materials for researchers to use. The Society also rescued the most significant portion of the battleground early in the twentieth century from being quarried into extinction.

During Peter's time as the Society's director (1981-2009), he was often involved with the annual ceremonies to commemorate the battle, and is grateful to the board for its support of those programs. He also created a brochure and section on the Society's web page that provided a concise history of the battle.

Sullivan County also deserves great credit for having included the battleground in its park system in 1955 and for having since expanded it beyond what the Society turned over. Most recently the county undertook major improvement efforts at the park that will benefit visitors for a long time to come. Kristen Porter and Bob Myers deserve great kudos along with the Sullivan County Legislature.

Alicia Batko gave more help than it is possible to thank her for. She was quick to share her own research and offer advice. She was responsible for finding the names of several of the men who were ultimately included in the list of known men in the battle and was the person who tracked down the known image of Ezekiel Lane, the only battle participant whose photographic image still survives.

Others who helped to make this new study of the battle possible are Alan Dampman, Laurie Ramie, John Masten, Sue Gardner, Michelle Figliomeni, Elizabeth Ahrens-Kley, Peter Blood, Joe Sherwood, Judy Gumaer Testa, Wendy Harris, Paul Joyal, Jess Novak, Betty Collins, Gerald DeGroat, Ted Sly, Tony and Louise Domingo and Rick Moran.

We are grateful to our readers who bear absolutely no blame for any mistakes that have come through the process but can take credit for the readability of the manuscript. The readers were Mary Tiebout, Frank Salvati, Fred and Barbara Johnson Weissman, and Nancy Vocci. Thank you for meeting the challenge that awaited you when you opened your folders with nine hundred sheets of paper to read and for agreeing to undertake this project.

We also want to express our gratitude to Rick Hibberd the designer of the cover and co-designer of the book. His web page can be found at *www.rhibberd.com.* When all three of us saw the cover for the first time our comments were almost all the same: "Holy Expletive." To get a book through the publishing process can often be a nightmare and Rick shed some beautiful moonlight upon us as the days went by.

The Navasing Longrifles have been friends and colleagues for many years and have been great supporters of this project, not only by gracing the cover of the book with their picture but in discussing the battle with us and participating in the annual commemorative programs at the battleground. If we have given birth to a baby in the form of this book, they are the proud uncles (and aunts).

Finally, if we have missed anyone, we are sorry. Your assistance was greatly appreciated.

About the Authors

Mark Hendrickson

Mark Hendrickson was born in Livingston Manor, New York, and has had a great interest in the Revolutionary War and the Battle of Minisink in particular. He is a graduate of the State University of New York at Delhi and a veteran of the United States Army, serving at Schofield Barracks, Hawaii. Since first learning of the battle after his service in the Army, he has made it his life's work to visit various archives to collect material on the battle and its participants. His material, presented here, is the culmination of nearly twenty years of research. Mark resides in Wake Forest, North Carolina.

Jon Inners

Jon Inners was a geologist with the Pennsylvania Geological Survey for thirty-one years, retiring as chief of the Geologic Mapping Division in 2004. He is currently a Conservation Volunteer at the Survey and an adjunct geologic field-camp instructor at SUNY Oswego. He and his wife Martha reside in Camp Hill, Pennsylvania.

Peter Osborne

Peter Osborne was born in Paterson, New Jersey, and was the executive director of the Minisink Valley Historical Society from 1981 to 2009. He has a Bachelor of Arts degree from Rutgers, the State University of New Jersey, and has served on the board of directors of the Orange County Historical Society, the Grey Towers Heritage Association and the Upper Delaware Scenic Byway.

During his directorship he was responsible for the Society's Fort Decker Museum of History, and the interpretation of its long and eventful history which included its destruction during the Brant raid into the Minisink in 1779. He has also been a participant for many years at the annual commemoration of the battle at the Minisink Battleground Park at Minisink Ford, New York.

Osborne has been published widely over the last two decades and has written books on the Civilian Conservation Corps and regional history. He is also the owner of the Pienpack Company. The mission of his company is to provide exciting journeys of discovery into our nation's history, and the Minisink Valley's regional history through presentations, lectures, demonstrations, motor coach tours and publications. The Pienpack Company has been providing programs to civic, historical, fraternal, church groups, seminars and meetings for thirty years.

About So Many Brave Men

The book was designed by Rick Hibberd and Peter Osborne, set in Adobe Garamond and printed in the United States of America. The cover of the book was created by Rick Hibberd using a photograph of members of the Navasing Longrifles that was taken by Peter Osborne at the Sentinel Rock on the Minisink Battleground Park. The photograph was shot at the annual commemoration ceremony on July 18, 2009, and tells the story of the many emotions that the men faced on the fateful day of the battle. The members of the Navasing Longrifles who are pictured left to right are: Dan Hogue, Jr., Kai Moessle, Tony Domingo, Kevin Williams, and Mike Roberts.

The Navasing Longrifles are a not-for-profit organization dedicated to promote, support and preserve our nation's rich heritage (including the sport of muzzle loading) through recreational, educational, and cultural avenues. These include but are not limited to encampments, exhibits, muzzle loading gun safety and competitions, historical impressions, museums, libraries and other related areas. For more information contact: Tony Domingo, President, 845-796-0938 or Michael Roberts, Vice President, 845-295-0481.

The authors are now working on a second volume of materials found in various archives across the country that pertain to the Battle at Minisink. They shed additional light on that dramatic day and the context of the events that were going on in the region at the time. For further information or to contact the authors, go to the publisher's web page at *www.pienpack.com.*

www.ingramcontent.com/pod-product-compliance
Lightning Source LLC
LaVergne TN
LVHW081958200926
848481LV00010B/2767
* 9 7 8 0 6 1 5 3 4 6 5 8 8 *